Advances in Intelligent Systems and Computing

Volume 1032

The series "Advances in Intelligent Systems and Computing" contains publications on theory, applications, and design methods of Intelligent Systems and Intelligent Computing. Virtually all disciplines such as engineering, natural sciences, computer and information science, ICT, economics, business, e-commerce, environment, healthcare, life science are covered. The list of topics spans all the areas of modern intelligent systems and computing such as: computational intelligence, soft computing including neural networks, fuzzy systems, evolutionary computing and the fusion of these paradigms, social intelligence, ambient intelligence, computational neuroscience, artificial life, virtual worlds and society, cognitive science and systems, Perception and Vision, DNA and immune based systems, self-organizing and adaptive systems, e-Learning and teaching, human-centered and human-centric computing, recommender systems, intelligent control, robotics and mechatronics including human-machine teaming, knowledge-based paradigms, learning paradigms, machine ethics, intelligent data analysis, knowledge management, intelligent agents, intelligent decision making and support, intelligent network security, trust management, interactive entertainment, Web intelligence and multimedia.

The publications within "Advances in Intelligent Systems and Computing" are primarily proceedings of important conferences, symposia and congresses. They cover significant recent developments in the field, both of a foundational and applicable character. An important characteristic feature of the series is the short publication time and world-wide distribution. This permits a rapid and broad dissemination of research results.

**** Indexing: The books of this series are submitted to ISI Proceedings, EI-Compendex, DBLP, SCOPUS, Google Scholar and Springerlink ****

More information about this series at http://www.springer.com/series/11156

Mirosław Siergiejczyk · Karolina Krzykowska
Editors

Research Methods and Solutions to Current Transport Problems

Proceedings of the International Scientific Conference Transport of the 21st Century, 9–12th of June 2019, Ryn, Poland

Editors
Mirosław Siergiejczyk
Faculty of Transport
Warsaw University of Technology
Warsaw, Poland

Karolina Krzykowska
Faculty of Transport
Warsaw University of Technology
Warsaw, Poland

ISSN 2194-5357 ISSN 2194-5365 (electronic)
Advances in Intelligent Systems and Computing
ISBN 978-3-030-27686-7 ISBN 978-3-030-27687-4 (eBook)
https://doi.org/10.1007/978-3-030-27687-4

This Springer imprint is published by the registered company Springer Nature Switzerland AG
The registered company address is: Gewerbestrasse 11, 6330 Cham, Switzerland

Preface

International Scientific Conference Transport of the 21st Century has its origins in the mid-1970s (1974) and is cyclized every three years. The message of this year's conference is to present the achievements of national and foreign research and scientific centers dealing with the issues of transport: rail, road, air, and sea in the technical and technological aspects, as well as organization and integration of the environment conducting research in the discipline civil engineering and transport.

Texts of articles submitted for the conference were accepted for publication in the conference materials on the basis of two independent reviews of both: members of the Scientific Committee of the conference, as well as recognized authorities in the field of civil engineering and transport. The reviewers were chosen in accordance with their competences and the principle of avoiding conflicts of interest—a personal relationship between the reviewer and the author, professional subordination, and direct scientific cooperation in the last two years before the review. Each review was concluded with a clear recommendation to accept or reject a reviewed article. Small or serious corrections to the submitted manuscripts with additional peer review cycles were also possible. About 40% of the finally accepted manuscripts for publication have been subjected to the re-verification procedure. In case of negative opinions, the submitted articles have been rejected. This situation occurred for about 30% of the articles sent. The review took place in accordance with the "single blind" procedure, so the authors did not know the names of people assessing their work. After receiving the required number of two positive reviews, the Organizing Committee sent such information to the authors, ending the evaluation phase of the articles. Before the final decision on qualifying for publication in the conference materials, the authors "responses to reviewers" comments were absolutely required. The entire process of submitting, reviewing, and accepting articles has been carried out electronically using the System Coffee platform.

We would like to thank all the authors for the presentation of their scientific and implementation achievements, as well as all participants for taking part in lively discussions during the conference. Special thanks to the reviewers for insightful and reliable evaluation of the submitted articles. We are also grateful to the Organizing Committee for the hard work of preparing and conducting the conference.

International Scientific Conference Transport of the 21st Century

Conference organized by thc Faculty of Transport under the auspices of His Magnificence Rector of Warsaw University of Technology and The Committee on Transport of the Polish Academy of Sciences Ryn, June 9–12, 2019

Conference Goal

The conference aims at presentation of achievements of both Polish and foreign scientific and research centers dealing with rail, road, air and marine transport issues in technical and organizational aspects as well as at integration of researchers and education environment in discipline of civil engineering and transport.

Conference Topics

- infrastructure of transport and communication engineering,
- construction and exploitation of transport means,
- logistics engineering and transport technology,
- organization and planning of transport, including public transport,
- transport traffic control systems,
- transport telematics and intelligent transport systems,
- smart city and electromobility,
- safety engineering and ecology in transport.

Scientific Committee

Chairwoman

Marianna Jacyna	Warsaw University of Technology, Poland

Members

Tomasz Ambroziak	Warsaw University of Technology, Poland
Ryszard Barcik	University of Bielsko-Biała, Poland
Marek Bartosik	Lodz University of Technology, Poland

Francesco Losurdo	University of Bari Aldo Moro, Italy
Jerzy Leszczyński	Warsaw University of Technology, Poland
Konrad Lewczuk	Warsaw University of Technology, Poland
Andrzej Lewiński	University of Technology and Humanities in Radom, Poland
Jerzy Lewitowicz	Air Force Institute of Technology, Poland
Zbigniew Lozia	Warsaw University of Technology, Poland
Mirosław Luft	University of Technology and Humanities in Radom, Poland
Witold Luty	Automotive Industry Institute, Poland
Bogusław Łazarz	Silesian University of Technology, Poland
Czesław Łukianowicz	Koszalin University of Technology, Poland
Zbigniew Łukasik	University of Technology and Humanities in Radom, Poland
Jerzy Manerowski	Warsaw University of Technology, Poland
Andrzej Massel	Railway Research Institute, Poland
Jerzy Merkisz	Poznan University of Technology, Poland
Agnieszka Merkisz-Guranowska	Poznan University of Technology, Poland
Jerzy Mikulski	University of Economics in Katowice, Poland
Mirosław Nader	Warsaw University of Technology, Poland
Andrzej Niewczas	Motor Transport Institute, Poland
Tomasz Nowakowski	Wroclaw University of Technology, Poland
Michał Opala	Warsaw University of Technology, Poland
Olexandr M. Pshinko	Dnipropetrovsk National University of Railway Transport, Ukraine
Dariusz Pyza	Warsaw University of Technology, Poland
Leszek Rafalski	Road and Bridge Research Institute, Poland
Adam Rosiński	Warsaw University of Technology, Military University of Technology, Poland
Barbara Rymsza	Road and Bridge Research Institute, Poland
Janusz Rymsza	Road and Bridge Research Institute, Poland
Iouri Semenov	West Pomeranian University of Technology Szczecin, Poland
Jarosław Sęp	Rzeszow University of Technology, Poland
Mirosław Siergiejczyk	Warsaw University of Technology, Poland
Marek Sitarz	WSB University of Dabrowa Gornicza, Poland
Jacek Skorupski	Warsaw University of Technology, Poland
Zbigniew Smalko	Air Force Institute of Technology, Poland
Leszek Smolarek	Gdynia Maritime University, Poland
Bogdan Sowiński	Warsaw University of Technology, Poland
Włodzimierz Stawecki	Institute of Rail Vehicles "TABOR", Poland
Anna Stelmach	Warsaw University of Technology, Poland
Adam Szeląg	Warsaw University of Technology, Poland
Janusz Szpytko	AGH University of Science and Technology, Poland
Grzegorz Szyszka	Institute of Logistics and Warehousing, Poland

Wojciech Ślączka	Szczecin Maritime University, Poland
Marcin Ślęzak	Motor Transport Institute, Poland
Geza Tarnai	Budapest University, Hungary
Tadeusz Tatara	Cracow University of Technology, Poland
Franciszek Tomaszewski	Poznan University of Technology, Poland
Piotr Tomczuk	Warsaw University of Technology, Poland
Kazimierz Towpik	Warsaw University of Technology, Poland
Mariusz Wasiak	Warsaw University of Technology, Poland
Wojciech Wawrzyński	Warsaw University of Technology, Poland
Adam Weintrit	Gdynia Maritime University, Poland
Sławomir Wiak	Lodz University of Technology, Poland
Andrzej Wolff	Warsaw University of Technology, Poland
Wiesław Zabłocki	Warsaw University of Technology, Poland
Irina Yatskiv	Transport and Telecommunication Institute, Riga, Latvia
Paweł Zalewski	Szczecin Maritime University, Poland
Krzysztof Zboinski	Warsaw University of Technology, Poland
Jolanta Żak	Warsaw University of Technology, Poland
Józef Żurek	Air Force Institute of Technology, Poland
Andrzej Żurkowski	Railway Research Institute, Poland

Organizing Committee

Chairman

Mirosław Siergiejczyk — Warsaw University of Technology, Poland

Secretary

Adam Rosiński — Warsaw University of Technology, Poland

Members

Amelia Chocholska	Warsaw University of Technology, Poland
Ewa Dudek	Warsaw University of Technology, Poland
Karolina Krzykowska	Warsaw University of Technology, Poland
Zbigniew Kasprzyk	Warsaw University of Technology, Poland
Mariusz Rychlicki	Warsaw University of Technology, Poland
Marek Stawowy	Warsaw University of Technology, Poland
Andrzej Szmigiel	Warsaw University of Technology, Poland

Contents

Research on Proper Integration Between an On-Board and a Trackside Control-Command and Signalling Subsystems

Dominik Adamski, Krzysztof Ortel, and Łukasz Zawadka(✉)

Instytut Kolejnictwa (Railway Research Institute), ul. Chłopickiego 50, Warsaw, Poland
{dadamski,lzawadka}@ikolej.pl

Abstract. Achieving the interoperability of the European rail system requires many measures to unify technical solutions as well as regulations in each Member State. However, there is a possibility of some incompatibilities between individual subsystems despite generating them in accordance with unified applicable requirements. It is possible that the interoperable rolling stock will not be able to move freely over the interoperable railway line due to some incompatibilities and differences in the versions of the installed firmware in the ERTMS/ETCS system devices. In connection with the above the correct integration of rail vehicles with track-side equipment should be examined by means of tests under operating conditions. The article presents compliance tests of the correct integration of the on-board subsystem with the track-side subsystem which are carried out by the Railway Research Institute.

Keywords: Interoperability · Signalling · ERTMS/ETCS · TSI · CCO

1 Introduction

Over the years railway systems in particular Member States of the European Union have been developed independently using diversified technical solutions. The diversification of the technical rules, power supply, signaling, etc. has prevented the free movement of people and goods through the Member States. Therefore, it was decided to draw up the so-called White Paper - 'Roadmap to a single European transport area towards a competitive and resource efficient transport system' - where harmonization of technical requirements was considered to be the key issue in rail transport. The main goal was to lead to a situation in which trains will be able to cross the borders of European countries freely without stopping, on the contrary as it is today. In order to reach this goal it is necessary to achieve interoperability of the European rail system which will consequently lead to the creation of a single European railway area. This requires the elimination of a number of technical, legal and administrative obstacles. Thus, the railway system has been divided into structural subsystems [2]: infrastructure, energy, the control-command and signalling and rolling stock as well as functional subsystems: maintenance, operation and traffic management and telematics applications for passenger transport and freight transport. This division allowed drawing up specific

M. Siergiejczyk and K. Krzykowska (Eds.): ISCT21 2019, AISC 1032, pp. 1–10, 2020.
https://doi.org/10.1007/978-3-030-27687-4_1

Technical Specifications for Interoperability (TSIs) for particular subsystems, i.e. the requirements that must be met in order to achieve full technical harmonization of the railway. Dedicated to the relevant subsystems, the TSIs have the same structure, where the essential requirements, basic parameters, interfaces with other subsystems, the scope of necessary checks and inspections in order to obtain EC verification certificates, etc. are described.

The article focuses on issues related to the structural control-command and signalling subsystem implemented by the Commission Regulation (EU) No. 2016/919 of 27 May 2016 on the technical specification for interoperability within the control-command and signalling subsystems of the rail system in the European Union, hereinafter referred to as "TSI CCS" (from control-command and signalling) [1].

According to the provisions of Directive 2008/57/EC of June 17, 2008 command and signalling subsystem is described as: "All the equipment necessary to ensure safety and to command and control movements of trains authorised to travel on the network". Moreover [2] divides command and signalling subsystem into two separate subsystems i.e. "Control-Command and Signalling On-board Subsystem" and "Control-Command and Signalling Track-side Subsystem". It is clearly visible that in spite of the common denominator "control-command and signalling", both subsystems are independent assessment subjects for their EC verification by notified bodies. It should also be remembered that apart from the above mentioned control-command and signalling layers of the subsystem so-called basic layer is also distinguished. It consists of track vacancy detection devices, turnouts, station, line equipment etc. The requirements for the basic layer are defined by national regulations and this is conditioned by the diversity of traffic regulations and the applied technical solutions in individual Member States like it is in Poland in [7]. The confirmation of meeting the TSI requirements by the subsystem is to receive the EC certificate of verification which, after carrying out the necessary checks and inspections is issued by the notified body [9].

2 ERTMS System

2.1 Characteristics of the ERTMS System

Regardless of the case under consideration, the European Rail Traffic Management System (ERTMS) is installed on interoperable railway vehicles and interoperable railway lines. ERTMS is defined to meet the TSI control-command and signalling requirements despite where it is implemented. The ERTMS system is classified as Class A and is divided into the European Train Control System (ETCS) and the Global System for Mobile Communications-Railways (GSM-R). The operation of the ETCS system is based on calculating and controlling the braking curves [4, 5]. The aforementioned curves depend on many factors relating both to the vehicle and to the track. When devising the system, it was assumed that these factors can be separated into track-dependent and vehicle-dependent. Thus, vehicle data includes information such as vehicle weight, maximum axle load, maximum speed, brake system parameters, etc.

Track data, as opposed to vehicle data (which is provided once before the train's departure), are received by the vehicle during the entire driving time. They change both in time (depending on the traffic situation) and in space (depending on the location of the vehicle). These data include, above all, an movement authority, which consists of the maximum distance that the vehicle can cover and the maximum speed as a function of the distance from the reference point. Together with the movement authority vehicle also receives another information which is determined by track-dependent factors affecting the braking curves counted by the system. Such information includes, for example, the track profile (up and down gradients), distances to neighboring balises and information about other track – vehicle transmission channels.

ETCS systems in both on-board and track-side variants appear in three levels and depending on the requirements also several baselines. Baselines correspond to the version of the technical specifications designed and validated for particular project. TSI 2016/919 [1] indicates three sets of specifications that may be applied during EC verification. From 1 January 2019 the set of specifications only number 2 and 3 are valid for on-board subsystem. That means that since that date only vehicles equipped with baseline 3.4.0 and 3.6.0 are able to obtain EC certificate of verification excluding some specific cases with derogations. Differences between baselines consist in introducing additional functionalities clarifying the requirements as well as improving the stability of the system operation. Baseline 3.4.0 comparing to baseline 2.3.0d introduces new functions such as: Passive Shunting mode, Limited Supervision mode, level crossing not protected, new specification for the Non Leading mode, track conditions for the power supply, track conditions for sounding the horn, virtual balise cover etc. It is worth to mentioned that baselines backwards and forwards compatibility is defined in [6]. The rules for assigning the appropriate baseline for the GSM-R system are analogous to those applicable to ETCS and GSM-R system can be implemented in two versions: baseline 0 and baseline 1. It is important to mention that project baseline is not related to the levels installed. The article focuses on on-board systems so detailed information regarding trackside subsystem baseline version management will be intentionally omitted.

ETCS is based on digital track-vehicle transmission. The transmission can be carried out by balises, short, medium or long loops, a digital radio channel or specialized transmission modules. Data describing the track and the vehicle are used to calculate static and dynamic speed profiles. The calculated profile is constantly compared to the current speed in the position function. The necessary location function is based on clearly distinguishable (through a unique number) and accurately located devices for point transmission (balises or loops).

The control and supervision functions always operate according to the same rules regardless of the channel in which information from the track was received. Figure 1 below shows the structure of the ERTMS/ETCS system with interfaces.

Figure 1 clearly indicates the division into: track-side devices, track-vehicle transmission and on-board vehicle devices.

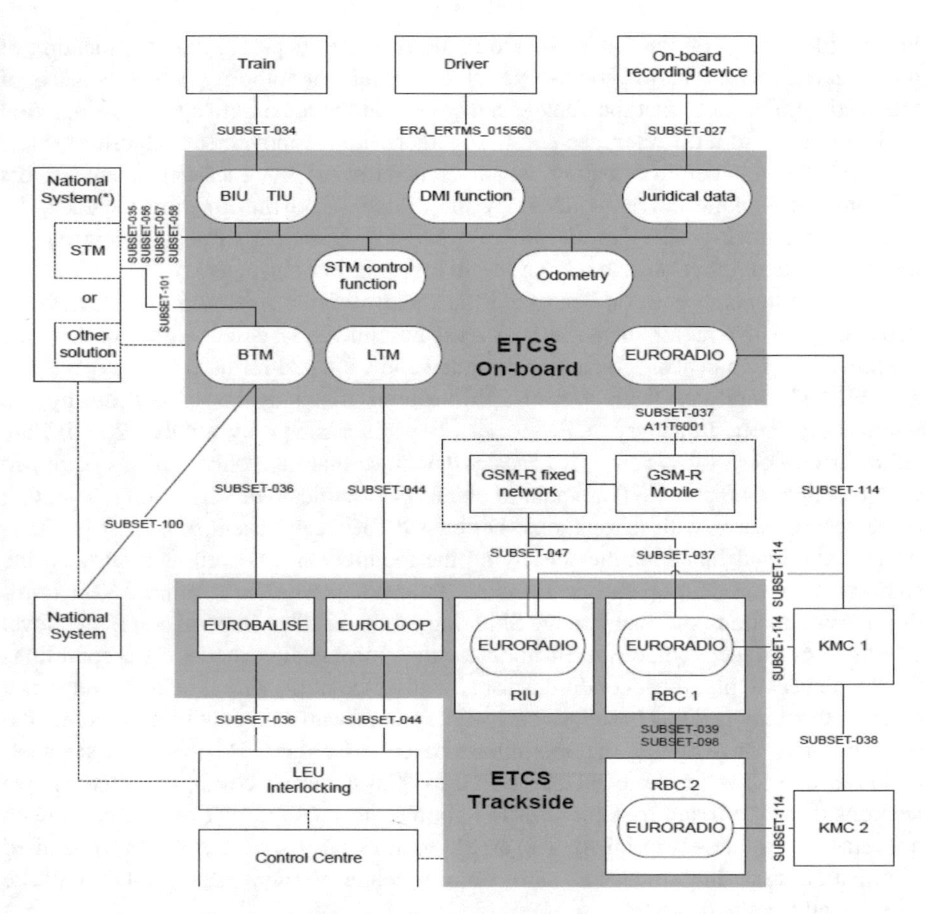

Fig. 1. The structure of the ERTMS/ETCS system with interfaces [4].

2.2 System Diversification

Depending on the specific project, both on-board and track-side equipment are provided by different manufacturers. In addition, individual hardware configurations may differ in software versions, versions of interoperability constituents or components used because a certain manufacturer may for example allow to use three types of modems or four types of sensors for odometry devices with their generic product [10]. Moreover, rolling stock manufacturers in consultation with manufacturers of on-board devices must jointly develop a dedicated ETCS/ERTMS system configuration specific to a particular vehicle type and the ETCS/ERTMS system itself must be properly implemented in order correctly communicate and interacts with train systems. ERTMS/ETCS trackside implementation is also subject to similar conditions. Challenges arise relating to the connection between the Radio Control Center (RBC) with relay interlockings (basic layer) or with neighbouring RBC, especially if it was built by another manufacturer.

Despite the introduction of standardization and unification of requirements relating to the control-command and signalling subsystem many issues still need to be clarified. Tests carried out in laboratories in simulated conditions as well as manufacturer's operational tests do not provide sufficient information on the correct integration of on-board equipment of the subsystem with track side equipment. This issue was already mentioned at the very introduction of [1] where it was stated: "Even a successful certification process cannot always exclude that, when an on-board CCS subsystem interacts with a trackside CCS subsystem, one of the subsystems repeatedly fails to function or perform as intended under certain conditions. This may be due to deficiencies in the specifications, different interpretations, design errors or equipment being installed incorrectly. A more coordinated way to perform compatibility tests should be introduced in order to help operators to take appropriate decisions."

Therefore Table 6.2 of the [1] clearly states the checks than need to be carried out when verifying the onboard subsystem regardless of the chosen evaluation module. One of the aspects that has to be assess is "Integration with Control-Command and Signalling Trackside Subsystems and other subsystems: tests under operational conditions". These checks must be carried out on the line under as many different operational conditions as reasonably possible e.g. line gradient, train speed, traction power, weather conditions etc. The scope of these tests should not include tests already carried out at earlier stages for example tests carried out on the interoperability constituents or in the laboratory.

3 Tests of a Proper Integration of Subsystems

3.1 Test Scenarios

Due to arisen situation the need to verify the correct integration of both subsystems in operational conditions was noticed and individual EU Member States were required by [1] to develop operational test scenarios. These tests are intended to check proper operation of the railway system in situations relevant to the ETCS and GSM-R and to demonstrate that the subsystems under examination are compatible with each other. Test scenarios created by each country's National Safety Authority (NSA) are to be sent to the European Railway Agency (ERA) in order to create a consolidated version of scenarios valid throughout the EU community in the future [11]. Currently scenarios elaborated by these entities are used in each Member State separately.

In Poland the body that defined the test scenarios was the Office of Rail Transport (UTK). They were published in the form of tables describing individual test cases along with possible driver's actions, expected system reaction and finally the behavior of the tested vehicle. Exemplary test scenario was presented on Fig. 2.

Every vehicle subject to EC verification, irrespective of the chosen evaluation module based on [3], must pass the tests of correct integration with the track-side subsystem. Such tests must be carried out by an entity having appropriate competences and qualifications. Most often it is a Notified Body (NoBo) or the Designated Body (DeBo) cooperating with it. The Railway Research Institute conducts tests on the compliance of on-board subsystems with track-side as NoBo or as DeBo depending on

Entry into the area equipped with ETCS L2 - on-board in L0 or STM SHP

No	Initial state	Action taken	Expected result
1	The vehicle is traveling on a railway line outside the GSM-R and ETCS area L2, on-board devices running with selected level 0 or STM SHP mode.	Drive over the balise group announcing the entry to the area GSM-R.	ETCS On-board equipment is establishing a connection with the correct radio network.
2	The vehicle is traveling on a railway line in the area of GSM-R and beyond ETCS L2 area, on-board devices running with selected level 0 or STM SHP mode.	Drive over the balise group establishing communications from RBC.	ETCS On-board equipment is establishing communication session with RBC, vehicle sends train data, vehicle receives parameters for reports, national variables and confirmation.
3	The vehicle is traveling on a railway line in the area of GSM-R and beyond ETCS L2 area after passing balise groups announcing the entry to the ETCS L2 area.	Drive over the balise group entering the ETCS area L2. Signal on the semaphore related to this group of balise - permitting.	Vehicle reports its position and obtains a permission for driving, changing on-board devices in ETCS level 2, FS operation mode, speed limit in accordance with the speed profile for the vehicle and line, sending information about changing train parameters.

Fig. 2. Exemplary test scenario.

the function performed in the given EC verification process. A similar situation takes place in the case of verification of track-side subsystems for which operational scenarios dedicated to specific implementations on specific lines are also provided by Infrastructure Managers. Issues related to the testing of trackside subsystems are not covered by scope of the article and therefore they will not be discussed.

3.2 Test Venues

The tests of proper integration of subsystems presented in the article are based on the knowledge and experience gained by the Railway Research Institute while carrying them out on the Polish ERTMS trackside infrastructure. The key issue that has been taken into account during the organization of tests was a selection of test fields depending on the version system and level. Since year 2016 Railway Research Institute has ETCS level 1 system installation on Test Track Centre near Żmigród. This system is compliant with the Functional Requirement Specification ERA/ERTMS/003204 ERTMS/ETCS FRS version 5.0 and [3]. The main devices of the ETCS level 1 system in the research area are: switchable and non-switchable balises, LEU encoders and the ETCS-L1_IK type simulator. At the beginning of 2019 the first ETCS L1 installation in Poland compliant with baseline 3.4.0 was introduced on the Test Track Center. It allows testing the compatibility of vehicles equipped with baseline 3.4.0 on-board devices with the infrastructure with the state of art system version.

In the case of level 2 ERTMS/ETCS tests it is necessary to conduct them on the line belonging to the Polish infrastructure manager PKP PLK, what determines a number of formal conditions that must be completed in order for the tests to take place. The procedure starts with specifying test track that meets the criteria for operational scenarios. Then technical and operational risk assessment is carried out for the tested vehicle, where all parameters influencing the tests are taken into account. Later on, in agreement with all parties involved in the study of a specific rolling stock, a temporary driving rules are created on the basis of which strictly specified track closures are introduced.

At present test of proper integration are carried out on the E-30 line Legnica Wsch. - Miłkowice - Chojnów along with adjacent lines. It is worth noting that the 296 railway line is not covered with the ERTMS/ETCS level 2 system, however, there are balises enabling the entry/exit to the system. Therefore this particular section of a track is used to verify level transitions scenarios during tests. At the Miłkowice and Chojnów stations the EBILock950 version 4 computer is installed. The essential part of the ERTMS/ETCS level 2 infrastructure is the Radio Block Center (RBC) which is an integral part of the complete INTERFLO 450 control system. RBC devices consist of two parts: a processing and executive module (RBU), an ISDN server that supports radio transmission. The RBC environment has external connections to: the interlocking system, the operator panel (CMI) and the GSM-R network. RBC's task is to supervise the movement of trains equipped with on-board ERTMS/ETCS level 2 devices. Supervision consists of sending telegrams to the vehicle with a movement authority (distance, permitted speeds, temporary speed limits, emergency braking order, etc.) and receiving from the vehicle radio information about the location of the vehicle (Fig. 3).

Fig. 3. Dispatch center with RBC (CMI on the right).

The balises are also an element of the trackside devices of the ERTMS/ETCS level 2 system but mainly they serve the positioning function of the train. The train equipped with ETCS on-board devices passing over the balises, reads the identification data of the balises and sends them to the radio control center. RBC has information about the location identifier and balise orientation (order of balises in a group) and on this basis determines the location and direction of travel. Balises are also used to calibrate devices responsible for measuring the distance (odometer). In the area of the test ground there are category A, E and B level crossings. The information of a state of level crossing devices are send to the LEU encoder which is connected with switchable balises installed in the track. The built-in devices comply with UNISIG specifications, including [3] and Subset 108 version 1.2.0 according to the Technical Specification of Interoperability for the Rail Traffic Control Subsystem.

3.3 Tests Results

The integration tests of a rolling stock with track-side infrastructure are carried out based on UTK operational scenarios and a set of additional checks developed by the Railway Research Institute. Train test runs are performed in fixed conditions in order to obtain a specific and expected results. Particular attention should be paid to the integration of the ERTMS/ETCS system with the Class B system, as it is the scope of a specific application for a given subsystem of a vehicle in a given country. During the tests, the correctness of the vehicle's transitions between levels and operating modes is checked, the response to changing the length of the Movement Authority (MA), speed control, reaction to the failure of individual components, etc. All indications on the Driver-Machine Interface (DMI) as well as the signals of semaphores are recorded using digital cameras. In addition, in the case of level 2 tests, it is coordinated from the level of RBC, where a representative of the Railway Research Institute is also present (Fig. 4).

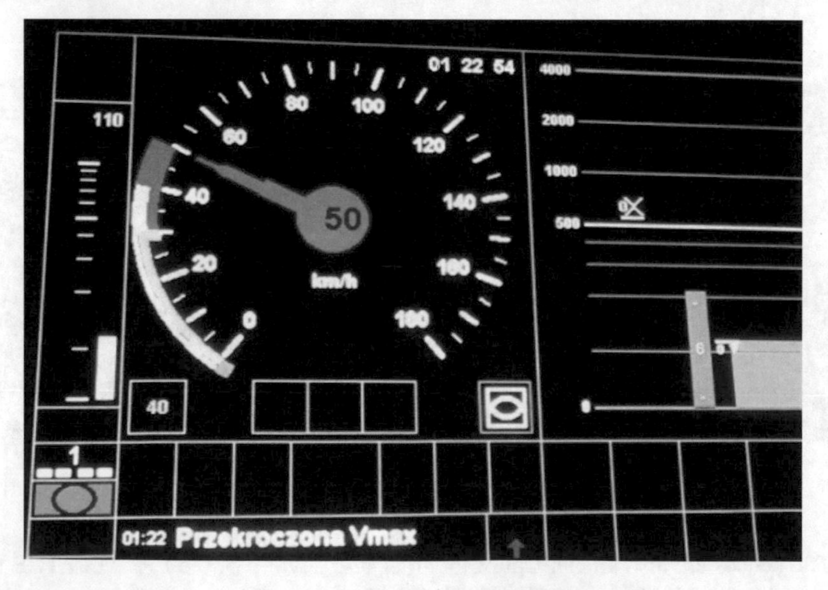

Fig. 4. DMI during test run [8].

The Juridical Recorder (JRU) is the source of all necessary data to evaluate the obtained results, in which all required train driving parameters are saved (see Fig. 5). In addition, in order to verify the correctness of telegrams read from the balises and RBCs it is also possible to use the so-called Logs obtained from the European Vital Computer (EVC).

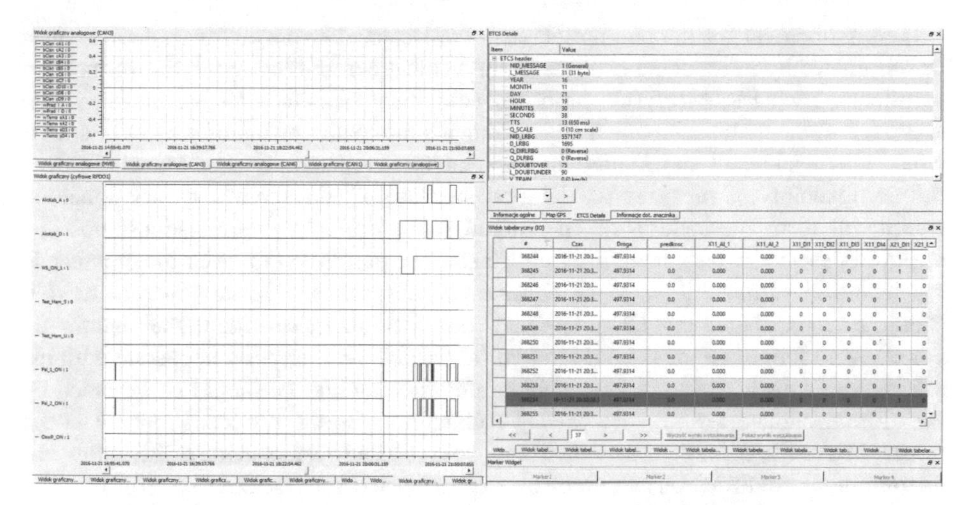

Fig. 5. Screen with raw data from JRU.

Data collected from the aforementioned sources, knowledge of the testing ground and visual observations made by qualified staff allow for an assessment of the correct integration of the on-board subsystem with track-side subsystem. In ambiguous cases, the assessment process is significantly extended, because it is necessary to analyze the error in detail in order to eliminate it. System errors that cause the TRIP - emergency vehicle stop, poor driving parameters (permitted speed reduced with no reason), problems with measuring the distance traveled, or measuring the distance from train antenna to the recently traveled balise are only a part of the difficulties encountered during discussed tests. In most cases, minor changes in the vehicle software are sufficient solution to the problems that arose but in extreme cases it is necessary to replace individual system components. Sometimes it is necessary to repeat whole test scope because necessary adjustments take time and solutions are not always obvious. For example, during the L2 tests, it turned out that the tested electric multiple traction unit that had two BTM components and EVC was incorrectly calculating distance between them. The result of this situation were problems with starting of a vehicle from known position after changing cab while RBC connection was still maintained. This issue was causing odometry errors and it was solved, after analyzing data from the vehicle and RBC telegrams, via software upgrade.

Observation of events and driving parameters in connection with the knowledge of individual subset requirements enables the analysis and subsequent assessment of the correct integration of on-board subsystem with trackside subsystem.

4 Conclusion

The issues of correct integration of the “Signalling On-board Subsystem” with the “Signalling Trackside Subsystem” discussed in the article are an important element in the approval of the rolling stock. This is the final verification of the vehicle before obtaining approval for the putting into service. Detected errors and irregularities must be eliminated by introducing changes to the vehicle before issuing the EC verification certificate. Each ERTMS/ETCS on-board equipment manufacturer has an individual approach to the manner in which he implements the requirements of subsets and TSIs what in the final result may require the implementation of additional measures to ensure full compatibility of subsystems. Therefore because of variance in national signalling equipment (e.g. interlocking's), operational rules, different interpretations, possible errors in design etc. tests mentioned above shall be performed in order to demonstrate the technical compatibility of the considered subsystems. The last but not least It should be remembered that laboratory and commissioning tests, due to their nature, do not allow to formulate the conclusion that the vehicle will properly cooperate with the infrastructure dedicated to it. Therefore commissioning checks should be extended by the aforementioned tests before placing rolling stock in the service. Probably in the future ERA will formulate a unified and consolidated test plan that will allow for unambiguous confirmation of compliance of the tested subsystems.

References

1. Commission Regulation (EU) 2016/919 of 27 May 2016 on the technical specification for interoperability relating to the 'control-command and signalling' subsystems of the rail system in the European Union
2. Commission Directive 2011/18/EU of 1 March 2011 amending Annexes II, V and VI to Directive 2008/57/EC of the European Parliament and of the Council on the interoperability of the rail system within the Community
3. Commission Decision (EU) 2010/713 of 9 November 2010 on modules for the procedures for assessment of conformity, suitability for use and EC verification to be used in the technical specifications for interoperability adopted under Directive 2008/57/EC of the European Parliament and of the Council
4. System Requirements Specification UNISIG SUBSET-026 version 2.3.0d
5. System Requirements Specification UNISIG SUBSET-026 version 3.4.0
6. System Requirements Specification UNISIG SUBSET-104 version 3.3.0
7. The list of the President of the Office of Rail Transport on the relevant national technical specifications and standardization documents, the use of which allows meeting the essential requirements relating to the interoperability of the rail system of 19 January 2017
8. Adamski, D., Ortel, K.: Problematyka współpracy pokładowego systemu ERTMS/ETCS z polską infrastrukturą; Autobusy, June 2018
9. Kycko, M.: Metodyka certyfikacji podsystemu sterowanie. Prace Naukowe Politechniki Warszawskiej, Warsaw (2016)
10. Pawlik, M.: Europejski System Zarządzania Ruchem Kolejowym, przegląd funkcji i rozwiązań technicznych – od idei do wdrożeń i eksploatacji. KOW, Warsaw (2015)
11. ERA homepage. https://www.era.europa.eu/activities/european-rail-traffic-management-system-ertms_en

Alternative Method of Diagnosing CAN Communication

Marcin Bednarek[1(✉)] and Tadeusz Dąbrowski[2]

[1] Faculty of Electrical and Computer Engineering, Rzeszow University of Technology, W. Pola 2, 35-959 Rzeszow, Poland
bednarek@prz.edu.pl

[2] Faculty of Electronics, Military University of Technology, gen. S. Kaliskiego 2, 00-908 Warsaw, Poland
tadeusz.dabrowski@wat.edu.pl

Abstract. The article discusses a suggestion for an alternative method of diagnosing communication in the CAN standard (*Controller Area Network*). Currently, a CAN communication bus is available not only in vehicles for which the standard was designed, but also in industrial plants. A CAN communication bus connects elements of, e.g. distributed control systems, vehicle onboard systems, remote intelligent I/Os of industrial controllers. The devices connected via the bus communicate in a broadcast mode. IDs (numbers) are assigned to messages and not devices. This enables simultaneous sending of a message to all awaiting recipients. A prerequisite for the correct functioning of the aforementioned systems is their proper operation, and primarily, maintaining the communication system in a state of fitness. Information regarding the sent variables are necessary to test the functioning of a data transmission system. It is the communication diagnostics that provides information necessary to formulate a diagnosis regarding the communication system condition. Based on the collected data sent in the messages, it is also possible to develop relevant remedial actions. In the case of CAN standard transmission, the diagnostic information can be acquired through intercepting and analysing the messages sent to the communication bus by communicating devices. The article suggests replacing often costly CAN bus analyser hardware solutions with a simpler one – an operator-diagnostic station – a computer equipped with an industrial CAN expansion card. By applying the right middleware, and the use of a standard visualization package, it is possible to monitor the data sent in CAN messages. This elaboration presents a concept of diagnosing communication using a standard SCADA visualization package, which is run on a diagnostic-operator station using middleware – a control window embedded directly in the synoptic image and a software gateway. This solution makes receiving (monitoring, visualization) and sending (testing) data frames to a CAN bus feasible.

Keywords: CAN standard · Diagnostics · Communication system

M. Siergiejczyk and K. Krzykowska (Eds.): ISCT21 2019, AISC 1032, pp. 11–20, 2020.
https://doi.org/10.1007/978-3-030-27687-4_2

1 Introduction

1.1 Considered System

The considered system is built of several actuating devices equipped with a CAN (Controller Area Network) interface [1–3] connected to a common communication bus (Fig. 1). The devices exchange data with each other creating a distributed control system. Originally, the CAN standard was developed to support connections and exchange data between equipment onboard vehicles [4]. However, it very quickly became clear that it satisfied the communication requirements for distributed control systems quite well [5]. The CAN standard is implemented in CANOpen [6–8] and DeviceNet [9, 10] solutions for connecting, controller devices, and sensors and actuators, respectively.

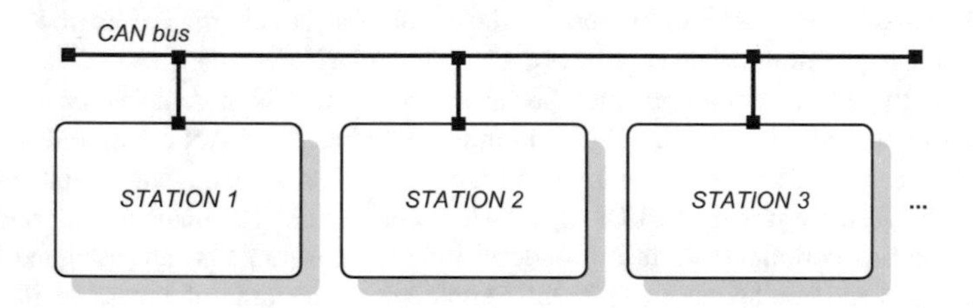

Fig. 1. Considered communication system.

1.2 CAN Basics

The CAN protocol, as mentioned previously, is based on bus topology. All devices hooked to a bus have the option to transmit or receive data. Master or slave devices are not distinguished. All stations have equal access to the transmission medium.

Each message has an 11-bit or 29-bit identifier, depending on the implemented standard (CAN 2.0A or CAN 2.0B, respectively). The data field contains a maximum of 8 bytes. Unlike the habits associated with computer networks, numbers/addresses are not assigned to devices. The adopted identification method is based on message identifiers. Access to the transmission medium is based on the CSMA/CA/AMP principle *(Carrier Sense Multiple Access with Collision Avoidance/Arbitration on Message Priority*). It involves the application of the so-called bit arbitration. Each device (station, node) can issue dominating or recessive bits, at the same time listening to what is going on within the bus. In the case of several stations simultaneously trying to broadcast a message, the devices keep sending a message until the internal supervision and therapeutic system does not signal a difference between the exposed bit value and the bus state. If one of the devices sends a recessive bit to the bus and the other one sends a dominating bit, the station with the lower priority “withdraws”. In the event of a simultaneous broadcast attempt by many devices connected to a medium, the node with a message with a lower identifier/address value is permitted to broadcast [1–3, 11, 12].

1.3 Need for Diagnostics

Due to the fact that the control software for the devices-stations connected to the bus are based on the data received from other stations, the communication system is one of the most important elements impacting the reliability of such a system. In systems, which are critical owing to their functions (mission critical systems) [13], a correctly functioning communication is a crucial element in ensuring the safety of process variable delivery between the devices. In the event of an unfitness within a communication system shown in Fig. 1, a system operator does not have the tools allowing to diagnose the communication state. Providing the operator with a preview of selected, send messages and an option to test-send certain data creates a possibility to take remedial action. Diagnostics is also possible in the event of complete fitness of a communication system. In such a case, the message sending/receiving tests are the elements of the operator's preventive actions, which ascertain him/her in the belief regarding full fitness or preventing the transition of the system into a state of unfitness.

The article proposes replacing an often very expensive solution of a ready protocol analyser (cf. cl. 2.1) with expanding the system with an operator station (e.g. computer) with an expansion card compatible with CAN and relevant middleware (cf. cl. 3.1).

2 Communication Diagnostics

2.1 Standard Solutions

In order to be able to diagnose a communication system state, the system should be equipped with a surveillance system (Fig. 2). Hardware or hardware and software protocol analyser are normally used for this purpose. They provide significant possibilities of analysing transmitted data. Selected possibilities are shown in cl. 2.2.

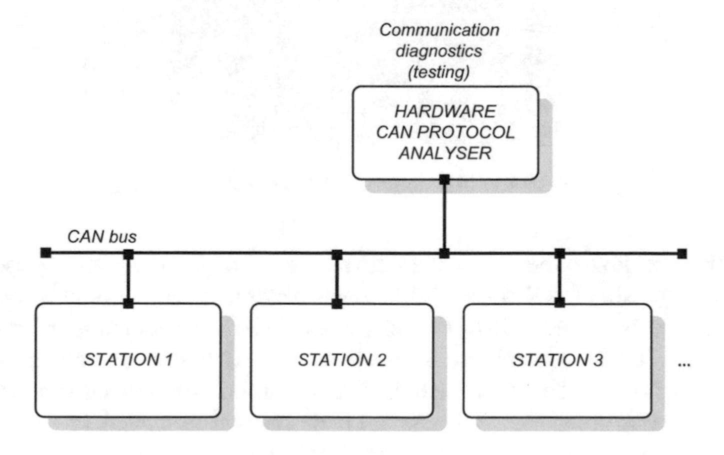

Fig. 2. Application of a hardware protocol analyser.

2.2 Overview of Available Solutions

Numerous CAN analyser solutions are available on the market, which enable a preview of the data sent to the CAN bus. They comprise a matching interface and dedicated software. The interface enables connection to the CAN bus via a USB port.

The solutions by Netronics Ltd. are worth mentioning at this point [14]. The analyser consists of a CAN-USB "Netronics Cando Interface" and dedicated software. The user has the option to use the available API.

Another similar software-based solution is a device by TKEngineering [15], the TKE CANtrace, which enables cooperation with, i.a., hardware interfaces by Kvaser, Vector Informatik GmbH and Peak System. Kvaser Leaf Light HS v2 interface can be used in this case [16].

There is also a hardware solution, which apart from the CAN-USB converter, has an option of Wi-Fi communication in the 802.11b/g/n standard. This is the Kvaser Black Bird device [16]. A built-in battery enables stand-alone use of the device and wireless communication facilitates contact with a device connected to an industrial CAN system. The manufacturer provides the libraries necessary to operate and program the device.

The available solutions also include the CanAnalyser by IXXAT. It cooperates with external cards (e.g. IXXAT USB-to-CAN v2 compact - Fig. 3) or expansion cards for PCs. In addition to the option of previewing the transmitted data, it is also possible to monitor, e.g. bus load trend images or errors [17].

Fig. 3. USB-to-CAN compact IXXAT hardware interface.

A synthetic review of the global literature indicates 3 major trends among the practical methods of diagnosing CAN buses. The first one relates to the use of built-in systems and small prototype systems [18, 19]. It is convergent with the aforementioned commercial solutions [14–17]. The second one is associated with utilizing software environments supporting diagnostics. This field includes the use of, i.a., specialized supporting environments (e.g. LabView [20], CANoe [21]) or original graphical interfaces [19] to diagnose CAN buses. The third one is oriented at safety-related diagnostics, regarding the detection and prevention of intentional intruder intervention [22–25], transmission security [24, 26, 27] or programming the behaviour of nodes based on transmission delay analysis [28, 29], as well as using additional microprocessor systems [30]. In contrast to the original, developed communication diagnostics solution presented herein, the previous applications of the SCADA visualization package in relation to CAN communication were limited to the aspects associated with the visualization and monitoring of the technological process [31], data acquisition and control [32, 33] or monitoring, failing to provide a possibility to preview the transmitted data.

3 Communication Diagnostics Using the SCADA Package

3.1 Proposed Solution

An alternative to the proposed CAN industrial network communication diagnostics method is using a standard SCADA visualization package for this purpose e.g. [34, 35]. Instead of a hardware-software protocol analyser, it is suggested to use an operator station, which successfully functions besides communicating devices in many industrial systems (Fig. 4). Such an application for the station results in a need to execute an additional coupling between the station and the CAN communication bus.

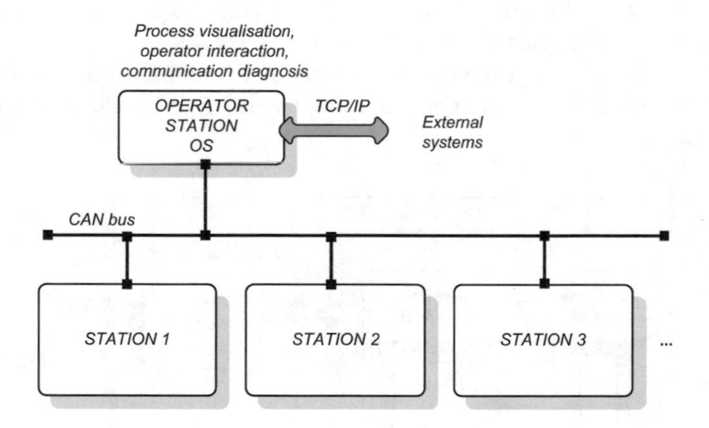

Fig. 4. Proposed solution – basics.

A hardware expansion card (Fig. 5a) [17] or a hardware CAN-USB converter (Fig. 5b) can be used for this purpose. The first applied solution is recommended for use in operator stations running on stationary or industrial computers. A CAN-USB converter is an alternative solution, which can be used as a coupler for the station and the CAN bus. References to the software libraries for this device are available at [36].

It should be stressed that the possibilities provided by the diagnostic application are limited to a few basic ones. However, its strengths should be borne in mind: versatility of the solution and application for monitoring/testing a standard SCADA package. The possible activities of a diagnostician using a SCADA software graphical interface with running software control window:

- reading the data field content of CAN frames with any identifier;
- reading the data field content of a frame with a set identifier;
- test-sending a frame with a specific identifier and set values of subsequent bytes;
- test-sending a remote data request frame.

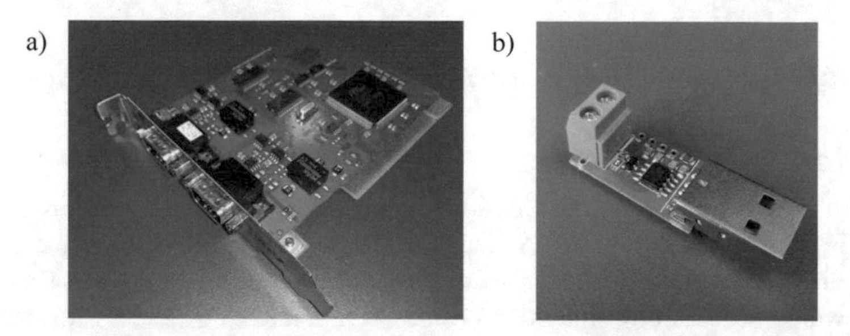

Fig. 5. CAN devices: (a) PC expansion card [17]; (b) CAN-USB converter [36].

The aforementioned functions enable monitoring (observing, aimed at identifying a status change) data field value changes within the received frames and use it to formulate a diagnosis regarding the correctness of sent data.

It is also possible to test the communication and response of attached devices to data received through sending a frame with set: identifier, data field byte number, data field byte value.

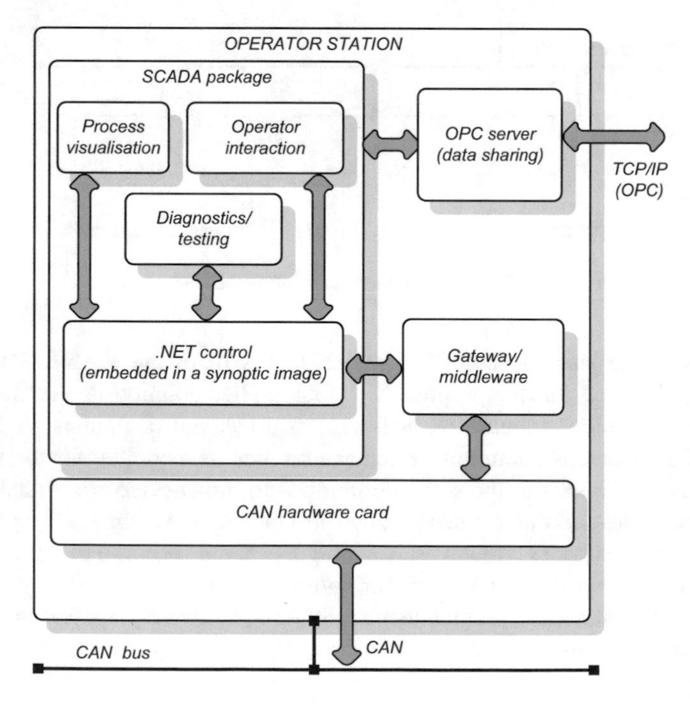

Fig. 6. Proposed solution.

Figure 6 shows a pictorial diagram of the suggested, alternative method for communication diagnostics. Middleware, otherwise called a gateway, directly communicates with the expansion card. Libraries supporting the CAN card are used for this purpose [17]. Next, the data is sent between the gateway and the control window with an interface embedded on the synoptic screen of the SCADA package. Using the synoptic image of the operator station, the diagnostician makes a decision regarding potential diagnostic activities (test-sending a message or observing the data field of a selected message). A message with a set ID is received after entering the corresponding frame number value into the ID field. Whereas sending a message involves filling appropriate, apart from the ID field, bytes of the data field to be sent.

3.2 Functional Test

The fragment of the conducted functional test of the SCADA package-based diagnostic system described herein involved performing the preparatory actions: hooking a CAN card to a communication bus connecting the central unit of an industrial controller (controller name not stated for ethical reasons) and a binary output expansion module; running the middleware and visualization package; creating a relevant synoptic screen with a constructed control window.

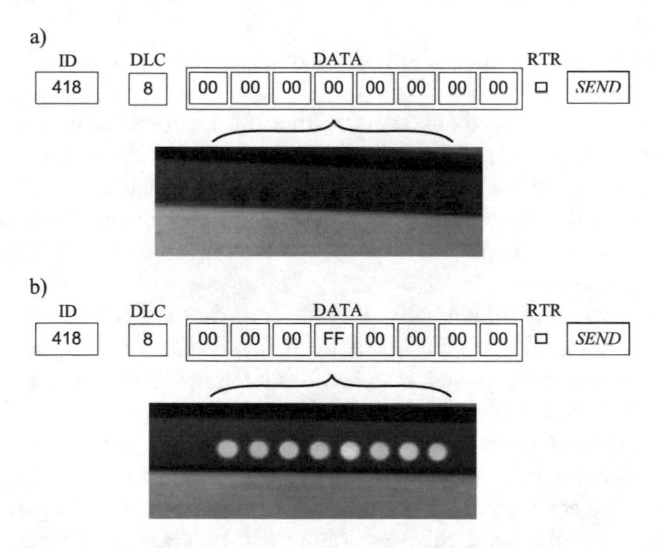

Fig. 7. Forcing test on binary outputs: (a) low states; (b) high states.

It was previously diagnosed that frame number 418 transmitted within the controller system contained information on the binary output settings of 16 subsequent channels (from channel no. 16 to 31). The data field value of frame number 418 was read. Zero values corresponded to low logic states on all sixteen channels of binary outputs. Data field values and a view of a front panel fragment of the controller output module with extinguished LEDs informing about low states on the outputs are shown in Fig. 7a. Next, the 0xFF value corresponding to the setting all channels in the high state was entered in an appropriate data field. Figure 7b illustrates frame 418 field value, sending which results in, i.a., illumination of high state indicator diodes next to relevant object binary outputs.

4 Conclusions

The article presents a concept of replacing an expensive CAN protocol analyser with a budget solution based on standard SCADA visualization software. Of course, the possibilities of such a diagnostic system are much poorer than standard specialized equipment, however, it is a sufficient tool in the case of diagnosing a communication system in terms of the correctness of data transmitted in the CAN frame data field subfields. Furthermore, by equipping it with an option to transmit a frame of a specific shape, we provide the diagnostician with a possibility to study the system response to a set forcing test.

It is worth to point out the advantages of the presented communication system diagnostic solution:

- the diagnostic system utilizes an operator station and existing SCADA visualization software with a control window, as well as ready, inexpensive hardware components (CAN card);
- enables reading a data field of any CAN frame sent along the bus in a transparent manner;
- provides an option to send any frame with a set data field value and number;
- all activities are performed from the visualization package synoptic image level, which can be placed among other operated operator images;
- transferring the diagnostic functions to a standard package creates a possibility for sending data through computer network connections to other hosts, hence, enables remote diagnostics;
- running the presented solution on a portable computer and equipping it with a USB-connected CAN card gives the freedom of diagnostics at any place, because the operator station can be moved anywhere – therefore, CAN communication can be diagnosed in, e.g. a vehicle or on a production line;
- the use of mobile devices provides the possibility to connect the device to a communication bus of a vehicle and transmit the information regarding any onboard equipment connected to the CAN network using WAN; it is of particular importance in terms of the constantly developing IoT (*Internet of Things*).

References

1. Bosch Homepage. http://www.bosch-semiconductors.com. Accessed 20 Apr 2019
2. Homepage. http://www.canbus.pl. Accessed 20 Apr 2019
3. ISO 11898-1:2015. Road vehicles - Controller area network (CAN) - Part 1: Data link layer and physical signalling (2015)

4. Bosch, R.: Vernetzung im Kraftfahrzeug. Gelbe Reihe Ausgabe 2007. Fachwissen Kfz-Technik. Elektrik und Elektronik für Kraftfahrzeuge [Networking in the motor vehicle. Yellow Series Edition 2007. Expertise Automotive Engineering. Electrics and electronics for motor vehicles], Robert Bosch GMBH, Plochingen (2007)
5. Morażyn, J.: CAN zdobywa popularność w przemyśle [CAN is gaining popularity within the industry]. ab2b Homepage. https://automatykab2b.pl/technika/37423-can-zdobywa-popularnosc-w-przemysle. Accessed 20 Apr 2019
6. CAN in Automation Homepage. http://www.can-cia.org. Accessed 20 Apr 2019
7. Moll, P.: Sieci CAN [CAN Networks]. Elektronika praktyczna **7**, 84–88 (2005)
8. HMS Technology Center Ravensburg Homepage. https://www.canopensolutions.com. Accessed 20 Apr 2019
9. Open DeviceNet Vendor Association. http://www.odva.org. Accessed 20 Apr 2019
10. Honeywell Smart Society Homepage. http://www.honeywell.com/sensing/prodinfo/sds/. Accessed 20 Apr 2019
11. Etschberger, K.: Controller Area Network: Basics, Protocols, Chips and Applications. IXXAT Press, Weingarten (2001)
12. Technical Documents CiA. https://www.can-cia.org/groups/specifications/. Accessed 20 Apr 2019
13. Storey, N.: The importance of data in safety-critical systems. Saf. Syst. **13**(2), 1–4 (2004)
14. Netronics CanBusAnalyser Homepage. http://www.cananalyser.co.uk/index.html. Accessed 20 Apr 2019
15. TKE CanTrace. https://tke.fi/products/can-bus/software/cantrace-can-bus-analyzer/. Accessed 20 Apr 2019
16. Kvaser Homepage. https://www.kvaser.com. Accessed 20 Apr 2019
17. IXXAT Homepage. https://www.ixxat.com. Accessed 20 Apr 2019
18. Dekanic, S., Grbic, R., Maruna, T., Kolak, I.: Integration of CAN bus drivers and UDS on Aurix platform. In: Proceedings of 2018 Zooming Innovation in Consumer Technologies Conference (ZINC), pp. 39–42. IEEE, Novi Sad (2018)
19. Song, Y., Wang, T., Xu, A., Wang, K., Yang, Z.: CAN based unified customizable diagnostic measure research and realization. In: Proceedings of 2012 UKACC International Conference on Control, pp. 825–829. IEEE, Cardiff (2012)
20. Drgoňa, P., Danko, M., Taraba, M., Adamec, J.: CAN BUS analyzer using modular instrumentation. In: Proceedings of 19th International Conference on Electrical Drives and Power Electronics (EDPE), pp. 336–340. IEEE, Dubrovnik (2017)
21. Hong, X., Chuan-guo, L.: Modeling and simulation analysis of CAN-bus on bus body. In: Proceedings of 2010 International Conference on Computer Application and System Modeling (ICCASM 2010), pp. V12-205–V12-208. IEEE, Taiyuan (2010)
22. Kyriakides, E., Polycarpou, M. (eds.): Intelligent Monitoring, Control, and Security of Critical Infrastructure Systems. Springer, Heidelberg (2015)
23. Moller, D.P.F., Haas, R.E.: Guide to Automotive Connectivity and Cybersecurity/Trends, Technologies, Innovations and Applications. Springer, Cham (2018)
24. Koscher, K., et al.: Experimental security analysis of a modern automobile. In: Proceedings of 2010 IEEE Symposium on Security and Privacy, pp. 447–462. IEEE, Berkeley/Oakland (2010)
25. Marchetti, M., Stabili, D.: Anomaly detection of CAN bus messages through analysis of ID sequences. In: Proceedings of 2017 IEEE Intelligent Vehicles Symposium (IV), pp. 1577–1583. IEEE, Los Angeles (2017)
26. Zhou, M., Ao, X., Wang, J.: Fault diagnosis of automobile based on CAN bus. In: Qi, L. (ed.) Information and Automation, ISIA 2010. Communications in Computer and Information Science, vol. 86, pp. 317–323. Springer, Heidelberg (2011)

27. Ying, X., Bernieri, G., Conti, M., Poovendran, R.: TACAN: transmitter authentication through covert channels in controller area networks. In: Proceedings of ACM ICCPS (ICCPS 2019), Montreal (2019)
28. Sagong, S.U., Ying, X., Bushnell, L., Poovendran, R.: Exploring attack surfaces of voltage-based intrusion detection systems in controller area networks. In: Proceedings ESCAR Europe (2018)
29. Pospíšil, T., Novák, J.: New method of CAN nodes health monitoring. In: Proceedings of International Conference on Applied Electronics, pp. 251–254. IEEE, Pilsen (2014)
30. Ying, X., Sagong, S.U., Clark, A., Bushnell, L., Poovendran, R.: Shape of the cloak: formal analysis of clock skew-based intrusion detection system in controller area networks. IEEE Trans. Inf. Forensics Secur. **14**, 2300–2314 (2019). Early Access
31. Yaolin, L., Xuange, P., Peipei, L.: Based on CAN bus wind generating set online monitoring and fault diagnosis system. In: Proceedings of 2008 IEEE International Conference on Sustainable Energy Technologies, pp. 195–197. IEEE, Singapore (2008)
32. Tcaciuc (Gherasim), S.-A.: A solution for the uniform integration of field devices in an industrial supervisory control and data acquisition system. Int. J. Adv. Comput. Sci. Appl. **9** (3), 319–323 (2018)
33. Liu, M., Guo, C., Yuan, M.: The framework of SCADA system based on cloud computing. In: Leung, V., Chen, M. (eds.) Cloud Computing, CloudComp 2013. Lecture Notes of the Institute for Computer Sciences. Social Informatics and Telecommunications Engineering, p. 133. Springer, Cham (2014)
34. Botland Homepage. https://botland.com.pl/pl/magistrala-can/8528-uccb-konwerter-usb-can.html. Accessed 20 Apr 2019
35. Wonderware (Part of AVEVA) Homepage. https://www.wonderware.com/. Accessed 20 Apr 2019
36. Vix Homepage. https://www.vix.com.pl/. Accessed 20 Apr 2019

Occupational Safety in a Medical Ambulances

Sylwia Bęczkowska(✉), Iwona Grabarek, and Katarzyna Mróz

Faculty of Transport, Warsaw University of Technology, Warsaw, Poland
bes@wt.pw.edu.pl

Abstract. A paramedic is a person performing a medical profession, authorised to provide health services in health care institutions, in particular to provide health services in a situation of direct, sudden threat to life or health. Scientific publications indicate a high rate of injuries among medical personnel while working in ambulances as well as difficulties in performing medical procedures due to the lack of unification of equipment in different ambulances. The problem of mismatch between the spatial structure of medical compartment in ambulances to the needs of rescuers and the methodology of its assessment is presented in the paper.

Keywords: Ambulance · Paramedic · Working conditions

1 Introduction

A paramedic is a person performing a medical profession, authorised to provide health services in health care institutions, in particular to provide health services in a situation of direct, sudden threat to life or health [1]. In accordance with the provisions of the Labour Code, the employer is obliged to ensure safe and hygienic working conditions. Due to the fact that the tasks of a paramedic include a wide range of duties, it is very difficult to identify the workplace of a paramedic, as he or she works wherever his or her assistance is necessary. The detailed scope of activities that a paramedic may perform independently or under the supervision of a physician is specified in the regulation [14]. Numerous publications contain information about high injury rates among medical personnel providing assistance in ambulances [7, 10, 15]. They also experience high levels of musculoskeletal trauma caused by lifting patients or other equipment [11]. A large part of this risk can be attributed to the design and arrangement of patient compartments, which are usually not designed to enable emergency services to perform patient care tasks while maintaining a stable seated position. The authors of the article limited their considerations to the workplace, which is a medical ambulance. The choice of the subject of the study was also dictated by the results of British and Polish scientists' research, which show that the time spent in the cabin of the ambulance during the performance of medical procedures is significant and amounts to about 44% of the active work intended to help the patient [5].

M. Siergiejczyk and K. Krzykowska (Eds.): ISCT21 2019, AISC 1032, pp. 21–30, 2020.
https://doi.org/10.1007/978-3-030-27687-4_3

2 Characteristics of Medical Ambulances

An ambulance is defined as a means of transport available to a place of sudden illness or accident, intended to provide assistance and transport sick or injured persons from the place of the incident to hospital, and often also used for medical and inter-hospital transport. Ambulances are operated by specially trained rescue teams and are part of the emergency aid system [13]. The transport of a sick or injured person takes place while ensuring maximum medical care during the transport of the patient. The actions which may be taken under qualified first aid are specified in Article 14, which mentions [1]:

- cardiopulmonary, non-instrumental and instrumental resuscitation, with administration of oxygen and use as indicated by an automated defibrillator
- stopping external bleeding and dressing wounds
- immobilisation of fractures and suspicions of fractures and sprains
- protection against cooling or overheating
- carrying out initial shock investigations
- use of passive oxygen therapy
- evacuation from the place of event of persons in a state of emergency health hazard
- mental support and initial medical segregation.

Therefore, an ambulance must be equipped with medical equipment that enables medical personnel to provide immediate assistance, including life-support [12]. According to the Polish Standard PN-EN 1789+A2 Motor Vehicles and their equipment, Road Ambulances distinguish 3 basic types of ambulances, namely: type A, B, C. Depending on the type of ambulance, its internal structure and equipment and medical materials are shaped in different ways. Type A ambulance is designed and equipped to transport patients whose lives are not seriously endangered.

The type B ambulance is designed and equipped to transport patients whose health condition requires only basic treatment and monitoring. A type C ambulance, called a “mobile intensive care unit”, is an ambulance whose design allows the transport of patients requiring advanced treatment and monitoring [18]. The ambulance is designed to transport patients requiring advanced treatment and monitoring. The medical space of the ambulance is equipped with necessary medical equipment and medical buildings (cabinets and shelves for small medical assortment). Proper spatial structure and location of medical equipment significantly facilitates rescue operations in life-threatening situations, shortens their time, which contributes to increased reliability of the lifeguard’s work. Today, most of the road ambulances in hospitals and medical services in the world are built on the basis of factory models of the world’s largest brands such as Mercedes, Renault, Citroen Peugeot, Fiat, Ford, Iveco, etc. Companies building road ambulances, use the bodies and chassis and adapt them to serve as road ambulances [17]. The PN-EN 1789+A2 standard specifies the basic requirements and recommendations for the vehicle and the medical space itself. Rescue ambulances must meet the following requirements: e.g. the height of the medical compartment must not be less than 175 cm, the height of the rear door must not be less than 170 cm. The configuration of the arrangement of the equipment depends on the customer and is based only on the experience and preferences of rescuers.

3 Hazard Identification at the Rescue Worker's Workplace

Regardless of the area of operation, rescuers are daily exposed to the forces of many dangerous, harmful and burdensome factors. The risks associated with the work of a paramedic are different in nature and affect occupational risks to different degrees. These include: communication accident, musculoskeletal overload, impact on moving parts, impact on fixed parts, fall at the same level, physical factors, chemicals and dust, biological factors, crushing, injury, psychosocial and work organisation factors. The source of many hazards is incorrect design and furnishing of the patient compartment, which may cause injuries to rescuers, e.g. by hitting stationary and moving parts of objects that are part of every ambulance. These objects usually have sharp, unprotected edges, are heavy and lack additional protection. In the United States, this problem was solved by designing a new spatial structure of the ambulance, which modified the security of cabinets, drawers, introduced safe locks for heavy equipment and elements that could cause injury to the rescuer's head [4]. In the medical compartment, rescuers are also exposed to injuries and cuts from the needles or knives they use to perform medical procedures. The reason for such a situation may be difficult access to medical equipment and materials as well as access to the patient from a sitting position, proper and safe for the rescuer while driving an ambulance. According to British researchers [20], approximately 48% of the 271 medical rescuers surveyed assessed the access to ambulance elements on a 5-degree scale at a bad level (only a very bad level is below). Also the seat comfort was negatively evaluated - 116 out of 271 people rated this feature at a bad level and 49 at a very bad level. Similarly, the backrest, which is a structural element of armchairs, affecting the comfort of sitting - very badly assessed 95 medical rescuers and 105 badly. The above mentioned factors indicate a lack of use of ergonomic principles in the process of designing interiors of medical ambulances. These problems have been described in [8] and [19]. In many countries, e.g. Austria, Germany or the USA, containers are mainly used as ambulances. Their dimensions make it possible to place medical equipment ensuring easy access both to the patient and to medical instruments.

Communications accidents caused by many external factors are a very big threat.

It is worth mentioning that rescuers do not fasten seat belts. According to Article 39 paragraph 1 of the on Road Traffic Law Act, road traffic participants are obliged to wear seat belts while driving, however paragraph 2 point 7 exempts from this obligation members of the emergency medical team providing assistance. In many cases, wearing a seat belt limits the performance of medical procedures and their absence unfortunately entails certain negative consequences, often resulting in injuries. This problem is also being raised in other countries and enforcement attempts are being made, such as in Finland [2].

Ambulance medical equipment includes gas cylinders, which incorrect mounting and operation may cause an explosion or fire and cause severe bodily injury, burns or even death. Similar injuries can result from damaged or unattached medical devices. Discomfort in the work of the rescuer also causes bad lighting, which can affect the probability of medical errors. The lighting system is also an inseparable part of the patient's compartment structure.

The presented analysis of the problem showed a number of threats, the source of which is the construction of the ambulance compartment. These threats may contribute to lowering the effectiveness of the work of medical rescuers and increasing the risk of musculoskeletal disorders, as well as a higher probability of an accident during the performance of medical procedures during transport of patients to hospital. Despite compliance with the provisions of the Polish Standard PN-EN 1789 [18], the problem remains unresolved. Too general provisions do not take into account the necessity to apply ergonomic rules in the design of medical compartments in ambulances. Currently, there is no standard or document that would specify the arrangement of ambulance equipment so as to ensure easy access to medical materials and devices used during the performance of medical procedures while driving, and on the other hand, to achieve unification of the interior of medical space, regardless of the type of car. This aspect is critical to the speed and reliability of the paramedic's work. A comprehensive solution to this problem requires a detailed analysis of the workload of rescuers performing various medical procedures in a moving ambulance.

4 Results of Pilot Studies and Research Directions

The development of a comprehensive solution should include an analysis of the existing condition, methods of workload assessment adequate to the specifics of the work of paramedics, formulation of ergonomic recommendations, the application of which will allow to adjust the spatial structure of the medical compartment to the capabilities and limitations of users. These actions will have a significant impact on increasing the safety of rescuers. The analysis of the existing situation in the field of safety of work of medical rescuers in ambulances was made on the basis of questionnaire and expert studies. A research form was developed which consisted of 18 questions of both closed and open nature. The questions concerned: participants' opinions on the vehicle fit, inconvenience involved in working in the ambulance, proposals and comments regarding the procedures performed and ambulance fit. The research was of a pilot nature and embraced 51 paramedics randomly selected from the Emergency Medical Rescue Service in Siedlce. The participants had different seniority, from 2 to 35 years, and therefore the average length of service was 14.3 years. The participants included both women and men. The group of women consisted of 5 people, which constituted 9.8% of all participants. The group of men embraced 46 people, or 91.2%. The distribution of seniority is shown in Fig. 1.

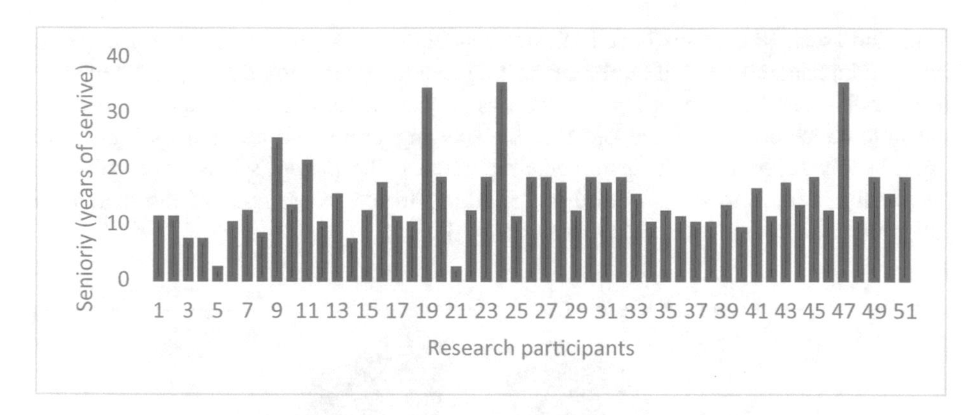

Fig. 1. Distribution of seniority among research participants

The tests were anonymous and were conducted with the consent of the Head of Emergency Medical Rescue Service in Siedlce. The obtained results were coded in the MS Excel 2010 program and developed statistically using the Statistica v. 8 program. The results obtained from the questionnaires revealed many irregularities present in ambulances. The article focuses on the discussion of the results concerning the safety of the rescuer's work. The survey raised, among others, the issue of position at work. Analysis of the responses showed that 71% paramedics perform medical procedures in a standing position. Only 17% of them worked in a sitting position (Fig. 2).

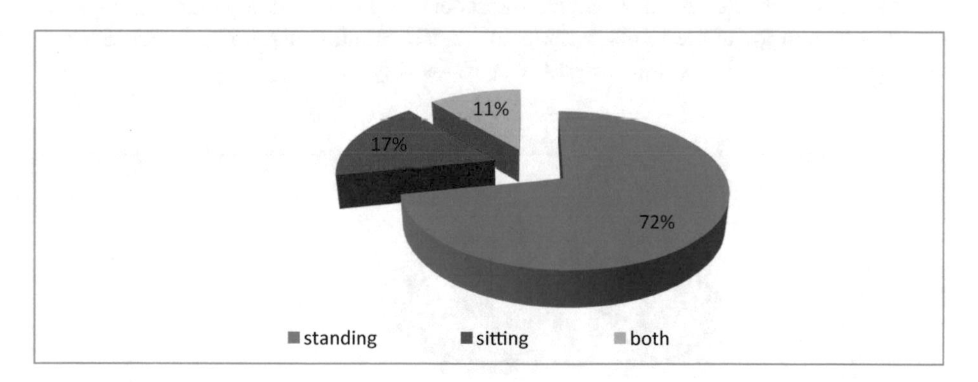

Fig. 2. Body positions adopted by paramedics preforming medical procedures

Standing position is particularly dangerous when the vehicle is moving, because the rescuer is exposed to many risks such as falls, injuries from stationary elements of ambulance equipment, needle-infected pricks or injuries. One of the questions on the position at work concerned the possibility of performing procedures in a standing position and how to prevent falling. Respondents indicated that there is no fall protection for standing position (50%). The other half indicated that such a protection exists and could be, for example, a safety net, which is installed in some vehicles next to the side seat. 18 rescuers have indicated a safety net as a protective net. 4 paramedics indicated a handrail located on the car ceiling as a fall protection device, and 3 medical rescuers indicated spreading legs far apart as a security feature. In addition, members of the ambulance crews pointed out that the existing protective net and railing do not provide any support for the paramedic standing at the patient's head. Difficulties in performing certain medical procedures resulting from the presence of the mentioned fall protection devices were also found (Fig. 3).

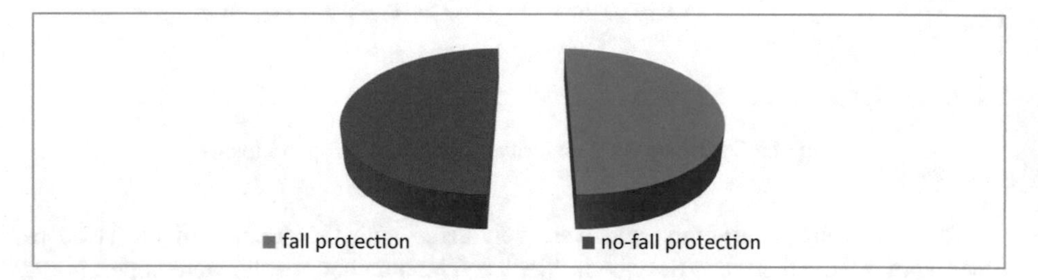

Fig. 3. Protection of a paramedic against falling

The standing position, additionally slightly inclined, is positioned unnaturally and keeping such position for a long time can cause weakening of the musculoskeletal system and diseases of the spine.

In the question about pain, as much as 56% respondents showed that their work is accompanied by back pain. 13% medical rescuers did not report any pain and 31% also reported pain in the ankle joints, wrist pain, bruises, contusions, limb cramps, aches of muscles, arms, shoulders, loins, and lower limbs (Fig. 4).

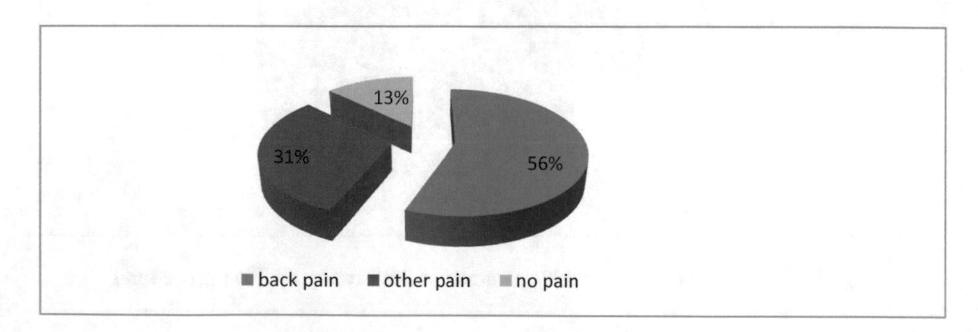

Fig. 4. Medical paramedic's ailments

Musculoskeletal ailments significantly reduce the comfort of work and may be the cause of making medical mistakes. The authors of the pilot studies also gained knowledge on the scope of modifications to the structure of medical ambulance, which would increase their safety and comfort. On this basis, a list of changes was created in order of the most important ones: increase of space, change of vehicle type to container, protection against falling, unification of equipment and ambulance installation, possibility to reach the patient from both sides, lack of sharp edges, access to equipment without touching the patient, description of the content of a given shelf/cupboard, better lighting. Rescuers in their surveys very often emphasized that it is necessary to modify the spatial structure - unification of ambulance equipment and buildings. The variety of arrangement makes it very time-consuming to get to know the "new interior", on average it takes about 33 min for rescuers, further increasing stress. Unification of the interior within at least one transport unit would significantly improve the work also in terms of its safety. A few years ago, paramedics of the Medical Air Rescue Service faced similar problems. This is why the purchase of helicopters was preceded by ergonomic research, which led, at the construction stage, to optimization of deployment of medical equipment and necessary medical materials in the helicopter medical compartment. Currently, all rescue helicopters are unified which results in identical placement of medical equipment in all machines. In their opinions, rescuers also stressed the need to equip medical compartments with protection against falls and injuries. This problem is often raised not only in the area of the unit whose rescuers were covered by the research. This may be confirmed by the analysis of the most frequent injuries of medical rescuers, which was carried out in one of the specialist hospitals of the Podkarpackie Voivodeship. Among the most numerous injuries were sprains or dislocations of joints, blows to the head, injuries, cuts. The highest number of injuries was recorded in lower limbs (32.5%) and upper limbs (22.5%) [19].

The problem of the abnormal spatial structure of ambulances and the nuisances resulting from them is present in many countries. Paramedics assessing ambulances operated in Malaysia [16] described the following characteristics of the ambulance as bad or very bad: total space available (60%), space above the head (40%), access to patient (43.3%), access to equipment in a sitting position (50%). Moreover, also in many foreign publications the problem of faulty solutions of ambulance spatial structure is discussed, which are characterized on the one hand by the lack of protection of equipment elements and on the other hand by their incorrect arrangement and thus the lack of access from a sitting position to necessary equipment and medical materials [3, 6, 9, 20]. The pilot studies carried out are the start of further studies related to occupational safety in medical ambulances. The results of the surveys showed that the necessary changes in the structure of the vehicle and its body should take into account ergonomic principles, because only in this way is it possible to better adapt the workplace to rescuers, guaranteeing safe work.

5 Research Methodology

The complexity of the presented issues related to the unification of patient compartments and adaptation of their equipment to the rescuers working in them requires a comprehensive approach and taking into account many aspects in the research, such as: the specificity of the rescuer's work, his anthropometric features, biological costs of work. A paramedic is often exposed to dangerous situations arising in an ambulance while saving the life and health of another person. Procedures performed on patients sometimes require lifting, staying in a forced, unnatural position, performing many monotype movements in insufficient lighting, in the presence of vibrations and noise affecting the paramedic. Additionally, rescuers are exposed to: stress, fatigue due to irregular working hours, work in constant tension or aggression on the part of patients. Adapting the spatial structure of the medical compartment of an ambulance to the specific needs of work and its unification will significantly minimize the risks occurring during the work of rescuers. This process requires the formulation of a comprehensive evaluation of the existing state, covering the following stages of research: Stage I - Selection of types of ambulances and analysis of performed medical procedures. Stage II - Assessment of existing solutions for the deployment of ambulance equipment from an ergonomic point of view (use of subjective, estimation and objective methods):

a. Developing a questionnaire/control list identifying the troubles accompanying the lifeguard during the assistance to the patient during transport to the hospital and carrying out research on a representative group of lifeguards.
b. Evaluation of the estimated physical effort for selected medical procedures,
c. Registration and analysis of the actual work of a rescuer - use of risk assessment methods related to musculoskeletal load, e.g. RULA (rapid upper limb assessment); REBA (Rapid Entire Body Assessment) and OWAS (Ovako Working Posture Analysis System).
d. Assessment of the body movement range of a rescuer based on the registration of motion kinematics by means of an inertial sensor system (MyoMotion), taking into account the anthropometric requirements.
e. Identification of the most involved muscles and their level of activity in a given ambulance body configuration - muscle tension measurements using surface electromyography (sEMG) during typical medical procedures.
f. Measurements in static and moving conditions (while driving) of muscle fatigue during the resuscitation procedure: manually by a rescuer and with the use of a massager (tests on phantoms).
g. Anthropometric verification of the ambulance medical compartment based on available simulation packages.
h. Development of a comprehensive (integrated) assessment based on the results of tests using heuristic techniques.

Stage III - Formulation of guidelines/recommendations for modification of medical compartment of an ambulance

a. Proposal of the concept of equipment deployment in the analyzed ambulance in the 3D program.
b. Ergonomic evaluation of the proposed solution

The interdisciplinary character and scope of the research proves the complexity of the problem.

6 Summary

The issue of safety in medical ambulances concerns both rescuers and patients. Ensuring work safety for rescuers is an important element determining the correct and effective performance of medical procedures by them, and their effect has a significant impact on the health of the rescued patient. Literature analysis showed revealed a widespread lack of unification of ambulance interiors, which affects many countries, not only Poland. However, actions are being taken in various research centers to formalize the requirements leading to the unification of ambulance equipment. Unification of operators' workstations is not a new problem and takes place in many means of transport, such as pilot cockpit or driver's cab. It promotes operator efficiency and reliability. The medical compartment is the place of work of a rescuer and at the same time the place of saving a patient's life. Ergonomic layout of medical equipment will ensure comfort for rescuers during their work, optimize the range and number of movements, reduce fatigue and, as a result, reduce the number of mistakes made during procedures. However, the change in the current state of affairs requires multifaceted research defined within the framework of the presented methodology of research.

References

1. Act of 8 September 2006 on state medical rescue service (consolidated text: Journal of Laws of 2017, item 2195)
2. Auvinen, T., Lisitsyn, D.: Study of paramedic staff safety comparing Greater Manchester and Finland. Thesis. https://www.theseus.fi/bitstream/handle/10024/137789/Thesis_2017_Paramedic_Staff_Safety_UK_VS_FIN_Teija_Auvinen_and_Dimitri_Lisitsyn.pdf?sequence=1/. Accessed 2017
3. Corbeil, P., Plamondon, A., Tremblay, A., Prairie, J., Larouche, D., Hegg-Deloye, S.: Measurement of emergency medical technician-paramedics' exposure to musculoskeletal risk factors, report R-944. Institut de recherche Robert-Sauvé en santé et en sécurité du travail, Montreal (2017)
4. Dąbrowska, A., Dąbrowski, M., Witt, M.: Bezpieczeństwo pracy personelu medycznego Zespołów Ratownictwa Medycznego. Anestezjologia i Ratownictwo **6**(4), 490–496 (2012)
5. Ferreira, J., Sue Hignett, S.: Reviewing ambulance design for clinical efficiency and paramedic safety. Appl. Ergonomics **36**, 97–105 (2005)
6. Gałązkowski, R., Binkowska, A., Samoliński, K.: Occupational injury rates in personel of emergency medical service. Ann. Agric. Environ. Med. **22**(4), 625–629 (2015)
7. Green, J.D., Ammons, D.E., Isaacs, A.J., Moore, P.H., Whisler, R.L., White, J.E.: Creating a safe work environment for emergency medical service workers. In: American Society of Safety Engineers' Professional Development Conference, Las Vegas Nevada (2008)
8. Kulczycka, K., Grzegorczyk-Puzio, E., Stychno, E., Piasecki, J., Strach, K.: Wpływ pracy na samopoczucie ratowników medycznych. Medycyna Ogólna i Nauki o Zdrowiu **22**(1), 66–71 (2016)
9. Levick, N., Grzebieta, R.: Crashworthiness analysis of three prototype ambulance vehicles. International Enhanced Safety of Vehicles (07-0249), Lyon, France, June 2007
10. Maguire, B.J., Hunting, K.L., Smith, G.S., Levick, N.R.: Occupational fatalities in emergency medical services: a hidden crisis. Ann. Emerg. Med. **40**, 625–632 (2002)

11. Maguire, B.J., Smith, S.: Injuries and fatalities among emergency medical technicians and paramedics in the United States. Prehospital Disaster Med. **28**(4), 376–382 (2013)
12. Olejnik, K., Nowacki, G., Woźniak, G.: Ocena obowiązujących wymagań w ambulansach pogotowia ratunkowego. Logistyka **4**, 8077–8082 (2015)
13. Pniewski, R., Pietruszczak, D., Ciupak, M.: Transport medyczny karetek pogotowia ratunkowego. Analiza czasów przejazdu. Autobusy-Technika, Eksploatacja, Systemy Transportowe **220**, 1092–1096 (2018)
14. Regulation of the Minister of Health of 20 April 2016 on medical rescue activities and health services other than medical rescue activities, which may be provided by a paramedic (Journal of Laws of 2016, item 587)
15. Reichard, A., Marsh, S., Moore, P.: Fatal and nonfatal injuries among emergency medical technicians and paramedics. Prehospital Emerg. Care **15**(4), 511–517 (2011)
16. Mohd Yusuff, R., Abidin, A.M.B.Z., Agamohamadi, F.: Task analysis of paramedics in the ambulance patient compartment. In: Advance Engineering Forum, vol. 10, pp. 278–284. Trans tech Publications, Switzerland (2013). ISSN: 2234-99X
17. Sobolewski, T., Posuniak, P.: Bezpieczeństwo pasażerów pojazdu medycznego podczas wypadku drogowego w świetle obowiązujących przepisów homologacyjnych. Logistyka **3**, 5855–5864 (2014)
18. Standard PN-EN 1789+A2: Mechanical vehicles and their equipment. Road ambulances. Polish Committee for Standardization (2014)
19. Wnukowski, K., Kopański, Z., Brukwicka, I., Sianos, G.: Zagrożenia towarzyszące pracy ratownika medycznego – wybrane zagadnienia. J. Clin. Healthc. **3**, 10–16 (2015)
20. Workers' Compensation Board of British Columbia: Evaluation of paramedics tasks and equipment to control the risk of musculoskeletal injury (2001)

Analysis of Elimination of Electromagnetic Disturbances at Power Ports of Railway Equipment

Kamil Białek(✉) and Patryk Wetoszka

Signalling and Telecommunication Laboratory, Railway Research Institute, Warsaw, Poland
kbialek@ikolej.pl

Abstract. The article presents the analysis of the occurrence of disturbances and the impact of the load of the selected railway device on the measurement result of conducted disturbances emission. The static converter used in railway vehicles (electric multiple units, electric locomotives) on the entry port ±3 kV DC was tested. The measurements were carried out using a specialized measuring station with a standardized hardware configuration. During the measurements, the exceeding admissible values in the frequency band of 150 kHz–300 kHz on both power terminals (+, −) were observed. In order to meet the requirements of EN 50121-3-2 [12] railway standard, and thus a reduction interference levels, an appropriate filter EMI at the EUT input was used. When selecting the filter taken account damping characteristic and other technical parameters such as: supply voltage, load current and operating temperature range. Installing the filter in accordance with the manufacturer's guidelines allowed to obtain the expected results. The article presents the effects of research in the form of tables containing specific values of electromagnetic levels of conducted disturbances and characteristics as a function of frequencies which were registered on the measuring device. The use of interference suppression systems is necessary in order to allow the use of electrical and electronic devices in the railway environment.

Keywords: Electromagnetic compatibility · Emission · Electromagnetic disturbances

1 Introduction

Railway equipment used and operated in the railway environment should meet standards of the electromagnetic compatibility (EMC). Meeting the normative requirements guarantees that the device will not interfere the operation of other electronic devices nearby and it will be resistant to the surrounding electromagnetic disturbances.

The article presents two examples of reduction of disturbances of the emission conducted by the selected railway equipment [6, 7].

In the railway environment, there is an increase in the number of impulse devices converting electrical energy, for example static converters, inverters, etc. The development of electronics in the field of semiconductor elements (the use of silicon carbide

M. Siergiejczyk and K. Krzykowska (Eds.): ISCT21 2019, AISC 1032, pp. 31–40, 2020.
https://doi.org/10.1007/978-3-030-27687-4_4

(SiC)) and the ferromagnetic material makes such devices with much smaller dimensions, weight and higher energy efficiency in respect to the traditional systems. Unfortunately, significant amounts of operated devices of this type, cause an increase in undesirable signals (disturbances), which have a negative impact on the work of other systems working nearby. In the railway environment, the following classification of disturbances depending on frequency is assumed: conducted disturbances from 150 kHz to 30 MHz and radiated disturbances from 30 MHz to 6 GHz [1, 2].

The mutual interactions of the device (system) and the environment are presented in Fig. 1. The interactions within a given system, which are often very complicated, are also illustrated. Electronic devices should be designed and constructed in such a way as not to interfere with the performance of their basic functions. Ensuring electromagnetic compatibility is a complex technical problem for many designers and constructors of electronic systems (devices, systems) [4, 5].

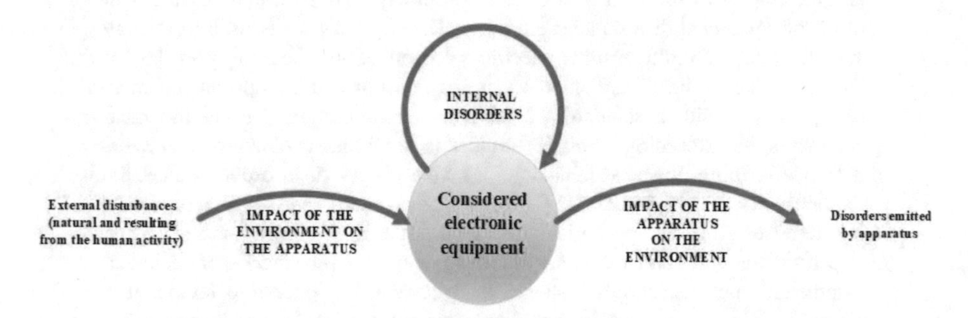

Fig. 1. The impact of disturbances in the electronic apparatus.

One of the rather specific sources of generating an electromagnetic field is the railway area. The problem of radioelectric disturbances occurring in the broadly understood railway environment turned out to be a difficult issue resulting from the multi-threaded subject matter. Problems of occurrence of disturbances in railway traffic control systems (SRK) have been well known for a long time, but in recent years their size and diversity have definitely increased. The reason is undoubtedly constant technological progress, miniaturization of systems and innovativeness of producers on the railway market [4].

Disturbances in the case of devices not directly supplied from low voltage networks get through capacitive, galvanic and inductive couplings. They are important for small distances between devices. The disturbances along the power cables - conductive disturbances - are much larger [2, 3].

Sources of conducted electromagnetic disturbances contain a symmetrical component of interferences - between lines and an asymmetrical component - between the conductor and the ground. Symmetrical disturbances penetrate very weakly.

The asymmetrical component of the signal is decisive. Excessive disturbances may affect the quality of the device or system in the railway environment, in particular, the functionality of srk devices. It is important, therefore, that the equipment under test (EUT - Equipment Under Test) does not emit disturbances of high values, as it may negatively affect safety in railway traffic. Testing equipment and systems for their compliance with the recommendations of the standard requires performing interference measurements and demonstrating that the device or system under test has a tolerable emission level in the frequency domain [4]. The hardware requirements and methodology for measuring the level of emission of conductive disorders are set by EN 55016-1-1 [9] and EN 55016-1-2 [10] standards. In contrast, emission limit values in the railway environment are described in specific EN 50121-3-2 [12] and EN 50121-4 [13] standards concerning electronic devices used on rolling stock as well as railway traffic and telecommunications devices – Fig. 2.

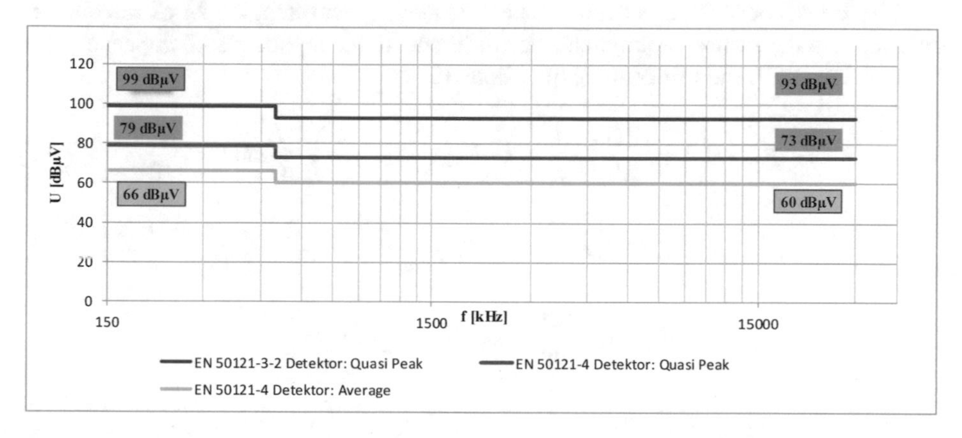

Fig. 2. Permitted limits for the emission of conducted disturbances required in the railway environment.

2 Analysis of the Occurrence of Conductive Disorders

Ensuring correct transmission, processing and use of electrical signals in the railway equipment require the existence of equipotential points and/or planes (potentials fixed in time, independent of the value of sink or source currents). These reference points or planes in a given subassembly, device or railway system constitute the so-called ground (also known as the grounds of signals). The name of grounding is fully adequate in case of the situation when a point or a given plane is on the earth potential (there is a physical connection with the earth). The points and planes located on the ground potential are safe for a human being. In practice, the reference points or planes are not equipotential – the grounding has low non-zero impedance (resistance). Then, the ground loops are formed in the electric and electronic circuits of railway equipment. If the signal source and the receiver (e.g. load – e.g. a battery pack in the railway vehicle) are connected to such a reference plane in the sufficiently distant points, then, a difference in potentials arises between these points, as a result of which the current flowing in this circuit occurs – Fig. 3. If multiple signal and load sources were connected to one plane or point, numerous, also common, return paths are formed, through which the currents flow from the loads to the sources. The result of such architecture is a mutual undesirable phenomenon – the interference effect of circuits. Then, the plane points – references have different potentials. If we consider two points, which are placed at the distance of 1 m on the metal plane (conductive housing of the railway vehicle) that constitutes a reference plane for operating signals, then, between these points there will be a difference in U potentials for a given f frequency caused by e.g. the impact of the unintentional magnetic field, which is generated by the power cable e.g. of the converter. The impedance Z between these two points can be recorded with the relationship (1) [1, 2]:

$$Z \approx R = R_{RF}(1 + \frac{ltg2\Pi l}{\lambda l}) \cdot \frac{l}{w} \tag{1}$$

The impedance value depends on the frequency (wavelengths λ) of interference occurring in the railway equipment. For different frequency ranges, Z impedance can be described with the use of the expressions (2–4)

$$Z \cong R_{RF}\left(1 + \frac{2\Pi l}{\lambda l}\right) \cdot \frac{l}{w} \cong k \cdot R_{DC}\left(1 + \frac{2\Pi l}{\lambda}\right) \cdot \frac{l}{w} dla\, l < \frac{\lambda}{10} \tag{2}$$

$$Z \cong R_{RF} \cdot \frac{l}{w} \cong k \cdot R_{DC} \cdot \frac{l}{w} dla\, l < \frac{\lambda}{20} \tag{3}$$

$$Z \cong R_{RF} \cdot \frac{2l}{w} \cong 2k \cdot R_{RF} \cdot \frac{l}{w} dla\, l \approx \frac{\lambda}{8} \tag{4}$$

$R_{RF}[\Omega]$ – surface resistance for the alternating current,
$R_{RF} = 0.26 \cdot 10^{-6}\sqrt{f}$ for copper,
$R_{RF} = 0.26 \cdot 10^{-6}\sqrt{\mu_w \cdot f/\sigma_w}$ for other metals,
σ_w – relative conductivity of the metal related to copper,
μ_w – relative permeability of the metal related to copper,
$\lambda[m]$ – wavelength corresponding to the frequency of the interference field,
$R_{DC}[\Omega]$ - surface resistance of the direct current,
k – number indicating the value R_{RF}/R_{DC}.

The relationships 1–4 will make it possible to estimate the impedance of the ground planes of various sizes, made of different metals for various frequencies of interference signals. The relationship 1 results in the fact that in practice, the impedance between two points on the equipotential plane may have different values. For the reference plane made of the metal with the following electrical properties ($\sigma_w = 10^{-1}$, $\sigma_w = 10^3$) and the dimensions of l = 70 cm, width w = 1 cm, Z value between extreme points on the plane for the interference frequency f = 100 kHz (λ = 3 km) can be estimated with the use of the relationship 3. The relationship shows that 1/λ = 2, $3 \cdot 10^{-4}$ < 1/20. The impedance Z for this case will be:

$$Z \cong R_{RF} \cdot \frac{l}{w} = 0,57\ \Omega \tag{5}$$

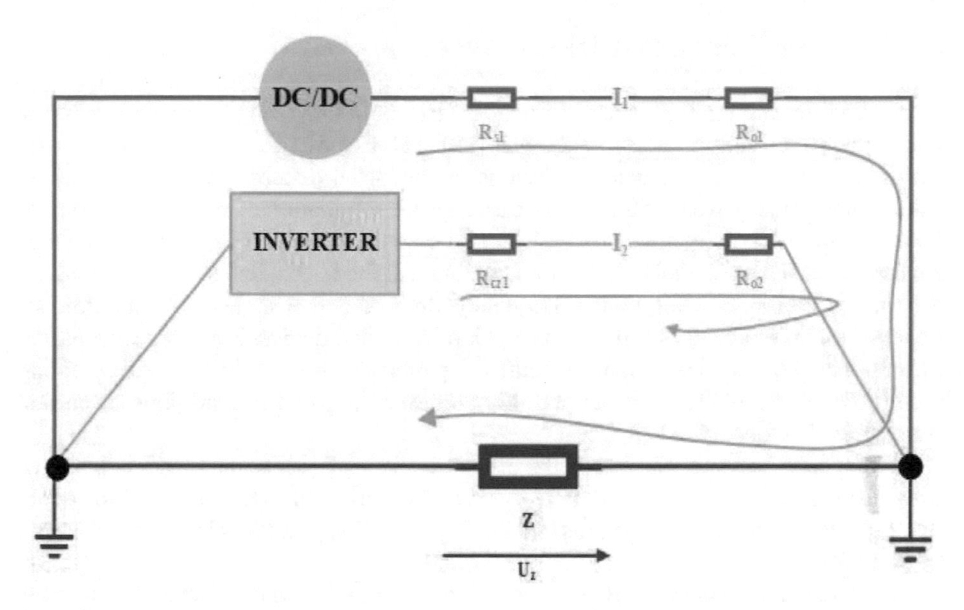

Fig. 3. The phenomenon of interference caused by the coupling of circuits with different levels of the signal on the common impedance Z of the ground plane.

R_{s1}, R_{cz1} – internal resistance of DC/DC converter source and inverter, R_{o1}, R_{o2} – load resistance DC/DC converter source and inverter.

The circuit No. 1 is the source of interference and traction inverter is an interference receiver. Interference signal is transferred by the common reference plane characterized by Z impedance [1, 2].

By analysing the circuit No. 1, U_z voltage will be:

$$U_z = E_1 \frac{Z}{R_{s1} + R_{o1} + Z} \tag{6}$$

The circuit No. 2 is a receiver of interference, the interference component on the resistance R_{o2}:

$$U_{zo2} = U_z \frac{R_{o2}}{R_{cz1} + R_{o2} + Z} \tag{7}$$

The voltage of interference for the circuit 2 will be:

$$U_{zo2} = E_1 \frac{ZR_{o2}}{(R_{s1} + R_{o1})(R_{cz1} + R_{o2})} \tag{8}$$

3 Research Conducted Disturbances

According to the methodology written in EN 55016-2-1 [11] standard and the requirements of railway EN 50121-3-2 [12] and EN 50121-4 [13] standards, the measurement of electromagnetic emissions of conducted disturbances is performed in the frequency range from 150 kHz up to 30 MHz. Examination of devices and systems in the scope of their compliance with the recommendations of the standard requires performing interference tests and demonstrating that the tested device or system has an acceptable emission level in the frequency domain [4]. The level of interference depends on the conditions of the target work of the tested device. For these reasons, the interference level measurements should be performed in clearly defined operating conditions of the EUT, as far as possible corresponding to the conditions of normal (target) exploitation [5, 8].

The paper presents the methods of elimination of electromagnetic conductive disturbances on the 3 kV power supply port of the static converter, designed to supply 3 × 380 V AC low voltage circuits and 24 V DC auxiliary circuits, including charging of batteries.

The main measuring device used for testing the emission of conducted disturbances of a static converter are a measuring receiver and a high-voltage measuring probe.

Research on the emission of conducted disturbances was carried out for research purposes using specialized measuring equipment – Table 1.

Table 1. Measuring apparatus.

Measuring device	Manufacturer	Type
Measuring receiver	Rohde & Schwarz	ESCI3
Measuring cable 5 m	HUBER SUHNER	RG58CU/11 N/11 N/005000
High voltage test probe	Schwarzbeck Mess Elektronik	TK 9420

The research concerned measurements of conducted disturbances of an asymmetrical component (between the cable and the earth) on the power supply port ±3 kV DC of the static converter. A multiple measurement of conducted disturbances emission was carried out, however, each course of electromagnetic disturbances as a function of frequency considerably exceeded the admissible values in the frequency band of 150 kHz–300 kHz. The results of the measurements carried out are presented below in the form of frequency characteristics - Figs. 4, 5 and Tables 2 and 3.

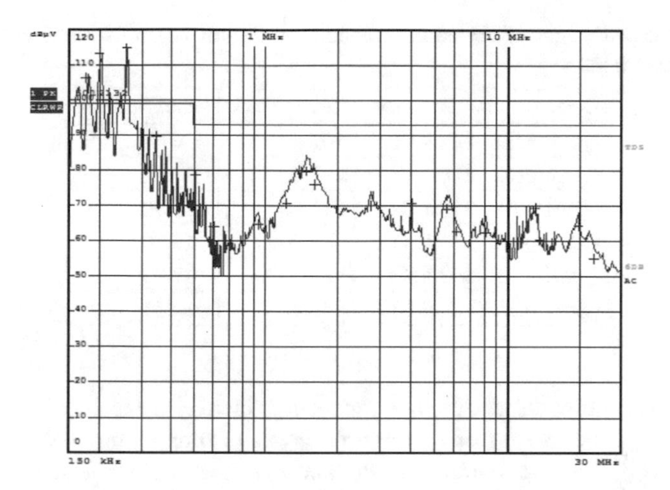

Fig. 4. Results of emission of conducted disturbances of a static converter on the +3 kV DC input port.

Table 2. Measured values of emission of conducted disturbances of a static converter on the +3 kV DC input port.

Detector	Frequency	Level dBμV	Delta limit dB
Quasi peak	17 kHz	106.41	7.40
Quasi peak	202 kHz	113.15	14.14
Quasi peak	258 kHz	114.79	15.79
Quasi peak	342 kHz	89.66	−9.33
Quasi peak	502 kHz	78.51	−14.48
Quasi peak	1.514 MHz	79.76	−13.24

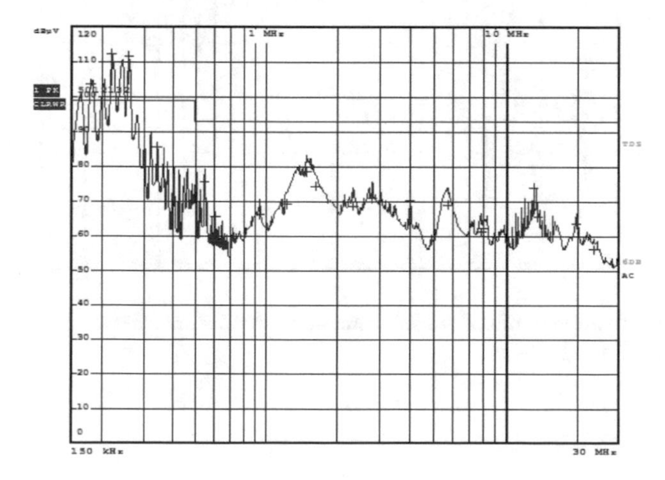

Fig. 5. Results of emission of conducted disturbances of a static converter on the −3 kV DC input port.

Table 3. Measured values of emission of conducted disturbances of a static converter on the −3 kV DC input port.

Detector	Frequency	Level dBμV	Delta limit dB
Quasi peak	182 kHz	103.67	4.66
Quasi peak	218 kHz	112.23	13.22
Quasi peak	258 kHz	111.70	12.70
Quasi peak	338 kHz	85.70	−13.29
Quasi peak	542 kHz	75.65	−17.34
Quasi peak	1.514 MHz	78.48	−14.51

In order to reduce the above-mentioned interference in the input waveform, and thus, to eliminate the excess of permissible levels resulting from EN-50121-3-2 [12] railway standard, it is important to use the anti-interference filter, which will direct the interference currents (of high frequency) to the common system ground point. It is also crucial to use the shortest connection to the system ground (impedance – earthing resistance), and also to avoid large areas of the ground loops. The ground loops should have the smallest possible area in order to minimize the impact of the interference electromagnetic fields on the cables. On Figs. 6, 7 and Tables 4, 5 present the results of testing the conducted disturbances static converter after applying the above mentioned solutions [1, 2].

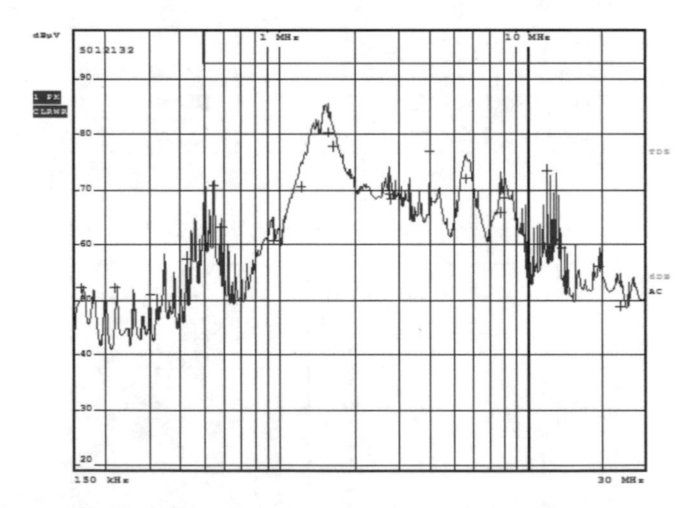

Fig. 6. Results of emission of conducted static converter disturbances on the +3 kV DC input port after proposed solutions.

Table 4. Measured values of emission of conducted disturbances of a static converter on the +3 kV DC input port after proposed solutions.

Detector	Frequency	Level dBμV	Delta limit dB
Quasi peak	542 kHz	70.81	−22.18
Quasi peak	1.57 MHz	80.48	−12.51
Quasi peak	1.634 MHz	77.99	−15.00
Quasi peak	4.018 MHz	77.05	−15.94
Quasi peak	5.626 MHz	71.95	−21.04
Quasi peak	8.082 MHz	68.63	−24.34

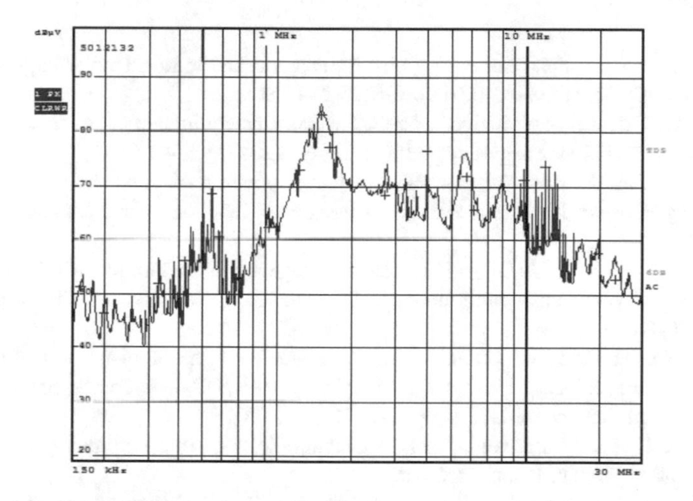

Fig. 7. Results of emission of conducted static converter disturbances on the −3 kV DC input port after proposed solutions.

Table 5. Measured values of emission of conducted disturbances of a static converter on the −3 kV DC input port after proposed solutions.

Detector	Frequency	Level dBμV	Delta limit dB
Quasi peak	542 kHz	68.73	−24.26
Quasi peak	1.246 MHz	73.07	−19.92
Quasi peak	1.53 MHz	83.16	−9.83
Quasi peak	1.642 MHz	77.31	−15.68
Quasi peak	4.018 MHz	76.59	−16.40
Quasi peak	12.058 MHz	73.69	−19.30

4 Summary

The measurements carried out made it possible to determine the frequency and magnitude of the amplitudes of the exceeded permissible levels. From these characteristics it was observed crossing the course of the input converter occur in the frequency range from 150 kHz to 300 kHz. The obtained results indicate the need for an anti-interference filter to reduce the level of emitted disturbances. The use of the EMI filter allowed to reduce the emissions of disturbances of the conducted static converter not only to the levels acceptable by the EN 50121-3-2 [12] standard, but significantly below the limit. In the device under consideration, the shortest connections to the system's mass were used - ground impedance. Earth loop in the test equipment and systems should have the smallest possible area to minimize the influence of disturbance of electromagnetic fields on internal wiring device.

References

1. Białek, K., Paś, J.: Analysis of the electromagnetic environment on a large railway are. Biuletyn WAT **67**(1), 53–63 (2018). ISSN 1234-5865
2. Białek, K., Paś, J.: Exploitation of selected railway equipment - conducted disturbance emission examination. Diagnostyka **19**(3), 29–35 (2018)
3. Dyduch, J., Paś, J., Rosiński, A.: Basics of the operation of transport electronic systems. Publishing House of Kazimierz Pułaski University of Technology and Humanities in Radom (2011)
4. Wetoszka, P., Paś, J., Laskowski, D.: Electromagnetic compatibility in selected electronic devices security systems - preliminary tests, pp. 1– 8. Elektronika, Publisher Sigma-NOT, August 2018
5. Krzykowski, M., Paś, J., Rosiński, A.: Assessment of the level of reliability of power supplies of the objects of critical infrastructure. In: IOP Conference Series: Earth and Environmental Science, vol. 214, pp. 1–9(2019). 012018
6. Donald, G.: Baker Electromagnetic Compatibility – Analysis and Case Studies in Transportation. Wiley, Hoboken (2016)
7. Montrose, M.I., Nakauchi, E.M.: Testing for EMC Compliance: Approaches and Techniques. Wiley, Hoboken (2014). IEEE
8. Soliman, S.A., Mantawy, A.H.: Modern Optimization Techniques with Applications in Electric Power Systems. Springer, Heidelberg (2012)
9. EN 55016-1-1:2010 Specification for radio disturbance and immunity measuring apparatus and methods – Part 1-1: Radio disturbance and immunity measuring apparatus – measuring apparatus
10. EN 55016-1-2:2014 Specification for radio disturbance and immunity measuring apparatus and methods - Part 1-2: Radio disturbance and immunity measuring apparatus - coupling devices for conducted disturbance measurements
11. EN 55016-2-1:2014 Specification for radio disturbance and immunity measuring apparatus and methods – Part 2-1: Methods of measurement of disturbances and immunity – conducted disturbance measurements
12. EN 50121-3-2:2016 Railway applications – Electromagnetic compatibility – Part 3-2: Rolling stock – apparatus
13. EN 50121-4:2016 Railway applications - Electromagnetic compatibility - Part 4: Emission and immunity of the signalling and telecommunications apparatus

Issues of Description of EMC Processes in Direct Current Catenary

Andrzej Białoń, Krzysztof Ortel, and Łukasz Zawadka(✉)

Instytut Kolejnictwa (Railway Institute), ul. Chłopickiego 50, Warsaw, Poland
{abialon,lzawadka}@ikolej.pl

Abstract. When examining engineering problems and physical phenomena, accurate results obtained in the laboratory calculations do not ensure compliance with actual results. Therefore, the results of analytical tests should be compared with experimental results. However, in the case of modern complex systems, including electromagnetic compatibility problems, which are characterized by numerous external influences, a direct comparison of a single calculation with a single experiment can often give absolutely unique results. The article refers to physical phenomena occurring in the traction network related to ensuring electromagnetic compatibility of the electric traction system with other railway devices. Characteristics of traction network (overhead and rail network) in terms of the possibility of calculating its parameters useful in the consideration of electromagnetic compatibility issues are presented. Replacement diagrams of traction networks (overhead and rail network) are shown. Calculation results obtained for networks with clustered and distributed parameters are shown. It was found that the treatment in calculations of the traction network with distributed parameters gives results very similar to those obtained in real conditions.

Keywords: EMC · Traction network · Noise

1 Introduction

Experimental researches are a basic information source for the study of electromagnetic processes in the traction energy system of the direct current electrified railways and for development of mathematical models of electric energy conversion and consumption processes, electromagnetic processes and interferences, as well as particular electrical devices [1]. The produced mathematical model enables conduction of variant theoretical research of the developed engineering measures and facilities, and identification of probable levels and extent of electromagnetic impact on the related low-power devices. The traction energy system electromagnetic compatibility (EMC) model is of probabilistic nature, while parameters of energy-exchange processes are statistical characteristics. Hence, the EMC researches are based on application of statistical methods of processing and analysis of experimental data obtained both from physical experiment and from mathematical simulation.

It is known [1] that at the study of engineering problems and physical phenomena, accurate (in regard of solution procedure) results obtained from some computation

M. Siergiejczyk and K. Krzykowska (Eds.): ISCT21 2019, AISC 1032, pp. 41–51, 2020.
https://doi.org/10.1007/978-3-030-27687-4_5

device do not ensure the agreement with actual ones. Therefore, analytical research results must be compared with experimental results. However, with the modern complex systems including electromagnetic compatibility problems, which are characterized by the numerous impact factors, the direct comparison of single calculation with the single experiment may often give absolutely non-typical results.

In such engineering systems, widely varying results may be obtained depending on combinations of these factors and random conditions. Therefore, disagreement of single calculation and single experiment results in some cases may not predictably characterize the theory incorrectness or calculation error, and similarly, single coincidence does not provide assurance [1].

In practice, in order to prove real rigor of engineering research, the comparison of calculation and experimental results for any complex system must be conducted taking into account the possible random variation in parameters, especially in the cases close to some extreme state, for example, at verification of harmonic component levels in the forced traction energy system operation mode. Such comparison must be conducted according to purposely designed method taking into account either variation of parameters obtained from experimental research of the real system or variation of parameters obtained from calculations provided that calculated and experimental results are represented and compared in the criterial form written in accordance with the theory of similarity. It is known that any mathematical simulation features the investigated process idealization to some extent. Therefore, estimation of validity of results obtained from the mathematical simulation of the traction energy system EMC problems must be conducted by comparison of theoretical and experimental data.

The solution of any problems with the use of probability theory in those cases where their statistical definition is used is only possible with the collection of proper statistical material based on the generous amount of experiments or observations. This involves origination of problems associated with the correct processing of statistical material and its transformation into format suitable for the further analysis. Population of the observed values of the above listed quantities is the primary statistical material and is called statistical or time series. Time series is an ordered sequence of observations made at specified instants. Series nature and influencing process structure determine an order of the said sequence. Usually, time series are formed based on discrete sequence of observations at specified regular intervals. Results of such measurements are represented as a continuous variable (random function), which follows some probabilistic law. The random process realization may also include observation errors. It should be noted that in many cases, besides the fluctuations and irregularities (overshoots), time series show some trends, which can be described by various models. Most common approaches to the trend identification are smoothing procedures: moving average and autoregression process models.

For characterization and description of random values, distribution laws are used, which can be set in various forms. With the knowledge of random value distribution law, its statistical characteristics can be determined. Note that both the selection of distribution law for investigated variable of numerical series and determination of statistical characteristics are conducted under the conditions of limited sampling, since large amount of measurements is virtually impossible under the real conditions. When processing such statistical material, it seems appropriate to address the issue of how to fit theoretical distribution curve for the obtained statistical series, which would only represent essential traits of statistical material, but not incidents resulting from the experimental data volume. In many cases, experimental data are well described by the normal probability distribution law, but it is also suitable for the use with random values not distributed according to the normal law, since [1]:

- quantity can be converted so that it follow normal distribution low;
- distribution of random values sum tends to normal law with the sample scope expansion up to ∞;
- error due to the use of statistical criteria based on assumption of normal distribution of experimental data is insignificant.

2 Approaches to Description of Electromagnetic Processes Occurring in the Direct-Current Traction Network

It is known that the traction energy system consists of multiple parts. Traction substations are static elements, while traction network is tens of kilometers long.

This fact supposes consideration of electromagnetic processes occurring in the traction network from two points: as circuits with the lumped parameters or circuits with the distributed parameters. The electrified railway power supply theory [1] assumes that traction network of both direct and alternating current is a line with the lumped parameters, and calculation of energy-exchange processes is carried out according to equivalent circuits, of which parameters are determined by multiplying the linear parameters (Ω/km, H/km …) by the respective length. At the same time, under the real-life conditions of traction energy system electric energy conversion and consumption, higher harmonic components circulate in the traction network, and these must be taken into account at the simulation and calculation [3, 4]. Appearance of the traction vehicle with the induction motors yet more complicates behavior of energy-exchange processes in traction network, since stand-alone inverter working frequency value is measured in kHz, and frequency band is considerably expanded.

It is known that any line should be considered as long one (i.e., with the distributed parameters) when its length equals to or exceeds current or voltage wave-length [1]. Wave-length is determined according to formula:

$$\lambda = \upsilon \cdot T = \frac{\upsilon}{f}, \tag{1}$$

where

$\boldsymbol{\upsilon}$ – phase velocity, υ = 2.5–10 m/s;
$\boldsymbol{T}$ – current or voltage harmonic period

As an example, let us consider characteristic harmonics within the range of 2 kHz accepted for traction energy system: for the six-pulse rectifier: min – 300 Hz, max – 2400 Hz; for the twelve-pulse rectifier: min – 600 Hz, max – 2400 Hz. Wave lengths make, respectively: 833 km, 417 km, 104 km. And given that working frequency value for voltage inverter of electric locomotive with induction motors makes up to 1.0 kHz, which means that spectrum frequency of harmonics generated into the traction network would be yet higher, and the fact that calculation of energy-exchange processes in traction network requires consideration of 7 inter-substation zones [5, 9] (with the average traction substations spacing $\boldsymbol{L} = 7 \times 16 = 112$ km), the fact becomes apparent that electromagnetic processes in traction network should be considered as for line with the distributed parameters.

Following equivalent circuit is commonly used for calculation of long lines (Fig. 1):

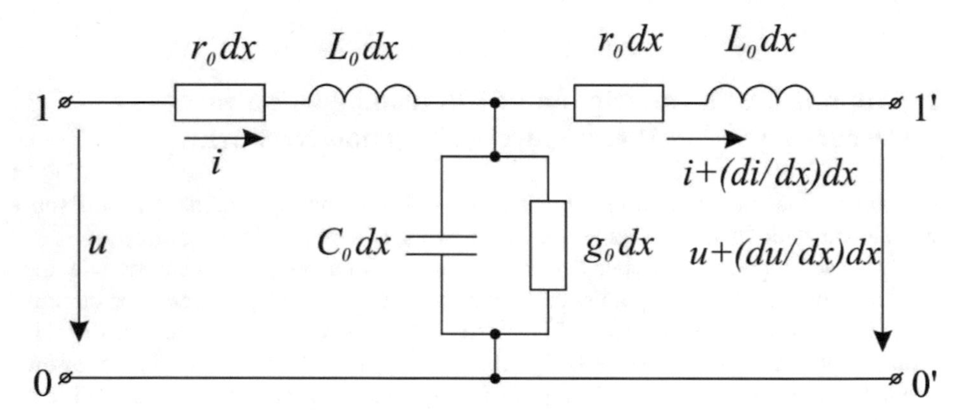

Fig. 1. Equivalent circuit of long line section

Such circuit is described by the system of differential equations, of which number is determined depending on elementary section length and calculation zone length. For example, for the nodal scheme of traction network power supply with the traction substations spaced 20 km apart, feeder zone length would be 10 km, and calculation zone length – 0.4 km. Hence, number of differential equations for calculation of electromagnetic processes in the overhead catenary of single track would make 10:0.4 = 25. At the same time, please note that calculation zone length of 0.2 km [6]. The shorter is length, the larger is number of calculation elements, and hence number of differential equations.

Traction network includes two components: overhead and rail systems, each of which has own features of equivalent circuit design. Feature of calculation of electromagnetic processes in OS is that main reactive elements (***L*** and ***C***) are concentrated at the traction substation. Since electric locomotive motors are connected directly to overhead system (unlike the alternating current electric locomotives), this enhances interrelation between the operation modes of electric locomotive and overhead system, which under the conditions of virtually continuous transient processes (electric locomotive operation modes changing, profile elements changing, pantograph bouncing, passing of rail joints, external power supply system operation modes changing, emergency processes, etc.) may result in occurrence of local switching overvoltage or oscillation processes in the interacting ‘traction vehicle – traction network’ pair.

Finite-element model of line with the distributed parameters [1] consisting of the series-connected elements shown in Fig. 1 is most commonly used for electromagnetic processes calculation and their simulation in overhead system. And it is assumed that the circuit parameters vary linearly, since surface effect phenomenon is not observed due to the small cross-section of the overhead system wires.

Rail system with copper equivalent of two rail strings of 800–1000 mm^2 has also much greater geometrical dimensions. When the time-varying current flows through it, its density is non-uniform over the rail section, which results in the necessary consideration of the surface effect phenomenon. In this case, longitudinal elements shown in Fig. 1 are as follows (Fig. 2):

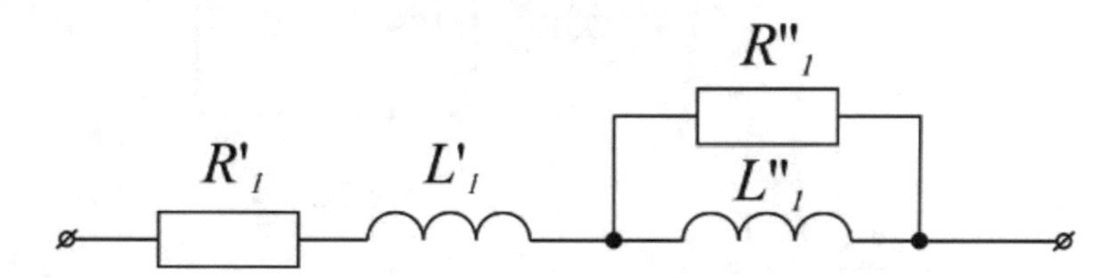

Fig. 2. Equivalent circuit of longitudinal rail system components

Taking all the foregoing into account, equivalent circuit of long rail line section would be as follows:

It should be noted that above proposed scheme applies to single rail. The mathematical simulation of rail system must be carried out for two rail strings with common conductor (earth), and equivalent circuit would be three-wire circuit. Figure 3 shows that the earth resistance is equal to zero. At the same time, scientific and technical literature [1, 4, 8], proposes to take into account ground path resistance, rail-to-rail mutual inductance, and conductivity and capacitance between rails (Fig. 4).

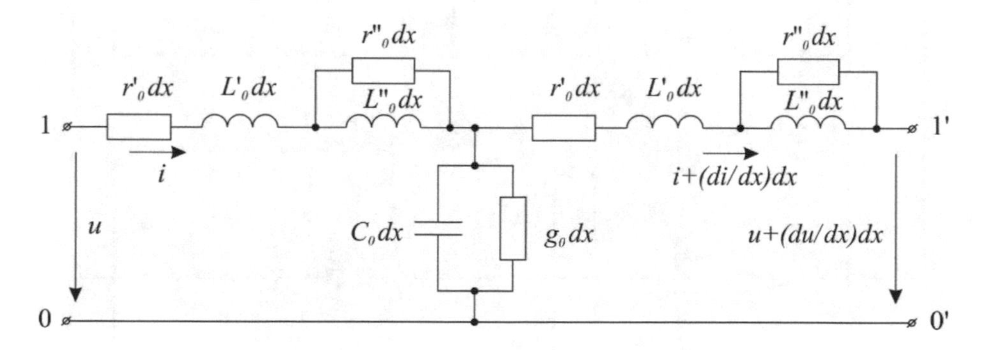

Fig. 3. Equivalent circuit of line section

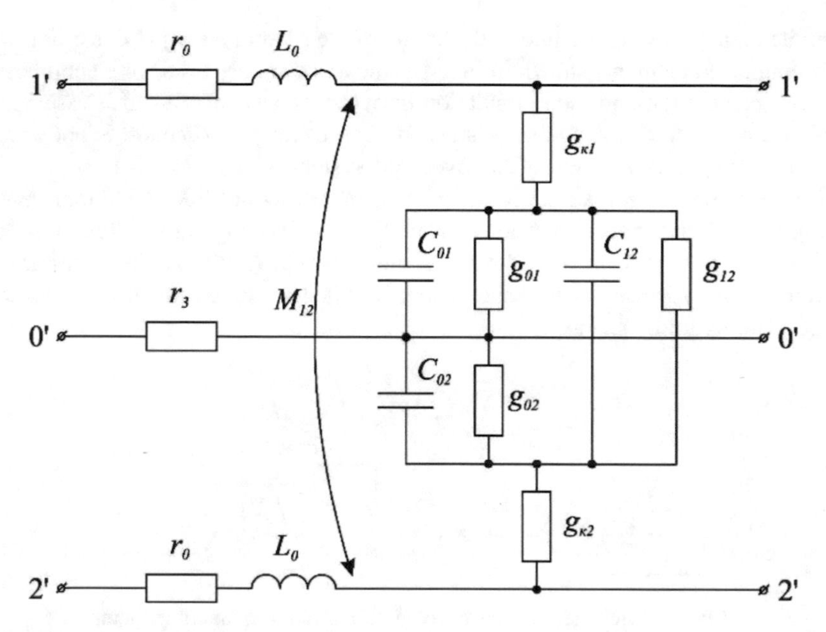

Fig. 4. Equivalent circuit of element taking into account earth resistance and rails interaction: $\boldsymbol{r}_3$ – earth resistance; $\boldsymbol{M}_{12}$ – rail-to-rail mutual inductance; $\boldsymbol{gk}_1$ and $\boldsymbol{gk}_2$ – rail-to-sleeper conductivity.

Summarizing all above-mentioned, following equivalent circuit must be used for the mathematical simulation of electromagnetic processes occurring in the direct-current traction network (Fig. 5):

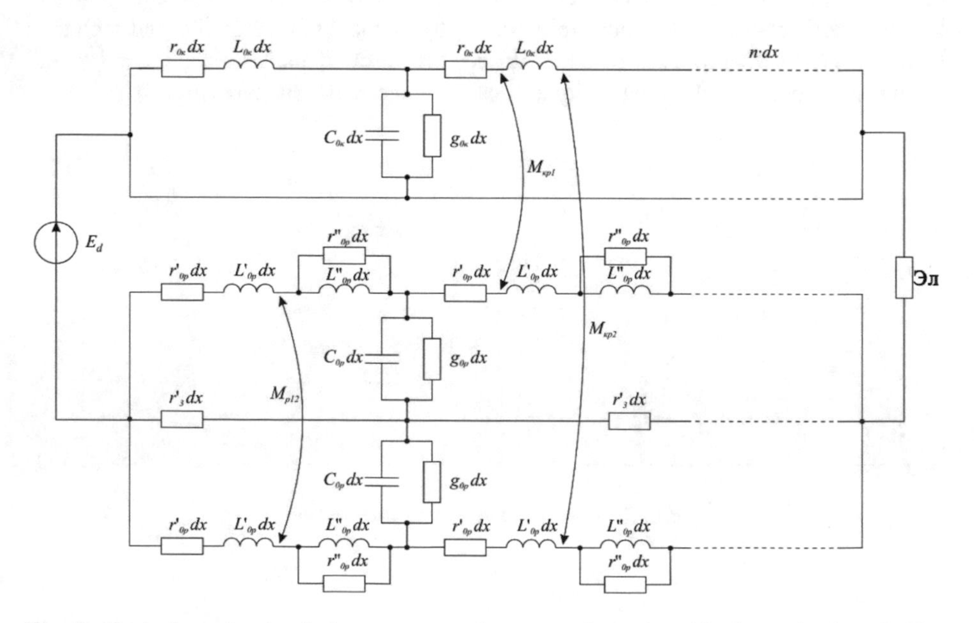

Fig. 5. Equivalent circuit of elementary traction network section (single track shown): E – voltage at the traction substation buses, Эл – traction network load (electric locomotive)

In consideration of the foregoing, Simulink-model of the traction network section will be as follows (Figs. 6 and 7):

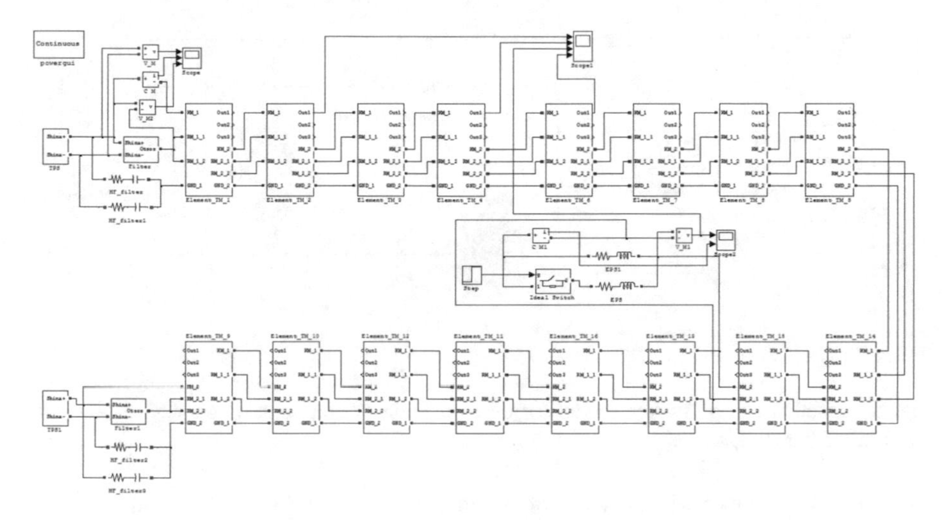

Fig. 6. Model of traction energy system with the distributed parameters

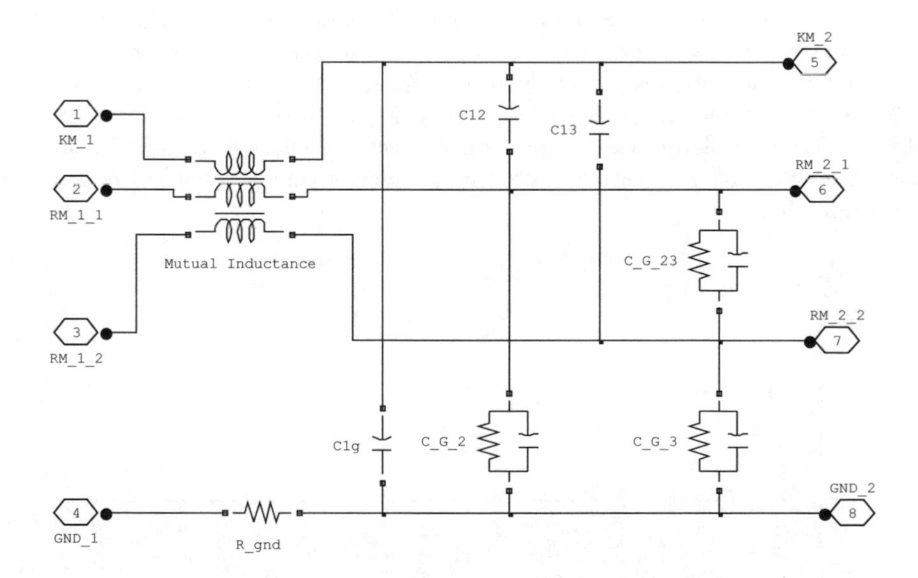

Fig. 7. Model of elementary traction energy system section with the distributed parameters

In order to determine the amplitude-frequency response of experimental section, model of traction energy system with the lumped parameters was also developed [2, 7, 8, 13] in the MATLAB Simulink package using output data of capacitance and inductance parameters of the traction substation smoothing filter circuits, initial capacitance, specific resistance of traction network and resistance of electric locomotive in various modes (Fig. 8).

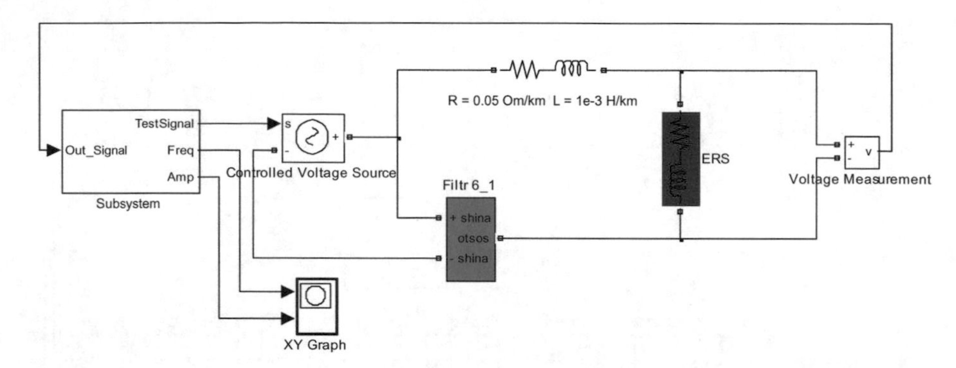

Fig. 8. Traction energy system model for determination of amplitude-frequency response with the lumped parameters

Following characteristics were obtained from simulation (Figs. 9 and 10). Their analysis shows that the electric rolling stock resistance (change of traction-coasting modes) does not change characteristic, in addition, it depends not much on the traction vehicle position within the section (0…20 km). Amplitude-frequency response shows the domination of harmonics with frequencies of 100 and 200 Hz over other ones, because within the frequency range of 100 ÷ 200 Hz, transmission factor exceeds 1.0 and may reach up to tens. Similar amplifying may also occur at frequency close to 400 Hz. These results provide support for existence of effect of electric locomotive operation mode on the harmonic content of traction network voltage determined experimentally.

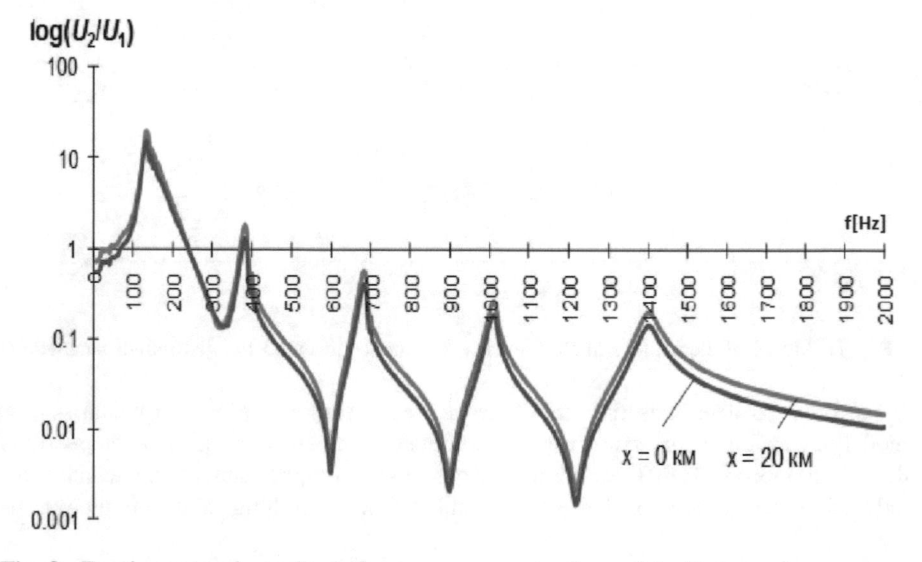

Fig. 9. Traction network amplitude-frequency response - the model with lumped parameters

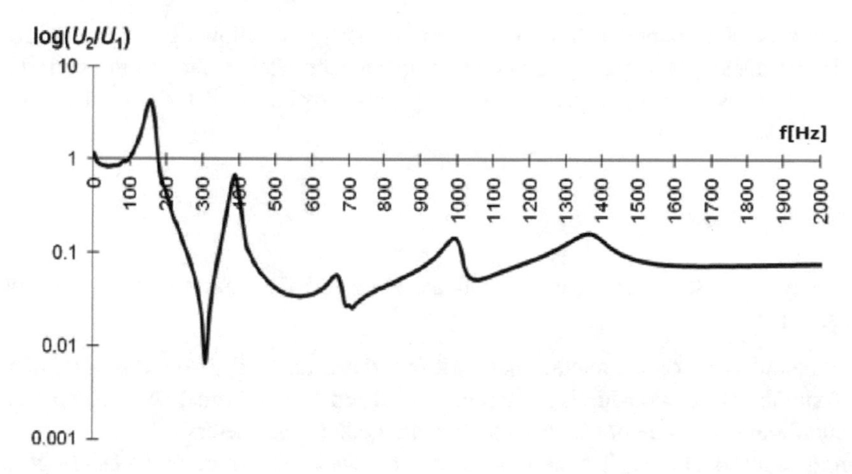

Fig. 10. Traction network amplitude-frequency response - the model with distributed parameters

Another aspect was study of possibility of the resonance phenomena [1, 10–12, 14] occurrence in the traction network at the electric energy transmission and consumption. The simulation was carried out with the electric locomotive current of 1000 A according to the above described method. An exemplary simulation result is shown in Fig. 11.

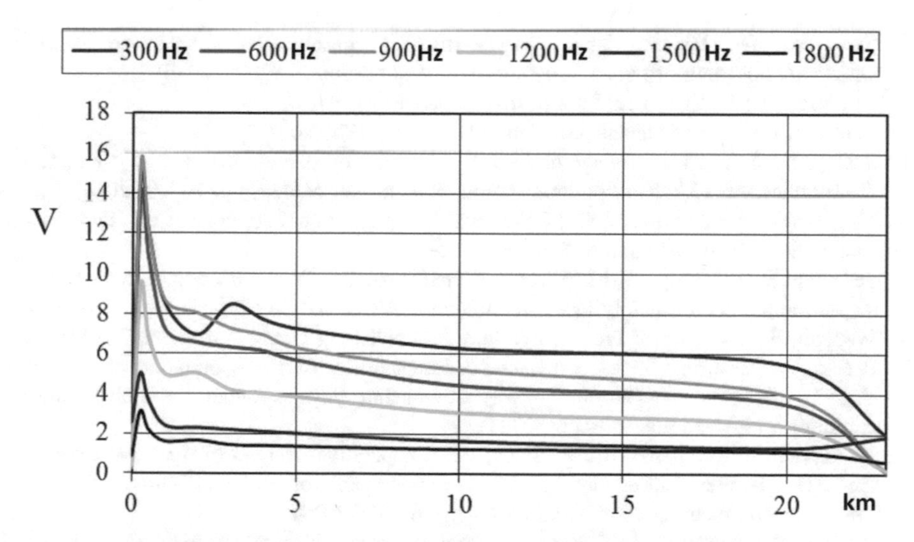

Fig. 11. Results of simulation of the harmonic voltage at locomotive pantograph vs. its position within the inter-substation zone (model with distributed parameters)

Analysis of obtained results allows us to highlight following point: due to the system parameters change, resonance phenomena occur in the traction network during the current consumption process [1, 13, 14], which results in deterioration of electromagnetic environment.

3 Conclusions

The analysis of statistical material and experimental data allows for the following conclusions:

- for calculations of the traction network (overhead and rail), from the point of view of suitability for considering phenomena related to electromagnetic compatibility, one should use network diagrams with distributed parameters;
- their analysis shows that the electric rolling stock resistance and vehicle position within the section does not change characteristic,
- for simulations of traction network parameters, the phenomenon of resonances should be taken into account,
- the results of calculations obtained in this way are similar to those obtained in real conditions.

References

1. Syczenko, W., Białoń, A.: Электромагнитная совместимость тягового электроснабжения постоянного тока при скоростном движении, Dnipro (2016)
2. Anderson, T.: Statistical analysis of time series. Mir (1976)
3. Bader, M.P.: Electromagnetic compatibility. UMK MPS (2002)
4. Szeląg, A.: Influence of voltage in 3 kV DC cafenary on traction and energy parameters of the supplied vehieles. Transportation electrification, No. 6/2013, pp. 74–79 (2013)
5. Mitrofanov, A.N.: Simulation of processes of prediction and control of train traction electrical energy consumption. Samara (2005)
6. Serbinenko, D., Khlopkov, M.: Electromagnetic processes in the traction network and their impact on the power quality indexes. Bulletin of VNIIZhT, No. 3/2003, pp. 23–34 (2003)
7. Wentzel, E.S.: Theory of Probability, 9th edn. Publishing House "Academia" (2003)
8. Białoń, A.: Ensuring the electromagnetic compatibility of new-generation direct current electric rolling stock with the railway automation devices under the Polish railways conditions. Ph.D. thesis, Moscow (2001)
9. Yandovich, V., Sychenko, V., Antonov, A.: Comparative analysis of overhead catenary in the EU Member States and Ukraine: arrangement of efficient current collection, Transportation electrification, No. 7/2014, pp. 67–77 (2014)
10. Byalon, A., Furman, J.: Research damping surges in contact network 3 kV DC. Електрифікація транспорту, No. 3/2012, pp. 35–39 (2012)
11. Sychenko, V., Bosiy, D.: Efficiency of electric power utilization in the electric traction systems, Modern technologies. System analysis modeling no. 48/2015, pp. 143–149 (2015)
12. Biedermann, N., Schütte, T., Elschner, K., Christoph Hinze, C.: Peak load management – reducing power peaks for a AC 15 kV 16,7 Hz railway, Elektrische Bahnen INT/2017, pp. 1–13 (2017)

13. Jefimowski, W., Szeląg, A.: Assessment of AC traction substation influence on energy quality in a supplying grid. Technical Transactions 12/2018 Electrical Engineering (2018)
14. Steczek, M., Szeląg, A.: Modification of the selective harmonic elimination method for effective catenary current harmonics reduction. In: 2015 International Conference on Electrical Drives and Power Electronics (EDPE), Tatranska Lomnica, Slovakia, pp. 394–401 (2015)

Designing of Transshipment Terminals for Selected Intermodal Transport Systems

Mariusz Brzeziński(✉) and Dariusz Pyza

Faculty of Transport, Warsaw University of Technology,
Koszykowa 75, Warsaw, Poland
mbrze94vp@gmail.com

Abstract. In this article the intermodal terminal designing method has been presented. The technologies of FlexiWaggon and CargoBeamer were put in the center of attention. Thanks to given assumptions the values of parameters in the proposed algorithm of the designing method have been calculated. Than the comparison of the results was possible. As a consequence it has been proved that modern *RO-RO* (*Roll-on/Roll-off*) technologies can be a reasonable alternative for traditional solutions such as road transport or *LO-LO* (*Lift-on/Lift-off*) technology.

Keywords: Intermodal transport · FlexiWaggon · CargoBeamer

1 Introduction

Over the years many authors were considering the subject of intermodal terminal designing. In the literature [1, 2, 10–12, 14–18] the meaning of intermodal terminals as a very important element of railway supply chain has been widely characterized. Their main functional, technical and organizational elements have been identified. Various variants of cargo fronts have been showed. In addition it has been specified how to localize the terminals and how to optimize their handling processes. There's a lot of statistic data concerning intermodal transport branch too. The practical procedure of container terminal designing method has been shown in the positions [3–5, 7–9, 13]. These literature positions contain also a case study involved with terminal projecting.

New intermodal solutions of transported commodities in the integrated loading units determine changes related with railway and road transport integration. This fact makes intermodal transport more flexible and more attractive for potential customers.

Mentioned before technologies have been described in more details in the headings [6, 18]. They constitute the transport solutions for combined transport. They are also concerning the following systems: Modalohr, CargoSpeed, MegaSwing, CargoBeamer and FlexiWaggon.

There are no currently literature available regarding terminal designing procedure in the *RO-RO* (*Roll-on/Roll-off*) transshipment systems. There are usually the already prepared parameters characterizing the given object. In addition there's lack of clear methodology to solve the algorithm. Therefore presented in the this article researches are synergy effect of researches contained in the headings [1, 3, 8, 11, 14, 16].

M. Siergiejczyk and K. Krzykowska (Eds.): ISCT21 2019, AISC 1032, pp. 52–62, 2020.
https://doi.org/10.1007/978-3-030-27687-4_6

For the purpose of scientific investigation from the collection of interesting *RO-RO* technologies there have been selected two: *FlexiWaggon* and *CargoBeamer*. The adaptation of equations from headings [2, 4, 9, 13] has been carried out. On this basis the new adjusted equations have been presented too. As a consequence these appliances made possible to formulate a new algorithm for the terminal designing. Finally the case studies have been accomplished.

To start the terminal designing procedure first it is mandatory to characterize their specifications. Technology of *FlexiWaggon* enables to hande full road sets, buses or other self-propelled units. Vehicle loading and unloading can be arranged by a truck driver what is a very important advantage. Wagons are being fitted with special ramps. Thus truck entry on wagon's platform and leaving from wagon is possible. The car's construction enables transport units which mass is not larger than 80t and length do not exceed 18,75 m. There is no other requirements regarding terminal construction. Hardened square is completely sufficient. The wagons are coupled with passenger wagon where vehicle drivers can rest or take a nap. Therefore the act *WE/561/2006* allows include working time into the *time of the rest*. In this case the *FlexiWaggon* solution can help to save some funds. On the other hand relatively long train cars are considerable disadvantage of this technology. It literally means a fewer train capacity - whereas the *FlexiWaggon's* cars are quite expensive.

CargoBeamer technology is a horizontal-parallel transshipment system for semi-trailers. Unit's loading and unloading is utterly automatic. All train composition can be loaded/unloaded during less than 15 min. Semi-trailers are settled in a special *pallets*. The *pallets* are conveyed into *wagon basket* by special beams installed underground. For effective *CargoBeamer* system functioning there are at least two terminals required – one in the beginning of carriage and second in the end of carriage. In addition the *CargoBeamer* system needs more surface than *FlexiWaggon* system.

2 FlexiWaggon and CargoBeamer Terminal Designing Procedure

2.1 Assumptions

The procedure of terminal transshipment system designing requires some kind of assumptions which determinate a technological-organizational solution.

First assumption it must have been taken is an annual flow of ITU (*Intermodal Transport Unit*) entering intermodal terminal handling system (ω_{RO}) [4].

The *ITU* flow extensiveness depends on few factors and can be estimated based on: (1) empirical experience (2) data from national or international statistical office (3) data coming from collecting fee system (e.g. *ViaToll*) (4) data and statistics from other intermodal terminals (5) market surveys and probes (6) micro- and macro-economic data (7) industry-service market analysis – coming both from regional and national market [3].

Having already the annual flow of *ITU* – now it is possible to figure out a daily terminal workload [8] (ω_{DZ}):

$$\omega_{DZ} = \left\lceil \frac{\omega_{RO}}{d_r} \right\rceil \left[\frac{ITU}{day} \right] \quad (1)$$

where: d_r – annual number of working days, ω_{RO} – total annual number of *ITU* entering terminal system.

Subsequently a percentage of *ITU* of the *c-type* (P_c) crossing terminal system should be determined in the total number of ITU.

Whereas:

$$\sum_{c \in \boldsymbol{C}} P_c = 1 \quad (2)$$

There the following designations are performing: (*a*) P_1 – percent of semi-trailers crossing terminal system in the total number of *ITU*, (b) P_2 – percent of road sets crossing terminal system in the total number of *ITU*.

Having P_c parameters it is obligatory to estimate the values of the *ITU* participation in the particular *terminal relation.*

$$\sum_{z \in \boldsymbol{Z}} U_z = 1 \quad (3)$$

where: z – *type* of *relation*.

There are the following interpretation of the terminal relations: (1) U_1 – *ITU* participation in the direct handlings of railway transit, (2) U_2 – *ITU* participation in the railway transit with *ITU* storage, (3) U_3 – participation in the *ITU* transshipments; from railway means of transport onto road means of transport, (4) U_4 – participation in the *ITU* transshipments; from railway means of transport onto road means of transport with ITU storage, (5) U_5 – participation in the *ITU* transshipments; from road means of transport onto railway means of transport with *ITU* storage, (6) U_6 – participation in the *ITU* transshipments from railway means of transport onto truck and *ITU* carriage to warehouse (*or cross-dock*), (7) U_7 – participation in the *ITU* transshipments from warehouse (*or cross-dock*) onto vehicle carrying *ITU* in order to handling onto railway means of transport.

Taking into account the previous assumptions the matrix of relations (*F*) takes the following form:

$$F = [F_{cz}]_{C \times Z} = \begin{bmatrix} F_{11} & F_{12} & F_{13} & F_{14} & F_{15} & F_{16} & F_{17} \\ F_{21} & F_{22} & F_{23} & F_{24} & F_{25} & F_{26} & F_{27} \end{bmatrix} [\%]$$

A subsequent formula (4) can help to figure out a virtual number of c – *kind ITU* taking part in *relation of type* – z.

$$G_{cz} = \omega_{DZ} \cdot \partial \cdot F_{cz} \left[\frac{ITU}{day} \right] \quad (4)$$

wherein: ∂ – peak factor (allowing prepare the terminal system for excessive number of ITU handling in some particular day of the year). A virtual number of *c-type ITU* taking part in *relation- z -* may be showed as *G-matrix.* In the considered case the matrix takes following form:

$$G = [G_{cz}]_{C \times Z} = \begin{bmatrix} G_{11} & G_{12} & G_{13} & G_{14} & G_{15} & G_{16} & G_{17} \\ G_{21} & G_{22} & G_{23} & G_{24} & G_{25} & G_{26} & G_{27} \end{bmatrix} [ITU]$$

Figure 1 shows the planned schema of *ITU* flow in the projecting terminal system.

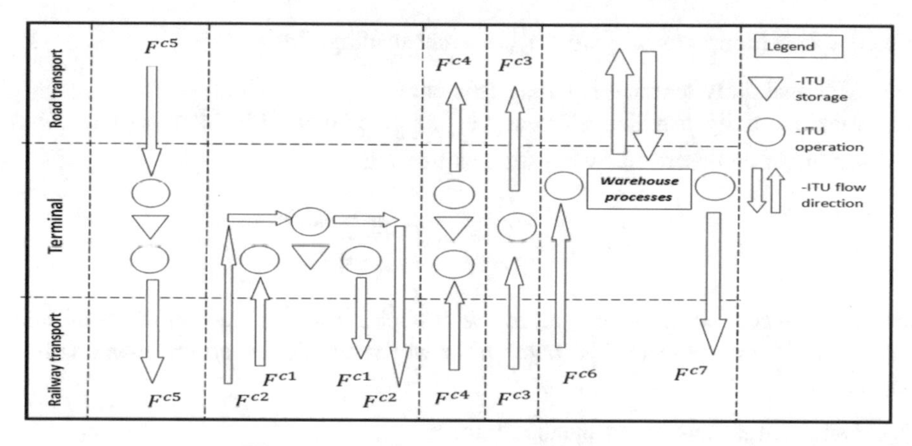

Fig. 1. The *ITU* flow directions in the considered intermodal terminal (Source: personal collection based on [4, 13])

2.2 Intermodal Terminal Designing Procedure

There has been formulated a following procedure for intermodal terminal designing both for FlexiWaggon (*further also marked as FW*) and CargoBeamer (*further also marked as CB*):

(1) First of all, it should be estimated the number of vehicles or semi-trailers which can be carried by a single train ($n_{FW/CB}$):

$$n_{FW/CB} = \frac{l^{stor}_{FW/CB} - l^{Loc}_{FW/CB} - l^{pas}_{FW/CB}}{d_{FW/CB}} \left[\frac{vehicles/semi\text{-}trailers}{train} \right] \quad (5)$$

where: $l^{stor}_{FW/CB}$ – length of *FW* or *CB* train, $d_{FW/CB}$ – length of *FW* or *CB* car, $l^{pas}_{FW/CB}$ – length of passenger car for truck drivers, $l^{Loc}_{FW/CB}$ – length of locomotive.

(2) Taking into consideration loading and unloading time a second step is estimating a train handling time ($T_{FW/CB}$):

$$T_{FW/CB} = t^{wj}_{FW/CB} + t^{op}_{FW/CB} + t^{prz}_{FW/CB} + t^{op'}_{FW/CB} + t^{leav}_{FW/CB}\ [min] \tag{6}$$

where: $t^{wj}_{FW/CB}$ – train entry time, $t^{op}_{FW/CB}$ – technical and trade clearance on terminal entry, $t^{op'}_{FW/CB}$ – technical and trade clearance before train departure, $t^{leav}_{FW/CB}$ – train departure time, $t^{prz}_{FW/CB}$ – semi-automatic or automatic ITU handling time.

In case of *FW*, the *ITU* handling time has the following form:

$$t^{prz}_{FW} = t^{lo}_{FW} + t^{unl}_{FW} \tag{7}$$

where: t^{unl}_{FW} – train unloading time, t^{lo}_{FW} – train loading time

(3) Maximal daily terminal handling efficiency
The maximal daily handling efficiency ($\tau^{trans}_{FW/CB}$) both for *FlexiWaggon* and *CargoBeamer* might be described by equation as beneath:

$$\tau^{trans}_{FW/CB} = \frac{T_{Wor}}{T_{FW/CB}} \cdot \varepsilon_{zm}\left[\frac{train\ handlings}{\frac{day}{rail\ track}}\right] \tag{8}$$

where: ε_{zm} – correction factor – related with track utilization rate, work organization, readiness of technical facility, *changes* in the work area, T_{Wor} – terminal working hours per day.

(4) Daily virtual number of handled trains
Having *ITU* flow on the terminal entry – now it can be calculated a virtual number of handled trains per day ($O^{Obs}_{FW/CB}$):

(a) for *FlexiWaggon* technology where; $c = 1 \wedge z = \{1, 2, 3, 4, 6\}$
(b) for *CargoBeamer* technology where; $c = 2 \wedge z = \{1, 2, 3, 4, 6\}$

$$O^{Obs}_{FW/CB} = \left\lceil \frac{\sum G_{cz}}{n_{FW/CB} \cdot \alpha_{zm}} \right\rceil \left[\frac{train\ handlings}{day}\right] \tag{9}$$

α_{zm} – reducing factor concerning utilization rate of train space.

(5) An average daily terminal efficiency utilization
Beneath it's been shown a very important parameter informing about daily terminal rate of capacity utilization. Low value of factor may testify about insufficient customer's business interesting. This parameter - for *FlexiWaggon* and *Cargo Beamer* technology has a similar form ($\omega_{Flex/CB}$):

$$\omega_{FW/CarBe} = \frac{O^{Obs}_{FW/CB}}{\tau^{trans}_{FW/CB}} \cdot 100\ [\%] \tag{10}$$

(6) Investment expenses for technology of *FlexiWaggon* and *CargoBeamer*
In each case of intermodal solutions the values of expenses are different. The level of investment expenditures is depending on infrastructure and terminal equipment specification. These investments are taking the following form (E_{Flex}/E_{CB}):

(a) for *FlexiWaggon* technology

$$E_{FW/CB} = E^{Infr}_{FW/CB} + E^{Sys}_{FW/CB} + E^{Wag}_{FW/CB}\ [€] \tag{11}$$

where: $E^{Infr}_{FW/CB}$ – expenses for *FW* or *CB* infrastructure (inclusive hardened buffer area, railway tracks and turnouts, lighting, designations, mechanisms installed underground – beams, engines etc.), $E^{Sys}_{FW/CB}$ – expenses for *FW* or *CB* system control expenses for system (inclusive computer control system, sensors), $E^{Wag}_{FW/CB}$ – expenses for wagons.

The *FlexiWaggon* and *CargoBeamer* investment expenditure for railcars ($E^{Wag}_{FW/CB}$) are product of multiplication of the number of acquired cars ($m^{Wag}_{FW/CB}$) and their unit price ($u^{Wag}_{FW/CB}$).

$$E^{Wag}_{Flex/CB} = m^{Wag}_{FW/CB} \cdot u^{Wag}_{FW/CB} \tag{12}$$

FlexiWaggon and *CargoBeamer* infrastructure expenses are sum of following components:

$$E^{Infr}_{FW/CB} = E^{Dr}_{FW/CB} + E^{Buf}_{FW/CB} + E^{Tor}_{FW/CB} + E^{Cros}_{FW/CB} = [(A^{Dr}_{FW/CB} \cdot W^{Dr}_{FW/CB} \cdot L^{Dr}_{FW/CB}) + (S^{Buf}_{FW/CB} \cdot H^{Buf}_{FW/CB}) + (NT^{Tor}_{FW/CB} \cdot Q^{Tor}_{FW/CB} \cdot P^{Tor}_{FW/CB}) + (n^{Cros}_{FW/CB} \cdot M^{Cros}_{FW/CB})][€] \tag{13}$$

where: $E^{Dr}_{FW/CB}$ – expenses for internal maneuvering routes, $E^{Buf}_{FW/CB}$ – expenses for buffer area, $E^{Tor}_{FW/CB}$ – expenses for loading/unloading railway tracks, $E^{Cros}_{FW/CB}$ – expenses for railway turnouts, $A^{Dr}_{FW/CB}$ – unit asphalt costs $W^{Dr}_{FW/CB}$ – road width, $L^{Dr}_{FW/CB}$ – total road length, $S^{Buf}_{Flex/CB}$ – unit cost of buffer area, $H^{Buf}_{Flex/CB}$ – buffer area square, $Q^{Tor}_{FW/CB}$ – unit cost of railway track, $P^{Tor}_{FW/CB}$ – length of railway tracks, $NT^{Tor}_{FW/CB}$ – number of railway tracks, $n^{Cros}_{FW/CB}$ – number of railway turnouts, $M^{Cros}_{FW/CB}$ – cost of singular railway turnout.

(7) Annual operational costs
Annual operational costs are tightly correlated with terminal reliable system prosperity. These cost are derivative of expenses and depends from: level of investments, random variables and terminal specification. In case of *FW* or *CB* technology, the operational costs ($K^{Ro}_{FW/CB}$) are sum of the following components:

$$K_{FW/CB}^{Ro} = K_{FW/CB}^{Ener} + K_{FW/CB}^{Infr} + K_{FW/CB}^{Sys} + K_{FW/CB}^{Nap} + K_{FW/CB}^{Opr} + K_{FW/CB}^{PHu} = [(r_{FW/CB} \cdot E_{FW/CB}) + (r_{FW/CB}^{Infr} \cdot E_{FW/CB}^{Infr}) + (r_{FW/CB}^{Sys} \cdot E_{FW/CB}^{Sys}) + (r_{FW/CB}^{Wag} \cdot E_{FW/CB}^{Wag}) + (r_{FW/CB} \cdot E_{FW/CB}) + (y_{FW/CB}^{PHu} \cdot Z_{FW/CB}^{PHu} \cdot NM_{FW/CB}^{PHu})] \left[\frac{€}{year}\right] \quad (14)$$

where: $K_{FW/CB}^{Ener}$ – terminal energy consumption costs, $K_{FW/CB}^{Infr}$ – infrastructure maintenance cost, $K_{FW/CB}^{Sys}$ – system control maintenance cost, $K_{FW/CB}^{Nap}$ – wagon maintenance cost, $K_{FW/CB}^{Opr}$ – borrowing and capital freezing cost, $K_{FW/CB}^{PHu}$ – human labor cost $r_{FW/CB}$ – percent of $E_{FW/CB}$, $r_{FW/CB}^{Infr}$ – percent of $E_{FW/CB}^{Infr}$, $r_{FW/CB}^{Sys}$ – percent of $E_{FW/CB}^{Sys}$, $r_{FW/CB}^{Wag}$ – percent of $E_{FW/CB}^{Wag}$, $y_{FW/CB}^{PHu}$ – number of terminal workers servicing *FW* or *CB* system, $Z_{FW/CB}^{PHu}$ – gross worker remuneration, $NM_{FW/CB}^{PHu}$ – number of working months per year.

3 FlexiWaggon and CargoBeamer System Terminal Designing – Case Study

Terminal projecting assumptions

To start a procedure of calculation for the algorithm above it is compulsory to make a basic assumptions concerning the *input* parameters. There a lot of *input* parameters can be distinguished among others: peak factor, flow extensiveness on the system entry, number of terminal working hours/days etc. (see Table 1). Most of the assumptions were adopted from headings [2, 13, 18].

Table 1. Terminal system *input* parameters

Parameter	Parameter unit	Value of parameter
ω_{RO}	ITU/year	24 000
$\partial/\varepsilon_{zm}/\alpha_{zm}$	–	1,2/0,85/0,8
T_{Wor}	h/day	16 (2 work shifts)
d_r	days	350
$l_{FW}^{stor}/l_{CB}^{skl}$ and d_{FW}/d_{CB}	m	600/600 and 34/19,4
$l_{FW}^{Loc}/l_{CB}^{Loc}$ and $l_{FW}^{pas}/l_{CB}^{pas}$	m	20/20 and 20/0
t_{FW}^{wj}/t_{CB}^{wj} and $t_{FW}^{leav}/t_{CB}^{leav}$	min	5/5 and 5/5
t_{FW}^{op}/t_{CB}^{op} and $t_{FW}^{op'}/t_{CB}^{op'}$	min	10/10 and 5/5
t_{FW}^{lo}/t_{FW}^{unl} and t_{CB}^{prz}	min	25/25 and 15
$E_{FW}^{Sys}/E_{CB}^{Sys}$	mln €	2/5
A_{FW}^{Dr}/A_{CB}^{Dr} and $S_{FW}^{Buf}/S_{CB}^{Buf}$	€/m^2	50/50 and 12/185
L_{FW}^{Dr}/L_{CB}^{Dr} and W_{FW}^{Dr}/W_{CB}^{Dr}	m	1800/1800 and 7/7
$H_{FW}^{Buf}/H_{CB}^{Buf}$	m^2	12 000/ 37 000

(*continued*)

Table 1. (*continued*)

Parameter	Parameter unit	Value of parameter
$NT_{FW}^{Tor}/NT_{CB}^{Tor}$ and $n_{FW}^{Cros}/n_{CB}^{Cros}$	–	1/2 and 1/1
$Q_{FW}^{Tor}/Q_{CB}^{Tor}$	€/m	1515/ 1515
$P_{FW}^{Tor}/P_{CB}^{Tor}$	m	900/900
$M_{FW}^{Cros}/M_{CB}^{Cros}$ and $u_{FW}^{Wag}/u_{CB}^{Wag}$	€/unit	70000/70000 and 330000/400000
$m_{FW}^{Wag}/m_{CB}^{Wag}$	–	20/30
$y_{FW}^{PHu}/y_{CB}^{PHu}$ and $NM_{FW}^{PHu}/NM_{CB}^{PHu}$	–	2/2 and 12/12
$Z_{FW}^{PHu}/Z_{CB}^{PHu}$	€/month	1000/1000
r_{FW}/r_{CB} and $r_{FW}^{Infr}/r_{CB}^{Infr}$ $r_{FW}^{Sys}/r_{CB}^{Sys}$ and $r_{FW}^{Wag}/r_{CB}^{Wag}$	%	3/3 and 5/3 6/6 and 6/4

Source: Personal collection

Based on Fig. 1 it has to be estimated percentage of *ITU* taking part in particular terminal *relation*:

$$[F_{cz}]_{C\times Z} = [F_{cz}]_{2\times 7} = \begin{vmatrix} 5 & 5 & 9 & 9 & 18 & 2 & 2 \\ 5 & 5 & 9 & 9 & 18 & 2 & 2 \end{vmatrix} [\%]$$

After the calculations and taking into consideration information contained in the literature - the couple of results have been posted in the Table 2.

Table 2. Intermodal technology comparison

Parameter	Traditional terminal	Cargo-Beamer	Flexi-Waggon
Train capacity [ITU/train]	30	29	16
Train handling time [h]	3–5	0,75	1,33
Maximal terminal handling efficiency [train handlings/day/rail track]	5–6	18	10
Investments for infrastructure [mln €]	4,65–8,15	10,3	2,28
Investment expenditures for rolling stock [mln €]	6,9	12	6,6
Investment expenditures for system control	Medium	High	Low
Total investment expenditures [mln €]	7–10	~27	~11
Carriage cost [€/km]	N/A	0,063	0,06
Operational costs [mln €/year]	1,00–2,3	~2,9	~1,4
Average rate of terminal efficiency utilization [%]	>40	11	30
Square for each single ITU [m²]	N/A	120	117
Terminal square [m²]	~50 000	~20 000	~17 000

Source: Adopted from [6, 13, 18]

The concept of terminal where *FlexiWaggon, CargoBeamer* and traditional *LO-LO* technologies are complementary to each other depicts Fig. 2.

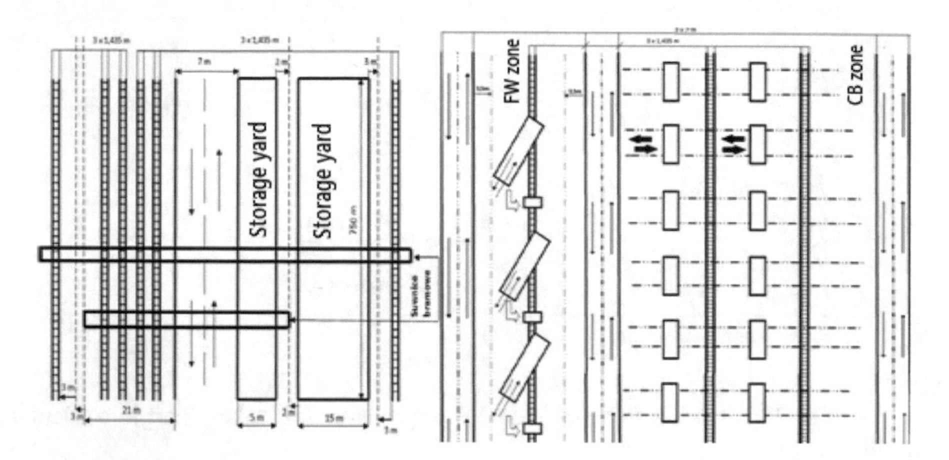

Fig. 2. The concept of integrated terminal. (Source: personal collection)

4 Conclusions

The main purpose of this article was to develop an algorithm of intermodal terminals designing for technology of *FlexiWaggon* and *CargoBeamer*. Thanks to calculations it has been proved that new intermodal technologies can compete both with *LO-LO* technology and with road transport. According to the Polish and European standards some assumptions regarding the input parameters have to had been taken on. Thus the algorithm presented in *point 2* could be accomplished.

In addition authors of this article wanted to draw attention to obstacles which crack down the development of these transshipment technologies. Literally it is investors uncertainty about their effectiveness and about their implementation cost level. To justify that unconventional railway transshipment systems can be a serious alternative for conventional solutions – the methodology involved with terminal designing have to had be shown. When the calculations have been carried out it turned out that intermodal solutions and their infrastructure investments are not such expensive as it may seem. On the other hand problematic may be a wagons acquisition. For example a cost of singular wagon is equal by around - 330 k € in case of *FW* and by around - 400 k € in case of *CB* [6]. Thus these wagons are more expensive than traditional one.

Contrary to *FW* and traditional railway technologies the *CB* technology is more expensive because of higher level of automation. *FlexiWaggon* technology has the lowest level of automation therefore investment expenses and operational cost are the lowest. The described in the papers *RO-RO* technologies need less surface than *LO-LO*. They are also characterized by a fewer train handling time what is a strong opportunity in the logistic of XXI century.

Calculations above shows also that technologies of *FW* and *CB* may be uneconomic in case of insufficient number of customers. Therefore terminal localization should be preceded by precise analysis. Nearby terminal should be localized logistics parks, manufactories or big cities to increase the annual ITU flow extensiveness [2].

Both technologies are characterized by shorter handling time because of process automation which allows to reduce a human factor. They are rather more effective for long distance (150–200 km) than for shorter [18].

New intermodal solutions are also environmentally-friendly what nowadays is very important. Their implementation can bring a relief to natural habitat and to overwhelmed infrastructure. In addition the development of *CB* and *FW* may be crucial in the concept of *SilkWay* between Europe and China.

Finally a human life would be easier if all the world put an importance on railway solutions.

References

1. Hipolito, T., Nabais, J.N., Botto, M.A.: Efficient operations at intermodal terminals using a multi-agent system. In: CONTROLO. Springer (2016). ISSN 1876-1119
2. Jacyna, M., Pyza, D., Jachimowski, R.: Intermodal transport. Designing of reloading terminals. PWN, Warsaw (2017). ISBN 978-83-01-19579-3
3. Jacyna, M., Pyza, D.: Role of intermodal terminals in the rail-road transport. Railway reports, vol. 59, nr. 169, pp. 15–27 (2016). ISSN 0552-2145
4. Jakubowski, L.: Technology of the cargo operations. Publishing House of the Warsaw University of Technology, Warsaw (2009)
5. Bose, J.W.: Handbook of Terminal Planning. Operations Research/Computer Science. Hamburg University of Technology. Springer (2010)
6. Klemenčič, M., Burg, R.: Data base and comparative analysis of CT and transshipment technologies for CT. University of Maribor (2017)
7. Alicke, K.: Modeling and optimization of the intermodal terminal Mega Hub. OR Spectr. **24** (1), 1–18 (2002). ISSN 0171-6468
8. Lee, B.K., Jung, B.J., Kim, K.H.: A simulation study for designing a rail terminal in a container port. IEEE, Monterey (2006). ISSN 1558-4305
9. Pyza, D., Jachimowski, R.: Designing of functional areas of intermodal terminals. Log. Transport **40**(4), 83–90 (2018). ISSN 1734-2015, Wroclaw
10. Pyza, D., Piątek, M.: Selected aspects of the design of intermodal terminals. Research papers of the WUT. Transport, no. 119, pp. 389–399 (2017). ISSN 1230-9265
11. Pyza, D., Piątek, M.: Intermodal terminals and their role in supply chain. Research papers of the WUT. Transport, no. 119, pp. 379–388 (2017). ISSN 1230-9265
12. Pyza, D.: Carriage Systems – Handling. Potential and Maintenance Issues. Publishing house of the WUT, Warsaw (2019). ISBN 978-83-7814-881-4
13. Pyza, D., Brzeziński, M.: The loadings service process rationalization in the intermodal terminals. Research papers of the WUT. Transport, no. 123, pp. 121–136 (2018). ISSN 1230-9265
14. Kostrzewski, A., Nader, M.: Analysis of the issue of intermodal terminals designing. Log. Mag. (2015). no. 2/2015, Poznań

15. Kostrzewski, A., Nader, M.: Semi - trailers and road sets transportation of the selected intermodal technology with horizontal transshipment. Research papers of the WUT. Transport, no. 111, pp. 287–299 (2016). ISSN 1230-9265
16. Pyza, D.: Optimization of transport in distribution systems with restrictions on delivery times. Arch. Transport **21**(3–4), 125–147 (2009). Polish Academy of Sciences Committee of Transport
17. Pyza, D.: Intermodal transport – technical, technological, organizational and functional conditions. The chosen issues of logistics, chapter IV, red. J. Feliks. AGH University of Science and Technology Press, pp. 168–179, Cracow (2016)
18. Savelsberg, E.: Innovation in European Freight Transportation: Basics, Methodology and Case Studies for the European Markets. Springer, Berlin (2008)
19. Steenken, D., Voß, S.: Container terminal operation and operations research - a classification and literature review. OR Spectr. **26**(1), 3–49 (2004). ISSN 0171-6468
20. https://combined-transport.eu/cargo-beamer. Accessed 21 May 2019

Study of the Microsleep in Public Transport Drivers

Rafał Burdzik(✉), Ireneusz Celiński, and Jakub Młyńczak

Faculty of Transport, Silesian University of Technology, 40-019 Katowice, Poland
{rafal.burdzik, ireneusz.celinski, jakub.mlynczak}@polsl.pl

Abstract. The article elaborates upon a study of the phenomenon of microsleep in drivers operating means of collective public transport on a regular basis. The microsleep phenomenon has been observed as a rather common one among drivers in general, however, in the case of professional drivers operating public transport lines, this is particularly dangerous. The consequences it may trigger are not limited to traffic accidents, but may even include road traffic disasters. The study addressed in the article was conducted using the eye tracking technique supported by auxiliary testing apparatus, including systems of the authors' own design and make. The study was conducted in operation of different means of collective transport on public transport lines in Poland. In one case, microsleep was studied in a driver of the EN57 passenger train, while the second case studied was that of a driver of the Solaris Urbino bus. The data retrieved from the measurements were processed using the SMI BeGaze 3.4 software. Results of the studies, subsequently processed in statistical terms, were used to formulate operational conclusions concerning the microsleep phenomenon and to define specific recommendations for drivers of means of collective public transport based on such grounds. The article also touches upon the matter of the values of parameters characterising the phenomena of microsleep and blinking, as previously commented in the literature of the subject.

Keywords: Microsleep · Public transport · Eye tracking

1 The Microsleep Phenomenon

Being a frequent and mass phenomenon, microsleep is definitely a serious social problem of the 21st century. The lifestyle changes observed over the last few decades, the voluntary extension of working hours, the increasing stress, fatigue, civilisation diseases, including eye diseases, cause that the phenomenon of microsleep should become a subject of special interest for researchers, especially when it is connected with the use of transport infrastructure. This problem appears to be all the more acute in the case of drivers operating means of public transport used by several to several hundred people at the same time. The phenomenon itself affects various social and professional groups in different ways, and it is age-dependent [15]. The Department of Transportation claims that ca. 16% of car accidents can be attributed to sleepiness.

M. Siergiejczyk and K. Krzykowska (Eds.): ISCT21 2019, AISC 1032, pp. 63–73, 2020.
https://doi.org/10.1007/978-3-030-27687-4_7

The 2013 statistics in this area are as follows: ca. 6 million accidents (ca. 1 million sleep related), 1.5 million injuries (ca. 300.000 sleep related), 30.000 fatalities (ca. 5000 sleep related) [17].

To investigate the microsleep (MS) problem, the authors studied drivers operating means of collective public transport in Poland. They assumed the parameters characterising microsleep and blinking, as previously defined in the literature of the subject, and analysed them in a critical manner. What they observed was symptomatic variability of the characteristics, hence the breakdown of observations into three intervals.

The microsleep (MS) phenomenon is related to short periods (spells) of sleep or sleepiness throughout a human being's 24-hour activity (including professional). This phenomenon manifests itself in external closing of eyelids, which may last from tenths of a second to several dozen seconds [1–5, 12–15]. Some sources set the lower limit of microsleep at 1 s, which may raise certain doubts. Most typically, microsleep examination is performed using the EEG technique, where the mean length of MS, as provided in the literature of the subject, ranges between 5 and 9 s [18]. Other EEG examinations imply that the minimum MS duration ranges at 0.5–10 s, while the maximum duration ranges between 14 and 30 s. The literature of the subject perceives this phenomenon more broadly, i.e. by also addressing the problem of dozing off, which has not been discussed in this article.

The MS phenomenon is accompanied by reduced concentration (no response to stimuli), and frequently also by a lapse of consciousness (the authors of this article have not addressed problems representing the fields of medicine and psychology, focusing primarily on the material data obtained by the ET technique). In the study elaborated in this article, MS was identified by ET. Every measured closing of eyes has been interpreted as a potential symptom of MS. The observed range of microsleep time variability has been illustrated by specific examples. On account of the divergence of the data retrieved from the literature of the subject as to the time of microsleep and blinking, an entire range of variability of the time in which those surveyed closed their eyes has been presented (this being an intentional simplification).

There are various interpretations of what is happening to a person affected by this phenomenon. According to different sources, such a person does not respond properly to environmental stimuli [6–11]. Most certainly, the perception of the external environment is completely or considerably reduced during microsleep due to the closing of eyelids (the perception channel is physically closed the more effectively, the longer the phenomenon lasts, while delays in data processing are no longer relevant).

With regard to driving vehicles, the foregoing has a direct effect on road safety [3–5, 17]. The microsleep phenomenon often occurs during monotonous and routine jobs in combination with fatigue resulting from previous activity [6]. And driving public transport vehicles repeatedly on the same routes in the transport network is certainly a monotonous activity [20, 21]. Microsleep is mainly studied using the following techniques: EEG, EOG, eye tracking, direct observation and other visual techniques. The application of visual techniques results from the fact that microsleep often manifests itself not only in the closing of eyelids, but also in specific movements of larger body parts. The phenomenon of short-term sleep can be observed in the case of systematic dropping and lifting of the head (alternately at a varying amplitude), drooping eyelids, turning away from sources of strong light, covering and rubbing eyes, yawning and other mimic gestures. On account of the number of the available techniques that can be used to observe this phenomenon and the large number of microsleep symptoms, there is no consensus as to the best examination method used in this respect. EEG is most commonly used as a procedure simple in technical terms.

The most dangerous aspect of this phenomenon is that those who succumb to it are often unaware of that. They are convinced that before being enquired about their condition they were fully aware. Ongoing vigilance is required while driving public transport vehicles. Therefore, it is reasonable to investigate this phenomenon with regard to the means of public transport.

2 Research Technique

The study addressed in this article was conducted as a pilot project for two cases in four public transport routes. In the first case, the work of a driver of the EN57 type passenger train was observed over a regular transport line (30 km). Examinations were performed while running in both directions on the same line. In the second case, the subject of observations was the driver of the Solaris Urbino 12 bus serving a large city on a regular basis (in which case, however, the passengers were limited to the research team and observers). These are two diametrically different cases in terms of the subject and object of the study. The study was a pilot project preceding more extensive population research. The train was always moving in a more or less straight line. There are only a dozen or so stations and stops, and more than a dozen level crossings on the railway route in question. Therefore, it is a case of driving monotony due to the railway network structure. As for the bus, its route changes directions at many points running through a city where high variability of traffic intensity and different road inclinations can be observed. Unlike on the railway, there are significantly more pedestrian crossings and vehicular traffic streams. In this case, a large number of other vehicles and pedestrians occupy the common elements of infrastructure at the same time. Consequently, the road network infrastructure should not make the driving monotonous in this case. What connects the two cases is the fact that both the bus driver and the train driver were tested after 4 h of operation starting from 6 a.m. (following initial fatigue). The tests were conducted in two series on explicitly defined routes while driving there and back (return run) with a short break of more than a dozen minutes between them. The equipment used for purposes of microsleep examination comprised SMI eye tracking glasses (60 Hz), an original system of cameras mounted in front of the bus and train driver, linear accelerometers and GPS receivers. The cameras detect the driver's gestures and the movement of his body parts (this aspect of the study will be described in another article).

Figure 1a illustrates the manner in which the eye tracking glasses were used during examinations, while Fig. 1b schematically depicts the testing apparatus.

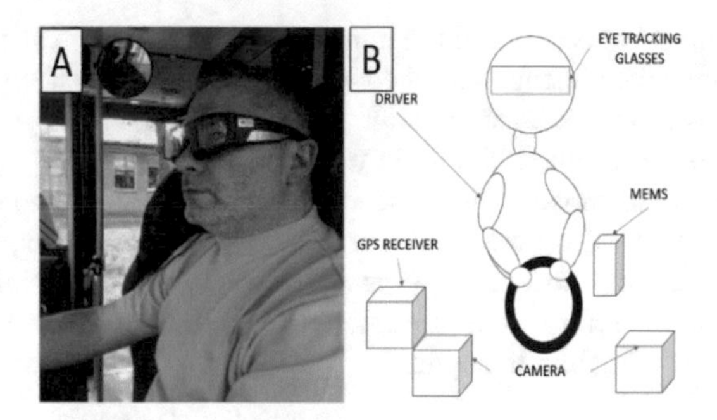

Fig. 1. (a) Eye tracking glasses (b) testing apparatus. *(Source: elaboration own)*

The ET glasses record parameters such as e.g. characteristics of the given surveyed person's blinking. In driving, they measure the duration of each blink in milliseconds. And in this article, the authors have taken even the shortest blink into consideration. What this device also measures is some other characteristics of eyes which may be correlated with blinking time. The camera set pointed at the bus and train driver's torso and head measures the movements made with individual body parts as well as the drivers' gestures. It is not the case that closing one's eyes is always necessarily connected with the microsleep phenomenon. Therefore, the examination of gestures and movements makes it possible to identify the actual microsleep periods. The measuring apparatus described in the paper does not constitute the entire research domain intended for purposes of observation of the phenomenon in question. In the future, this equipment will be complemented with an air quality meter, a CO_2 m, light intensity sensors and biometric measurement apparatus (temperature, pulse).

3 Empirical Data

A vehicle running 10 km/h in uniform motion travels ca. 3 metres over one second. At a speed of 70 km/h, it covers 20 m in 1 s. It is a distance within which 4 passenger cars or a bus can be encountered. Consequently, the microsleep phenomenon may involve safety of 4 up to several dozen persons. Figure 2 shows the eyelid closing times measured in all four cases (2x bus driver, 2x train driver). The first two cases presented concern the bus driver (driving in two directions in a return run), while the other two pertain to the train driver (return run).

For illustrative purposes, the authors have assumed that each eyelid closing is a potential manifestation of the microsleep phenomenon, which is obviously a considerable simplification of the problem – still necessary at this stage of presentation. There are specific conclusions which can be drawn from the observation of the microsleep of the bus and train drivers (Fig. 2a–h). A cursory analysis of Fig. 2a–d implies that the bus driver blinked less frequently and for a shorter time than the train driver.

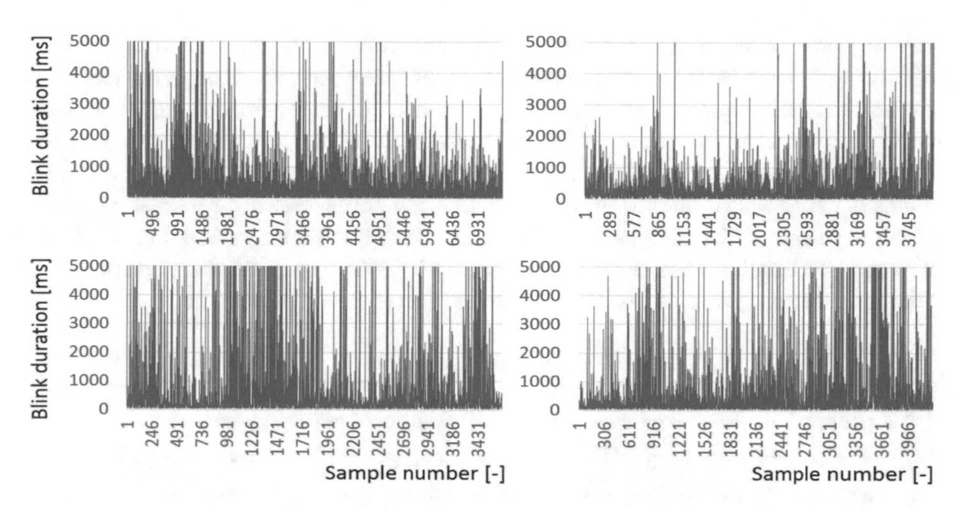

Fig. 2. Blink duration: (a) case1 (b) case2 (c) case 3 (d) case4. *(Source: elaboration own)*

In all the cases analysed, high dynamics of this phenomenon was observed, and it exceeded half a minute in several cases. At a speed of 50 km/h, over this period (30 s) the car driver can travel a distance of ca. 300 m (speed reduction was taken into account in this case), while the train driver closed his eyes for a longer time (Table 1). The train driver showed much higher variability in terms of this property. An interesting fact is that the time value observed most frequently is identical in all four cases (which is attributable to the physiological activity of blinking). Table 1 shows the numerical characteristics of the data set for the bus and train driver's blinking.

Table 1. Blink duration characteristics.

Object	Case1	Case2	Case3	Case4
Average [ms]	438.81	436.89	814.71	696.62
Mode [ms]	33	33	33	33
Median [ms]	199	183	183	199
Std [ms]	840.11	1,065.60	1,950.14	1,560.64
Max [ms]	20,098	36,120	33,009	38,948

An analysis of the mean value exceeding 400 ms implies that the data also comprise observations of microsleep. (Source: elaboration own)

4 Analysis

Figure 3 presents cumulative values of the time of eyelid closing for both cases (a-b – bus driver, c-d – train driver). Every distortion of the curve provided indicates a potential increase in the microsleep time while the person surveyed was moving. What one can observe in these diagrams is that the increase in the microsleep time is linear for short periods of time only. The process of fatigue (those surveyed were initially exhausted by several hours of work starting from 6 a.m.) manifests itself after just a dozen or so minutes (as soon as after 10 min in the bus driver case)! An even though the bus driver stayed constantly attentive (showing small variations in characteristics) and maintained the capacity to effectively perceive the road traffic scene, he succumbed to microsleep immediately after arriving at the depot. The train driver, on the other hand, displayed constant variations of attentiveness throughout the entire return run. These variations were irregular over the whole period of driving. During a break of a dozen or so minutes between consecutive runs, it was the train driver who managed to regenerate more efficiently before further driving (longer linear increase in blinking time).

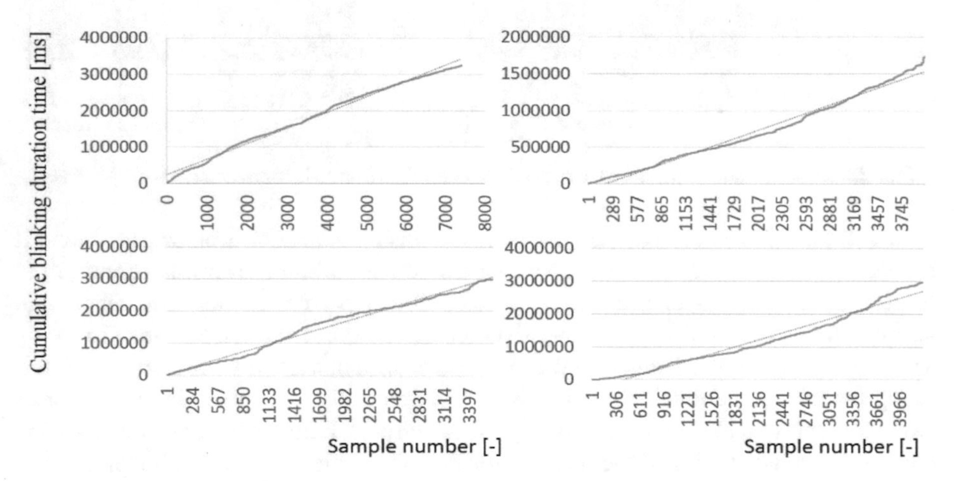

Fig. 3. Cumulative blinking duration time, fatigue graph: (a-b) bus driver (c-d) train driver. *(Source: elaboration own)*

For the sake of comparison, Fig. 4 presents a fragment of the characteristics from Fig. 3 in the part common to both persons examined, showing small differences in the duration of a typical driving time resulting from the self-regulation of traffic which has an impact on the large variation in driving time (also due to a different return route of the bus driver, while the train driver always traversed along the same route).

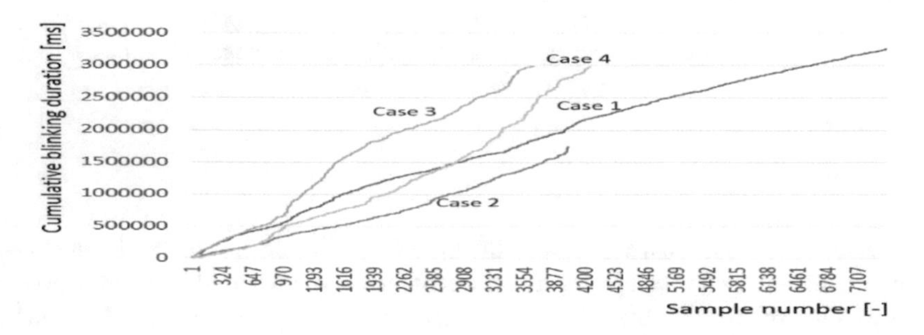

Fig. 4. Cumulative blinking duration time, fatigue chart (comparison). *(Source: elaboration own)*

Figure 4 illustrates the differences in the characteristic curves of the potential microsleep process. It clearly shows that the train driver differs significantly from the bus driver in this respect (monotony of the route). What the authors did not study was the microclimate inside the vehicle driver's cab and the local illumination conditions. Figure 5 shows the cumulative distribution functions of blinking time.

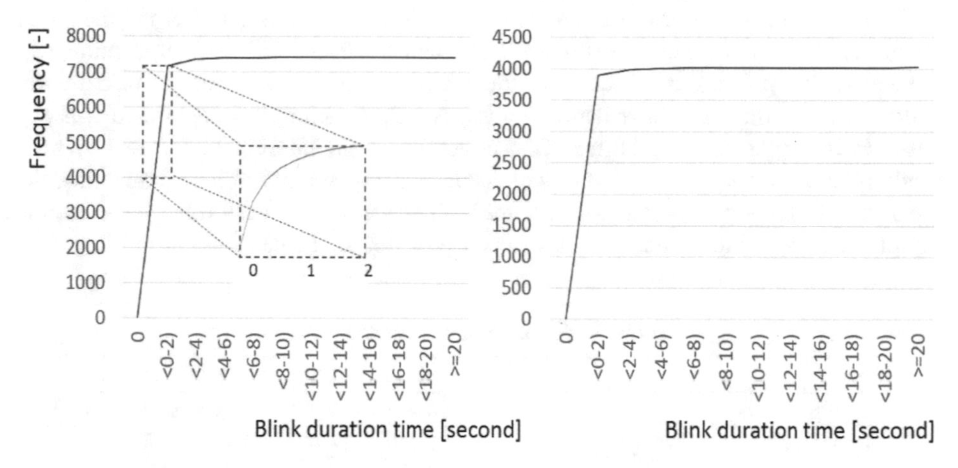

Fig. 5. Blink duration time distribution (bus driver only). *(Source: elaboration own)*

In all four cases, a decided majority of observations are distributed over the first interval, i.e. 0–2 s. The number of blinks increases rapidly in this range of values. Above this interval, the eyelid closing increases in duration, and the number of closing actions grows slowly in a non-linear manner. Table 2 summarises the function forms of the cumulative distribution functions observed in cases in interval < 0,2).

Table 2. Cumulative distribution function of blink duration time.

Object	Case1	Case2	Case3	Case4
<0-2)	y = 1385,6ln(x) + 4263,4; R^2 = 0,943	y = 734,37ln(x) + 2365,8; R^2 = 0,942	y = 577,92ln(x) +2047; R^2 = 0,973	y = 698,22ln(x) + 2380,7; R^2 = 0,976

(Source: elaboration own)

Table 2 illustrates the fact that the number of blinks in the interval from zero to 2 s grows logarithmically (R^2 values have been provided). However, it grows most rapidly between 0 and ca. 1.2 s. These cases should be distinguished from other cases where the increase in the number of blinks in individual intervals, (1.2-4) and above 4, is smaller. Consequently, with reference to these data the authors claim that the microsleep phenomenon should be studied in intervals of (400;1,200 ms), (1,200;4,000 ms) and above 4,000 ms. The cases of microsleep from the latter interval account for 37–176 incidents per an hour of driving, which is relatively low (although it is still 2 per minute). The main problem in this respect is the change in the time of the driver's response, which is different in cases of a very short microsleep. Therefore, one should concur with other researchers claiming that the microsleep problem begins when it is longer than 1 s.

Blinking is an essential physiological activity which manifests itself in rapid movement (different ranges of values are provided in the literature: 100 ÷ 400 ms) or closing shut of eyelids [15, 16, 19]. The aim of this reflex is to create a tear film and moisten the eye (which dries more intensely while driving on account of the increased attentiveness). An average person typically blinks 10 ÷ 17 times per 1 min [16]. However, this process is affected by the illnesses one has suffered or is currently suffering from, the state of irritation, stress, etc. Some sources recognise microsleep when it is longer than 1,000 ms. Consequently, in an interval of 0.5 ÷ 1 s, both blinking and short-term microsleep may occur. For the sake of further calculations, the mean of 14 blinks per minute was assumed. The calculations performed to determine the parameters of individual tests have been provided in Table 3.

Table 3. Blink duration characteristics.

	Object	Case1	Case2	Case3	Case4
1	Testing time [s]	4,425.49	2,479.47	4,057.05	4,513.60
2	Number of observations [-]	7,418	4,027	3,663	4,265
3	Mean of blinks (14*Δt)	1,033	579	947	1,054
4	Number of observations above 1,000 ms [-]	738	388	642	704
5	Number of observations above 400 ms [-]	2,031	1,034	1,153	1,333
6	Number of observations above 4,000 ms [-]	68	37	176	154
7	Percentage of average blinks [%]	14%	14%	26%	25%

(Source: elaboration own)

Table 3 provides information on the number of blinks which should statistically occur in driving (line 3). For a bus driver, there is an excess number of observations. Lines 4 ÷ 6 state the number of blinks over 400 ms, 1 s. and 4 s. An analysis of Table 3 implies that 3/4 of all recorded blinks should theoretically be related to the microsleep phenomenon. At the same time, the probability of a long microsleep is lower.

Figure 6a provides a comparison of times of blinking and visual fixation (< 1,000 ms). The relevant plots have been superimposed according to the testing time curve (case 2, rough comparison). The same has been done in Fig. 6b with regard to the time of blinking and saccadic movements. Falling into microsleep should not be associated with long gazing at the objects within the traffic scene. More or less the same can be said about performing intense saccadic movements while falling into microsleep at the same time. This is contradictory for physical reasons. Intense gazing may actually involve moistening of eyes at the most.

In Fig. 6a, the (1) marks the period of examination where long times of eyelid closing correspond to very long fixation times (ca. 0.5 min). It is hardly imaginable that a vehicle driver should gaze at the surrounding so intensely while falling into microsleep at the same time (ca. 2 s). Probably it is no microsleep. The (2) marks the point where the microsleep could come to ca. 1 s upon simultaneous fading of fixation.

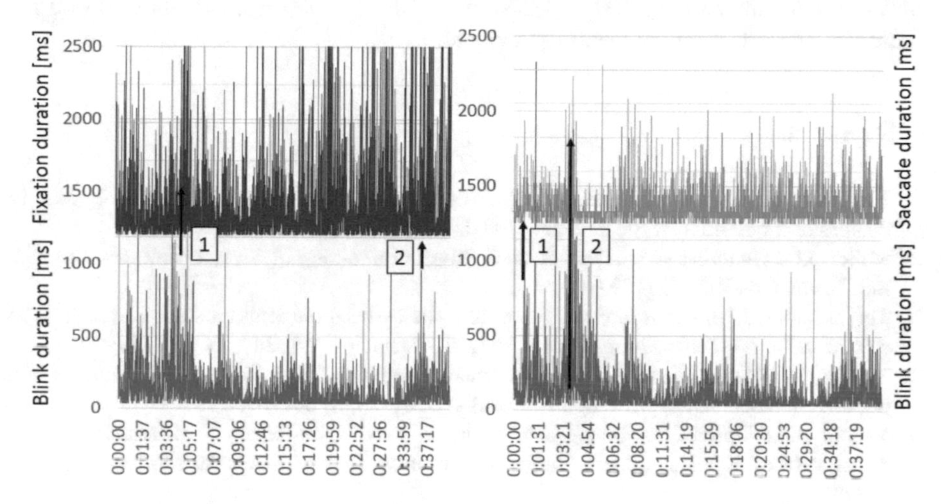

Fig. 6. Rough comparison: (a) blinking and visual fix. times (b) blinking and saccadic mov. Times. *(Source: elaboration own)*

It can be a microsleep. Interpretation of the data provided in Fig. 6b is similar. In either case, the point is to reach a conclusion such as the following: while falling into microsleep, the driver cannot intensely examine the surrounding at the same time.

5 Conclusions

The problems addressed in the paper involve processing of BIG DATA sets. Even if one uses special tools intended for handling of the ET data (BeGaze), processing of the results requires much time and effort. This calls for automation of the process. A single test run (ca. 40') entails the necessity to process data sets containing millions of data items, which is remarkably time-consuming. And this all the more laborious if different explanatory variables (visual fixations, saccadic movements, hot spots, AOI) available in the input data for the ET technology are assumed for purposes of the analysis, as demonstrated in this article. Such relationships exist locally for small intervals. Owing to these relationships, one can tell the difference between blinking and microsleep related processes which actually take place. With regard to the microsleep phenomenon itself, the conclusions derived from the pilot observations may be of purely working nature only. For the sake of generalisation, the authors will run further tests on a more numerous sample and under different infrastructure conditions.

However, one can be certain that the microsleep phenomenon does take place in the analysed samples. Moreover, its initial symptoms in a bus driver exposed to initial fatigue surface as soon as after a dozen or so minutes of driving, and the microsleep phenomenon is also conditional to the monotony resulting from the surrounding infrastructure. One can also observe a high variability of the characteristics analysed. Further studies should concentrate on explaining the differences between the data analysed in the intervals of 400 ÷ 1,200 ms, 1,200 ÷ 4,000 ms, and above 4,000 ms where the microsleep phenomenon is evident.

References

1. Hakkanen, H., Summala, H., et al.: Blink duration as an indicator of driver sleepiness in professional bus drivers. Sleep **22**(6), 798–802 (1999)
2. Miller, J.C.: Quantitative analysis of truck driver EEG during highway operations. Biomed. Sci. Instrum. **34**, 93–98 (1997)
3. Teran-Santos, J., Jimenez-Gomez, A., et al.: The association between sleep apnea and the risk of traffic accidents. Coop. Group Burgos-Santander. Med. **340**, 847–851 (1999)
4. Young, T., Blustein, J., et al.: Sleep-disordered breathing and motor vehicle accidents in a population based sample of employed adults. Sleep **20**(8), 608–613 (1997)
5. Vespa, S.R.: Options for changes to hours of service for commercial vehicle drivers. Transport Development Centre, Safety and Security, Transport Canada, Montreal, pp. 3–5 (1998)
6. Torsvall, L., Akerstedt, T., et al.: Sleep on the night shift: 24-hour EEG monitoring of spontaneous sleep, wake behavior. Psychophysiology **26**(3), 352–358 (1989)
7. Torsvall, L., Akerstedt, T.: Sleepiness on the job: continuously measured EEG changes in train drivers. Electroencephalogr. Clin. Neurophys. **66**(6), 502–511 (1987)
8. Miller, J.C.: Batch processing of 10,000 h of truck driver EEG data. Biol. Psychol. **40**(1–2), 209–222 (1995)
9. Lorenzo, I., Ramos, J., et al.: Effect of total sleep deprivation on reaction time and waking EEG activity in man. Sleep **18**(5), 346–354 (1995)
10. Haraldsson, P.O., Carenfelt, C., et al.: Clinical symptoms of sleep apnea syndrome and automobile accidents. ORL J. Otorhinolaryngol. Relat. Spec. **52**, 57–62 (1990)
11. Corsi-Cabrera, M., Ramos, J., et al.: Changes in the waking EEG as a consequence of sleep and sleep deprivation. Sleep **15**(6), 550–555 (1992)
12. Risser, M.R., Ware, J.C., Freeman, F.G.: Driving simulation with EEG monitoring in normal and obstructive sleep apnea patients. Sleep **23**(3), 393–398 (2000)
13. Young, T., Palta, M., Dempsey, J., Skatrud, J., Weber, S., Badr, S.: The occurrence of sleep-disordered breathing among middle-aged adults. N. Engl. J. Med. **328**(17), 1230–1235 (1993)
14. Amit, P., Boyle, L., Tippin, J., Rizzo, M.: Variability of driving performance during microsleeps. In: Proceedings of the Third International Driving Symposium on Human Factors in Driver Assessment, Training and Vehicle Design. (2005)
15. Schiffman, H.R.: Sensation and Perception. An Integrated Approach. Wiley, New York (2001)
16. UCL LONDON, Institute of Neurology. http://news.bbc.co.uk/2/hi/health/4714067.stm. Accessed 16 Mar 2019 (2005)
17. https://www.transportation.gov/. Accessed 16 Mar 2019 (2014)
18. Golz, M., Sommer, D., Krajewski, J., Trutschel, U., Edwards, D.: Microsleep episodes and related crashes during overnight driving simulations. In: Proceedings of the 6-th International Driving Symposium on Human Factors in Driver Assessment. Iowa (2011)

19. Golz, M., Sommer, D., Chen, M., Trutschel, U., Mandic, D.: Feature fusion for the detection of microsleep events. J. VLSI Sig. Proc. Syst. **49**, 329–342 (2007)
20. Jacyna, M., Wasiak, M., Lewczuk, K., Kłodawski, M.: Simulation model of transport system of Poland as a tool for developing sustainable transport. Arch. Transp. **31**, 23–35 (2014)
21. Siergiejczyk, M., Pas, J., Rosinski, A.: Issue of reliability–exploitation evaluation of electronic transport systems used in the railway environment with consideration of electromagnetic interference. IET Intell. Transp. Syst. **10**(9), 587–593 (2016)

Comparison of Energy Consumption of Short and Long City Buses in Terms of Assessing the Needs for e-Mobility

Rafał Burdzik[1(✉)], Łukasz Konieczny[1], Robert Jaworski[2], Dariusz Laskowski[3], and Rafał Polak[4]

[1] Faculty of Transport, Silesian University of Technology, 40-019 Katowice, Poland
{rafal.burdzik, lukasz.konieczny}@polsl.pl
[2] Automotive Expertise Office, Częstochowa, Poland
jaworski.rzeczoznawca@gmail.com
[3] Military University of Technology, Warsaw, Poland
[4] Transbit, Warsaw, Poland
rafal.polak@transbit.com.pl

Abstract. The article presents the results of simulation studies, the aim of which was to compare comparative energy consumption of short and long city buses in terms of assessing the needs for e-mobility. Currently, a number of e-mobility activities are observed, the largest of which has a dynamic change in the bus fleet structure and a significant, defined in subsequent years, increase in the percentage share of zero-emission buses (electric as well as hybrid and hydrogen). Decisions regarding the exchange of fleets should be preceded by proper analyzes, including the requirements and energy consumption of individual bus lines with the assumption of a certain number of passengers and the selection of a bus with proper capacity. The methodology for calculating energy consumption and determination of the required engine power presented in the article enables dedicated analyzes depending on driving profiles and specific bus parameters.

Keywords: Energy consumption · Public transport · Standardised On Road Test Cycles

1 Introduction

Current activities in the field of environmental transport responsibility are oriented at negative impacts in cities [1, 2]. An important means of transport in urban agglomerations are buses, which in Poland are still powered mostly by self-ignition engines. These engines produce significant amounts of particulate matter, nitrogen oxides and excessive noise. Therefore, actions were taken to change the structure of the fleet of collective transport carriers. In addition to the specific growth rate of the share of zero-emission buses (electric), an increase in the number of hybrid buses and CNG-powered buses is also observed [3, 4].

When choosing an electric or hybrid drive system, pay attention to the appropriate battery or other energy storage system. For this purpose, account shall be taken of:

M. Siergiejczyk and K. Krzykowska (Eds.): ISCT21 2019, AISC 1032, pp. 74–83, 2020.
https://doi.org/10.1007/978-3-030-27687-4_8

vehicle weight, required propulsion power, vehicle operating range, maintenance organization and costs [5, 6]. The paper [7] presents a cost-benefit analysis of hybrid and electric city buses in fleet operation. The paper [8] analyzes the TTW (tank-to-wheel) energy conversion efficiency of a series plug-in hybrid electric bus. The TTW process is characterized by the recuperation and fuel-to-traction efficiencies, which are quantified and compared for two optimization-based energy management strategies.

In addition, the current requirements of public transport should be taken into account. Currently, the "Mobility as a Service – MaaS" approach should be distinguished, in which the needs of residents are met by means of one service that combines the offer of many carriers, navigation systems and payment technologies - using a dedicated application (Fig. 1). It is assumed that MaaS development will improve both city traffic and contribute to reducing the pressure on own means of transport. For the effective functioning of MaaS, it is necessary to provide a number of technical and organizational parameters, including universal access to 3G/4G/5G mobile networks and a high level of telecommunications network quality, visual information systems and mobile search engines providing secure, dynamic and up-to-date information about travel options, data and platform management systems, integration of transport, energy and ICT infrastructure, properly designed infrastructure of integrated transfer nodes, taking into account parking spaces for bicycles and cars used in the car-sharing model and non-cash payment systems [9–11].

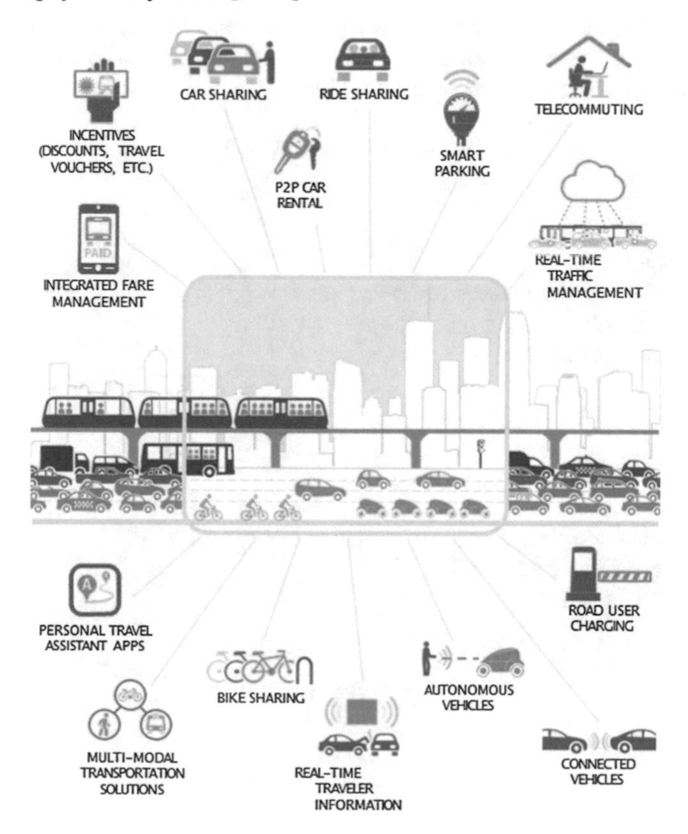

Fig. 1. Future mobility options [10]

2 Components of Energy Consumption in Driving

The article presents an approach to estimating energy consumption of traffic as the sum of traffic resistance. A model based on four main resistance groups was adopted.

Air resistance is the force created by the difference in dynamic pressures that act on the surface of the vehicle. It consists of: profile resistance (around 58%), depending on the shape of the longitudinal sections of vehicles; interference resistance (approx. 14%) caused by such vehicle parts as: mirrors, handles, wheels and other elements; frictional resistance (about 10%); resistance of the cooling and ventilation system (about 10%); inductive resistance (approx. 8%) caused by turbulence of air streams on the sides of the vehicle. The size of the air resistance depends on the front face of the vehicle, air density, aerodynamic coefficient and speed squared.

The rolling resistance is a force which results from the deformation during the cooperation of the wheel surface with the ground, internal friction which occurs in the tires and the adhesive force between the tire and the surface. The value of rolling force depends on the type of surface and tires as well as the weight of the vehicles. Its size is influenced by the dimensions of the drive wheel, air pressure in the tires and the speed of movement.

The resistance of the hill should be taken into account when riding on hill clearance. It is a component of the vehicle's weight, which depends on the inclination of the route covered. Inertia resistance is a force that counteracts the change in speed at which a vehicle moves at a given moment. The force occurs during accelerations and braking. These resistances are calculated as follow [12]:

$$F_a = 0,047 \cdot c_x \cdot A \cdot v^2, \quad (1)$$

where:

F_a – Air resistance;
A – the front face of the vehicle [m^2];
c_x – drag coefficient;
v – vehicle speed [km/h].

$$F_t = G \cdot \cos\alpha \cdot f_0 \cdot \left(1 + k \cdot v^2\right), \quad (2)$$

where:

F_t – rolling resistance [N];
G – the gravity of the vehicle [N];
f_0 – coefficient of friction;
k - coefficient.

$$F_w = G \cdot tg\alpha, \quad (3)$$

where:

F_w - resistance of the hill [N];
G - vehicle gravity [N];
α - inclination angle of the slope.

$$F_b = m \cdot a \cdot \delta, \tag{4}$$

where:

F_b - inertia force [N];
m - vehicle mass [kg];
a - acceleration [m/s^2];
δ - coefficient of reduced masses.

3 Research Method

As part of the study, simulations of driving were carried out for specific cycles representing the most common traffic conditions of city buses. The speed of city buses depends on external factors such as other road users, road surface, and weather conditions. The bus traveling on the appropriate section of the road is characterized by irregularly changing speed changes (acceleration, braking, driving at a constant speed). The speed profiles of each module are distinguished by their range and the intensity of changes in speed and duration of individual stages of motion. Analyzing the urban movement, the modules are short and divided by the time of stoppage. We divide the travel cycle taking into account the speed profile on:

(a) single-segmented:

– trapezoidal,
– triangular,

(b) multi-segment complex:

– trapezoidal (SORT cycles),
– mixed (Japanese, Manhattan Cycle, Californian).

The first aim of SORT (Standardised On Road Test cycles) methodology was to provide the bus sector a standardised way to compare the energy consumption of different buses. It is very important that bus operators get an objective information about especially the energy consumption which is one of the most important key factor for TCO (Total Cost of Ownership) and it doesn't matter if we speak about classic driven or e-driven buses. For now 12 years, International Association of Public Transport UITP has been publishing a standardised SORT protocol for comparing the energy consumption of buses. Starting with 12 m diesel buses, it has been enlarged to all bus sizes (from midi-buses to articulated or double-decker vehicles), using either liquid or gaseous fuels, and to non-plug-in hybrids.

There are three driving cycles that represent the given traffic categories:

- SORT 1 in city centers - Heavy Urban,
- SORT 2 in cities - Easy Urban,
- SORT 3 on suburban routes - Easy Suburban.

The SORT 1 cycle - Heavy Urban Cycle is used to simulate bus driving in the city center. The assumptions of these simulations are the average vehicle speed of 12.1 km/h. Acceleration of individual profiles is successively: 1.03 m/s^2, 0.77 m/s^2, 0.62 m/s^2. The delay for each is 0.8 m/s^2.

The SORT 2 cycle - Easy Urban Cycle is used to simulate bus travel in the city. The assumptions of these simulations are the average vehicle speed of 18 km/h. Acceleration of individual profiles is successively: 1.03 m/s^2, 0.62 m/s^2, 0.57 m/s^2. The delay for each is 0.8 m/s^2.

Cycle 3 - Suburban Cycle is used to simulate the travel of buses outside the city. The assumptions of these simulations are the average vehicle speed of 25.3 km/h. Acceleration of individual profiles is successively 0.77 m/s^2, 0.57 m/s^2, 0.46 m/s^2. The delay for each is 0.8 m/s^2.

Due to the increasing share of electric and hybrid buses, UITP introduced new standards dedicated to this type of city buses. The aim of the project "E-SORT" is to measure, in a reproducible way, the energy consumption and the Zero Emissions range of a bus. This E-SORT protocol applies to 2 types of vehicles: Full electric vehicles (These are vehicles that use electricity from the grid as the only energy source) and Hybrid Vehicles that can be externally re-charged (Plug-in Hybrid Vehicles – PHEV).

Every test has to be considered in term of reliability of transport system [13, 14].

The calculations were carried out for a 12-m long and a short (articulated) 18-m long bus. It was assumed that both buses have the same drive units (Table 1).

Table 1. Basic parameters of the tested buses.

Parameter	12-m	18-m
Body type	Low-floor city bus	Low-floor city bus
Length [mm]	12 000	18 000
Width [mm]	2 550	2 550
Height [mm]	3 350	3 350
Mass [kg]	18 000	28 000
Engine power [kW]	230 (2000 rpm)	230 (2000 rpm)

4 Result and Analysis

The article presents the results of calculations of energy consumption of individual cycles of driving cycles and the demand for traction motor power determined on this basis. Each of the profiles was characterized by sections of the time of moving with a given speed or with constant acceleration in a given unit of time, hence the driving profiles were presented as functions of distance and time. The results are shown in the tables and figures below. This makes it possible to compare energy consumption for a 12-m long and 18-m long buses (Figs. 2, 3, 4, 5 and Tables 2, 3, 4, 5).

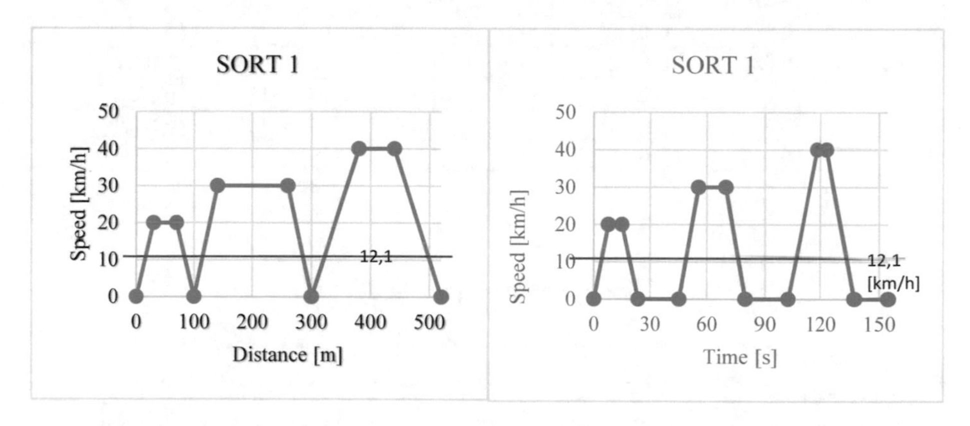

Fig. 2. SORT 1 – profiles.

Table 2. Energy consumption and demand for power calculation for 12-m long bus – SORT1.

	Time [s]	Acceleration [m/s^2]	Start speed [km/h]	End speed [km/h]	Distance [m]	Energy consumption [N]	Demand for power [kW]
Profile 1	7,6	1,03	0	20	30	21857,882	95,87
	7,2	0	20	20	40	2020,082	12,47
	8,6	−0,8	20	0	30	−13387,918	−51,89
Stop	22	0	0	0	0	0	0
Profile 2	10,2	0,77	0	30	40	16988,27	74,02
	14,4	0	30	30	120	2158,07	19,98
	10	−0,8	30	0	40	−13249,93	−58,89
Stop	22	0	0	0	0	0	0
Profile 3	16	0,62	0	40	80	14237,537	79,1
	5,4	0	40	40	60	2296,337	28,35
	14,1	−0,8	40	0	80	−13111,663	−82,66
Stop	17	0	0	0	0	0	0

Table 3. Energy consumption and demand for power calculation for 18-m long bus – SORT1.

	Time [s]	Acceleration [m/s^2]	Start speed [km/h]	End speed [km/h]	Distance [m]	Energy consumption [N]	Demand for power [kW]
Profile 1	7,6	1,03	0	20	30	33879,502	148,59
	7,2	0	20	20	40	3020,702	18,65
	8,6	−0,8	20	0	30	−20947,298	−81,19
Stop	22	0	0	0	0	0	0
Profile 2	10,2	0,77	0	30	40	26252,415	114,39
	14,4	0	30	30	120	3183,215	29,47
	10	−0,8	30	0	40	−20784,785	−92,38
Stop	22	0	0	0	0	0	0
Profile 3	16	0,62	0	40	80	21931,017	121,84
	5,4	0	40	40	60	3355,817	41,43
	14,1	−0,8	40	0	80	−20612,183	−129,94
Stop	17	0	0	0	0	0	0

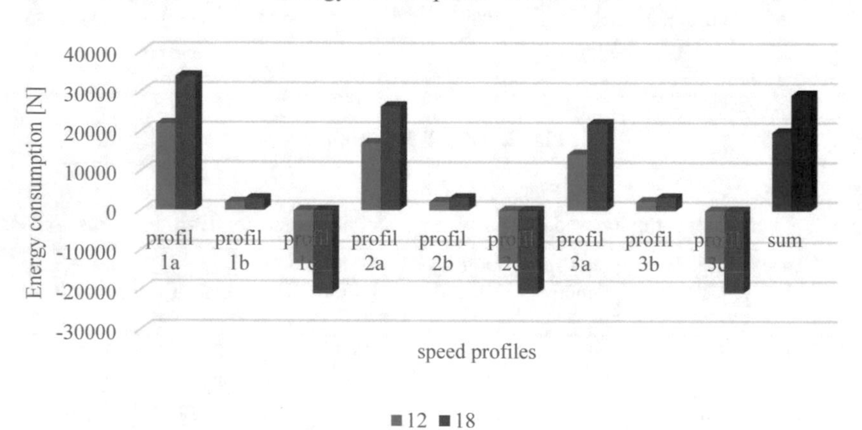

Fig. 3. Comparison of energy consumptions for 12-m long and 18-m long buses, SORT 1 – profiles.

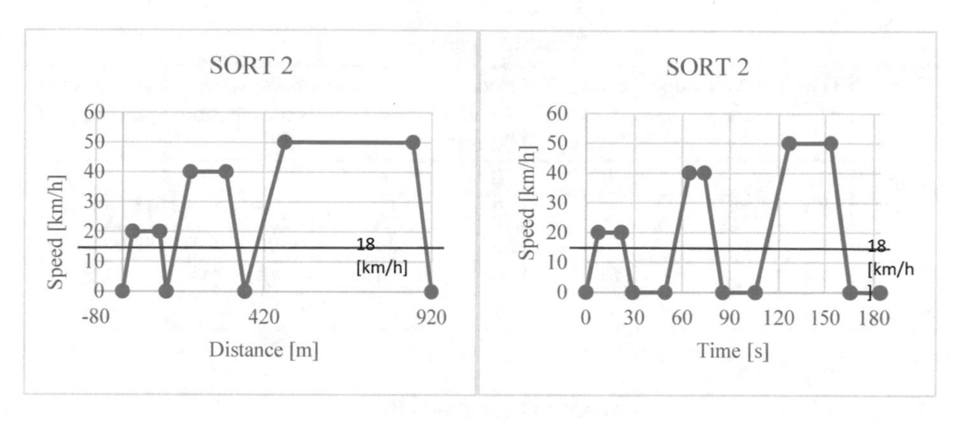

Fig. 4. SORT 2 – profiles.

Table 4. Energy consumption and demand for power calculation for 12-m long bus – SORT2.

	Time [s]	Acceleration [m/s²]	Start speed [km/h]	End speed [km/h]	Distance [m]	Energy consumption [N]	Demand for power [kW]
Profile 1	7,6	1,03	0	20	30	21857,882	95,87
	14,4	0	20	20	80	2020,082	12,47
	7	−0,8	20	0	20	−13387,918	−42,5
Stop	20,3	0	0	0	0	0	0
Profile 2	15	0,62	0	40	70	14237,537	73,82
	9,5	0	40	40	106	2296,337	28,47
	11,8	−0,8	40	0	56	−13111,663	−69,14
Stop	20,3	0	0	0	0	0	0
Profile 3	20,4	0,57	0	50	119	13572,964	87,97
	27,4	0	50	50	380	2596,764	40,02
	12,1	−0,8	50	0	59	−12813,236	−69,42
Stop	18,1	0	0	0	0	0	0

Table 5. Energy consumption and demand for power calculation for 18-m long bus – SORT2.

	Time [s]	Acceleration [m/s²]	Start speed [km/h]	End speed [km/h]	Distance [m]	Energy consumption [N]	Demand for power [kW]
Profile 1	7,6	1,03	0	20	30	33879,502	148,59
	14,4	0	20	20	80	3020,702	18,65
	7	−0,8	20	0	20	−20947,298	−66,5
Stop	20,3	0	0	0	0	0	0
Profile 2	15	0,62	0	40	70	21931,017	113,72
	9,5	0	40	40	106	3355,817	41,6
	11,8	−0,8	40	0	56	−20612,183	−108,69

(continued)

Table 5. (*continued*)

	Time [s]	Acceleration [m/s²]	Start speed [km/h]	End speed [km/h]	Distance [m]	Energy consumption [N]	Demand for power [kW]
Stop	20,3	0	0	0	0	0	0
Profile 3	20,4	0,57	0	50	119	20775,589	134,66
	27,4	0	50	50	380	3698,389	56,99
	12,1	−0,8	50	0	59	−20269,611	−109,82
Stop	18,1	0	0	0	0	0	0

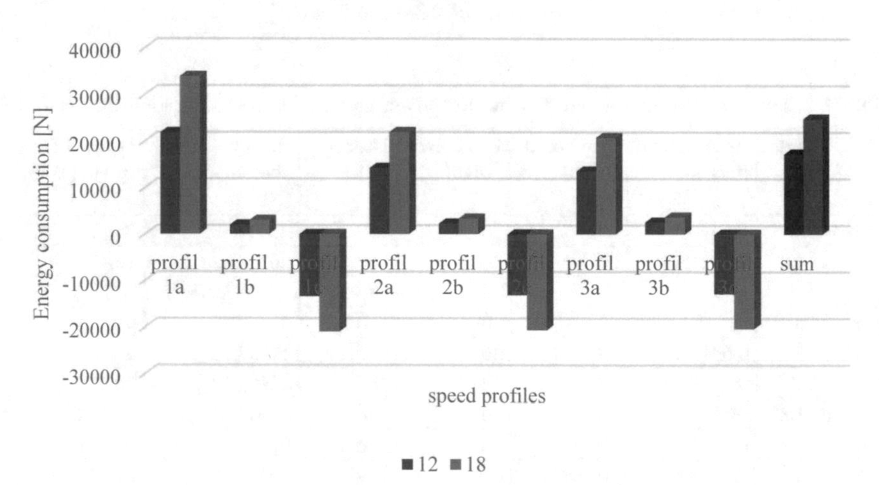

Fig. 5. Comparison of energy consumptions for 12-m long and 18-m long buses, SORT 2 – profiles.

5 Conclusions

The presented results of simulation studies on energy consumption of bus traffic in urban cycles allow estimation of energy consumption for specific vehicle. In addition, based on the knowledge of the drive transmission parameters, including general efficiency, they enable calculation of the required traction motor power. As a result, it enables proper selection of the vehicle for specific transport tasks, which may be an important decision-making factor in the planning of "electrification" of urban city rolling stock or design of charging infrastructure. Accurate and precise profiling of the routes of particular lines, taking into account the level of terrain, sections between stops and average staple speeds, will allow to increase the accuracy of calculations. Also results can be source of information for energy management strategy, as it was presented for fuel cell hybrid bus in paper [15]. In addition, the presented calculations can also be the basis for determining fuel consumption and the emission volume of harmful compounds in the case of buses powered by an internal combustion engine.

References

1. Jacyna, M., et al.: Noise and environmental pollution from transport: decisive problems in developing ecologically efficient transport systems. J. VibroEng. **19**(7), 5639–5655 (2017)
2. Jacyna, M., Merkisz, J.: Proecological approach to modelling traffic organization in national transport system. Arch. Transp. **30**, 31–41 (2014)
3. Wojewoda, P.: The methodology of selecting an internal combustion engine for a selected configuration of the hybrid drive of a city bus. Ph.D. Dissertation, Rzeszów (2012)
4. Burdzik, R., Folęga, P., Konieczny, Ł., Jaworski, R.: E-mobility - the challenge of the present. Prace naukowe Politechniki Warszawskiej **118**, 17–29 (2017)
5. Zajler, W.: The use of hybrid drive in combat vehicles. High-speed tracked vehicles, nr.1 (21). OBRUM, Gliwice (2005)
6. Goethem, S., Koornneef, G., Spronkmans, S.: Performance of battery electric buses in practice. energy consumption and range, TNO report, nr 033.27092/01.43 (2013)
7. Lajunen, A.: Energy consumption and cost-benefit analysis of hybrid and electric city buses. Transp. Res. Part C: Emerg. Technol. **38**, 1–15 (2014)
8. Hu, X., et al.: Energy efficiency analysis of a series plug-in hybrid electric bus with different energy management strategies and battery sizes. Appl. Energy **111**, 1001–1009 (2013)
9. Hensher, D.A.: Future bus transport contracts under a mobility as a service (MaaS) regime in the digital age: are they likely to change? Transp. Res. Part A: Policy Pract. **98**, 86–96 (2017)
10. Fishman, T.D.: Digital Age Transportation: The Future of Urban Mobility. Deloitte University Press, Deloitte Development LLC (2012)
11. Kamargianni, M., et al.: A critical review of new mobility services for urban transport. Transp. Res. Procedia **14**, 3294–3303 (2016)
12. Domek, M.: Resistance of vehicle motion and minimization of rolling resistance. Drives Control **12**, 96–100 (2017)
13. Siergiejczyk, M., Krzykowska, K., Rosinski, A., Grieco, L.A.: Reliability and viewpoints of selected ITS system. In: 2017 25th International Conference on Systems Engineering (ICSEng), pp. 141–146. IEEE (2017)
14. Siergiejczyk, M., Pas, J., Rosinski, A.: Assurance of the electromagnetic compatibility in the chosen transport telematic systems. Arch. Transp. Syst. Telematics **6**, 38–41 (2013)
15. Gao, D., Jin, Z., Lu, Q.: Energy management strategy based on fuzzy logic for a fuel cell hybrid bus. J. Power Sources **185**(1), 311–317 (2008)

Personal Rapid Transit – Polish Concept

Włodzimierz Choromański(✉) and Iwona Grabarek

Warsaw University of Technology, Koszykowa 75, 00-662 Warsaw, Poland
prof.wch@gmail.com

Abstract. In the last period in the world you can observe a very large interest in vehicles and autonomous transport systems. To the level of L5 autonomy (according to the SAE classification), it seems that we still have very far away (in the technological sense) and this technology according to various forecasts will be achieved in the perspective of 15–20 years with regard to motor vehicles. It is difficult to precisely define today which technical, IT and legal standards will be finally adopted. The authors of this work propose to define the concept of an autonomous vehicle a little differently. It will be understood as a car or track vehicle that can partially move without the participation of the driver and who can autonomously (without the participation of the driver) choose a route from the starting point to the destination. The concept of a track starts with a debatable character if, for example, it is determined by an IT system. The authors argue that a system with such features combines the features of systems: an autonomous vehicle at level L4 b and an extended PRT system. Polish cities (Rzeszów, Katowice, Łódź) have signed a letter of intent regarding the implementation of this system in Poland.

Keywords: Personal rapid system · Modeling and simulation · Polish concept of PRT

1 Introduction

The problems of urban transport are the subject of intense research and constant search for new solutions [1, 2, 5, 9]. We are looking for solutions that:

- They will limit individual transport for public transport or for other organizational forms of individual transport;
- They will be ecological (they will use electric drive, they will not emit pollutants and greenhouse gases, they will undergo the process of recycling and remanufacturing);
- They will be safe and adapted to carry disabled people;
- They will be material-saving and energy-saving (through energy recovery systems and implementation of the "on demand" transport concept).

Undoubtedly, such features have the Personal Rapid Transit idea, which was born in the '50 s in the USA [6]. PRT is a transport combining the features of individual and collective transport. It consists of many vehicles (usually 3–4 persons) completely remotely operated (without the participation of the driver) realizing the "door to door" transport, i.e. from the initial stop to the final stop without intermediate stops. Vehicles

M. Siergiejczyk and K. Krzykowska (Eds.): ISCT21 2019, AISC 1032, pp. 84–93, 2020.
https://doi.org/10.1007/978-3-030-27687-4_9

move on a light aboveground infrastructure (see Fig. 1) easily scalable i.e. easily expandable with new network elements (lines, stops, etc.). Between two stops (initial and final), the connection is redundant (the control system can choose the optimal route - bypassing e.g. blockages on the lines). Because vehicles are placed on a light aboveground infrastructure (about 5 m above the ground) there is no collision with other vehicles by surface transport. To date, no PRT system with all its functions has been implemented. Rather, pilot solutions (Ultra system at Heathrow airport, PRT company - 2getthere or British-Swedish-Korean Vectus system).

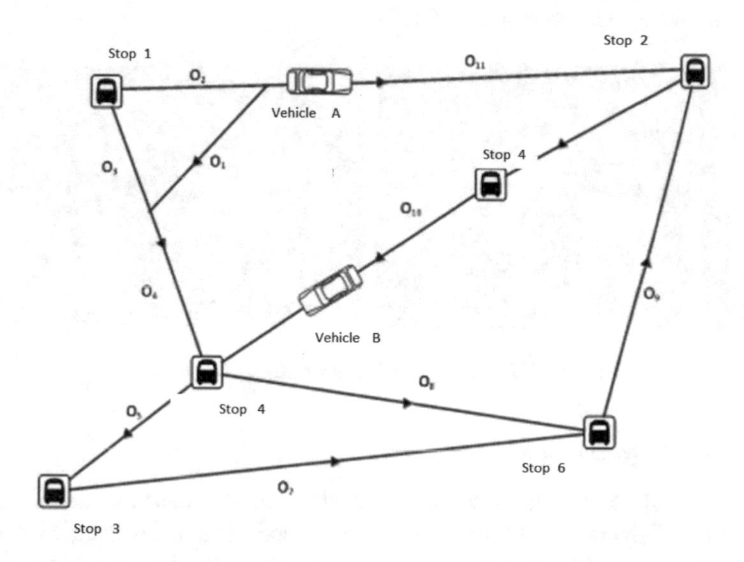

Fig. 1. An example of a PRT network

The implementation of the PRT system requires answers to at least a few questions:

- What is the broadly understood efficiency of the transport of this system (including its capacity) in comparison with other modes of transport?
- What are the costs of building this system and its operating costs in comparison with already used means of transport?
- What is the reliability and safety of this system compared to existing systems?
- How to implement system management and control algorithms?

We do not have an answer for most of these questions. There are too few systems implemented (apart from the above-mentioned demonstrators). However, computer simulation tests are carried out and forecasts (often very different) are given.

2 Polish PRT System

The Warsaw University of Technology developed (in the sense of: a 1:1 scale vehicle solution and a model for the validation of management and control concepts on a scale of 1: 4) PRT system - Prometheus. In terms of individual assemblies of this complex system, the following have been developed: control and management system and algorithms, drive systems, mechanical systems, system of changing direction, hierarchic control system, vehicle design and passenger interface as well as computer simulation software for dynamic and traffic simulation Fig. 2 presents the vehicle of polish Prometheus PRT sytem on a scale of 1:1.

,

Fig. 2. Vehicle of polish Prometheus PRT system

2.1 Technical Description

Detailed technical specifications of the Polish solution are presented, among others, in [1, 3, 8]. The Polish solution is characterized by, among others: an electric drive system using a linear motor (the active part of the engine is on the vehicle); the original solution (patent) of changing the direction of movement on turnouts using the mechanism placed on the vehicle (the so-called passive switch); a modern energy recovery system using a supercapacitor; non-contact energy transfer to the vehicle; hierarchical, three-level management and control system; precise vehicle identification systems; control and management algorithms. Some elements are shown in Fig. 3.

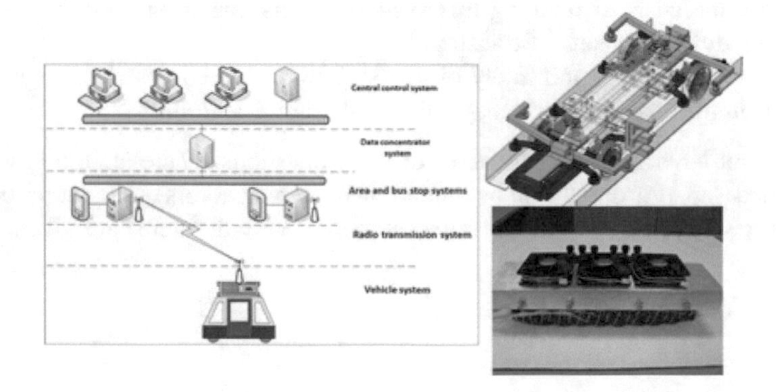

Fig. 3. Examples of the elements of PRT developed at Warsaw University of Technology (hierarchical control system, passive switch, linear motor)

2.2 Modeling and Simulation

The general simulation model of the system will be described in the following [7]. This model consists of the few groups of objects.

PRT vehicles moving in the PRT network are characterized by the following properties: number of seats for passengers (usually 4–5 seats), vehicle priority (some vehicles can be distinguished and privileged), vehicle performance (in particular acceleration, delay and maximum speed). Communication routes represent elements of the track system on which PRT vehicles move. The basic unit of communication routes is the Q_i segment (see Fig. 1). It is a part of the track connecting directly two separate points of the network - stations, intersections and depots. Each segment is described by several parameters, such as: length segment; the maximum speed allowed for the vehicle to travel on it; direction of vehicles moving; priority (it represents the category of the communication route, e.g. main bus access segment).

Another important element of the PRT network are the stations. It is a place where passengers order vehicles and wait for them, get on and off. In the system, we distinguish two types of stations: FIFO and with bays. The depot is the place from which vehicles start their traffic and to which they return for service or garage purposes. The depot is characterized by the number of parking spaces.

An important element in the simulation of PRT networks are passenger flows characterized by traffic between individual stations. The PRT vehicle motion simulator uses the model in the form of a complex cellular automaton. Cellular automata are the structures described by the cellular mesh and their states, transitions and rules of these transitions. Each simple cellular automaton consists of a n-dimensional regular, discrete cell grid, each cell is the same (it is a copy of the previous one), the entire grid space must be occupied entirely by cells arranged side by side. Each of them has one state from a finite set of states. The evolution of each cell follows the same strictly defined local rules, which depend only on the previous state of the cell and on the state of the finite number of cells - neighbors. Evolution occurs at discrete time intervals, simultaneously for each cell. In the cellular automaton, the cell is a finite state machine.

For the problem of simulating PRT traffic, the adaptation of cellular automata to a more complex structure, which is a directed graph representing the infrastructure of the PRT network, has been proposed. The computational model is a directed graph in which the nodes are communication nodes, while the edges are sections, or communication routes.

At the moment when the structure of the graph reflects the topology of the PRT network, the dynamic elements (i.e. vehicles, depots, intersections, logical modules) receive their initial conditions (including position, speed, weight). After determining and defining all the components of the cell automaton, you can go to the description of rules describing the evolution of the machine in time. As mentioned, all cells change their state synchronously, realizing a single iteration step time. The implementation of the each step can be divided into several phases:

- Checking pass rules - in this phase the current state of the cell and neighbor cell status are checked;

- Neighbor checking - it is investigated whether any of the neighboring cells does not enter the state in which the next step of the iteration is inconsistent with the assumed principles of the automaton's evolution (e.g. one vehicle will cross over to the other). Such states will be called conflicts.
- In this phase, all existing conflicts should be eliminated according to pre-defined rules (e.g. speed reduction);
- Checking boundary conditions - the cells that are on the edges of the cellular automaton are checked (e.g. the vehicle enters the depot). In this case, such vehicles should be removed from the machine (vehicles in the depot do not participate in the evolution of the machine until they leave it).
- Checking the number of iterations - if it is an automaton with a finite, predetermined life cycle, then in this phase it is checked whether the end of the evolution of the machine can occur.

2.3 Description of Traffic Management and Control Rules

The main task of the simulator is the ability to analyze various vehicle control algorithms in terms of optimizing the network capacity, i.e. the number of passengers transported to the destination within a set period of time. In the adopted solution, the model's logic is divided into the following modules: route calculation module, free vehicle control module, vehicle interlocking module at the intersection, vehicle allocation module for order fulfillment [4].

In the first layer of the traffic control process in the PRT network one can distinguish two most important algorithms, which are responsible for effective traffic simulation - a single vehicle traffic control algorithm and an algorithm for selecting the shortest route. One of the basic assumptions of the PRT system simulator is the fact that the simulation takes place in discrete time. This means that the status of all objects in the system is refreshed every unit of time (simulation step).

Similarly, with sections that are divided into smaller pieces, corresponding to the shortest distance that PRT vehicles will be able to overcome in a unit of time. In practice, this means that each episode contains a list of points on which vehicles can move. It is therefore not possible for the vehicle to be between these defined points.

The second most important algorithm is the algorithm for selecting the shortest path. This algorithm is executed every time the vehicle sets off (transports passengers from point A to point B, the empty wagon returns from the garage to the station, etc.). It should be noted that the route can be updated at any time (as a result of the changing network situation). The proposed algorithm is the version of the shortest path search algorithm using the Dijkstra method. Because the classic Dijkstra algorithm is very general and has many assumptions, the algorithm that has been implemented in the PRT network traffic simulation system has been extended for additional checks and conditions. The next sections will describe the individual modules responsible for traffic management and control.

At the stage of PRT vehicle control and management, two optimization tasks are carried out. First of all, for a given network structure (network topology, number of available vehicles, distribution of streams, etc.) the problem is to choose the optimal route of the vehicle. At the same time, the term optimal route is understood as a route that minimizes the function of the destination, which is the travel time of the vehicle.

The second problem of optimization of traffic in the PRT network is to find the optimal configuration of selected network elements, which maximizes the transport efficiency of the entire system. The concept of transport capacity of the PRT network is understood as the number of passengers served during the survey, who did not wait for the vehicle longer than 300 s (5 min).

For the needs of research, the travel time of the vehicle $T_{\_travel\ time}$ was defined as the sum of the theoretical time of the $T_{\_reference}$ run and the delay time $\Delta_{\_delay}$. $T_{\_reference}$ is the shortest possible journey time between two stations (start and destination), assuming there are no traffic jams on the route, and the vehicle is traveling at the maximum possible speed (taking into account the maximum permissible speed on the segments). In this case, acceleration and deceleration are ignored. $\Delta_{\_delay}$ is the waiting time for a vehicle, delays caused by traffic jams on the route and the time needed for acceleration and deceleration.

$$T_{travel} = T_{reference} + \Delta_{delay} \tag{1}$$

$$\Delta_{delay} = t_1 + t_2 + t_3 + t_4 \tag{2}$$

$$\delta = \frac{\Delta_{delay}}{T_{reference}} \tag{3}$$

δ – coefficient of deviation of travel time
t_1 – waiting time for the vehicle
t_2 – boarding and disembarking time
t_3 – the delay time resulted from moving at a speed lower than the maximum allowed on a given road section;
t_4 – time of accelerations and delays

Coefficients defined by Eqs. (1) and (3) can be understood as some measures of transport efficiency.

2.4 Example Results

The analyzed example concerned the PRT network as per the drawings on Fig. 4. The network referred to the interchange hub, which was to connect metro stations with other eight transport hubs (bus lines, tram lines and suburban train). The charts (see Fig. 5) show very clearly and a satisfactory result was obtained. The number of depots handled depends strongly on the number of vehicles. However, we draw attention to the disturbingly high rate of deviation of travel time. Travel time becomes significantly less than the delay time.

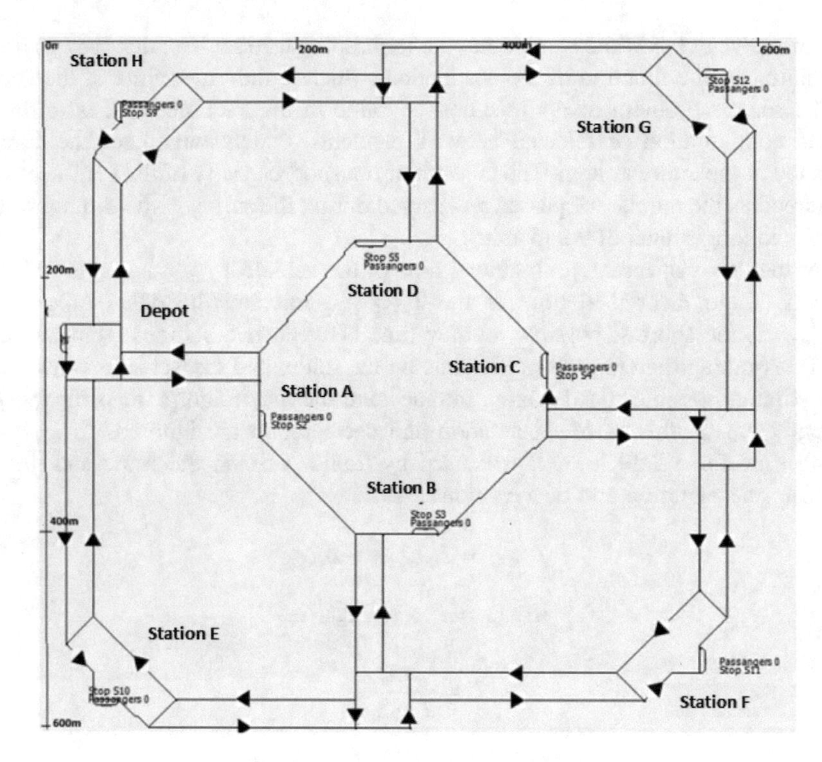

Fig. 4. The analyzed PRT network

a)

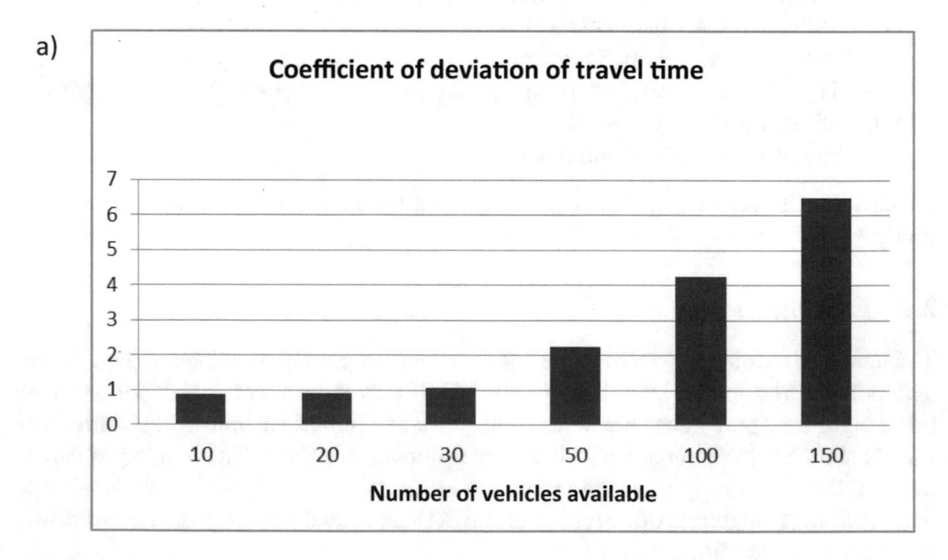

Fig. 5. Charts showing: changes (a) deviations of travel time and (b) number of passengers handled during waiting time less than 300 s as a function of the number of PRT vehicles (traffic density)

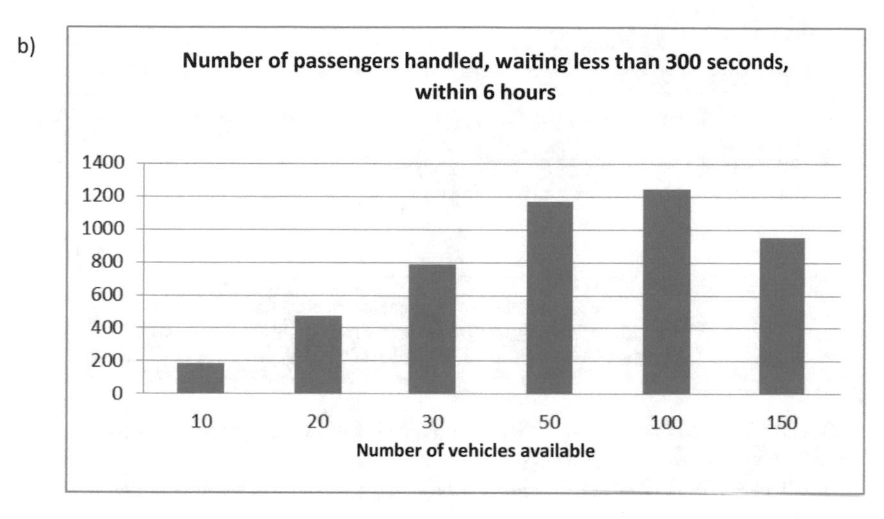

Fig. 5. (*continued*)

3 The Hybrid, Modular and Automated System for Sustainable Urban Transport Prometheus – Warsaw Technical University Concept

Currently, work on the new system is underway at Warsaw University of Technology. The concept of the system was presented by Choromański [2]. The new system integrates three technologies: PRT technology, autonomous and electric vehicle technology. The basis for the construction of the new system is the PRT system and (not shown in this paper) and the eco-car also developed at the Warsaw University of Technology (see Figs. 6 and 7). The basic modification of the PRT system, as compared to the one already developed, consists in the fact that PRT vehicles can also move on roads which they share with other vehicles. In this case, they use a special (not mechanical track) similar to map technology adapted to autonomous vehicles.

The eco-car can be used as an electric car move on all roads or move on the road in the same technology as the PRT vehicle. In this case, the limitation is the lack of maneuvers of overtaking and bypassing. The new system is characterized by extremely high scalability and flexibility. It can be used to implement very specific individual tasks.

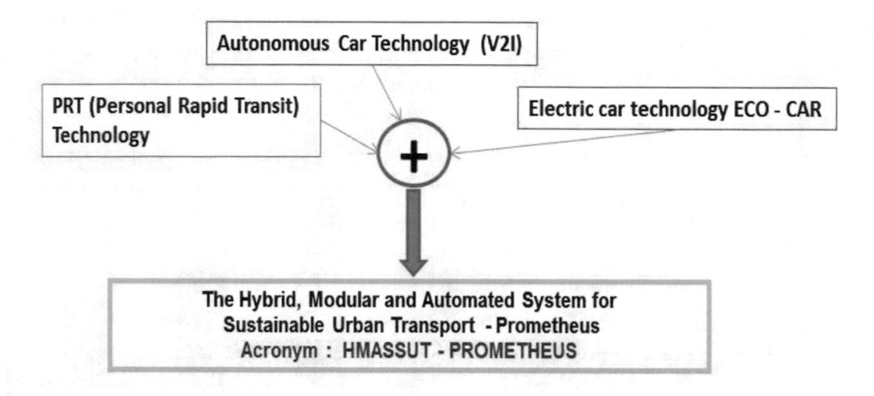

HYBRID – because, referring to the electric drive, it does not eliminate the use of others (CNG, hydrogen fuel drives, fuel cells)
MODULAR – because it enables adaptive system design, using all its modules or only some
AUTOMATED – because it uses technologies allowing traveling in "driverless" mode

Fig. 6. Polish autonomous transport system

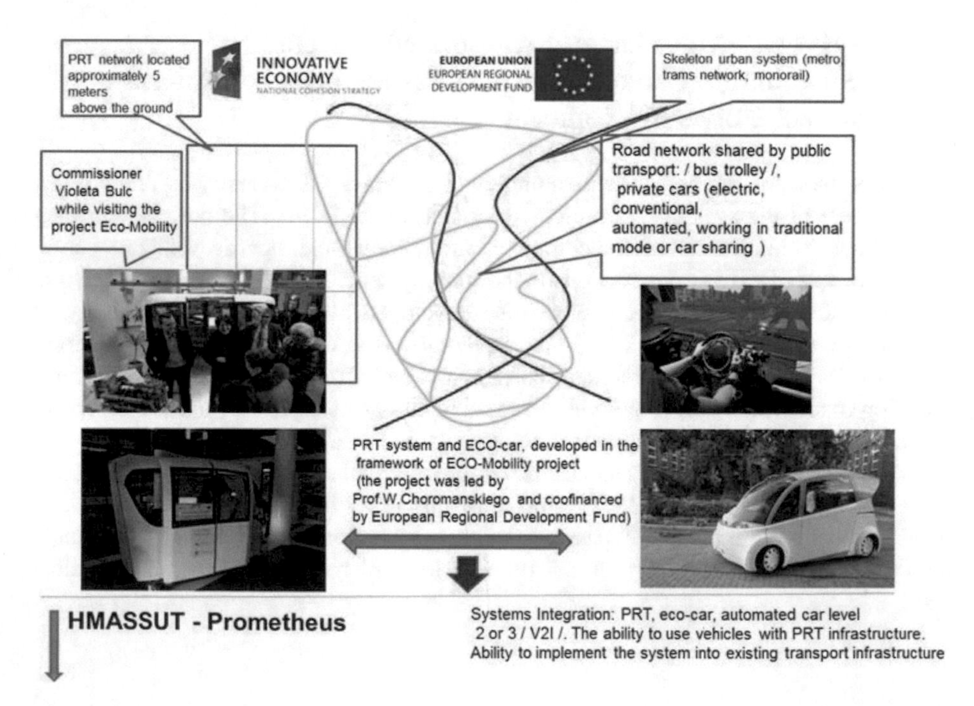

Fig. 7. The polish hybrid, modular and automated system for sustainable urban Transport – Prometheus

4 Summary

The PRT system presented and its extension to Hybrid, modular and automated sustainable urban transport system - Prometheus is an interesting solution for urban transport. It practically meets all the requirements of sustainable urban transport. At the same time, computer simulation tests should determine the suitability of the system for a specific solution. All solutions presented in the article should be treated as a supplement (of a local character) to mass public transport systems (monorail, metro).

Acknowledgements. The article is based partly on the results of the ECO-Mobility project WND-POIG.01.03.01-14-154/09, co-financed from European Regional Development Fund within the framework of Operational Programme Innovative Economy and research conducted as part of the statutory work of Department of Information Technology and Mechatronics in Transport at the Faculty of Transport.

References

1. Choromański, W.: Transport Systems PRT, Transport and Communication Publishers, Warsaw (2016)
2. Choromański, W.: The concept of the Electric and Autonomous Vehicles System for Urban Transport, - an unpublished report available at the Warsaw University of Technology, Transport Department, Warsaw (2017)
3. Choromański, W., Grabarek, I.: Autonomous vehicles in urban agglomerations. In: Urban and Regional Transport, pp. 12–16, vol. 11 (2018). ISSN 1732-5153
4. Daszczuk, W., Choromański, W., Mieścicki, J., Grabski, W.: Empty vehicles management as a method for reducing passenger waiting time in Personal Rapid Transit. IET Intell. Transp. Syst. **9**(3), 231–239 (2015)
5. Heinrichs, D.: Autonomous driving and urban land use. In: Maurer, M., Gerdes, J., Lenz, B., Winner, H. (eds.) Autonomous Driving. Springer, Berlin, Heidelberg (2016)
6. Irving, J.H., Olson, C.L., Buyan, J., Bernstein, H.: Fundamentals of Personal Rapid Transit, Lexington Books (1978)
7. Kozłowski, M., Choromański, W.: PRT simulation research. Arch. Transp. **27**(3–4), 95–102 (2013)
8. Kozłowski, M., Choromański, W., Kowara, J.: Analysis of dynamic properties of the PRT vehicle-track system. Bull. Polish Acad. Sci. **63**(3), 1–8 (2015)
9. Piao, J., McDonald, M., Hounsell, N., Graindorge, M., Graindorge, T., Malhene, N.: Public views towards implementation of automated vehicles in urban areas. Transp. Res. Procedia **14**, 2168–2177 (2016)

ATM Functionality Development Concept with Particular Analysis of Free Route Airspace Operations in Poland

Ewa Dudek(✉)

Faculty of Transport, Warsaw University of Technology, Koszykowa 75, 00-662 Warsaw, Poland
edudek@wt.pw.edu.pl

Abstract. This article concerns Air Traffic Management (ATM) development concept arising from the constant and undeniable development of air transport. In Europe, the Single European Sky Air Traffic Research and Development (SESAR) project and Single European Sky (SES) idea are still valid and up to date. As a consequence, in 2014 one of the Commission Regulations introduced a project called 'Pilot Common Project', which identified the ATM functionalities to be implemented as part of the European Air Traffic Management Master Plan. The defined deployment target dates are approaching that is why it was decided to analyse the issue of ATM development. The paper starts with a short introduction giving the background identification and describing ATM functionalities listed in the 'Pilot Common Project'. Then, due to the closest deployment date, particular attention was paid to the Flexible Airspace Management and Free Route functionality. In the following part of the publication the idea of free route airspace operations was presented – its definition, goal, important dates, boundaries as well as restrictions. Then risk analysis of its implementation using Failure Modes and Effects Analysis (FMEA) was carried out. The obtained results were summed up and conclusions drawn in the last paragraph of the paper.

Keywords: Air Traffic Management (ATM) · Free routing · Risk analysis

1 Introduction

1.1 Background Identification

The development of air transport in European Union as well as in Poland is constant and undeniable. This fact may be confirmed, for example, by the growing number of air operations in Poland (263028 in 2013 versus 341199 in 2017) or the growing number of attended passengers (24 982 623 in 2013 versus 39 972 247 in 2017) during the last few years (see Fig. 1) [15]. However, in other UE countries the situation is similar.

At the same time, regardless of the significant growth in performed operations, involved carriers, passenger or operating aircrafts, it is still important to maintain the same high level of safety in accordance with ICAO standards (e.g. ICAO Annex 19 [9]), EU regulations, Eurocontrol specifications as well as Polish National Civil Aviation Safety Program [15] statements. Safety and quality assurance was subject of other

M. Siergiejczyk and K. Krzykowska (Eds.): ISCT21 2019, AISC 1032, pp. 94–103, 2020.
https://doi.org/10.1007/978-3-030-27687-4_10

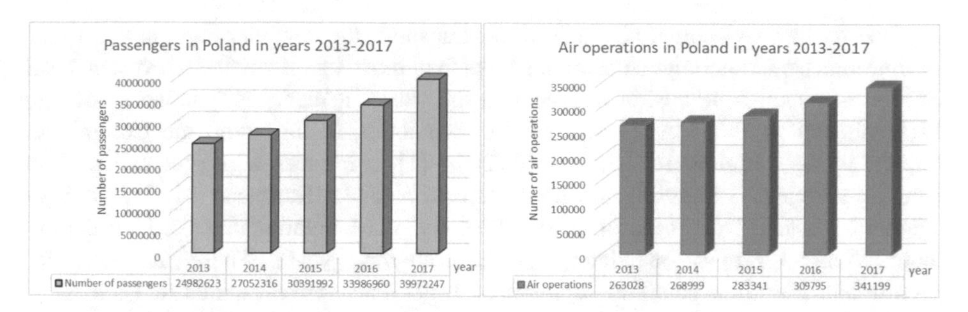

Fig. 1. Number of passengers attended and air operations performed in Poland in years 2013–2017 [15]

author's publications in general [e.g.: 3, 16], for selected systems [e.g. 4] or cases [e.g. 11]. The issue of safety in Air Traffic Management (ATM) is also subject to other authors analysis both in Poland [e.g. 17] and in Europe [e.g. 2]. Permanent growth, cooperation and integration, capacity problems, information sharing and traffic and flow management are also the background for the Single European Sky Air Traffic Research and Development (SESAR) project and Single European Sky (SES) idea aiming at (among others) modernization of the entire ATM in order to provide by 2030 "air traffic management infrastructure that will enable the safe and environmentally friendly operation and development of air transport" [1].

1.2 ATM and Its Functionalities Identification

Air Traffic Management (ATM) based on [8] may be defined as dynamic, integrated management of air traffic and airspace – safely, economically and efficiently – through the provision of facilities and seamless services in collaboration with all parties and involving airborne and ground-based functions. Primarily ATM consists of three main activities:

- Air Traffic Flow and Capacity Management (ATFCM) – a service established to assure safety, efficiency, orderliness and optimization of traffic flows with the maximum use of the air traffic control capacity, declared by the relevant ATS centres,
- Airspace Management (ASM) – planning function with the main objective to maximize the use of available airspace thanks to dynamic time division and airspace segregation between different categories of participants, based on short-term needs, nowadays the efforts concentrate on the idea of flexible use of airspace,
- Air Traffic Services (ATS) – a general term, representing services, which regulate and assist aircrafts in real-time to ensure their safe operations, which in practice means assistance to: prevent collisions between aircraft, provide advice for the safe and efficient conduct of the flight, conduct and maintain an orderly flow of air traffic and notify concerned organizations of and assist in search and rescue operations. The general term ATS hides four services: air traffic control service (ATC), air traffic advisory service, flight information service (FIS) and alerting service.

Due to the cross-border character of air transport, the new ideas and projects must be implemented not only on national level but most of all on the European level. Europe has one of the busiest airspaces in the world. It manages a network covering 11,5 mln km^2 with 63 ATM centres [5]. Remaining in this trend the Commission implementing Regulation (UE) No 716/2014 [1] introduces a project called 'Pilot Common Project' supporting the implementation of the European Air Traffic Management Master Plan. The project identifies six ATM functionalities, which deployment should be made mandatory in the following years. These functionalities, accompanied by their goal and deployment target dates, are presented in Table 1.

Table 1. Identified ATM functionalities to be implemented as part of the European Air Traffic Management Master Plan (own work based on [1]).

No	ATM functionalities name	ATM functionalities goal	Deployment target date
1.	Extended Arrival Management and Performance Based Navigation in the high density Terminal Manoeuvring Areas	to improve the precision of approach trajectory as well as facilitate the traffic sequencing at an early stage, thus allowing reducing fuel consumption and environmental impact in descent/arrival phase	January 1st 2024
2.	Airport Integration and Throughput	to improve runway safety and throughput, ensuring benefits in terms of fuel consumption and delay reduction as well as airport capacity	January 1st 2021 or 2024 (depending on the sub-function)
3.	Flexible Airspace Management and Free Route	to enable a more efficient use of airspace, thus providing significant benefits linked to fuel consumption and delay reduction	January 1st 2018 for Direct Routing (DCT) or January 1st 2022 for Free Route
4.	Network Collaborative Management	to improve the quality and the timeliness of the network information shared by all ATM stakeholders, thus ensuring significant benefits in terms of Air Navigation Services (ANS) productivity gains and delay cost savings	January 1st 2022
5.	Initial System Wide Information Management	to bring significant benefits in terms of ANS productivity	January 1st 2025
6.	Initial Trajectory Information Sharing	to improve predictability of aircraft trajectory for the benefit of airspace users, the network manager and ANS providers, implying less tactical interventions and improved deconfliction situation with a positive impact on ANS productivity, fuel saving and delay variability	January 1st 2025 or 2026 (depending on the sub-function)

The first ATM functionality to be implemented, based on the deployment target date, is no 3 - Flexible Airspace Management and Free Route. The issue of Free Route implementation has already been subject to others authors publications. The study of its implementation in European airspace can be found in [13]. The assessment of free route deployment in the north European airspace, based on aircrafts' separation and airspace complexity with reference to the SESAR project can be found in [12] for example. However, it is also worth mentioning the publications from outside of our region. In [18] the implementation in Manila FIR is described with the main goal to increase the local airspace capacity and efficiency in Philippines. These are just a few examples, but free route airspace implementation with similar problems to be solved concerns also Poland. That is why, in the following part of this article particular attention is paid to its deployment in our country. First, the idea of free route airspace operations is described and then risk analysis of its implementation is carried out.

2 Free Route Airspace

Free Route Airspace (FRA) according to its definition [19] is a specified airspace within which users may freely plan a route between a defined entry point and a defined exit point, with the possibility to route via intermediate (published) waypoints, without reference to the ATS route network, subject to airspace availability. Within this airspace, flights remain subject to air traffic control. In other words, the idea of free routing is to allow airspace users to fly as close to their preferred trajectories as possible, so that their flight routes are the shortest, the quickest and the most efficient. At the same time, it is expected to obtain significant cost reduction – thanks to reducing the number of nautical miles travelled and burning less fuel. In the transition period, while proceeding from fixed routes network into the free route airspace, it is possible to allow users to plan their flights based on the published Direct Routes (DCTs). Which means to determine the available entry/exit points for certain traffic flows, through the publication of DCTs.

In Poland the free route airspace, known as POLFRA, is based on the rules described above. Although the Commission Regulation [1] expects all European countries to remove all fixed waypoints above Flight Level (FL) 310 till January 1st 2022, the Polish Air Navigation Services Agency (PANSA) aimed at implementing POLFRA already in 2019. The task was deployed successfully and the notification to airspace users concerning the implementation of POLFRA together with the background information was published in Aeronautical Information Circular AIC03/18 [14] effective from September 13th 2018. Mentioned document announces implementation of POLFRA on February 28th 2019. Consequently, all necessary POLFRA information were published in AIP Polska two AIRAC cycles before implementation, which means from January 3rd 2019.

POLFRA allows users to freely plan their flight trajectories from flight level FL095 (which is 9 500 ft) to the cruising altitude (up to FL660) between any defined in Flight Information Region (FIR) Warszawa entry and exit point, with an option to route via intermediate points, if necessary. However, it must be noted that free route airspace rules for civil aviation are not applicable in:

- Terminal Manoeuvring Areas (TMA),
- Controlled zones around airports (CTR) and Control Areas (CTA),
- Uncontrolled airspace.

Figure 2 shows the general POLFRA map with the preliminary boundaries, while Fig. 3 shows POLFRA boundaries for a specific flight level. As an example FL095-115 was selected.

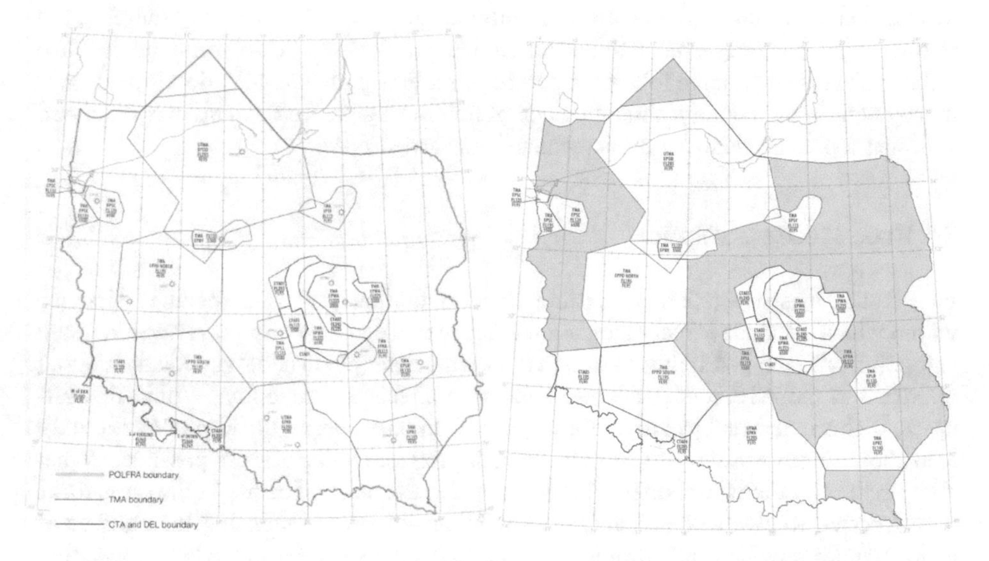

Fig. 2. POLFRA boundaries [19]

Fig. 3. POLFRA area FL095-115 [19]

The restrictions for free routing in polish airspace in practice mean that aircrafts departing from airports located within FIR Warszawa are obliged to plan their flights via the existing ATS route network from the end of the runway to a published departure connecting point into free route airspace (see exemplary POLFRA connectivity schemes on Figs. 4 and 5). After that point they may plan their trajectories according to free route airspace rules. While arriving to the FIR Warszawa airports, the rules are equal and fixed ATS network is obligatory from the POLFRA arrivals connecting points for each airport.

FRA/POLFRA deployment has a number of benefits: additional airspace capacity, greater resilience with possibility to re-plan the flight route in case of unfavourable weather and cost savings. The Polish Air Navigation Services Agency (PANSA) estimates potential track mile savings around 4500 nautical miles per day, equivalent to over 1 650 000 miles per year across all aircrafts operating in Polish airspace [6]. However, the risks of its implementation must also be considered and the level of safety managed in accordance with the Safety Management System (SMS) approach, available in ICAO Annex 19 [9]. That is why, taking into account the proactive

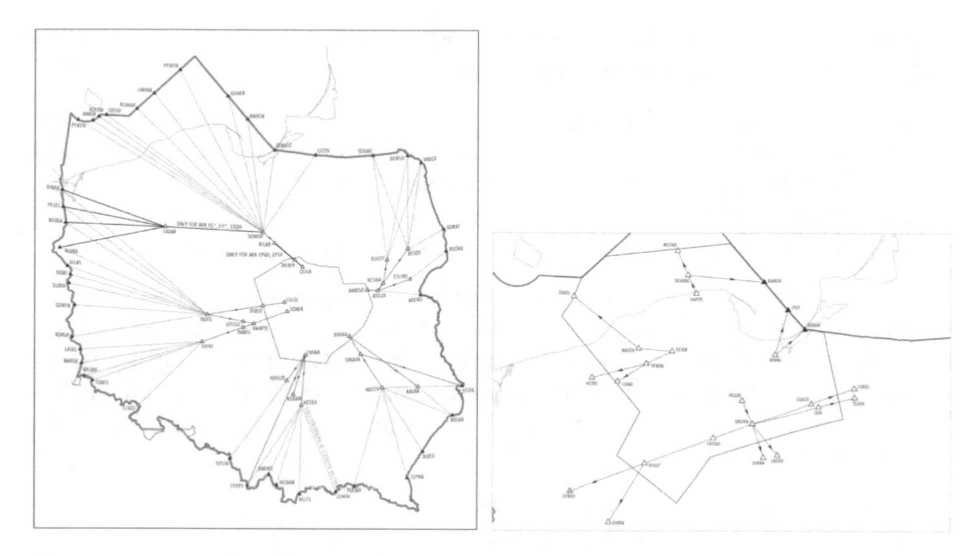

Fig. 4. POLFRA EPWA/EPMO departure connecting points [19]

Fig. 5. POLFRA EPGD departure connecting points [19]

approach to safety assurance in the following part of this article a risk analysis concerning implementation of POLFRA is carried out.

3 Failure Mode and Effects Analysis for POLFRA

Safety management standards, in particular those described for safety management systems (SMS) [7], demand that this management is systematic, transparent and proactive. Adjective "systematic" stands here for acting in accordance with continuous improvement cycle Define-Measure-Analyse-Improve-Control (DMAIC) as well as Plan-Do-Check-Act (PDCA) approach. "Transparency" should be meet thanks to clear management and incompatibilities recording rules. While "proactive security management" means that special emphasis should be placed on preventing events by identifying threats and overseeing and implementing risk mitigation measures before a risky event occurs. Early detection of errors and non-compliances and prevention of their consequences before they occur is therefore the basis of the required approach. Taking all this into account, it was decided to carry out a risk analysis for free route airspace implementation. Although the deployment date has already passed (on February 28th 2019), it is still possible to implement minor changes and most of all systematic and transparent safety analysis must be carried out in a cyclic way.

Risk assessment techniques are described in one of the ISO standards - ISO 31010 [10]. The review of those technics allowed selection of Failure Modes and Effects Analysis (FMEA) method to be applied in the considered case. FMEA analysis is a technique used to identify ways in which components, systems or processes can fail and not fulfil their tasks. FMEA defines [10]:

- all potential causes of damage to individual system components (the non-compliance mode determines what should be supervised or functions incorrectly),
- the effects that these incompatibilities may have on the system,
- the mechanism of failure,
- how to avoid the incompatibilities and/or mitigate their impact on the system.

With such an analysis carried out, at the end of the FMEA process a list of potential failure modes, the failure mechanisms and their impact on all highlighted stages or components of the process/product is obtained. It is also possible to identify the consequences of listed incompatibilities and their impact on the system as a whole.

The implementation process as a whole in this first risk analysis was divided into three main stages:

1. aeronautical data and information preparation,
2. operational use of published aeronautical information,
3. compatibility,

Concentrating first of all on analysing data and information preparation and final use. The deployment of POLFRA does not require changes to the existing material assets such as infrastructure or aeronautical systems' components, and even if - the required changes are minor. However, the change of airspace structure and functioning require publication of modified data and information (such as departing/arrival connecting points, FRA boundaries for specific flight levels, etc.), which must be available for all airspace users and are published in Aeronautical Information Publication AIP Polska. That is why the risk analysis carried out in this paper focuses on them. An important issue are also arrangements with other (neighbouring) countries, according to the cross-border character of air transport. Last but not least, the relationships with other ATM functionalities and systems were considered.

The results obtained are shown in Table 2. However, it must be taken into account that the results obtained, as in each risk analysis, are subjective. Nevertheless, they can still constitute a basis for safety and risk management.

In this first FMEA risk analysis the results obtained have descriptive form only. In the future it is possible to classify each of the identified failure modes according to its criticality based on the Risk Priority Number (RPN) calculation. However, to make it possible it is necessary to define the rating scales for all three assessed criteria of the RPN number: consequence of failure, likelihood of its appearance and ability to detect the problem and only then for each failure mode assign them the right value. This will be subject of author's further work.

Table 2. FMEA analysis of POLFRA implementation [own work].

No	Component's/stage's name	Potential failure/incompatibility	Possible effect	Potential cause	Corrective actions
1	**Aeronautical data and information preparation**				
1a	Determination of entry, exit and intermediate points	Incorrect determination of entry, exit and intermediate points	Creation of incorrect data and charts; serious incident involving two (or more) aircrafts	Lack of verification procedures; lack of people competence; misunderstanding; human error	Implementation of verification procedures; careful study of the documentation; designation of the competent person
		Incompatibility in AIP Polska publication	Lack of information publication in a correct form, place and time; operational use of incorrect data; aeronautical accident involving two (or more) aircrafts; aircraft's destruction	Incompatibilities in the aeronautical data and information chain; no stage verification; ignorance of rules and requirements; human error	Data and information verification; implementation of verification procedures on subsequent stages
1b	Determination of departure/arrival connecting points	Incorrect determination of departure/arrival connecting points	Creation of incorrect data and charts; serious incident involving two (or more) aircrafts	Lack of verification procedures; lack of people competence; misunderstanding; human error; ignorance of TMA boundaries	Implementation of verification procedures; careful study of the documentation; designation of the competent person
		Incompatibility in AIP Polska publication	Lack of information publication in a correct form, place and time; operational use of incorrect data; aeronautical accident involving two (or more) aircrafts; aircraft's destruction	Incompatibilities in the aeronautical data and information chain; no stage verification; ignorance of rules and requirements; human error	Data and information verification; implementation of verification procedures on subsequent stages
1c	Publication of information in different forms	Inconsistency between published written information and published charts	Operational use of incorrect data, serious incident involving two (or more) aircrafts	Lack of verification procedures; lack of people competence; human error	Data and information verification; designation of the competent person
1d	DCT (Direct Routing) determination	Incompatibility in DCTs publication	Incorrect route planning through restricted areas	Lack of verification procedures; lack of people competence; human error	Data and information verification; designation of the competent person
2	**Operational use of published aeronautical data and information**				
2a	Route planning	Route planning through areas where POLFRA is not applicable: DA, TMA, CTR	Aeronautical accident involving two (or more) aircrafts; aircraft's destruction	Incompatibilities on the 1st stage - aeronautical data and information preparation; ignorance of the current situation; human error	Implementation of the verification procedures on the 1st stage; detection of conflicts in planned routes
		Entry into POLFRA area from airports located in FIR Warszawa from other point then defined in departure connecting point	Traffic congestion around defined connecting points; aircrafts separation minima infrindgment; serious incident involving two (or more) aircrafts	Incompatibilities on stage 1b; human error	Implementation of the verification procedures on the 1b stage; good knowledge of published points; correct route planning
		Exit from POLFRA area into the airport's CTR through other point then defined in arrival connecting point	Safety risk; aeronautical accident involving two (or more) aircrafts; aircraft's destruction	Incompatibilities on stage 1b; human error	Implementation of the verification procedures on the 1b stage; good knowledge of published points; correct route planning
2b	Traffic management	Traffic congestion around defined entry/exit points	Serious incident involving two (or more) aircrafts; negative effects for the environment; additional use of fuel	Incorrect determination of entry, exit and intermediate points; errors on stage 1a; incorrect route planning	Implementation of the verification procedures on the 1a stage; correct route planning; correct traffic management
		Aircrafts separation minima infrindgment	Serious incident involving two (or more) aircrafts	Traffic congestion around defined entry/exit points; incorrect route planning; ATC error; human error	Correct traffic management
		Conflicts in planned routes	Urgent need to re-plan the route; delays, dissatisfaction; if not detected - aeronautical accident involving two (or more) aircrafts; aircraft's destruction	Incompatibilities on the 1st stage - aeronautical data and information preparation; lack of conflict detection tool	Implementation of conflict detection tool
3	**Compatibility**				
3a	Cross-border arrangements	Implementation delays	Incomplete application of the expected effects; dissatisfaction	Indolence; lack of competences	Proactive approach and active operation
		Lack of compatibility of POLFRA significant points and DCTs located on FIR Warszawa boundaries with other countries arrangements (Baltic FAB, Germany, Ukraine, Scandynavia)	Impossible deployment of FRA on international level; incomplete application of the expected effects	Lack of necessary arrangements; indolence	Active operation
3b	Interdependencies and compatibility with other ATM functions/systems	Lack of compatibility with other ATM functions and systems (such as SWIM, ATFCM, etc.)	Impossible full deployment of FRA on national and international level	Lack of necessary arrangements; indolence	Active operation; need for synchronization; implementation of other required functionalities

4 Conclusions

The development of ATM functionalities gives an opportunity to improve throughput and productivity, to reduce fuel consumption and impact of air transport on the environment. However, at the same time implementation on new functions is a challenge and must not influence the level of air operations' safety. That is why it is necessary to manage the level of safety in the entire ATM system in a systematic, transparent and proactive way.

In this article, special attention was paid to Free Routing, implemented in Poland from February 28th 2019. The FMEA risk analysis of its deployment, carried out in the paper, gave back a list of potential incompatibilities, their possible effects (up to serious aeronautical accidents), potential causes and corrective actions proposed for the subsequent, indicated process components. The elaborated results indicate first of all the necessity to implement verification procedures in the entire aeronautical data and information chain. Then, it was noted that: no rules and requirements should be ignored, overall and current operational situation must be well known to allow responsible trajectory planning, human errors are one of the most frequent, unpredictable and dangerous errors (which is also confirmed statistically).

In the future the FMEA analysis may be extended by the quantitative assessment – calculation of the Risk Priority Number (RPN) for example, to make the results more precise. The risk analysis may also be carried out for other ATM functionalities listed in the Pilot Common Project.

References

1. Commission Implementing Regulation (EU) No 716/2014 of 27 June 2014 on the establishment of the Pilot Common Project supporting the implementation of the European Air Traffic Management Master Plan
2. Di Gravio, G., Patriarca, R., Mancini, M., Constantino, F.: Overall safety performance of the Air Traffic Management system: the Italian ANSP's experience on APF. Res. Transp. Bus. Manag. **20**, 3–12 (2016)
3. Dudek, E., Kozłowski, M.: The concept of a method ensuring aeronautical data quality. J. KONBiN **1**(37), 319–340 (2016)
4. Dudek, E., Kozłowski, M.: The concept of the Instrument Landing System – ILS continuity risk analysis method. In: Jerzy, M. (ed.) Management Perspective for Transport Telematics. Communications in Computer and Information Science, pp. 305–319. Springer (2018)
5. European ATM Master Plan, Edition 2012. www.atmmasterplan.eu
6. IATA and PANSA: Airspace Strategy for Poland, Edition 1, November 2018
7. ICAO Doc. 9859, AN/474, Safety Management Manual, Third Edition (2013)
8. ICAO Annex 15 to the Convention on International Civil Aviation, Aeronautical Information Services, International Civil Aviation Organization, July 2013
9. ICAO Annex 19 to the Convention on International Civil Aviation, Safety Management, International Civil Aviation Organization, July 2013
10. ISO 31010:2010 standard, Risk management – Risk assessment techniques (2010)

11. Kozłowski, M., Dudek, E.: Risk analysis in air transport telematics systems based on aircraft's Airbus A320 accident. In: Jerzy, M. (ed.) Smart Solutions in Today's Transport. Communications in Computer and Information Science, vol. 715, pp. 385–395. Springer (2017)
12. Nava Gaxiola, C.A., Barrado, C., Royo, P., Pastor, E.: Assessment of the North European free route airspace deployment. J. Air Traffic Manag. **73**, 113–119 (2018)
13. Nava-Gaxiola, C., Barrado, C., Royo, P.: Study of a full implementation of Free Route in the European Airspace. In: IEEE/AIAA Digital Avionics Systems Conference, Institute of Electrical and Electronics Engineers, pp. 1–7 (2018)
14. Polish Air navigation Services Agency, Aeronautical Information Service: AIC 03/18 effective from 13th September 2018
15. Polish National Civil Aviation Safety Program, Civil Aviation Authority of the Republic of Poland (2016)
16. Siergiejczyk, M., Kozłowski, M., Dudek, E.: Diagnostics of potential incompatibilities in aeronautical data and information chain. In: Diagnostyka, vol. 18, no 2, pp. 87–93 (2017)
17. Skorupski, J.: About the need of a new look at safety as a goal and constraint in Air Traffic Management. Procedia Eng. **187**, 117–123 (2017)
18. Xie, Z., Aneeka, S., Lee, Y.X., Zhong, Z.W.: Study on building efficient airspace through implementation of free route concept in the Manila FIR. Int. J. Adv. Appl. Sci. **4**(12), 10–15 (2017)
19. www.pansa.pl/_site/index.php?menu_lewe=POLFRA&lang=_pl&opis=general_info. Accessed 12 Apr 2019

A Simulation-Based Approach for the Conflict Resolution Method Optimization in a Distributed Air Traffic Control System

Anrieta Dudoit[1] and Jacek Skorupski[2(✉)]

[1] Vilnius Gediminas Technical University, Vilnius, Lithuania
anrieta.dudoit@vgtu.lt

[2] Warsaw University of Technology, Warsaw, Poland
jacek.skorupski@pw.edu.pl

Abstract. The observed increase in air traffic volume leads to the development of methods like "free flight" aiming at better use of airspace. The expected consequence of the widespread implementation of this concept is the transfer of some air traffic management tasks to the aircraft. The aim of the research was to analyze the conflict resolution method in the airspace with the distributed air traffic control system (DATCS). The 3HC method was proposed, in which one aircraft bypasses the collision point by making a triple heading change. To determine parameters of the method, a conflict model implemented as a dynamic, stochastic, colored Petri Net (DSCPN) was used. Simulation experiments have shown that it is possible to select the parameters of the method that guarantee safety while minimizing the distance covered by the maneuvering aircraft. The starting point of these maneuvers should be selected as late as possible, i.e. when boundary separation is reached. In case of disturbances, however, the distance in which the collision point is omitted should be increased. The applied model and the simulation approach allow determining the relationship between this distance and the starting point of the maneuvers. It has been shown that the 3HC method can be used in DATCS for both free flight systems, where some of air traffic controller functions is performed by the aircraft crew as well as when it is delegated to an autonomous system of both manned or unmanned aircraft.

Keywords: Air traffic management · Distributed control system · Petri nets · Discrete events simulation · Conflict detection and resolution

1 Introduction

In recent years, there has been a constant increase in air traffic volume. Forecasts suggest continuation of this trend. Unfortunately, the current organization of airspace and air traffic control (ATC) make it impossible to handle the expected traffic. Therefore, the concepts of better use of the airspace emerges, for example through so-called "free flights", which take place independently of the air route network, and directly connect the departure and destination airports.

M. Siergiejczyk and K. Krzykowska (Eds.): ISCT21 2019, AISC 1032, pp. 104–114, 2020.
https://doi.org/10.1007/978-3-030-27687-4_11

Changes in the organization of air traffic must cause changes in the ATC system. Their essence is replacing the existing centralized surveillance and decision making with the distributed air traffic control system (DATCS). Depending on the development of this concept, some or all safety ensuring tasks can be handed over to aircraft. In this concept, it is important to choose the right conflict detection and resolution (CD&R) algorithm, allowing automation of this process. Of course, all these functions must be supported by appropriate computer systems.

One of the possible conflict resolution algorithms, called the triple heading change (3HC) method, was tested in this paper. To adapt this method to automation and application in DATCS, it is necessary to specify some of its parameters. It is also important to analyze how the uncertainty as to the aircraft position or its movement characteristics affects the change of these parameters.

It is worth noting that the considerations presented in this article apply to both DATCS where the ATC roles will be taken over by the aircraft crews as well as to the automatic CD&R systems. In the latter case, such systems can be installed both on manned aircraft and unmanned aerial vehicles.

1.1 Distributed Air Traffic Control System Concept

The starting point for the research has been the concept of "free flight" [5]. It uses flight trajectories which are directly connecting input and output points of the analyzed airspace [12]. These aims at increasing the airspace capacity and reduce congestion. However, increasing the number of entry points of aircraft leads to a greater complexity of traffic. Thus, the air traffic controller (ATCo) workload will also increase. The problem may be solved by handing over the separation task to aircraft [3]. In this case, we are dealing with a distributed ATC system.

An example of the DATCS is the airborne separation assurance system (ASAS), which enables aircraft crews to maintain separation from nearby aircraft [2]. In this system, the decision-making process is fully automated – it is accomplished by a suitable supporting system. This allows for elimination of the ATCo, although during the transition phase, the system can be used to support the traditional ATC system [11].

To properly implement the DATCS concept, the applied automatic inference systems must have information resources that are comparable to the knowledge of the air traffic controller. This will require a significant improvement of the previously used ground and on-board flight surveillance devices [16].

1.2 Conflict Detection and Resolution Systems

At present, TCAS (Traffic Collision Avoidance System) is commonly installed and used on aircraft. It is used to detect conflicts in the short-time horizon and to generate a solution consisting in changing the aircraft flight level. In this article, we analyze the problem of ensuring collision-free flight in a much longer time horizon than the TCAS. Additionally, we focus on maneuvers in the horizontal plane. This solution is more difficult to implement and requires more time. However, in practice, there are situations when such an approach is necessary.

The DATCS concept includes many elements, among which the CD&R subsystem has a significant role. There are many CD&R algorithms, their review can be found in [10, 17]. Some of the directions of CD&R development are: design of collision-free routes [1], application of technology to many aircraft [4] and to unmanned aircraft systems [18]. Relatively little space is devoted to the additional optimization criteria of solving the conflict. This article deals with the latter issue. The analysis of the 3HC method includes the criterion of minimizing the flight distance.

It should be noted that CD&R systems consist of two main elements – conflict detection and conflict resolution. This paper deals with this latter aspect. We assume that conflicts are reliably detected and classified by another element of the system.

1.3 The Concept of Work

The proposed method has been called triple heading change (3HC) method and is described in more detail in Sect. 2.1. Considering the available technical means of detecting neighboring traffic, it is possible to detect the conflict relatively early in the DATCS. However, the early start of maneuvers to avoid the collision causes a significant extension of the flight distance and may induce new conflicts with other aircraft.

Therefore, the 3HC method was analyzed to find the proper starting point for maneuvering, which will ensure the separation defined in international regulations [7] and, at the same time, as short flight distance as possible. This approach is in line with new ATC concepts in which the safety criterion is treated as a constraint, and optimization is made considering economic or environmental criteria [14].

The model of the collision situation implemented as DSCPN (Dynamically Stochastically Coloured Petri Net) was used. In the case study, two aircraft move with such parameters that a mid-air collision (MAC) would occur if there is no action. The analysis of the 3HC method both without and with disturbances was presented. This paper extends the existing research by using a microscale model, which allows for discrete event simulation and state space analysis, including disturbances and uncertainty as to the location of aircraft and their movement parameters.

2 Model of a Collision Situation to Analyze the Conflict Resolution Method

2.1 The General Concept of 3HC Conflict Resolution Method

The 3HC conflict resolution method involves three aircraft heading changes to bypass the collision point. We will consider the case when both aircraft A and B involved in the conflict are in the same distance from the collision point. This is a symmetrical problem and the choice of the aircraft to perform maneuvers is arbitrary. We choose that aircraft A maneuvers to resolve the conflict (Fig. 1).

The general algorithm of the 3HC conflict resolution method is as follows:

1. At point $A_1(x_{A_1}, y_{A_1})$, the on-board collision detection system notices a collision with aircraft B located at point $B_1(x_{B_1}, y_{B_1})$ with the collision point designated as $O_s(x_{O_s}, y_{O_s})$.
2. Aircraft A constantly calculates separation with aircraft B. Technically this task is easy considering the requirement to equip both aircraft with TCAS devices for monitoring the surrounding traffic.
3. When the distance between aircraft reaches the so-called boundary separation d_g, i.e. at point $C(x_C, y_C)$, aircraft A makes a turn in the horizontal plane and starts a flight towards point $E(x_E, y_E)$ located on a line perpendicular to the initial flight path of aircraft A at a distance R from the collision point O_s. This distance is dependent on the position of the C point, the headings and flight parameters of both aircraft, and is selected by simulation in such a way as to ensure the minimum separation d_s required by international regulations [7].
4. After reaching point E, aircraft A makes a second change of heading, starting the flight towards point A_2, i.e. the point lying symmetrically to the A_1 in relation to the collision point O_s.
5. After reaching point A_2, aircraft A returns to its original heading.

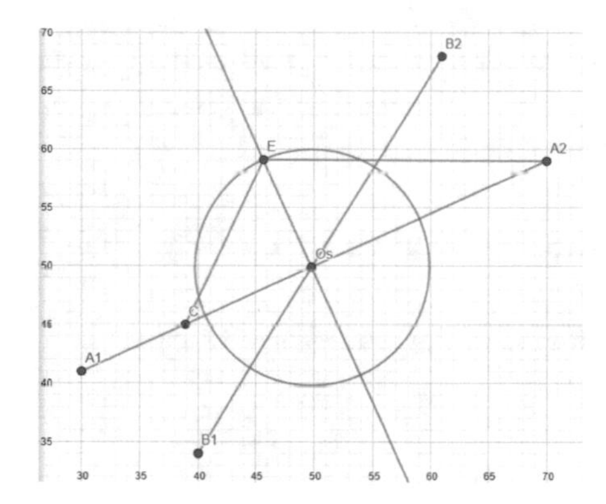

Fig. 1. The idea of the 3HC conflict resolution method

2.2 General Concept of Determining Parameters of the 3HC Method

Parameters of the method are determined in two stages. Firstly, it is checked if for actual input data (conflict detection point A_1, start of maneuvers point C, distance R) the conflict is really resolved, and the separation was not violated. If so, the value of Δl criterion is calculated. This criterion is understood as the extension of the distance flown by aircraft A compared to the straight flight. We will attempt to minimize it.

$$\Delta l = d(A_1, C) + d(C, E) + d(E, A_2) - d(A_1, A_2) \tag{1}$$

where d is the distance between points, for instance for $A_1(x_{A_1}, y_{A_1})$ and $A_2(x_{A_2}, y_{A_2})$

$$d(A_1, A_2) = \sqrt{(x_{A_1} - x_{A_2})^2 + (y_{A_1} - y_{A_2})^2} \quad (2)$$

Considering the modeling goals, it is possible to make simplifications in modeling aircraft movements while solving a conflict situation:

1. The aircraft will be represented as point-mass.
2. Initial positions of aircraft A_1 and B_1 at the time of conflict detection, their velocities V_A, V_B and initial headings k_{A_0}, k_{B_0} are known.
3. In the moment of changing the heading, there are no transitional phenomena, causing that in reality aircraft move along an arc during the heading changes.

Using these simplifications and designations, we can specify that the collision point O_s has the following coordinates

$$x_{O_s} = \frac{x_{A_1} \cdot \mathrm{tg}(k_{A_0}) - x_{B_1} \cdot \mathrm{tg}(k_{B_0}) + y_{B_1} - y_{A_1}}{\mathrm{tg}(k_{A_0}) - \mathrm{tg}(k_{B_0})} \quad (3)$$

$$y_{O_s} = \frac{(x_{A_1} - x_{B_1}) \cdot \mathrm{tg}(k_{B_0}) + y_{B_1} - y_{A_1}}{\mathrm{tg}(k_{A_0}) - \mathrm{tg}(k_{B_0})} \mathrm{tg}(k_{A_0}) + y_{A_1} \quad (4)$$

As already mentioned, point $C(x_C, y_C)$, in which the aircraft A starts the avoiding maneuvers will be determined by simulation depending on the assumed value of the boundary separation d_g. The coordinates of the point $E(x_E, y_E)$, in which the second phase of conflict resolution ends are determined as

$$x_E = x_{O_s} - \sqrt{R^2 \mathrm{tg}^2(k_{A_0}) + 1}, y_E = y_{O_s} - \frac{1}{\mathrm{tg}(k_{A_0})} x_E \quad (5)$$

and the aircraft A heading after the first change is equal

$$k_{A_1} = \mathrm{arctg}\left(\frac{y_E - y_C}{x_E - x_C}\right) \quad (6)$$

Coordinates of point $A_2(x_{A_2}, y_{A_2})$, where avoiding maneuvers are finished are

$$x_{A_2} = 2x_{O_s} - x_{A_1}, y_{A_2} = 2y_{O_s} - y_{A_1} \quad (7)$$

and aircraft A heading after the second change is equal

$$k_{A_2} = \mathrm{arctg}\left(\frac{y_{A_2} - y_E}{x_{A_2} - x_E}\right) \quad (8)$$

2.3 Petri Nets

Petri nets are a convenient tool for modeling and simulation of traffic processes in transport, especially when a discrete simulation of many dynamic and concurrent processes is necessary [6, 9, 13, 15]. The general form of colored Petri net for conflict resolution method analysis is as follows:

$$S_{CR} = \{P, T, A, M_0, \tau, X, \Gamma, C, G, E, R, r_0, B\} \tag{9}$$

where:

P – set of places,
T – set of transitions $T \cap P = \emptyset$,
$A \subseteq (T \times P) \cup (P \times T)$ – set of arcs,
$M_0 : P \rightarrow \mathbb{Z}_+ \times R$ – marking which defines the initial state of the system,
$\tau : T \times P \rightarrow \mathbb{R}_+$ – function determining the static delay that of activity (event) t,
$X : T \times P \rightarrow \mathbb{R}_+$ – random time of carrying out an activity (event) t,
Γ – finite set of colors which correspond to the possible properties of tokens,
C – function determining what kinds of tokens can be stored in a place: $C : P \rightarrow \Gamma$,
G – function which determines the conditions for a given event to occur,
E – function describing properties of tokens that are processed,
R – set of timestamps (also called time points) $R \subseteq \mathbb{R}$,
r_0 – initial time, $r_0 \in R$.
$B : T \rightarrow \mathbb{R}_+$ – function determining the priority of an event.

CPN Tools 4.0 package was selected for creating and researching colored Petri nets model. It is very convenient for this type of analysis – largely due to the transparency and coherence of the model, and especially the integrated simulation engine that enabled a precise analysis of the internal processes in investigated 3HC method [8].

2.4 DSCPN Approach to Modeling Air Traffic

The DSCPN model was used to analyze the 3HC method. The net has a hierarchical structure, consisting of three levels, corresponding to the stages of conflict resolution and heading changes. In the CPN Tools 4.0 environment such structures are built as the so-called "pages". Figure 2 presents the page representing the second phase of conflict resolution, that is when the aircraft flights between points *C* and *E*.

To synchronize the hierarchy levels, so-called "fused places" have been used. They are marked with labels in the lower left corner, and the most important are:

- *A* (labeled *AC A*) and *B* (labeled *AC B*), which store information about the dynamically changing position of both aircraft and other traffic parameters,
- *MA* and *MB*, in which the detailed actual trajectory of the aircraft is stored,
- *DDist* (labeled *Dist*), containing the flight distance in relation to a straight flight,
- *M*, that contains the smallest observed separation between aircraft.

From a functional point of view, the most important transitions are:

- *RSep*, in which the actual separation between aircraft is calculated,
- *Move A* and *Move B*, in which the next aircraft's positions are calculated using the *newpos*() procedure,

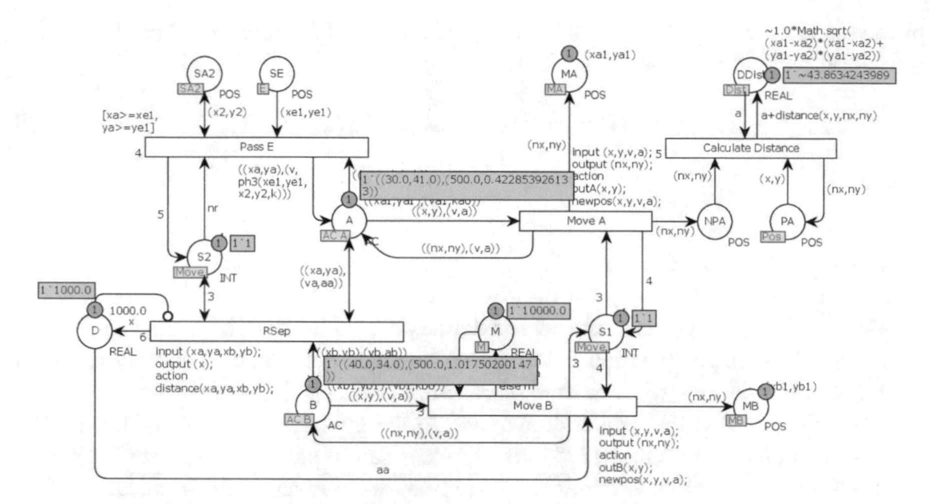

Fig. 2. The page of the model implementing the second phase of conflict resolution

- *Calculate Distance*, which calculates the additional distance flown by the aircraft A because of the conflict resolution procedure,
- *Pass E*, in which termination conditions of this phase of the algorithm are checked.

The model was verified by comparing two graphs. The first graph was of real aircraft trajectories which were calculated from initial coordinates. And the second graph was of trajectories which were obtained from subsequent simulation steps (positions stored in places *MA* and *MB*).

3 Simulation Analysis of 3HC Conflict Resolution Method

3.1 Method Analysis in the Ideal Variant

There are several parameters in the 3HC method that should be determined. The first one is the location of point *C*. We will define it indirectly, by setting the boundary separation d_g, at which aircraft A starts maneuvering. We will assume that radar separation of 5 NM is used, therefore $d_g \geq 5[\mathrm{NM}]$.

The second parameter to be determined is the radius *R*. It is selected by simulation, using the conflict model implemented in the DSCPN. Separation in the second (on section $C - E$) and third (on section $E - A_2$) phase of conflict resolution depends on *R*. It was chosen in such a way that the additional distance Δl flown by aircraft A is as small as possible. The results of these experiments are shown in Table 1.

Table 1. Parameters of the 3HC method for ideal conditions

d_g[NM]	5.0	6.0	7.0	8.0	9.0	10.0	11.0	12.0
R[NM]	8.7	9.5	10.2	10.8	11.3	11.8	12.2	12.6
Δl[NM]	5.44	5.72	6.0	6.28	6.41	6.69	6.69	6.83

In the situation of no disturbances, the most advantageous strategy is to start the maneuvers as late as possible, when we adopt the additional distance Δl of the aircraft A as the criterion. It also expresses the time, economic and environmental criterion.

3.2 Analysis of the Method Under Disturbed Conditions

The strategy defined in Sect. 3.1 for $d_g = 5$ NM seems to be risky, because even minor disturbances caused by meteorological factors or inaccurate navigation, can result in a separation violation. For a better understanding of this issue, a simulation experiment was carried out, in which the traffic was disturbed by a change in the aircraft speed in the range [490, 510] kt, which is up to 2% of the nominal value of 500 kt. At some speed combinations, for the parameters presented in Table 1, separation violation occurs. We can prevent this by increasing the radius R, which also increases the distance flown. The exemplary results for $d_g = 6$ NM are shown in Table 2.

Table 2. Results for disturbed conditions $d_g = 6$ NM, $R = 9.5$ NM

V_A, V_B[kt]	Δl[NM]	Separation violation	New R[NM]	New Δl[NM]
490,490	5.72	No	-	-
490,500	5.82	No	-	-
490,510	5.82	No	-	-
500,490	-	Yes	10.5	6.83
500,500	5.72	No	-	-
500,510	5.72	No	-	-
510,490	-	Yes	11.4	7.84
510,500	-	Yes	10.4	6.71
510,510	5.72	No	-	-

Even increasing the boundary separation d_g to 6 NM which means a 20% safety buffer does not prevent a separation violation in cases when the speed of aircraft A is greater than of aircraft B. In extreme cases, for $V_A = 510$ kt and $V_B = 490$ kt it is necessary to increase the radius R from 9.5 NM to 11.4 NM, which increases Δl by 2.12 NM, and this means 37% increase compared to the case without disturbances.

Results of experiments for different values of boundary separation are presented in Table 3. Again, also in the case of disturbances, the smallest additional distance Δl is observed when the aircraft A begins conflict resolution maneuvers as late as possible. However, to minimize the possibility of a separation violation, it is necessary to introduce a safety buffer consisting in the adoption of a larger radius R.

Table 3. Parameters of the 3HC method for disturbed conditions ($V_A = 510$ kt i $V_B = 490$ kt)

d_g[NM]	5.0	6.0	7.0	8.0	9.0	10.0	11.0	12.0
R[NM]	10.6	11.4	12.1	12.7	13.2	13.7	14.2	14.6
Δl[NM]	7.56	7.84	8.13	8.27	8.55	8.69	8.98	8.98

Simple analysis with popular spreadsheet package shows that R values follow the logarithmic trend line, which makes it possible to formulate the practical dependence

$$R = 4.5 \cdot ln(d_g) + 3.23 \tag{10}$$

The coefficient of determination is 0.9995, but the dependence (10) is valid only for the headings k_{A_0} and k_{B_0}. For other values, these relationships differ by a constant.

3.3 Selection of the 3HC Method Parameters

In the case of disturbances, we considered changes in the speed of both aircraft. Then, it is better to increase the radius R, but begin maneuvers when the smallest possible boundary separation is reached. However, it is worth noting that speed changes of 2% requires to increase the R parameter by 20%, and this increase Δl by almost 40%.

Further research is needed, in which other types of disturbances will be analyzed. Then, increase of the boundary separation may be necessary. However, the point C is determined by the actual observed distance between aircraft. Thus, even in the case of disturbances, conflict avoidance maneuvers start when the aircraft are separated by at least d_g. Therefore, for the 3HC method, we recommend the following parameters: $d_g = 5$ NM and $R = 10.6$ NM, according to the relation (10). For an ideal case, the increase in distance Δl is then 7.66 NM. The resulting trajectories of both aircraft are shown in Fig. 3a, and the separation versus time is presented in Fig. 3b.

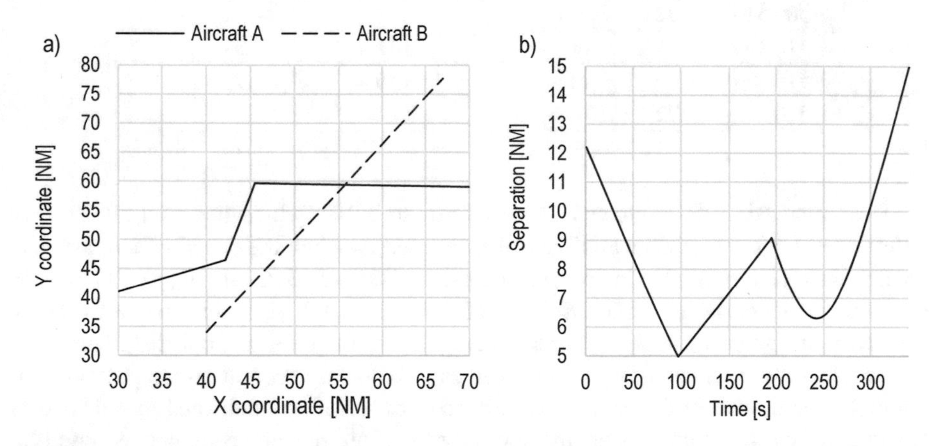

Fig. 3. (a) Aircraft trajectories for the recommended parameters of the 3HC method. (b) Separation during conflict resolution by the 3HC method.

4 Summary and Conclusions

The presented method of analysis uses Dynamic, Stochastic, Colored Petri nets to create a microscale model of air traffic conflict situation. It allows for: (i) modeling the dynamics of aircraft, (ii) detailed tracing of the flight trajectory to check separation conditions fulfillment and (iii) analyzing other traffic optimization criteria, characterizing the adopted 3HC conflict resolution method.

Simulation experiments with the created model allowed determining the point on the trajectory, where the maneuvers aimed at solving the conflict should start. Also, other parameters of the method, such as the distance at which the collision point should be passed. The approach using DSCPN model allows for the practical determination of 3HC method parameters for the assumed level of aircraft movements uncertainty. The analysis of the 3HC method indicates its usefulness in solving air traffic conflicts, especially when maneuvers in the vertical plane are not possible.

In the subsequent stages of the research, other disturbances will be analyzed. Also, a comparison will be made of the 3HC method with other conflict resolution methods based on maneuvers in a horizontal plane, for example using the Dubins trajectory.

References

1. Cafieri, S., Rey, D.: Maximizing the number of conflict-free aircraft using mixed-integer nonlinear programming. Comput. Oper. Res. **80**, 147–158 (2017)
2. Eurocontrol: Principles of Operation for the Use of Airborne Separation Assurance Systems, Brussels, Belgium (2001)
3. Friske, M., Ehrmanntraut, R.: Modeling a responsibility-transfer service for the delegation of tasks in airborne separation procedures and ground automation. Aerosp. Sci. Technol. **9**(6), 533–542 (2005)
4. Hao, S., Cheng, S., Zhang, Y.: A multi-aircraft conflict detection and resolution method for 4-dimensional trajectory-based operation. Chin. J. Aeronaut. **31**(7), 1579–1593 (2018)
5. Hoekstra, J., van Gent, R., Ruigrok, R.: Designing for safety: the 'free flight' air traffic management concept. Reliab. Eng. Syst. Saf. **75**(2), 215–232 (2002)
6. Huang, Y., Chung, T.: Modelling and analysis of air traffic control systems using hierarchical timed coloured Petri nets. Trans. Inst. Meas. Control **33**(1), 30–49 (2010)
7. ICAO: Procedures for Air Navigation Services - Rules of the Air and Air Traffic Services. DOC 4444-RAC/501. International Civil Aviation Organisation, Montreal (2007)
8. Jensen, K., Kristensen, L., Wells, L.: Coloured Petri Nets and CPN tools for modelling and validation of concurrent systems. Int. J. Softw. Tools Technol. Transf. **9**(3–4), 213–254 (2007)
9. Kovács, A., Németh, E., Hangos, K.M.: Modeling and optimization of runway traffic flow using coloured Petri Nets. In: International Conference on Control and Automation (ICCA), pp. 881–886 (2005)
10. Kuchar, J., Yang, L.: A review of conflict detection and resolution modeling methods. IEEE Trans. Intell. Transp. Syst. **1**(4), 179–189 (2000)
11. Płanda, B., Skorupski, J.: ASAS as a new concept of ensuring air traffic safety. In: Załęski et al. (eds.) Selected Aspects of Transport Safety, pp. 152–160. Polish Air Force University Publishers, Dęblin (2005). (In Polish)
12. Prevot, T., Battiste, V., Palmer, E., Shelden, S.: Air traffic concept utilizing 4D trajectories and airborne separation assistance. In: AIAA Guidance Navigation and Control Conference, Austin, TX (2003)
13. Skorupski, J.: The risk of an air accident as a result of a serious incident of the hybrid type. Reliab. Eng. Syst. Saf. **140**, 37–52 (2015)

14. Skorupski, J.: About the need of a new look at safety as a goal and constraint in air traffic management. Proc. Eng. **187**, 117–123 (2017)
15. Skorupski, J., Florowski, A.: Method for evaluating the landing aircraft sequence under disturbed conditions with the use of Petri nets. Aeronaut. J. **120**(1227), 819–844 (2016)
16. Skybrary: Airborne Separation Assurance Systems (2019). https://www.skybrary.aero/index.php
17. Tang, J.: Review: analysis and improvement of traffic alert and collision avoidance system. IEEE Access **5**, 21419–21429 (2017). Article number 8052481
18. Yu, X., Zhou, X., Zhang, Y.: Collision-free trajectory generation and tracking for UAVs using Markov decision process in a cluttered environment. J. Intell. Robot. Syst. **93**(1–2), 17–32 (2019)

Premises for Developing an IT Network Design for Railway Transport in Poland

Stanisław Gago[1] and Mirosław Siergiejczyk[2(✉)]

[1] Railway Institute, Warsaw, Poland
[2] Faculty of Transport, Warsaw University of Technology, Warsaw, Poland
msi@wt.pw.edu.pl

Abstract. The objective of the research paper is to present the influence of IT and logistics system global development on the planning and expansion of the ICT network for the purposes of railway transport in Poland. The PKP PLK Company is currently in the process of building an ICT network, primarily for the needs of the ERTMS system.

The authors believe that a railway ICT network should be planned in a way, so as to satisfy the current and future demands of all railway Companies in terms of data transmission, enabling the ICT expansion of these Companies onto other types of transport.

Keywords: Design · ICT network · Business areas in the railway industry

1 Introduction

The developing IT techniques and technologies are currently the major and necessary tool for conducting business activities. A particularly important aspect is the utilization of all possibilities of the IT technologies in companies with a network structure. Such companies include almost all railway companies, e.g., PKP PLK, carrier companies (passenger and cargo), as well as other rail and road transport companies.

Transport dynamics and cooperation with other institutions and companies result in a situation that transport cannot fall behind the global development trends in the field of ICT solution application. Even if only due to the environment and cooperation with this environment, transport cannot be the IT system "museum". A common feature of IT systems, especially the ones developed for the needs of transport is that their operation covers large areas, while at the same time, they have to use data transmission. The authors believe that a data transmission network is one of the main components of the current IT systems [5, 16].

Contemporary IT technologies are characterized by increasing functionality, efficiency and decreasing investment and operating costs.

Transport is one of the logistics chain elements, which also includes:

- a multifunctional logistic services system,
- an integrated ICT system.

In order for a logistics hub to be operating correctly, these systems must transmit valid information between each other. The aforementioned systems are increasingly

M. Siergiejczyk and K. Krzykowska (Eds.): ISCT21 2019, AISC 1032, pp. 115–123, 2020.
https://doi.org/10.1007/978-3-030-27687-4_12

better interconnected and are constantly developing. Current research in the field of logistics is associated with searching for a system solution, which would enable increasing the implementation effectiveness of processes and the development of logistics, while simultaneously obtaining economic, social and environmental balance. This solution is to be the so-called "Physical Internet". The Physical Internet (PI) is a connected communication infrastructure, using standardized modular container units for transport and exchanging data regarding the ongoing logistics processes.

The development and management of the IT technology will be significantly impacted by – practically already under development – the Internet of Things-IoT (applications), which will enable better utilization of IT and ICT technologies through making available new functionalities, increasing the effectiveness, automation and also increasing the bit rate of data, allowing to make decisions in real-time.

The Internet of Things will be used for managing, collecting and transmitting data between systems or devices equipped with computers connected to the Internet. This is a direction, which provides an opportunity for the construction of Intelligent Transport Systems (ITS) and will also support railway companies in the field of providing transport services. The Internet of Things is a solution supporting and optimizing the acquisition of data burdened with fewer errors. Without implementing innovative solutions in transport companies in the near future, it will be difficult to improve economic results and keep these services on the market.

The Internet of Things (IoT) is a development area for numerous innovations, which can support not only transport companies, but also the client and his/her business. This network has the best chance to be integrated and involved in the cooperation with various networks and IT systems. It is particularly vital in the railway industry because of its nature, which applies to the entire field covered by its services and involves moving means of transport, people and cargo.

Rail transport in Poland should keep up with the development of rail transport in Europe and also with the development of other land transport sectors. One of the prerequisites for this development is employment of modern ICT systems (Cloud Computing, Big Data, etc.), which enable the cooperation of railway Companies between each other but also with their foreign partners, and the introduction of new IT services for their clients, thus ensuring a more complete satisfaction of their transport demand. In order to implement the above, it is necessary to develop an appropriate ICT network [5, 13, 15].

The PKP ICT network should be compatible with the railway ICT networks within the EU both hardwired, as well as wireless (frequency band, technology), so that it is able to cooperate with ICT networks of other Railway Administrators.

2 Assumptions for Developing an ICT Network Design for PKP

Prior to commencing the development of an ICT network design, first of all it is imperative to identify and determine the expectations of future network users in various time perspectives:

- during the first period after commissioning the network
- within several years of operation (3–5 years)
- in a longer term (e.g. approx. 10 years)

The adopted assumptions and premises regarding the network being developed should be consulted with experts familiar with the state of the art of constructing ICT network, who should evaluate the feasibility of the assumptions both in terms of technical, as well as economic-wise implementation.

- It is currently assumed that a Holding of PKP Companies (2020) will be established soon, therefore, the designed PKP IT network should satisfy all business and operating needs of the Companies comprising the PKP Holding in terms of data transmission. see Raport kolejowy 6/2018 str.10 (Railway report 6/2018) p. 10.
- Each Company should be able to receive a guaranteed privacy and security of their data sent within the network.
- The PKP IT network should be open to cooperation with companies working with PKP, e.g. PKP Energetyka.
- The PKP IT network should be open to cooperation with the systems (networks) of major railway customers, e.g. modal transport (ports), owners of large sidings (mine sidings, sidings of large chemical plants, etc.),
- The PKP IT network should be open to cooperation with carriers not included in the PKP Holding, e.g. passenger traffic – municipal railway transport (SKM), freight traffic, e.g. – Lotos.

Three basic business areas can be distinguished within the railway industry:

- railway infrastructure,
- goods traffic,
- passenger traffic.

Each of these areas has own business priorities, which do not overlap and are linked with various points within a railway network:

- infrastructure – e.g. railway junctions and lines, viaducts, bridges, train safe travel systems, etc.
- passenger traffic – e.g. stations in cities, traction vehicles, wagons, carriage sets, etc.
- goods traffic – e.g. freight stations, sidings, locomotives, freight wagons, etc.

Many railway companies and enterprises conduct their business activity within the aforementioned areas:

1. The largest railway network operator within the Infrastructure area is PKP PLK. Apart from this Company, also other railway infrastructure operators are active: PKP SKM, PKM, PKP LHS, PKP WKD, etc.
2. Several dozen companies are registered in the field of cargo transport, with the leader being PKP Cargo, followed by Lotos Kolej, CTL Logistics, PKP LHS, etc.
3. The passenger transport area has a dozen or so registered companies: the biggest carriers include PKP Intercity, Przewozy Regionalne, Koleje Mazowieckie.

Railway transport companies (passenger and goods) conduct their activity throughout the entire Poland or only within a limited area, e.g. within a specific province or even just a specific railway line.

An ICT network should satisfy all the requirements within the aforementioned areas in terms of topology and ICT traffic over an appropriate time period [13–15].

Each enterprise wanting to stay on the market must develop along with technical progress in the field of a given company. This principle also applies to rail companies, which also have to catch up with the technical development by introducing new services arising from:

- Development in the field of increasing goods and passenger transport, size of the handled area, increasing train speed, operating new facilities, e.g. the planned Central Communication Port, etc.
- Development of infrastructure and vehicle use and maintenance (operation) technologies. Currently both the infrastructure, as well as the mobile equipment are subject to the "point" inspection of their operation efficiency, that is, for example, every specified period (quarterly, yearly, etc.) or by the quantity of completed work (travelled specific number of kilometres, etc.). The introduction of the Internet of Things (IoT) technology will enable, through installing appropriate "sensors", constant monitoring critical parameters of the railway infrastructure (rail track wear, strength of bridges, viaducts and the contact system condition, etc.) and mobile equipment (wear of traction vehicle and wagon wheels, etc.). It will be possible to automatically diagnose and maintain objects, evaluate the condition of rail tracks and the electric traction based on the large amount of collected data [6, 17, 18].
- Development of monitoring systems, i.e.
- monitoring the safe travel of trains,
- monitoring railway areas (parking stations, warehouses, depots, etc.),
- monitoring the widely-understood railway infrastructure (train stations, platforms and other station infrastructure),
- monitoring EU border stations (passenger traffic, goods traffic), continuous passenger train (wagon) and freight shipment monitoring.
- Introducing new services for rail customers, e.g. dynamic timetables, ticket sales (travel or shipment freight fees), monitoring "railway" areas – e.g. yards outside train stations.
- Cooperation with large railway clients and their IT systems, e.g. ports, international transport ("Silk Road", "Physical Internet PI").
- Development of IT technologies among railway clients, in neighbouring railway Administrators (e.g. PKP Cargo conducts business activity in several Central-European countries), etc.
- Development of ICT equipment and network monitoring technologies.

All of the aforementioned factors will impact the amount of data transmitted through the ICT network. In the longer term, it is also considered to introduce systems improving the safety of train travel through [3, 7, 16, 17]:

- automatic decisions regarding train speed based on weather data and object detection,

- autonomous travel, which stops trains, decreases speed or directs trains onto other routes depending on the situation,
- train travel support systems through constant monitoring of drive unit operation (energy-saving operation, timetable adherence, etc.).

For the aforementioned reasons, the designed ICT network should be open to the continuous growth of transmitted data volume.

An ICT network for the purposes of railway transport, which is organically adapted to cooperate with numerous IT systems, at the same time taking into account the European trends and standards, should also be susceptible to satisfying the requirements regarding the cooperation (integration) with networks of economy sectors impacting the humanity "survival" possibility in critical situations, e.g. extensive failures of power grids, failures of heating networks, water supply, network hazards for railway transport, fuel market, etc. Whereas IT systems developed for the purposes of railway transport should cooperate with the railway ICT network (networks), which is to be equipped with various data transmission technologies, owing to its evolutionary (and not revolutionary) development and the multitude of IT systems used in the field of transport.

It can be stated with a high degree of probability that the core network for the purposes of railway transport will be a network based on TCP/IP protocols, the so-called "Internet of Things" *(IoT)*. This network has the best chance to be integrated and cooperate with various networks and IT systems. It will ensure connections between Data hub clouds and users equipped with any computer stations or mobile devices, which will be connected with service servers through a broadband Internet connection. It is particularly important in the railway industry because of its nature, which applies to the entire field covered by its services and involves moving means of transport, people and cargo.

Therefore, the issues associated with increasing the resistance and survivability of the network, i.e., the ability to ensure the required service quality even in the event of attacks, large-scale disasters or other failures are becoming crucial [9, 17].

Summing up, it can be stated that one should strive for developing an IT network platform which would integrate various types of information on, e.g.:

- monitoring ground infrastructure
- monitoring vehicle condition
- monitoring train traffic
- monitoring train travel conditions
- monitoring power and supply demand
- monitoring passenger traffic volume for individual routes
- information regarding planned maintenance
- geospatial information
- and other.

3 Logical Model of a Network

According to the definition, IT network engineering should be a complete process, which matches business needs to the available technology in order to provide a system maximizing the success of an organization. An IT network for railway transport needs in Poland should cover many Companies, which constitute separate business entities that can compete and cooperate with each other, at the same time looking after their business interests (data confidentiality) and very often using the same infrastructure. When creating an IT network for railway transport, one can conclude that it is a "network of networks", since each of the users of this network will have an "own", isolated IT network with an option to cooperate with other IT networks (railway or public).

Each company should define the location and size of expected data streams, at least for the main routes for this company, the data types, processes and users, who gain access to the data or impact their change. This will apply to both the backbone network of a given company, as well as local area networks (LAN).

Collecting the information from the future users of a given IT network will be used to determine the requirements and structure of a future IT network for railway transport in Poland. The most important aspect is developing a logical concept of the IT network model, which would show the basic components, divided by function and system structure. Whereas the physical model would show the equipment and specific technologies, as well as their implementation method

The network logical model shall include such features as [1, 2, 8]:

- network susceptibility to practically constant development and changes,
- network security, which means appropriate protection of the network against hacker attacks, viruses, etc.,
- network resistance to equipment and software failures, damage, human errors, disasters,
- operational continuity after a disaster,
- supporting time-dependant applications (supporting real-time applications, e.g. control, VoIP, etc.),

and take into account existing limitations, such as:

- limited budget for network construction,
- personnel assigned to construct and operate the network,
- limited network construction time,
- "policy" of the Companies in relation to ICT application.

A network logical model shall determine appropriate features of the planned network, such as:

- traffic stream size in individual (basic) routes,
- location of data sources and sinks, and the location of data bases and warehouses,
- offered connection data rates,
- properties of data traffic streams (priorities, traffic classes, real-time applications, etc.),

- requirements regarding the quality of network-provided services (QoS).

In order to calculate whether the offered connection rate is sufficient, in addition to knowing the traffic stream volumes, one should know:

- the number of data sources and sinks,
- average idle time between sent packets,
- time required to send a message after gaining access to a medium.

As mentioned before, the network logical model engineering stage deals with the general architecture, sizes, shapes and network interconnectivity [8].

Network topology engineering should cover:

- Network scalability – the possibility to introduce new applications, handle increase data traffic (traffic increase percentage to be handled by the designed network), network expansion possibility.
- Availability – time for which the network is available to users, often expressed as a percentage of uptime or mean time between failures (MTBF) and the mean time to failure (MTTF).
- Security – protection of the ability of the organization to conduct its activity without interfering intruders, which gain access to the equipment, data or operation in an improper way and damage it. Particular security risks are that:
  - data can be intercepted, analysed, changed or deleted,
  - user passwords can be breached,
  - device configuration can be modified.
- Management – managing errors, configuration, performance and security.
- Usability – ease with which users can gain access to the network and its services.
- Adaptability – ease of adapting networks to faults, changing traffic patterns, additional business or technical requirements, new business practices and other changes.

A very important feature of a network is its capacity, which comprises the following common factors [2, 8]:

- available band
- bandwidth
- bandwidth use
- offered traffic volume
- precision (adapting to traffic demand)
- delay and delay changes
- response time.

The throughput is impacted by such factors as:

- packet size
- intervals between send packets
- number of packets per second in packet transmitting devices
- client access time to the processor, memory and HD
- server access time to the processor, memory and HD
- network design (queues, contentions, delays)
- protocols

- distances
- errors
- time of day, etc.

4 Conclusions

Creating a logical model of an IT network as a base to develop a physical model for this network first of all involves understanding the requirements of the ordering party, secondly, the possibilities of contemporary ICT networks, and thirdly, the size of the budget and time assigned by the ordering party to the implementation of the network in question. Designing IT networks for the railway in Poland involves additional engineering complication due to the different activity of railway companies – various business fields (e.g. passenger traffic, goods traffic), various activity coverage (regional transport, IC transport, foreign activity – PKP Cargo), various activity areas (infrastructure, transport, industry-related companies – e.g. telecommunications, IT). Each company has own confidential interests, which would be available only to itself. At the same time, these companies operating within the railway industry have to cooperate, e.g. PKP IC and PKP PLK. Furthermore, each company has different requirements regarding the services provided within the network. This would result in an ICT network for the purposes of railway companies comprising numerous logical networks, with access only for specific companies. It is obvious that the logical networks in question would operate based on one physical model of an IT network. Developing a physical model, as mentioned previously, requires the creation of a logical model of the network, taking into account all logical networks of individual railway companies – a "logical network of networks".

As shown by experience, familiarity with IT technology varies from company to company, as do the interests and requirements of individual enterprises. The prerequisite for the creation of a network logical model taking into account the requirements of individual companies is an explicit interpretation of the issues regarding network functionality and the services provided by the network.

The authors believe that it requires the establishment of a team composed of experts familiar with railway, IT and ICT issues, which would develop a set and interpretation of issues associated with networks for individual railway Companies.

This team should create logical models of ICT networks for individual Companies, the so-called VPN, using previously developed surveys as a reference.

The VPN network models for individual Companies should constitute a base for the creation and development of a network logical model for the purposes of the railway in Poland, which would be the backbone for a physical model of an IT network for the purposes of railway transport in Poland.

References

1. Anderson, D.J., Brown, T.J., Carter, C.M.: System of Systems Operational Availability Modeling. Sandia National Laboratories, October 2013
2. Bauschert, T., Büsing, Ch., D'Andreagiovanni, F., Koster, A.M.C.A., Kutschka, M., Steglich, U.: Network planning under demand uncertainty with robust optimization. IEEE Commun. Mag. **52**(2), 178–185 (2014)
3. Calle-Sánchez, J., Molina-García, M., Alonso, J.I., Fernández-Durán, A.: Long term evolution in high speed railway environments: feasibility and challenges. Bell Labs Tech. J. **18**(2), 237–253 (2013)
4. Chołda, P., Følstad, E.L., Helvik, B.E., Kuusela, P., Naldi, M., Norros, I.: Towards risk-aware communications networking. Reliab. Eng. Syst. Saf. **109**, 160–174 (2013)
5. Gago, S.: ICT in the Polish railway industry. Problemy Kolejnictwa (Railw. Rep.) (179) (2018). (in Polish)
6. Hiraguri, S.: Recent research on application of ICT for railway. QR RTRI **58**(4), 254–258 (2017)
7. Hiraguri, S.: Recent research and development of signaling and telecommunications technologies. QR RTRI **56**(3), 160–163 (2015)
8. Sterbenz, J.P.G., Çetinkaya, E.K., Hameed, M.A., Abdul Jabbar, A., Qian, S, Rohrer, J.P.: Evaluation of network resilience, survivability, and disruption tolerance: analysis, topology generation, simulation, and experimentation. Springer Science + Business Media, LLC, Heidelberg, 7 December 2011
9. Pawlik, M., Siergiejczyk, M., Gago, S.: European rail transport management system mobile transmission safety analysis. In: Walls, L., Revie, M., Bedford, T. (eds.) Risk, Reliability and Safety: Innovating Theory and Practice. Taylor & Francis Group, London (2017)
10. Qiu, S., Sallak, M., Schön, W., Cherfi-Boulanger, Z.: Epistemic parametric uncertainties in availability assessment of a Railway Signalling System using Monte Carlo simulation. Taylor & Francis Group, London (2014)
11. Saito, Y., Motoyoshi, S., Konishi, T., Matsuura, K., Yokoyama, A.: Innovative changes for track maintenance by using ICT. This is an open access article under the CC BY-NC-ND license 2017 The Authors. Published by Elsevier B.V. (2017)
12. Sansò, B., Soriano, P.: Telecommunications Network Planning. Springer, Heidelberg (2012)
13. Siergiejczyk, M., Gago, S.: GSM-R system operation in the Polish railway. Logistics 4/2015. (in Polish)
14. Siergiejczyk, M., Gago, S.: Issues of the GSMR system security in terms of the rail transport. Logistics No. 6/2012. Eds. ILiM, Poznań (2012). (in Polish)
15. Siergiejczyk, M., Gago, S.: Selected problems of reliability and security of information transmission in the GSM-R system. Problemy Kolejnictwa [Railw. Probl.] (162) (2014). (in Polish)
16. Siergiejczyk, M., Pawlik, M., Gago, S.: Safety of the new control command European System. In: Nowakowski, T., et al. (eds.) Safety and Reliability: Methodology and Applications. Taylor & Francis Group, London (2015)
17. Takikawa, M.: Innovation in railway maintenance utilizing information and communication technology (smart maintenance initiative). Japan Railw. Transp. Rev. **67**, 22–35 (2016)
18. Zhu, E., Crainic, T.G., Gendreau, M.: Scheduled service network design for freight rail transportation. Publ.: Oper. Res. **62**(2), 383–400 (2014)

Modeling of Adjustable Muffler in the Exhaust System of an Internal Combustion Engine

Andrzej Gagorowski(✉)

Warsaw University of Technology, Koszykowa 75, 00-662 Warsaw, Poland
agag@wt.pw.edu.pl

Abstract. The article presents model studies of the exhaust system of an internal combustion engine based on own concept of a regulated reactive muffler. A structure with a regulated input in the expansion chamber was proposed and the influence of adjustment degrees on the effectiveness of acoustic energy attenuation was determined in the simulation research process. In the modeling process of the mentioned structure, the basic acoustic properties of the system were mapped and the calculation process was carried out for selected real exploiting conditions of the internal combustion engine. For the mapping of real non-linear phenomena during exhaust gases flow, the boundary element method and volumetric flows were used as well as the specialized calculation package CFD (Computer Fluid Dynamics)-AVL AST. Comparisons of the test results of various degrees of regulation allowed determining the suitability of the proposed solution and evaluating the benefits in shaping the acoustic energy by determining the transmission loss.

Keywords: Exhaust system · Adjustable muffler · Gas-dynamic flow model

1 Introduction

The exhaust system of a motor vehicle (in combination with an internal combustion engine) has the features of a fluid system [4, 8], in which exhaust gases play the main role. When we design a specific structure, we influence the changes of dynamic interactions (e.g. by changing the pressure) in the process of energy transport. In the car exhaust system the basic structure is constituted by structural elements that create flow and transport channels as well as devices that convert chemical, thermal, mechanical and acoustic energy. Exhaust systems are a particularly important area among fluidized systems. This importance is connected with their basic functions, which apart from ensuring the correct work of the engine, include exhaust gases cleaning and noise reduction as well as heat dissipation to some extent. We deal here with an open system, in which energy flows into the environment in a shaped form. The issue of environmental pollution by exhaust gases and noise from motor vehicles is currently one of the basic problems in the world [6, 7]. The subject of research in this article concerns the reduction of noise emission to the environment by shaping the acoustic energy in the exhaust system of an internal combustion engine. For this purpose, an adjustable muffler system was proposed.

M. Siergiejczyk and K. Krzykowska (Eds.): ISCT21 2019, AISC 1032, pp. 124–132, 2020.
https://doi.org/10.1007/978-3-030-27687-4_13

When testing the exhaust system from the point of view of noise reduction, its entire structure should be taken into account, in which the main devices performing the above task are mufflers. The use of a muffler of appropriate design determines not only effective reduction of acoustic emission to the environment, but also the correct work of the engine. Various solutions can be found in the literature, but only a small part of them finds practical use in mass-produced vehicles. Among the various proposals structures with constant parameters dominate. Two main structural variants of exhaust mufflers can be distinguished, differing essentially in the method of creating flow resistance and shaping the acoustic energy. In the absorption variant, the material with sound-absorbing properties plays the leading role [10], through which exhaust gases are passed, while the cellular structure of reactive muffler forces the flow through a selected system of channels and chambers, shaping the acoustic waves by reflection or interference method [9]. The combination of the absorption and reactive systems in one integrated muffler gives often a good effect [15]. The normative limitation of the permissible noise emission to the environment from the exhaust system forces the use of muffler structures, and in the case of road vehicles the systems are subject to the approval process [3, 14].

In the engine - exhaust system structure we can distinguish different types of transported and processed energy (thermal, chemical, mechanical, acoustic), and changes of the parameters of its particular forms during the flow happen very quickly. The above facts indicate the complexity of the occurring processes and the difficulty in conducting a correct analysis for specific exploitation conditions. Therefore, modeling of energy transport and transformation processes during gas-dynamic flows is often presented in various degrees of simplification. In the work [5] authors analyze selected methods and techniques of modeling mufflers of exhaust systems, including models based on the theory of linear acoustics. During the flow through successive muffler structures, dynamic, local changes in temperature, pressure, velocity and exhaust gas density occur. Taking into account changes in these parameters over time, their description is possible by using non-linear Navier - Stokes or Euler equations [11], which in turn forces the use of numerical methods when solving them.

The article presents a proposal for the construction of an exhaust system muffler with an adjustable structure with an extended input, and the process of its modeling is presented on the basis of the boundary element method with using numerical procedures that allow to solve nonlinear Euler equations. Its effectiveness was determined by determining the transmission loss of the system with the assumed regulation levels (with the assumed settings). The model studies of acoustic gas flow in the exhaust system, with taking into account the proposed structure, were based on a specialized calculation package CFD (Computer Fluid Dynamics) - AVL AST, dedicated to automotive applications.

2 System with Adjustable Structure

The subject of the research is the construction of a muffler with an adjustable structure with variable input extension (Fig. 1). The system belongs to the group of reactive systems and enables the change of the internal structure by step-by-step controlling the extension of the input in the range limited by the length of the main expansion chamber. According to the theory of gas dynamics [13], changes in the internal structure of the system cause variable flow of the liquid medium (in this case of exhaust gases), among others by creating a pressure difference on the border of specific areas. From the point of view of acoustic energy analysis [2, 12], we deal with changes in flow resistance and acoustic pressure as well as the speed and flow direction. As a result, the acoustic energy passing through the system can be reduced or increased by incorporating into the basic system - successive acoustic discontinuities with different resistance. These discontinuities include, for example, all changes in cross-section and other geometric parameters like inlet or outlet channels. In the case of the proposed structure, a variable length extension element was added to the suppression inlet channel. As a result, the exhaust gases flow for a certain period of time through a narrow channel of regulated length instead of flowing directly into the main expansion chamber.

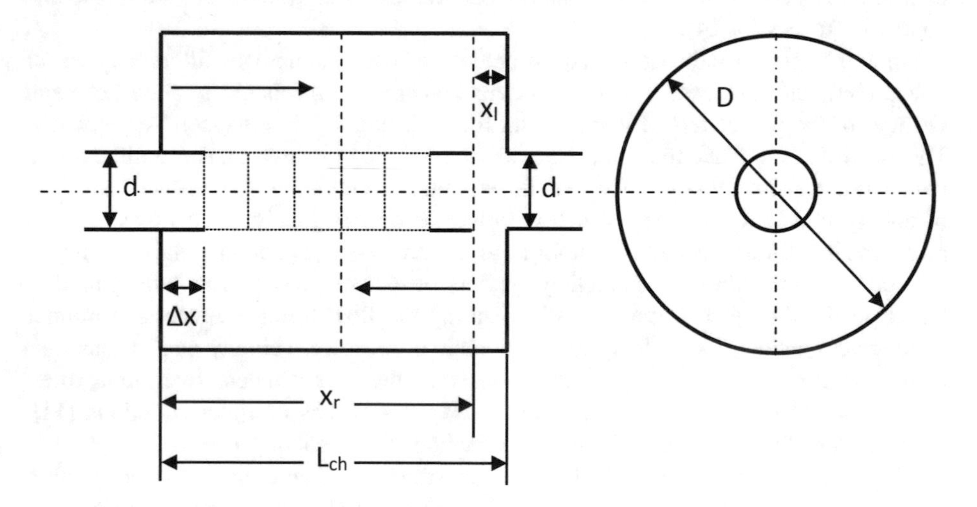

Fig. 1. System with adjustable structure

As parameters with fixed values, the diameter of section of the main chamber (D), the diameter of the inlet and outlet channel (d), and the length of the main chamber (L_{ch}) were adopted. The adjustable parameters are the length of the input extension (x_r), the distance between the end of the extension and the beginning of the outlet channel (x_l). The control of the exhaust gases flow takes place by step-by-step change of the inlet channel in the expansion chamber. The value of the regulation in the system is described by the parameter Δx denoting the minimum assumed increase or reduction of the length of the extension. The change of the input extension in the i-th step of the control is described by dependencies (1):

$$\begin{cases} x_r(i) = x_r(i-1) \pm k\Delta x \\ \Delta x \leq x_r \leq x_{\max} \\ 0 < x_l < L \\ \Delta x > 0 \end{cases} \tag{1}$$

where $k\Delta x$ (k = 1, 2 …) is a multiple of the used, minimal value of the length change of the extension in the proposed system.

3 Muffler Model in the Engine - Exhaust System Structure

The models of exhaust gases flow in the engine - exhaust system structure with a muffler, as mentioned in the introduction, were built by using the method of boundary elements. The presented approach allowed determining the basic acoustic properties of the system and its behavior at selected operating conditions of the internal combustion engine. Numerical calculations were made in the AVL AST (Advanced Simulation Tools) software environment, including the module for calculations of acoustic flows - AVL BOOST. In order to eliminate the influence of other devices on the flow of acoustic energy, the model does not include other elements of the exhaust system (e.g. catalyst). Figure 2 presents a built model of connections which implemented the previously described muffler models with adjustable structure.

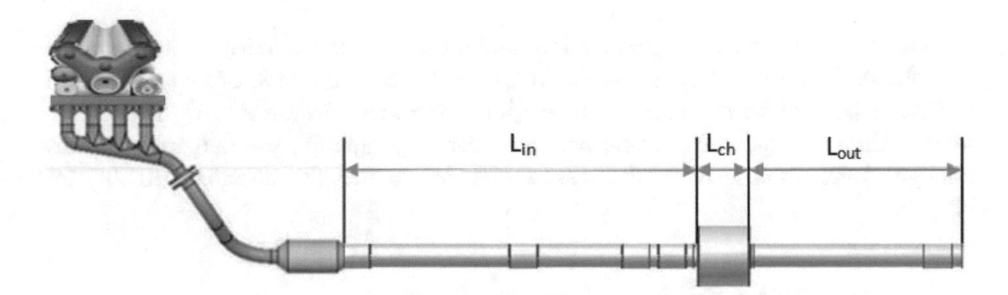

Fig. 2. The model of exhaust flow in the engine - exhaust system structure

While selecting the parameters of individual elements of the proposed structure, the limitations resulting from the construction requirements of motor vehicles were guided, including exhaust systems whose location and dimensions result from many premises and design assumptions. Table 1 presents the main geometric parameters of the exhaust system, including:

- length of the exhaust system in front of the muffler (L_{in} - inlet part),
- length of the exhaust system behind the muffler (L_{out} - outlet part),
- length of the muffler chamber (L_{ch}),
- diameter of the main section of the muffler (D),
- diameter of inlet and outlet channel (d).

Table 1. Geometric parameters of the exhaust system

L_{in}[mm]	L_{out}[mm]	d[mm]	D[mm]	L_{ch}[mm]
750	450	51	130	110

In the model based on the method of boundary elements, numerical procedures were used for Euler's equations [11] describing local pressure changes in particular cross-sections (A_n) of the exhaust system. The boundary conditions were determined taking into account the studies described in [6] and [9]. Table 2 presents selected reference parameters (initial) and boundary conditions used in the calculation for the duty cycle of a four-stroke internal combustion engine. It was assumed that at the moment of starting the opening of the exhalation valve at the beginning of the exhaust system, an overpressure impulse is created. This impulse travels along the system. The regulation of the fuel mixture composition is determined by the coefficient expressing the stoichiometric A/F ratio.

Table 2. Selected reference parameters and boundary conditions

Reference speed [rpm]	A/F [-]	Reference pressure [bar]	Reference temperature[C]	Wall temperature[C]
1000.0	14.5	1.013	24.85	23

During the gas flow, the temperature and pressure in the exhaust system increases (in relation to the initial values given in Table 2). The magnitude of these changes is strictly dependent on the stage of the engine's operation. As a result of the processes taking place, changes in acoustic energy occur very rapidly. The dynamics of these changes were confirmed by the results obtained, which are described in the next chapter.

4 Studies of the Engine - Exhaust System - Analysis of Transmission Loss

The calculations made, taking into account the model described in the previous chapter, allowed to determine, among others, values of the pressure and velocity of local exhaust flows in specific cross-sections of the system engine - exhaust system, with different input extension settings. The determined acoustic parameters made it possible to assess the attenuation of the system taking into account the changes in the length of the extension. The analysis covered 10 systems (Table 3) with various steps of extended inlet control taking into account dependence (1) presented in chapter 2.

Table 3. Selected cases of input extension control

Adjustment	x_r[mm]	x_l[mm]	Δx[mm]
xr1	10	100	10
xr2	20	90	10
xr3	30	80	10
xr4	40	70	10
xr5	50	60	10
xr6	60	50	10
xr7	70	40	10
xr8	80	30	10
xr9	90	20	10
xr10	100	10	10

The evaluation of the effectiveness of suppression reactive systems from the point of view of noise reduction is carried out using various indicators [1, 2]. In the work, to analyze the impact of regulation on the level of transmitted acoustic energy, a basic indicator was used to determine transmission suppression in the system, namely *TL* (Transmission Loss). The suppression of transmission determines the differences between the acoustic power levels of the sound falling on the muffler W_{in} and propagated from the muffler W_{out} [2, 6]:

$$TL = 10\log\frac{W_{in}}{W_{out}} \tag{2}$$

The acoustic power level in a particular A_l cross-section of the system can be defined as follows:

$$W_i = \int\limits_A \frac{|p^2|}{2\rho c} dA_i \tag{3}$$

Where

- p - pressure amplitude in the analyzed system cross-section,
- ρ - the exhaust gas density,
- c - speed of sound.

The obtained results for selected control cases characterized in Table 3 are shown below. To show the differences between the various levels of adjustment, the comparison of transmission loss for different cases in the whole analyzed frequency domain is presented in Figs. 3a and b.

The results from simulation tests show the strong non-linearity of local changes in the acoustic energy level at a given degree of regulation and the actual working conditions of the internal combustion engine, which in turn translates into a large variation of the *TL* level. Analyzing the course of the *TL* parameter in a function of frequency, it can be concluded that the use of both minimal and maximal expansion as well as intermediate steps of extended inlet adjustment causes beneficial changes in terms of attenuation of acoustic energy. Figure 3a shows a comparison of three cases with minimal adjustments – 10 mm, 20 mm and 30 mm (level of adjustment xr1 and xr3 - Table 3). The system with these extensions gives improvement in the entire tested frequency band, but detailed differences can be noted by analyzing individual frequency ranges, for example the adjustment xr2 (20 mm) brings a significant improvement in the middle and upper part of the considered spectrum.

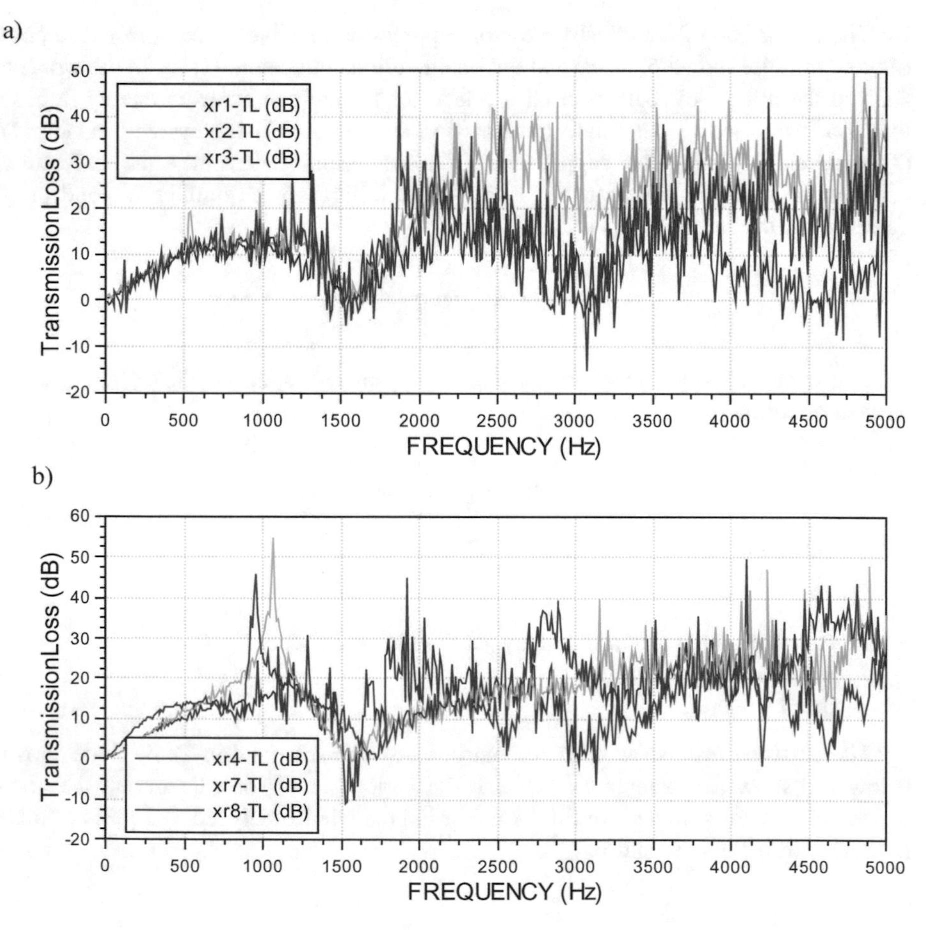

Fig. 3. Comparison of transmission loss (a) adjustment/xr1, xr2, xr3/, (b) adjustment/xr4, xr7, xr8/

Other beneficial effects were also obtained for other cases including intermediate control steps ($30 < x_r < 100$ mm). Comparing, for example, systems with the 40 mm, 70 mm and 80 mm extensions (Fig. 3b), we can notice differences in the resonant areas in terms of both band width and suppression values.

Comparing all structures with particular levels of regulation of the input extension, it can be clearly stated that the proposed system enables shaping both the attenuation level and the frequency response, which will allow adapting the system to the specific vehicle structure and the requirements of manufacturers or users.

5 Conclusion

The article presents a proposal for the construction of the muffler of the exhaust system with a regulated structure, with variable extension of the input as well as the method of its testing using the boundary element method and CFD technique. Taking into consideration numerical procedures based on the Euler's equations allowed determining changes in acoustic pressure and determining the effect of real gas-dynamic phenomena in the process of exhaust gases flow on the shaping of acoustic energy at various levels of adjustment. The system has been analyzed in the field of transmission suppression taking into account the frequency band characteristic for internal combustion engines of motor vehicles. The model tests of the system with the regulation of input extensions showed that due to even small changes in the internal structure of the muffler, large differences in the frequency range as well as a specific noise reduction in the terms of specific frequencies can be obtained. Profitable results were obtained for all applied control stages from the point of view of obtaining the appropriate level of suppression in a particular frequency range. The results show very high dynamics and non-linearity of changes as well as significant differences in the level of transmission loss depending on the frequency band. The study made for the basic rotational speed are a reference point for further analysis of acoustic processes at higher engine performance parameters and obtaining wider conclusions for building a prototype of the system.

Thanks to small dimensions and simple regulations, the proposed solution can be easily adapted to specific needs. The system can optionally be an additional muffler in an existing system or an initial/final section. However, it should be remembered that the final effect of the applied solution will be strictly dependent on the construction of the specific vehicle model.

The research has been conducted in the international program: "AVL AST University Partnership Program" (AVL, Graz, Austria).

References

1. Bender, E.K., Brammer, A.J.: Internal combustion engine intake and exhaust system noise. J. Acoust. Soc. Am. **58**, 22–30 (1975)
2. Bies, D.A., Hansen, C.H.: Engineering noise control theory and practice, 3rd edn. Spon Press - Taylor & Francis Group, London and New York (2003)

3. Commission Directive 2007/34/EC of 14 June 2007 amending, for the purposes of its adaptation to technical progress, Council Directive 70/157/EEC concerning the permissible sound level and the exhaust system of motor vehicles
4. Fister, W.: Fluidenergiemaschinen. Springer-Verlag, Berlin (1984)
5. Gągorowski, A., Melon, A.: Selected aspects of modelling mufflers for exhaust systems of vehicles. J. Kones Inst. Aviat. **20**(2), 97–103 (2013)
6. Gągorowski, A.: Study selected structures of exhaust systems mufflers in AVL environment. Scientific works of Warsaw University of Technology. Transport, Oficyna Wydawnicza Politechniki Warszawskiej, vol. 112, Warsaw, pp. 101–110 (2016)
7. Gozalo, G.R., Morillas, J.M.B.: Analysis of Sampling Methodologies for Noise Pollution Assessment and the Impact on the Population. Int. J. Environ. Res. Public Health **13**(490), 1–18 (2016)
8. Saleh, J.M.: Fluid Flow Handbook. McGraw-Hill, New York (2002)
9. Ji, Z., Ma, Q., Zhang, Z.: Application of the boundary element method to predicting acoustic performance of expansion chamber mufflers with mean flow. J. Sound Vibr. **173**, 57–71 (1994)
10. Jones, P.W.: Prediction of the acoustic performance of small poroelastic foam filled mufflers: a case study. Acoust. Aust. **38**(2), 73–79 (2010)
11. Lunev, V.V.: Real Gas Flows with High Velocities. CRC Press, Boca Raton (2009)
12. Rienstra, S.W., Hirschberg, A.: An Introduction to Acoustics. The extended and revised edition of IWDE 92-06. University of Technology, Eindhoven (2018)
13. Rathakrishnan, E.: Applied Gas Dynamics. Wiley, Singapore (2010)
14. Regulation (Eu) No 540/2014 of the European Parliament and of the Council of 16 April 2014 on the sound level of motor vehicles and of replacement silencing systems, and amending Directive 2007/46/EC and repealing Directive 70/157/EEC
15. Selamet, A., Lee, I.J., Huff, N.T.: Acoustic attenuation of hybrid silencers. J. Sound Vib. **262**, 509–527 (2003)

Virtual Reality Technologies in the Training of Professional Drivers. Comparison of the 2D and 3D Simulation Application

Kamila Gąsiorek(✉), Ewa Odachowska, Arkadiusz Matysiak, and Małgorzata Pędzierska

Transport Telematics Center, Motor Transport Institute, Warsaw, Poland
kamila.gasiorek@its.waw.pl

Abstract. During recent years virtual reality (VR) technology has been gaining popularity in many fields, including the acknowledgement given in professional driver training. Currently, the implementation of methodologically correct, ICT-based (including both the driving simulators and e-learning) training is both resource- and time-demanding. It is also strongly connected with high personnel costs. Due to their low implementation costs and impact made on a trainee, VR-based tools could play a complementary role in the training process and could support the real life training, as well as provide an alternative for traditional educational platforms. The article introduces the comparative analysis on how the virtual reality systems influence the simulation quality in relations to the 2D screens used in low-class simulators. The dynamic parameters of driving, intensity of the simulation sickness symptoms occurrence, immersion level, registered during three research drives were analyzed in a group of 30 drivers. Both the benefits and inconveniences deriving from the characteristics of certain virtual environment projection methods were specified.

Keywords: Immersive virtual reality · Driver training · HMD

1 Introduction

Modern virtual reality technologies (Virtual Reality – VR) are the solutions that support many industries today, including the transport industry. Due to the very rapid development of technologies and devices allowing to create a feeling of being complete transferred to a virtual environment, the use of this type of equipment in the process of training drivers seemed only a matter of time.

Due to the low implementation costs and the scope of impact on the trainee, these tools can be used to complement real-world training and an attractive alternative to traditional e-learning platforms.

At the moment, virtual reality technologies are primarily used to train in modelling safe driving habits, especially for young drivers [7, 9]. The VR based training programs are effective in improving the ability of inexperienced drivers to predict hidden dangers and prepare them to function in the real traffic [1, 8]. They are also useful for training professional drivers [4, 10, 11]. Special software allows drivers to participate in remote

M. Siergiejczyk and K. Krzykowska (Eds.): ISCT21 2019, AISC 1032, pp. 133–142, 2020.
https://doi.org/10.1007/978-3-030-27687-4_14

training sessions under the supervision of a driving instructor from anywhere and gives the opportunity to undergo part of practical training, including driving in special conditions and additional training sessions.

The research has confirmed the effectiveness of using VR to model the behaviour when transferring control in autonomous vehicles [15] as well as to study frustration, anger and behaviour of the driver when driving in adverse road conditions [2]. Virtual reality is also a useful solution for analyzing the behaviour of other road users, such as cyclists or pedestrians [14], as well as training of drivers of other professions, for example forklift drivers [8].

These technologies are gaining more and more recognition in the training environment, mainly due to the high level of drivers' involvement (immersion). The use of VR goggles can potentially separate participants to a greater extent from the real world compared to the use of flat screens of a low-end simulator [12, 20]. However, a particular attention is paid to the possibility of simulation sickness occurring during a virtual reality session [3, 13, 21]. In the context of the above-mentioned advantages, as well as limitations of the existing solutions, and also due to the very rapid development of virtual reality technologies and devices, the need to analyze the impact of VR tools available on the Polish market is justified.

The findings and analyzes presented are the result of the ICT-INEX project implemented by an international consortium under the Erasmus + program. The aim of the project is to increase the availability and effectiveness of the training for drivers and candidates for drivers using modern ICT tools (Information and Communication Technologies). As part of the research at the Motor Transport Institute, tests have been conducted using the VR Oculus and HTC VIVE Pro goggles, aimed at determining the possibilities of using VR technology in driver training.

2 Methodology

2.1 Participants

The study involved 30 drivers, including 15 men aged 20–29 years old and 15 men aged 50–65 years old. All participants of the study were active drivers, have had a driving license for a minimum of 3 years and travelled annually during this period about 3,000 km. Drivers were recruited to the test by telephone through an external research company. They received remuneration for their participation in the study.

2.2 Equipment

The study used two different methods of virtual environment presentation, i.e. virtual reality goggles - HTC Vive Pro and Oculus Rift and screen projection.

Conventional 2D Display (PC)

A 40-inch LCD screen with a resolution (1960 × 1080) was used to present the virtual environment using the screen projection method. The screen displayed the image at 60 Hz. During the drives, the drivers could observe the virtual environment and change

Fig. 1. Left: A scene in virtual reality presented by means of a conventional screen projection. Right: The same virtual reality scene is displayed using the HTC Vive Pro goggles.

the viewing point of the scene with the help of the steering wheel. The monitor was placed approximately 120 cm from the driver's seat (Fig. 1).

HMD Oculus Rift
In comparison to a conventional screen projection, where the image of the simulated environment is limited to the screen size and visible only from the simulator's front, head mounted displays (HMDs) generate high-quality stereoscopic image. The user has the opportunity to look around, including the movement of the head in the image displayed in the mirrors. Oculus Rift goggles used in the study consisted of two OLED displays with an effective resolution of 2160 × 1200 (1080 × 1200 for each eye) with a 90 Hz refresh rate. The device generates an image with an angular size of 110°, while monitoring the user's head movements for the full angle. The system also includes sound effects that are important for achieving the immersion effect, i.e. generates 3D surround sound.

HMD HTC Vive Pro
HTC Vive Pro goggles, like the Oculus Rift goggles, use two displays, whose task is to create a three-dimensional image from two two-dimensional ones. HTC Vive Pro is characterized by a resolution of 2880 × 1600 (1440 × 1600 for each eye) with dual-OLED display. The refresh rate and the field of view in this goggle model are 90 Hz and 110°, respectively. Goggles also generate 3D surround sound.

2.3 Study Protocol (Experimental Procedure)

The research experiment was carried out in accordance with the research methodology involving the use of virtual reality developed at the Institute of Motor Transport. The subjects were familiarized with the purpose of the study and informed about the procedure of the experiment and the possible adverse effects of participation in the research using VR. Next, the participants filled out a short questionnaire containing demographic and behavioural questions about road traffic, which were important due to the purpose of the study. Information on gaming experiences was also collected, such as the frequency of games and their type, previous experiences with virtual reality and augmented reality.

Each participant of the study before taking part in the test drive had an adaptive drive on a simulator, during which he had the opportunity to become accustomed to steering in the virtual reality. Adaptation was also aimed at eliminating people who are particularly sensitive and who feel the symptoms of simulation sickness. Next, the drivers proceeded to drive through three research scenarios. The driver's task during each scenario was to drive through the same straight road section. The section consisted of the local road of the built-up area, the extra-urban road and the motorway. During two drives, the HTC Vive and Oculus Rift goggles were used. However, the third drive took place using the 2D screen (see Fig. 2). A random assignment of the research scenarios order was applied. After each drive, the participants were asked to fill in the questionnaires containing questions about factors such as immersion, usability and simulator sickness. A detailed description of the methods used to assess these factors can be found in point 2.4. The total duration of the study was 1 h.

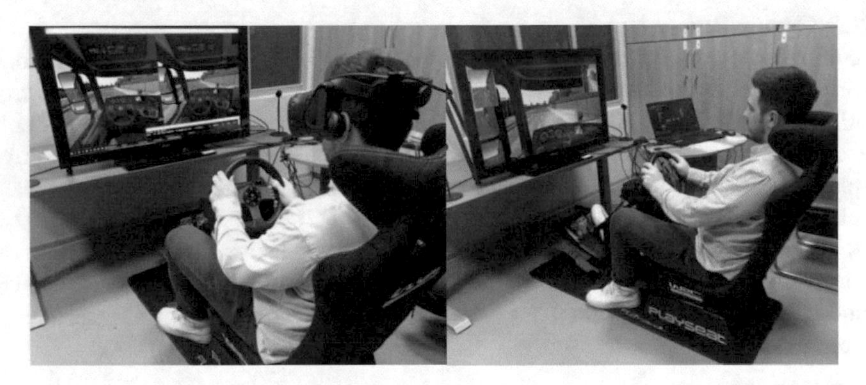

Fig. 2. Participants during research drives. On the left, a drive using the HTC Vive Pro goggles, on the right, drives using a 2D display.

2.4 Evaluation Measures

After each research drive and drivers' interaction with each virtual reality projection method (2D screen, Oculus Rift, HTC Vive Pro) the quality of the experiment was assessed, i.e. perceived usability, immersion level, intensity of the simulator sickness symptoms occurring and the dynamic driving parameters.

To evaluate the perceived usefulness, a questionnaire was created for the needs of the survey. The respondents were asked to assess the difficulty level of using a given projection method, the degree of realism of the simulation and the level of comfort of particular aspects of the simulation and the ease of operation, such as comfort of use, visual comfort (smoothness of looking around the scene), contrast and depth level, image sharpness, comfort of steering a vehicle, control of the situation in the virtual reality and well-being during the drive. The questions were based on a 6-point Likert scale. In addition, the respondents were asked about the usefulness of using goggles for driver training.

To evaluate the level of immersion during the experiment, the Immersion Questionnaire was used, which is the Polish adaptation of The Immersion Questionnaire presented by Strojny and Strojny [6, 18]. The Polish version of the questionnaire consists of 27 test questions, in which answers can be provided on a five-point Likert scale.

The RSSQ questionnaire was used to determine the degree of impact of the simulator sickness on the person examined. This method evaluates 28 symptoms in a four-point scale. Symptoms are grouped into three categories: symptoms of nausea, oculomotor symptoms, and disorientation symptoms.

Additionally, the following vehicle movement parameters as well as driver behaviour and response ones were recorded in the study: drive time, vehicle speed, number and time of speed deviations, number of brake pedal operations, average fuel consumption during the trip and number of collisions. All parameters listed above were recorded using the driving simulator.

3 Results and Discussion

As expected, the results obtained on the levels of immersion in the specific simulations indicate that both HTC Vive Pro and Oculus Rift goggles increase the sense of immersion in the virtual reality. Statistically significant differences were found between each group. The largest immersion experience was recorded in the Oculus Rift goggles (95.6 points). While the level of immersion in the case of HTC Vive Pro amounted to 91.9 pts, while driving without goggles (screen projection) - 86.1 pts. There were also statistically significant differences in the immersion experience when driving without goggles between a group of young people (up to 30 years of age) and a group of older people (over 50 years). Thc young people during this simulation experienced a significantly lower level of immersion than older ones (76.9 points vs. 95.3 points). The graph below (Fig. 3) provides a summary of all immersion levels.

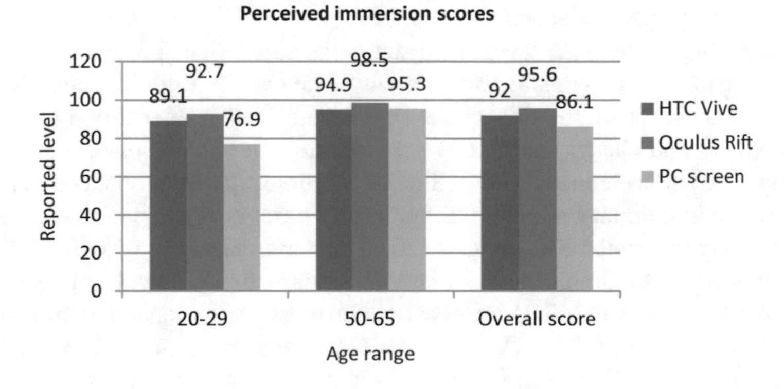

Fig. 3. Immersion experiment in the group of young, older people and in total, observed during 3D and 2D simulations.

Almost every aspect of usability was perceived slightly more positively in relation to the traditional 2D screen than virtual reality systems (Fig. 4). Statistically significant differences were found in the level of declared comfort (HTC Vive Pro - 3.9 vs. screen projection - 4.7) and perceived image sharpness between each of the VR systems and screen projection (HTC Vive Pro - 3.7, Oculus Rift - 4.2, screen projection - 4.7). Simulations in virtual reality goggles, however, better reflected real driving in real road conditions. Much lesser realism of the simulation the drivers felt during the trip without goggles (3.7) compared to Oculus Rift (4.4). Greater realism was also recorded in the case of HTC Vive (4.0), but this difference was not statistically significant.

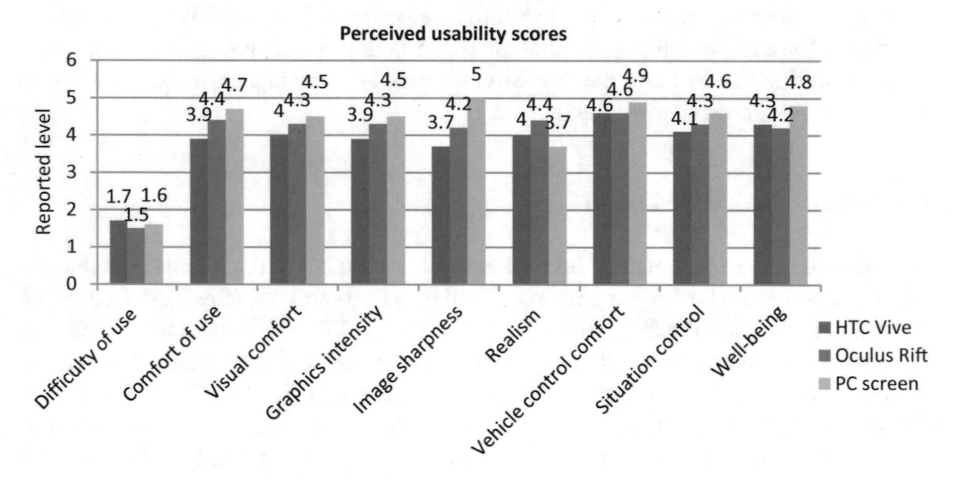

Fig. 4. Perceived utility of individual virtual reality devices and screen projection.

Despite slight differences in the level of utility, 83% of the surveyed drivers believe that virtual reality-based tools can be used for driver training. 43.3% of respondents preferred Oculus Rift goggles, 26.7% HTC Vive Pro goggles, while 30% of drivers preferred to use projection screens.

The worrying result is the level of impact of the simulation sickness on the persons tested in the case of driving using virtual reality. The overall result, defining the overall impact of the sickness on the driver was significantly higher after driving in the HCT Vive Pro (21.2) and Oculus Rift (21.3) goggles than after the simulation in which the projection screens were used (8.9). Similar relationships were noticed during the analysis of the individual symptoms. Significantly more symptoms of nausea occurred after simulation in Oculus Rift goggles (20.4) than after screen projection (8.9). After the 2D simulation, the number of oculomotor symptoms (8.1) was noticeably lower than in the Oculus Rift (17.7). However, the most symptoms of confusion were recorded after driving in the HTC Vive Pro (21.8) goggles compared to the 2D simulation (5.6). It is worth noting that there were no statistically significant differences between the two types of goggles used in the study. The results are presented in Fig. 5.

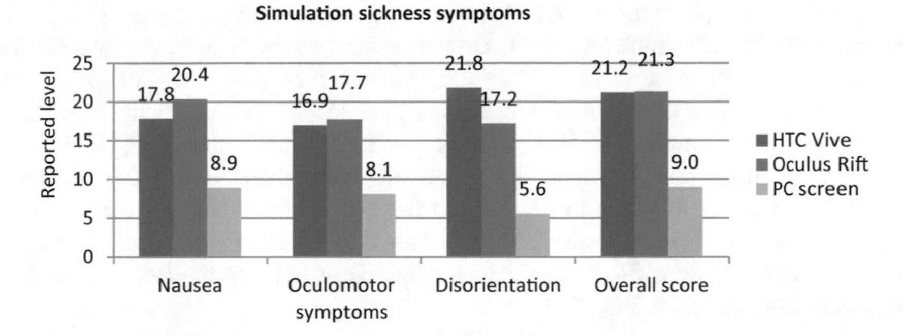

Fig. 5. Simulation sickness symptoms and total sickness impact on subjects observed after 3D and 2D simulations.

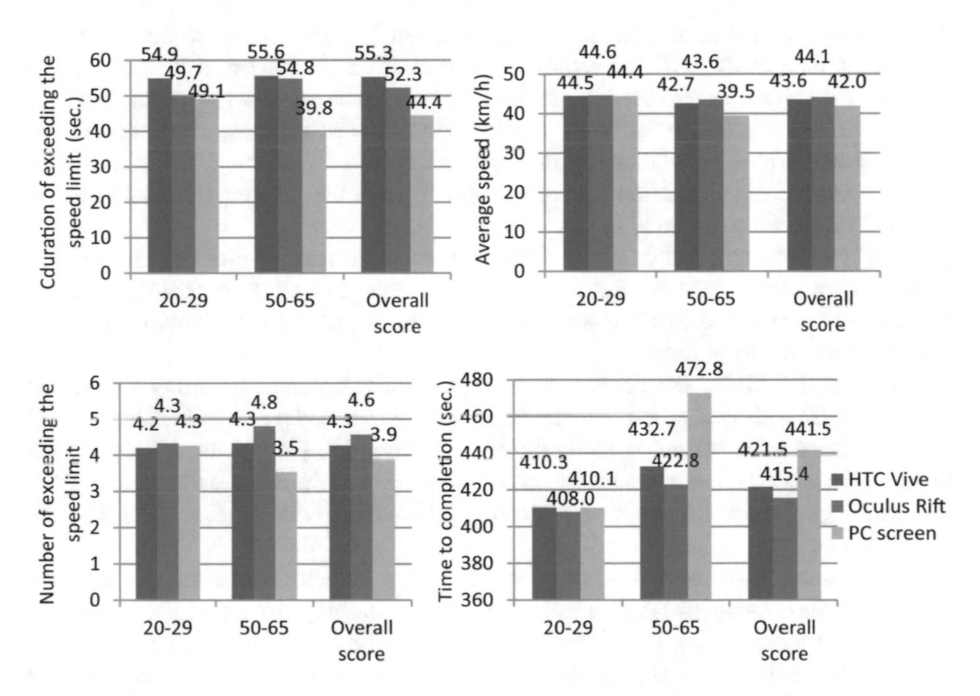

Fig. 6. Parameters of the vehicle movement in the simulation, behaviour and driver's reaction during 3D and 2D simulations.

Analysis of dynamic driving parameters did not show statistically significant differences in the number of collisions, brake pedal pressing or average fuel consumption between 3D simulations and 2D simulation. Drivers drove faster while travelling in goggles (43.6 km/h in HTC Vive Pro, 44.1 km/h in Oculus Rift) compared to 2D simulation (41.9 km/h). Therefore, detailed analyzes were carried out into age groups. It was shown that during 2D simulations, older people (above 50 years of age) were driving slower (39.5 km/h) than young people (44.4 km/h).

Speeding took place more often during simulations in Oculus Rift goggles (4.6) in comparison with 2D simulation (3.9). However, the longest time of speed deviations was recorded in the case of HTC Vive Pro (55.3 s) in comparison with the 2D simulation (44.4 s). The drivers took the longest time to pass 2D simulations (441.5 s), compared to the simulations in HTC Vive Pro (421.5 s) and Oculus Rift (415.2 s). The longest route time was shown in the group of older drivers during 2D simulation (472.8 s), It was significantly longer than in the case of young drivers (410 s). There were also statistically significant differences between the older and younger drivers in the number of collisions (9 vs. 2 collisions respectively) during the Oculus Rift drive. The results are shown in Fig. 6.

4 Conclusion

The research aimed to determine how virtual reality systems affect the quality of simulation in respect to 2D screens in low-level simulators showed that during 3D simulations, drivers experienced higher level of immersion and driving a virtual vehicle was it was not a problem for them. Additionally, the test drives using virtual reality were considered as simulations with a greater level of realism. This is important because, during the simulation, the driver should have the greatest impression of the 'reality' of the situation he is in.

The results in terms of comfort of use and the subjective level of sharpness of the picture were more advantageous in the case of conventional screen projection. No differences were found in other aspects, nor in the perceived usefulness of the individual virtual reality devices.

During 3D drives, the drivers were significantly faster than during the screen projection. The highest speed and the highest speed exceedances, indicating a lot of quick changes of acceleration made during the drive, i.e. jerking, were noted in the Oculus Rift goggles. It seems that this may be the result of the 3D simulation giving the opportunity to "be in" the situation, which provides a more certain sense of control over the situation.

There was a high percentage of people reporting symptoms of simulator sickness after the drives using goggles, which may be a disturbing aspect in the context of conducting training using this device. In addition, when using the goggles, the difficulty in sharpening the image, disorientation and nausea symptoms were much more common ($p < 0.05$). The symptoms of nausea and oculomotor ones observed mainly after driving in the Oculus Rift goggles were probably caused by a narrower field of vision, as well as a noticeable higher speed. The binocular field of view in the human reaches in the horizontal plane 180° and 120° in the vertical axis [22]. In the virtual reality technology, the field of vision plays a key role. It is strongly associated with simulator sickness. Maneuvering and observation of peripheral areas, due to the narrower field of view of the driver, required more movements and head twists, thus also adjusting the posture to the image. This is important because current studies show that the probability of simulator sickness symptoms occurring may increase due to the different severity of the head movements [16, 17]. In addition, the virtual environment may cause inadequate stimulation, because inter alia, the image resolution may not correspond with the real world and with what the perceptual system is used to [19].

The field of view can also be associated with the disorientation symptoms that were especially observed after the drive in the HTC Vive goggles. A narrower field of view can cause confusion for the goggle user as soon as he moves faster in the virtual environment [5].

References

1. Agrawal, R., Knodler, M., Fisher, D., Samuel, S.: Advanced virtual reality based training to improve young drivers' latent hazard anticipation ability. In: Proceedings of the Human Factors and Ergonomics Society Annual Meeting. SAGE Publications (2017)
2. Agrawal, M., Vemuri, K.: A heterogeneous traffic virtual-reality simulator to study irritation/anger and driving behavior under adverse conditions. In: Proceedings of British HCI 2018, Belfast, UK. BCS Learning and Development Ltd. (2018)
3. Blissing, B., Bruzelius, F.: Exploring the suitability of virtual reality for driving simulation. In: Proceedings of the Driving Simulation Conference 2018, France (2018)
4. City Car Driving. http://citycardriving.com/features. Accessed 25 Apr 2019
5. Harvey, C., Howarth, P.: The effect of display size on Visually-Induced Motion Sickness (VIMS) and skin temperature. In: Proceedings of VIMS 2017, pp. 96–103 (2007)
6. Jennett, C., Cox, A.L., Cairns, P., Dhoparee, S., Epps, A.: Measuring and defining the experience of immersion in games. Int. J. Hum.-Comput Stud. **66**, 641–661 (2008)
7. Lang, Y., Liang, W., Xu, F., Zhao, Y., Yu, L.: Synthesizing personalized training programs for improving driving habits via virtual reality. In: IEEE Virtual Reality (2018)
8. Hupont, I., Gracia, J., Sanagustín, L., Gracia, M.A.: How do new visual immersive systems influence gaming QoE? A use case of serious gaming with oculus rift. In: 2015 Seventh International Workshop on Quality of Multimedia Experience (QoMEX) (2015)
9. Project Toyota TeenDrive365. http://www.teendrive365inschool.com. Accessed 25 Apr 2019
10. Project Eco2Trainer. http://www.eco2trainer.se/en/ecodriving-utbildning-education. Accessed 25 Apr 2019
11. Press release about Cargo Dynasty. http://docslide.us/documents/cargo-dynasty-press-release.html. Accessed 25 Apr 2019
12. Read, J., Saleem, J.: Task performance and situation awareness with a virtual reality head-mounted display. In: Proceedings of the Human Factors and Ergonomics Society Annual Meeting, vol. 61, no. 1, pp. 2105–2109 (2017)
13. Ricaud, B., Lietar, B., Joly, C.: Are virtual reality headsets efficient for remote driving? In: Proceedings of the International Conference on Road Safety Simulation, RSS 2015, Orlando, United States, October 2015
14. Sobhani, A., Farooq, B.: Impact of smartphone distraction on pedestrians' crossing behaviour: an application of head-mounted immersive virtual reality. J. Transp. Res. Part F: Traffic Psychol. Behav. **58**, 228–241 (2018)
15. Sportillo, D., Paljic, A., Ojeda, L., Fuchs, P., Roussarie, V.: Light virtual reality systems for the training of conditionally automated vehicle drivers. In: IEEE Virtual Reality, Reutlingen, Germany, March 2018
16. Stoffregen, T.A., Smart Jr., L.J.: Postural instability precedes motion sickness. Brain Res. Bull. **47**(5), 437–448 (1998)

17. Stoffregen, T.A., Hettinger, L.J., Haas, M.W., Roe, M.M., Smart, L.J.: Postural instability and motion sickness in a fixed-base flight simulator. Hum. Factors **42**(3), 458–469 (2000)
18. Strojny, A., Strojny, P.: The immersion questionnaire – Polish adaptation and empirical verification of the scale. In: Homo Ludens, vol. 1, no. 6, pp. 171–185 (2014)
19. Tiiro, A.: Effect of visual realism on cybersickness in virtual reality. University of Oulu (2018)
20. Walch, M., et al.: Evaluating VR driving simulation from a player experience perspective, pp. 17–25 (2017)
21. Weidner, F., Hoesch, A., Poeschl, S., Broll, W.: Comparing VR and non-VR driving simulations: an experimental user study. In: 2017 IEEE Virtual Reality (VR), pp. 281–282. IEEE (2017)
22. Youngblut, C., Johnston, R., Nash, S., Wienclaw, R., Will, C.: Review of virtual environment interface technology. Institute for Defense Analyses (1996)

Evaluation of Exhaust Emissions in Real Driving Emissions Tests in Different Test Route Configurations

Wojciech Gis[1], Maciej Gis[1], and Jacek Pielecha[2](✉)

[1] Motor Transport Institute, Jagiellonska Street 80, 03-301 Warsaw, Poland
[2] Poznan University of Technology, Piotrowo Street 3, 60-965 Poznan, Poland
jacek.pielecha@put.poznan.pl

Abstract. The article presents an analysis of exhaust emissions in vehicles with gasoline engines according to the most recent research procedures in RDE tests. The compliance of test parameters (on chosen test routes) was evaluated against the requirements of the standard and dynamic parameters were determined characterizing the trips (among others – the product of speed and positive acceleration). Time ranges of respective trips were analyzed in coordinates – vehicle's speed vs acceleration; based on that a matrix was developed that allows for comparing trips not only based on averaged parameters (e.g. average speed, stoppage time, etc.) but also all conditions of the vehicle's operation during the road test. The comparison of those parameters for tests carried out along various test routes and the identification of discrepancies provided grounds for the next step of the evaluation in which exhaust emissions in respective test phases (urban, rural and motorway parts) were compared. The result is a methodology for determining exhaust emissions in road tests of various route configurations in which the exhaust emissions intensity matrix for the vehicle and basic trip parameters, e.g. speed profile, are known.

Keywords: Exhaust emissions · Real driving emissions · Gasoline engine

1 Introduction

The exhaust emission standards are defined for the purpose of controlling vehicle pollution emissions globally. In most regions also the limits of CO_2 emissions have been determined, which is directly associated with fuel consumption. The values of exhaust emissions are measured in laboratory conditions in the agreed approval test. This part of the vehicle approval process determines its "environmentally friendly features" and is identical for all passenger cars. The driving test corresponds to "the most likely" road conditions and the same tests conducted for all vehicles justify comparisons of the exhaust emission results between them [6, 7, 10]. However, presently more attention is given to road tests (which is reflected in the proposed EU regulations), known as Real Driving Emissions, performed with the use of portable test equipment Portable Emission Measurement System [1]. The most recent studies on the exhaust emissions of vehicles in traffic conditions performed with the use of portable measurement equipment reflect the actual environmentally-friendly status of the

M. Siergiejczyk and K. Krzykowska (Eds.): ISCT21 2019, AISC 1032, pp. 143–153, 2020.
https://doi.org/10.1007/978-3-030-27687-4_15

vehicles [9, 12, 15]. Much attention is given to the possibility of using those studies for calibrating drive units to reduce pollution emissions not only during the tests but also within the entire range of the engines' operation.

The authors of the publication [8] pointed that future RDE studies, presently simulated in various studies can bring an increased road emission of nitrogen oxides in vehicles. To limit the above risk, they recommended necessary changes in the software of the vehicles' controllers, claiming that such changes would prove successful in the case of vehicles equipped with gasoline engines. Vehicles equipped with Diesel engines will require expenditure to increase the effectiveness of exhaust processing outside the engine, owing to new methods of reducing NO_x content [13].

The same conclusions have been drawn by the authors of the article in [11] which RDE exhaust emissions were compared with the use of PEMS-type analyzers and COPERT programme. It was found that for the speed range of 20–120 km/h, the calculations in COPERT programme are approx. 10% greater for such parameters as fuel consumption and HC emission. Whereas regarding road emissions of NO_x – according to COPERT programme the values are smaller by ca. 30%. Comparisons of exhaust emissions in Euro 5 vehicles conducted in a laboratory on a chassis dynamometer in various trip tests also proved the results characterized previously.

As regards the accuracy of measurements in real driving conditions the end result depends on the operating conditions of the vehicle and the engine (among others the speed of the driving vehicles, road surface, the driver's predisposition and his/her driving style and other factors determining road traffic). Those conditions are unpredictable and can have significant impact on the results of pollution emissions measurement. Data presented inter alia in publications [14] prove that the thermal condition of the vehicle (engine), average speed, driving dynamics and road inclination have the greatest impact on the achieved results of pollutant emissions.

Starting from 2019 the approval process for the new type of passenger cars in the European Union comprises a procedure for the measurement of exhaust emissions in real traffic conditions. Regulation 2017/1151/EC [5] and 692/2008/EC [2] regarding RDE studies come as an answer to the results of studies concerning increased emissions of NO_x in vehicles equipped with gasoline engines, even though those vehicles complied with the permissible standards in laboratory conditions (Table 1).

Table 1. RDE test requirements in Europe (Type Approval, PCM) [3–5].

2015	2016	2017	2018	2019	2020	2021	2022
Euro 6b			Euro 6d-Temp		Euro 6d		
Research and concept phase			Conformity Factor (CF) $CF_{NOx} = 2.1$, $CF_{PN} = 1.5$		$CF_{NOx} = 1.5$, $CF_{PN} = 1.5$		

Those requirements do not specify the route of the road test for measuring exhaust emissions. The variability of a particular parameter (e.g. the share of respective route portions) should not have significant impact on the end results of road emissions. Those arguments were the primary objective of the article. The article discusses the evaluation of differences in road emissions of hazardous substances of two vehicles (A and B) with a gasoline engine of different ecological parameters (production year 2016 and 2019) and with different mileages. The tests were conducted on two test routes (X and Y, complying with RDE), characterizing with different parameters.

The purpose of the articles was to evaluate the differences in road emissions in vehicles in two test variants:

- Variant 1: determining road emissions in RDE tests for vehicle A on test route X; determining road emissions in RDE tests for vehicle B on test route Y.
- Variant 2: determining driving parameters (route Y) with the use of two-dimensional characteristics of the time share of the vehicle's operation in speed and acceleration coordinates and thereafter simulating the trip of vehicle A along route Y; determining driving parameters (route Y) with the use of two-dimensional characteristics of the time share of the vehicle's operation in speed and acceleration coordinates and thereafter simulating the trip of vehicle B along route X.

2 Methodology of Tests

2.1 Test Objects

The test objects were passenger cars featuring drive units characterising with parameters shown in Table 2. The vehicles were equipped with gasoline engines and compliant with Euro 6 and Euro 6d-Temp toxicity standards. Despite differences in the weight of the vehicles and engine displacement, their common feature comprised similar operating parameters (power and torque).

Table 2. Characteristics of the tested objects.

Parameter	Vehicle A	Vehicle B
Cylinder number, arrangement	4, in series	4, in series
Displacement [cm^3]	1991	1591
Emission standard	Euro 6	Euro 6d-Temp
Max. power [kW] at [rpm]	135/5500	132/5500
Max. torque [Nm] at [rpm]	300/1200–4000	265/1500–4500
Vehicle curb weight [kg]	1570	1465
Mileage [km]	60,000	200

2.2 Test Routes

The test route was chosen in such a way as to ensure that it conformed to the requirements of the European Commission laid down in the regulation, in particular taking into account its division into urban, rural and motorway portions. The maximum driving speed – despite considerable changes in the course of the route – differed between the trips by 3.2 km/h (the difference is 2.5% in relation to the lower speed).

2.3 Test Equipment

A portable Semtech DS analyzer was used for the measurement of exhaust emissions from vehicles. It allowed for measuring CO, HC and NO_x. In terms of benchmarking and quality control, zero-span checks were performed before and after each measurement. Linearization of the equipment was carried out every three months. Post-processing plausibility checks were made on all data, focusing on CO_2, to ensure that the data collected were realistic.

3 Analysing the Similarities of Test Routes

The comparison of the speeds on test routes points to differences in their passage (Fig. 1). Considerable variances were recorded for all compared parameters: accelerations: 36% (route X) and 28% (route Y), constant driving speed: 12% (route X) and 37% (route Y), braking: 35% (route X) and 23% (route Y), stoppage: 17% (route X) and 11% (route Y).

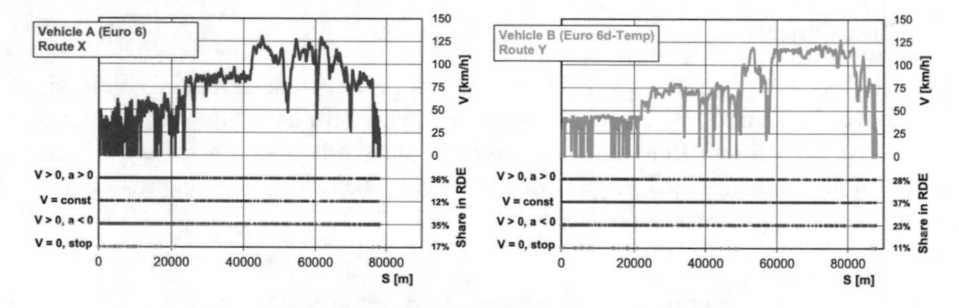

Fig. 1. Comparison of speeds on routes X and Y in vehicles A and B.

The average speed along route X was 45.8 km/h, and for route Y – 50.4 km/h. The most important phase was to verify the accuracy of road tests specified in the Regulations, in particular in terms of the duration of respective RDE test phases and their shares in the entire test, duration of the test, share of stops in the urban part and driving dynamics in respective portions of the RDE test. All parameters were verified with the use of relevant procedures and no deviations from the required values were found (Table 3).

Road tests characterize with high non-repeatability of the road conditions even if the formal requirements presented above are fulfilled. The definition of the performance correctness of road tests, described with relative positive acceleration or the product of speed and positive acceleration cannot be considered enough in proving the comparability of results achieved in road tests.

However, the possibility of comparing the results of exhaust emissions in road tests is correct in the case of a considerable similarity of trips during the road tests. For the comparison to be reliable an analysis was conducted regarding the similarity of the trips. For this purpose, characteristics were developed concerning the share – in terms of duration – of respective trip phases against the coordinates of the vehicle's speed and its acceleration – for vehicle A during road test along X route and vehicle B along Y route. The comparison shows (Fig. 2) there is a similarity between the trips.

Table 3. Validation of trip characteristics in accordance with EU RDE requirements.

Trip characteristics	Route X Vehicle A	Route Y Vehicle B	Requirement Route X/Route Y	Valid?
Urban distance [km]	25.9	29.9	>16	OK
Rural distance [km]	27.2	28.0	>16	OK
Motorway distance [km]	25.0	29.9	>16	OK
Total trip distance [km]	78.1	87.9	>48	OK
Urban distance share [%]	33.1	34.1	29–44	OK
Rural distance share [%]	34.9	31.9	33 ± 10	OK
Motorway distance share [%]	32.0	34.0	33 ± 10	OK
Urban stop time [%]	23.3	19.8	6–30	OK
Total trip duration [min]	99.2	104.7	90–120	OK
Cold start requirements				
Cooling temperature [°C]	63.0	67.0	<70	OK
Max vehicle speed [km/h]	49.7	41.9	<60	OK
Stop time [s]	91	36	<90	OK
Idling after ignition [s]	5	6	<15	OK
Overall trip dynamics				
Urban 95th percentile $V{\cdot}a_+$ $[m^2/s^3]$	10.8	12.9	<17.7/<18.2	OK
Rural 95th percentile $V{\cdot}a_+$ $[m^2/s^3]$	15.1	13.8	<24.9/<24.2	OK
Motorway 95th percentile $V{\cdot}a_+$ $[m^2/s^3]$	16.4	15.4	<27.1/<27.2	OK
Urban RPA $[m/s^2]$	0.21	0.14	>0.137/>0.131	OK
Rural RPA $[m/s^2]$	0.09	0.07	>0.047/>0.061	OK
Motorway RPA $[m/s^2]$	0.10	0.07	>0.025/>0.025	OK

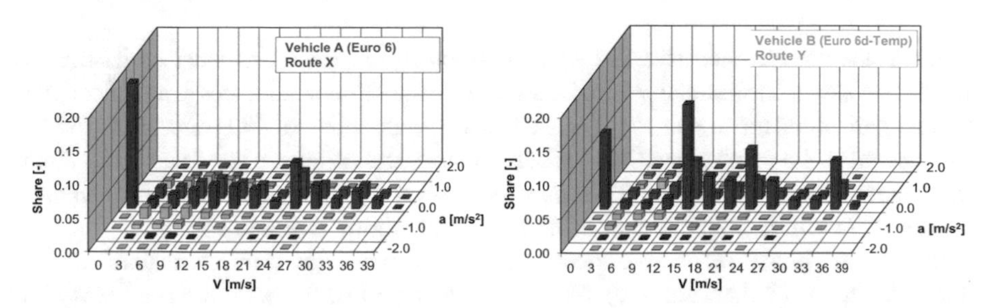

Fig. 2. Characteristics of the share of the vehicle's operating time against speed – acceleration coordinates for vehicles A and B along routes X and Y.

The comparison of the conditions in which the vehicles were driven can be made based on the two-dimensional characteristics of the vehicle's operation. For this purpose, relevant shares of the vehicle's driving time are compared based on two-dimensional features. In result there are two columns of data that may be used to determine a regression equation (y = ax + b), in which the coefficient of determination (R^2) provides a benchmark allowing for the trips to be compared. The compared data are similar if the slope is close to 1 (it is assumed that intercept is equal to 0). Such a comparison is presented in Fig. 3 which compares the entire trip (for vehicle A along route X and vehicle B along route Y), and the urban, rural and motorway portions. The achieved values of the coefficient of determination are different in different vehicles in the same phases of RDE tests, which means that the shares of the vehicles' operation in relevant conditions of speed and acceleration are not the same.

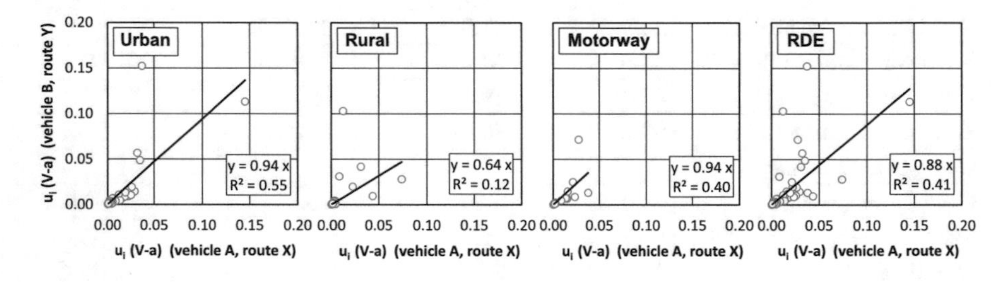

Fig. 3. Comparison of the consistency of trips (vehicle A along route X and vehicle B along route Y) with the use of line correlation of the shares of the operating time in V–a ranges.

The achieved results of the comparison point to different conditions of the vehicle's operation in different road tests even though in both cases the procedural requirements under the regulations concerning driving parameters and consistency of the driving dynamics were fulfilled. The comparison emphasises only the dissimilarity of the respective driving phases, but it does not affect the correctness in determining road emissions of hazardous substances.

4 Analysis of the Achieved Results of Exhaust Emissions

Based on the measurements of the concentration of hazardous exhaust emissions and the exhaust flows the intensity of emissions of respective substances was determined. The comparison of those values reveals comparable intensity of CO_2 emissions for both vehicles and at the same time different behaviour of the exhaust after-treatment systems for the remaining substances. This may be caused by smaller engine displacement, however with similar operating parameters of the engine maintained.

In the next phase respective measurements were assigned with trip features (U – urban, R – rural, M – motorway). Next, for the same features average road emissions were determined with the use of the window method, determined based on the identified road emission of carbon dioxide in the approval test. Other factors were considered too, regarding – among others – the tolerance range (±25% and ±50%) against the curve of carbon dioxide emission determined based on WLTC test phases. Results of road emissions were achieved in respective parts of RDE test and in the entire road test (Fig. 4).

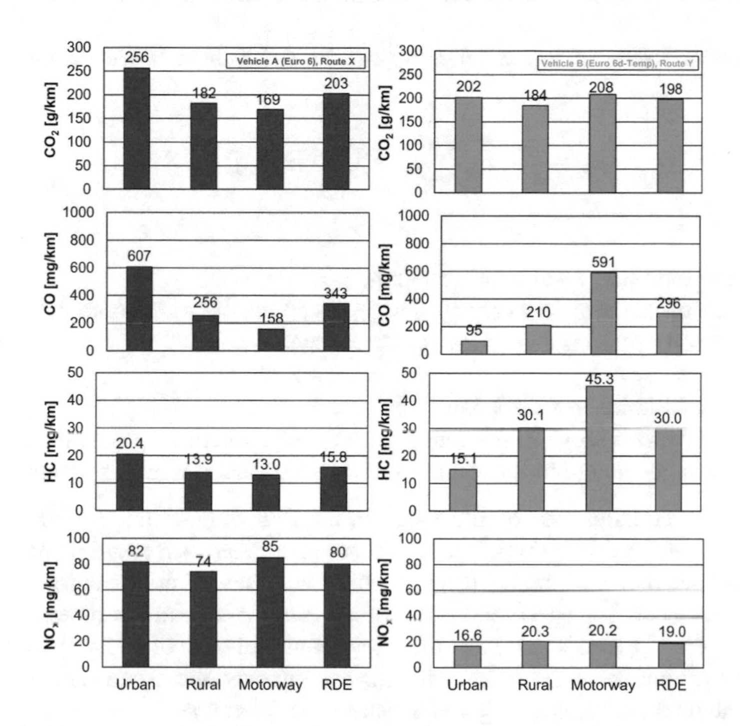

Fig. 4. Comparison of road emissions of respective hazardous substances for vehicles A and B along routes X and Y, taking into account respective urban, rural and motorway portions.

The comparison of those data shows that:

- road emission of CO_2 in the entire RDE test is comparable (a difference of 2.5%),
- road emission of CO is greater by 14% for the vehicle of a lower emission class,
- road emission of HC is twice greater for the vehicle with a higher emission class,
- road emission of NO_x is 4 times greater for the vehicle with a lower emission class.

The presented results of road emissions of all hazardous substances do not exceed the conformity factor (CF) which according to the legislation is equal to 1.5. This means that in the event of:

- road emission of carbon oxide for which the normative value is 1000 mg/km, the limit of 1500 mg/km has not been exceeded,
- road emission of hydrocarbons for which the normative value is 100 mg/km, the limit of 150 mg/km has not been exceeded,
- road emission of nitrogen oxides for which the normative value is 60 mg/km, the limit of 90 mg/km has not been exceeded.

Attention should be given to the fact that the permissible limits were not exceeded in any phase of RDE test and as an ecological result in the entire test.

5 Simulation of Exhaust Emissions in Other Test Routes

The results of RDEs in a particular trip can be determined based on the following relation:

$$b_k = \sum_{i=1}^{N} \sum_{j=1}^{M} u(i, j) \cdot E_k(i, j) \cdot t \cdot S^{-1} \quad (1)$$

where:

b_k – road emission k-substance, mg/km,
u – share of the vehicle's driving time in speed-acceleration coordinates,
E_k – intensity of emission of k-substance, mg/s,
t – test time, s,
S – linear distance travelled, km,
i – driving speed range in the range from 0 to N, where N is maximum value,
j – acceleration range in the range from 0 to M, where M is the maximum value.

The above equation can be illustrated in the following manner: with a particular characteristic of the driving time's share (in speed–acceleration coordinates) in the test and the characteristics of the intensity of emission of any of the hazardous substances (averaged in respective speed and acceleration ranges) the sum is determined of the products of driving time's share and the intensity of emission of a respective substance in relevant driving fields. Next, knowing the test time and the distance travelled by the vehicle, the road emission of a given substance is determined.

Hence, determining road emission for a particular vehicle along a different measurement route is achieved through changed emission characteristics of the vehicle ($E_{k(V,a)}$), whereas the parameters of the route itself remain unchanged (share of the driving time, test duration, test distance). This approach involves imprecision as the same driving windows are not filled in in the parametrised route and the characteristics of emission intensity are not identified. However, the assumption of the possibility of those simulations will reveal the imprecision of the results in the form of relative changes in road emissions of respective hazardous substances.

For the purpose of the abovementioned tasks the characteristics of the vehicles' driving time in speed–acceleration coordinates were used (presented earlier – Fig. 2) and respective exhaust emissions in those vehicles. The remaining parameters – duration and length of the measurement routes – are presented in Table 3.

The analysis of those data proves that the characteristics of CO_2 emissions are very similar. The greatest emission intensity occurs with the increasing speed of the vehicle and also within ranges of the greatest acceleration. Different are the results regarding the CO emissions – in the vehicle with a greater engine displacement they are the greatest in the low speed ranges and high acceleration, while in the vehicle with a smaller engine displacement – increased CO emissions occur when the driving speed and the acceleration are high. The characteristics of the HC emissions (unburned fuel) are similar to the characteristics of CO_2 emissions; however, these differ in the level of the achieved values. The intensity of NO_x emissions is approximately 5 times greater (for maximum driving speeds) in the vehicle of a lower emission class (Euro 6) compared to a vehicle of Euro 6d-Temp emission class.

Based on those data the values of road emissions were determined on routes along which road studies had not been performed, i.e.: for vehicle A – road emissions were determined of respective hazardous substances along route Y; for vehicle B – road emissions were determined of respective hazardous substances along route X.

Thereafter, the resulting values were compared. Figure 5 presents the relative differences with reference to real results.

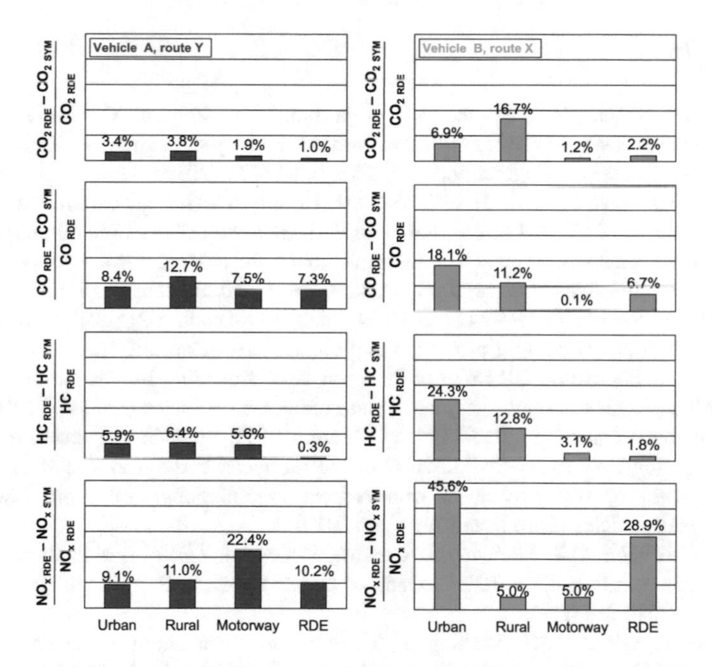

Fig. 5. Relative differences (absolute values) of road emissions of respective hazardous substances achieved in simulated trips of vehicle A along route Y and vehicle B along route X.

The comparison of the resulting data shows that the results of simulation along other measurement routes than those performed in reality do not differ much. The very small values of relative differences are for the final emissions in RDE tests a confirmation of the possibility of estimating road emissions in road tests, where only the driving speed parameters in any RDE test are known.

6 Summary

The article presents possibilities of evaluating exhaust emissions for different vehicles along different test routes, consistent with the requirements of RDE procedure. It has been found that despite differences in the speed profiles of tests, they are sufficient for evaluating the exhaust emissions in passenger cars. Verified has also been the possibility of estimating the emissions of hazardous substances along other test routes (compliant with RDE procedure), knowing only the speed profile and also the characteristics of intensity of respective exhaust emissions. It was proved that such approach is prone to a minor error (up to 10%), which does not affect the decision whether (or not) the permissible exhaust emission criterion has been fulfilled.

References

1. Clenci, A., Sălan, V., Niculescu, R., Iorga-Simăn, V., Zaharia, C.: Assessment of real driving emissions via portable emission measurement system. In: IOP Conference Series: Materials Science and Engineering, vol. 252, p. 012084 (2017)
2. Commission Regulation (EU) 692/2008 of 18 July 2008 implementing and amending Regulation (EC) 715/2007 of the European Parliament and of the Council on type-approval of motor vehicles with respect to emissions from light passenger and commercial vehicles (Euro 5 and Euro 6) and on access to vehicle repair and maintenance information (2008)
3. Commission Regulation 2016/427 of 10 March 2016 amending Regulation No 692/2008 as regards emissions from light passenger and commercial vehicles (Euro 6) (2016)
4. Commission Regulation 2016/646 of 20 April 2016 amending Regulation No 692/2008 as regards emissions from light passenger and commercial vehicles (Euro 6) (2016)
5. Commission Regulation 2017/1154 of 7 June 2017 amending Regulation 2017/1151 supplementing Regulation No 715/2007 of the European Parliament and of the Council on type-approval of motor vehicles with respect to emissions from light passenger and commercial vehicles (Euro 5 and Euro 6) (2017)
6. Gis, W., Pielecha, J., Waskiewicz, J., Gis, M., Menes, M.: Use of certain alternative fuels in road transport in Poland. In: IOP Conference Series: Materials Science and Engineering, vol. 148, p. 012040 (2018)
7. Korniski, T., Gierczak, C., Wallington, T.: Laboratory evaluation of the 2.5 inch diameter Semtech® exhaust flow meter with gasoline fueled vehicles. In: Sensors 4th Annual SUN Conference, Ann Arbor (2007)
8. Kousoulidou, M., Fontaras, G., Ntziachristos, L., Bonnel, P., Samaras, Z., Dilara, P.: Use of portable emissions measurement system (PEMS) for the development and validation of passenger car emission factors. Atmos. Environ. **64**, 329–338 (2013)
9. Kruczynski, S.W., Gis, W., Zin, D.: Effect of bioethanol addition to motor gasoline on selected engine operating parameters. Przem. Chem. **97**, 2185–2188 (2018)
10. Lijewski, P., Merkisz, J., Fuc, P., Ziolkowski, A., Rymaniak, L., Kusiak, W.: Fuel consumption and exhaust emissions in the process of mechanized timber extraction and transport. Eur. J. Forest Res. **136**, 153–160 (2017)
11. May, J., Favre, C., Bosteels, D.: Emissions from Euro 3 to Euro 6 light-duty vehicles equipped with a range of emissions control technologies. Association for Emissions Control by Catalyst, London (2013)

12. Pielecha, J., Andrych-Zalewska, M., Skobiej, K.: The impact of using an in-cylinder catalyst on the exhaust gas emission in real driving conditions tests of a diesel engine. In: IOP Conference Series: Materials Science and Engineering, vol. 421, p. 042064 (2018)
13. Pielecha, J., Magdziak, A., Brzezinski, L.: Nitrogen oxides emission evaluation for Euro 6 category vehicles equipped with combustion engines of different displacement volume. IOP Conference Series: Materials Science and Engineering, vol. 214, p. 012010 (2019)
14. Pielecha, J., Merkisz, J., Markowski, J., Jasinski, R.: Analysis of passenger car emission factors in RDE tests. In: E3S Web Conference, vol. 10, p. 00073 (2016)
15. Suarez-Bertoa, R., Astorga, C.: Impact of cold temperature on Euro 6 passenger car emissions. Environ. Pollut. **234**, 318–329 (2018)

Method of Planning the Work of Conductor Crews Taking into Account the Polish Conditions

Piotr Gołębiowski(✉)

Faculty of Transport, Warsaw University of Technology, 00662 Warsaw, Poland
pgolebiowski@wt.pw.edu.pl

Abstract. The aim of the article is to develop a method of planning the work of conductor crews, taking into account the Polish conditions - mainly in terms of compliance with working time regulations by employees. The paper describes the problems of scheduling conductor crews' work in Polish conditions. The mathematical model of assigning conductor crews to tasks in Polish conditions was developed based on Task Assignment. Moreover, an algorithm was developed for a method of assigning conductor crews to tasks using a block diagram. In this method it is assumed that a timetable is given, based on which tasks performed by the employees are then developed. For such tasks, personnel is assigned to service - train managers and, where required, conductors.

Keywords: Railway transport · Crew scheduling · Task assignment

1 Introduction

People play a very important role in rail transport. In the direct transport process, the most important function from the point of view of the railway operator is played by the train crews. A conductor crew can be defined as [18, 47, 48] an employee or crew of employees of a railway undertaking who operate a train on a specific section of its journey in the scope of technical and operational activities. A conductor crew may be the train manager himself or the train manager together with one or more conductors.

The aim of the article is to develop a method of planning the work of conductor crews taking into account the Polish conditions - mainly in terms of compliance with working time regulations by employees. In order to achieve that goal, a review of literature, both Polish and English, was carried out. The review covered manners of formulating the problem and methods of solving it. Next, the problems of planning conductor crews' work in Polish conditions were characterised. The ***MPDKPL*** (from Polish: ***M*** – model, ***P*** – planning, ***D*** – crews, ***K*** – conductor, ***PL*** – Poland) mathematical model of assigning conductor crews to tasks in Polish conditions was developed based on Task Assignment. Moreover, an algorithm was developed for a method of assigning conductor crews to tasks using a block diagram. It indicated data, decision variables, limitations together with boundary conditions and the indicator of the solution's quality assessment as elements of the optimisation task. Using a block diagram, an algorithm for assigning conductor crews to tasks was developed.

M. Siergiejczyk and K. Krzykowska (Eds.): ISCT21 2019, AISC 1032, pp. 154–163, 2020.
https://doi.org/10.1007/978-3-030-27687-4_16

2 Problems of Planning the Work of Conductor Crews in Literature

In Polish literature very little attention is paid to the issue of organizing the work of conductor crews. Apart from official documents from railway undertakings [18, 47, 48] there are no specific guidelines on how to organise the work of the train crew. The only one is the Labour Code Act [53], which applies to most social groups in Poland. In scientific publications it can be found that the problem of planning the work of conductor crews is related to the organization of railway traffic [23, 41]. The paper [43] analyses the way of estimating the degree of losses incurred by railway undertakings as a result of undesirable events in railway traffic, including losses resulting from the work of conductor crews. The article [55] analyses the operational efficiency of passenger rolling stock also from the point of view of problems related to the organisation of work of conductor crews.

In English literature there are many publications related to the subject of conductor crew work planning, formulated and solved with the use of mathematical modelling [7, 12–14, 17, 19, 50–52, 54]. Many variations of the problem were considered, among others, problems with the frequency of train operation by conductors [28] and problems related to the tactical level of timetable development [49].

The literature also describes methods of solving the problem of planning the work of the conductor crews. Algorithms used to solve the problem [6, 11, 15, 21] were characterized and examples were presented of solutions for different management boards, e.g. for Italian national railways [10, 11], Dutch railways [1–3, 22, 40], German railways [5], Taiwanese railways [29], Swedish railways [4], Chinese railways [16] and European rail freight companies [35, 39] - e.g. DB Schenker [37] (these publications essentially refer to the problem of traction crew scheduling [29]). In particular, the literature describes the application of the column generation method for the large scale problem [2, 35, 37, 46], Branch and Bound method [54], iterative division [1], pairing [5], dynamic programming [24], formic algorithm [29], division of sets [45] and graphs [36], hybrid solutions in the form of a combined algorithm of simulated annealing with constructive heuristics [26], Taboo Search method and total number programming [25], as well as relaxation of Lagrange and column generation [2]. An interesting concept seems to be the use of the fuzzy set method to solve the problem of planning the work of conductor crews [27]. The concepts of applications dedicated to solving problems [16, 37, 38, 42] were also presented.

On the basis of the literature review it should be concluded that there are no publications concerning the planning of the work of conductor crews relevant to the Polish conditions - taking into account the specific operation of the Polish transport system.

3 Problems of Planning the Work of Conductor Crews in Polish Conditions

A single train is operated in three stages:

– acceptance of the train (starting work on the train and preparing it for the road),

- driving the train (consisting of two phases: driving and stopping [8]),
- the handover of the train (completion of work on the train).

Adequate time must be devoted to the above mentioned activities. The basic time of a train service is called a turn. It may be represented by the formula:

$$\forall poc \in \boldsymbol{POC} \quad T_t^{poc} = t_{opr}^{poc} + t_p^{poc} + t_j^{poc} + t_z^{poc} + t_{opo}^{poc} \tag{1}$$

where:

- T_t^{poc} – duration of train-service turn ($poc \in \boldsymbol{POC}$, $\boldsymbol{POC}$ - set of serviced trains),
- t_{opr}^{poc} – waiting time for the *poc* train to be accepted,
- t_p^{poc} – *poc* train acceptance time,
- t_j^{poc} – *poc* train driving time,
- t_z^{poc} – *poc* train handover time,
- t_{opo}^{poc} – waiting time after handing over the *poc* train,

It should be noted that the turn also includes the empty runs undertaken by the conductor crew as passengers for the purpose of accepting the first train. It is impossible to distinguish between the time of accepting the train and the time of handing over the train in that situation.

When structuring the service for conductor crews, it needs to be taken into account that it must begin and end in the same place, i.e. in the hub of the conductor crews. The members of conductor crews must be familiar with the local conditions on the routes they operate (the so-called route knowledge) and at the points of operation they operate.

In Polish legislation there are no dedicated laws setting out the rules concerning the working time of conductor crews. There is only the Labour Code Act [53], which applies to most social groups in Poland.

4 Mathematical Model of Planning the Work of Conductor Crews in Polish Conditions

The ***MPDKPL*** model of assigning conductor crews to tasks in Polish conditions can be expressed in the following form:

$$\boldsymbol{MPDKPL} = \langle \boldsymbol{P}, \boldsymbol{RJ}, \boldsymbol{Z}, \boldsymbol{PRZ} \rangle \tag{2}$$

where: ***P*** - set of employees, ***RJ*** - train timetable, ***Z*** - set of tasks for employees, ***PRZ*** - allocation of employees to perform tasks.

In order to search for train sequences that constitute a service task and to assign crew members to tasks using the Task Assignment formula, it is necessary to formulate an optimisation task consisting of the following parts:

I. Identification of Data

- $\boldsymbol{P} = \{1, \ldots, p, \ldots, P\}$ - a set of employee numbers, where p is the employee number and P is the number of employees,

- $a(p)$ - function of employee with number p, if $a(p) = 1$ then employee with number p is train manager and if $a(p) = 2$ then employee with number p is conductor,
- $\boldsymbol{T} = \{1, \ldots, t, \ldots, T\}$ - a set of numbers of types of schedule days (e.g. type 1: weekdays from Monday to Friday except holidays), where t is the number of type of schedule day and T is the number of types of schedule days,
- $\boldsymbol{RJ}(t) = \{1, \ldots, poc(t), \ldots, POC(t)\}$ - timetable for schedule day type t, where $poc(t)$ - train number for schedule day type t and $POC(t)$ - number of trains for day type t,
- $b(poc(t))$ - parameter defining whether a conductor should be allocated to a given train, if so $b(poc(t)) = 1$, otherwise $b(poc(t)) = 0$,
- $c(poc(t))$ - parameter defining the number of conductors assigned to operate a train $poc(t)$, $c(poc(t)) \in \mathbb{N}$,
- $d(poc(t))$ - parameter defining whether train is equipped with devices for closing door together with an audible signal, if so $d(poc(t)) = 1$, otherwise $d(poc(t)) = 0$,
- $\boldsymbol{ERJ}(t) = \{(po(po(t)), ko(ko(t)))\}$ - a set of train routes of individual trains $poc(t)$ for a schedule day type t, where $po(poc(t))$ - beginning of the train route $poc(t)$, and $ko(poc(t))$ - end of train route $poc(t)$,
- $e(p, poc(t))$ - parameter specifying whether the *p-number* worker has route knowledge to operate train $poc(t)$ - if so, $e(p, poc(t)) = 1$, otherwise $e(p, poc(t)) = 0$,
- ord - length of the settlement period, not longer than 4 months, expressed in days, $ord \in \boldsymbol{ORD}$,
- $t1(p, ord)$ - working time of employee with number p in the settlement period ord,
- ort - length of the settlement period, not longer than 4 months, expressed in weeks, $ort \in \boldsymbol{ORT}$,
- $t2(p, ort)$ - working time of the employee with number p in settlement period ort,
- $t(\mathbf{Z}_n)$ - duration of task n ($\mathbf{Z}_n$ is a decision variable),
- $lsw(ord)$ - number of public holidays (including Sundays) in settlement period ord,
- $ldw(p, ord)$ - number of employee with number p days off in settlement period ord,
- $k(p, z)$ - cost of the task performed by the employee p.

II. Identification of Decision Variables

- vector of decision variables $\mathbf{Z}$ - process of building tasks based on available trains - assignment of trains to specific tasks,

$$\mathbf{Z} = [\mathbf{Z}_n] \tag{3}$$

$\mathbf{Z}_n$ – decision variable interpreting the sequence Z_n of the trains $poc^m(t)$ serviced in a given task (n is the task number and N is a number of tasks, $n \in \mathbb{N}$), where m is the consecutive number of the train comprising the task number n, and M is the number of trains comprising the task, $m \in \mathbb{N}$:

$$\mathbf{Z}_n = \langle poc^1(t), \ldots, poc^m(t), \ldots, poc^M(t) \rangle \tag{4}$$

– decision variables matrix **X**:

$$\mathbf{X} = [x(p, z)] \tag{5}$$

$x(p, z)$ – decision variable interpreting the assignment of the employee with the number p to the task with the number z, $x(p, z) \in \{0,1\}$, $x(p, z) = 1$ - if the employee with the number p was assigned to the task with the number z, otherwise $x(p, z) = 0$.

III. Identification of Limitations

– per type of decision variable $x(p, z)$ to take values of 0 or 1,
– only one task z can be assigned to each employee p,
– only one employee p can be assigned to each task z,
– for each train (task z) an employee p must be assigned to act as the train manager ($a(p) = 1$),
– for selected trains (tasks z) an employee p needs to be assigned to act as the conductor ($b(poc(t)) = 1$),
– for selected trains (tasks z) an appropriate number of employees p acting as conductors should be assigned ($c(poc(t))$ and $d(poc(t))$),
– the service must begin and end at the same place,
– at least one employee p assigned to a specific task should have knowledge of the route ($e(p, poc(t))$),
– the average daily working time of an employee p in a settlement period *ord* should not exceed 8 h,
– the average weekly working time of an employee p in a settlement period *ord* should not exceed 40 h,
– the average weekly working time of an employee p taking into account overtime in a settlement period *ord* should not exceed 48 h,
– to maintain the daily rest period for each employee p,
– to maintain the weekly rest period for each employee p,
– to maintain an adequate number of days off in the settlement period for each employee p,
– not to exceed the maximum daily working time of an employee p working under in an equivalent working arrangement,
– to maintain every fourth Sunday off for every employee.

IV. Identification of the Target Function: minimizing the costs of the conductor crew system, which can be described as the sum of the products of the values of decision variables interpreting the assignment of the employee to the task and cost $k(p, z)$ of the employee's p work related to the task z:

$$F(\mathbf{X}) = \sum_{p \in \boldsymbol{P}} \sum_{z \in \boldsymbol{Z}} x(p, z) \cdot k(p, z) \longrightarrow \min \tag{6}$$

5 Method of Planning Work of Conductor Crews in Polish Conditions

The proposed method of planning work of conductor crews in Polish conditions is presented in Fig. 1.

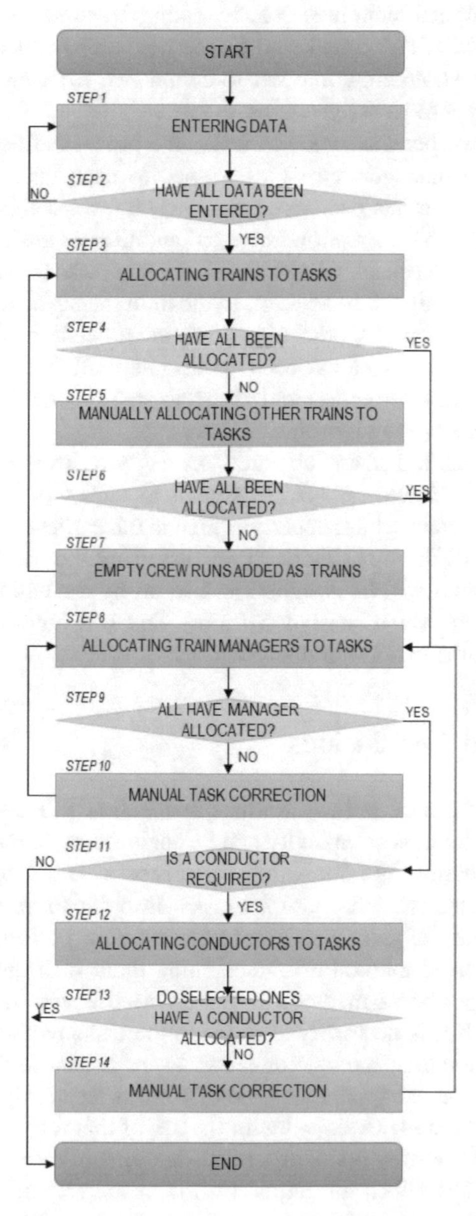

Fig. 1. Method of planning work of conductor crews in Polish conditions (source: own work)

The planning of conductor crews' work can be presented in 14 steps. In step one, all the necessary data must be entered. Then ensure that all data have been entered. If not, return to step 1, otherwise go to step 3 where the assignment of trains to tasks takes place. In that step, a sequence of trains which can be operated by one person is selected from the train set. When assigning, the restrictions arising from labour law must be observed. A selected optimization algorithm should be used for this purpose.

It may happen that not all trains are assigned to particular tasks. Check this in step 4. If you have not assigned them in step 5, try adding them to the existing tasks. If this has not been successful (check under step 6) add new trains that will symbolise empty runs undertaken by the train crew in order to commence working on the train (step 7). Next, go back to step 3 to re-assign trains to tasks.

Once all trains have been assigned to tasks, the process of assigning staff to tasks must begin. The train manager must be assigned to each task first (step 8). Step 9 checks whether a manager has been assigned to each task. If this is not the case, you should make a manual task correction (step 10) and then re-assign managers to tasks.

Otherwise, you need to check whether there are any tasks to which a conductor was supposed to have been assigned in addition to the train manager. If yes (step 11), assign conductors to the tasks (step 12), otherwise finish the procedure. Then make sure that a conductor has been assigned to all selected trains (step 13). If this is not the case, you should repeat manual task correction and return to step 8, where you assign managers to tasks again. Otherwise, the algorithm is complete.

The order in which trains are operated on a given task will be determined as follows. All trains that have not been assigned to task will be placed in a set of unassigned trains. The train which departure time is the earliest will be taken from this set. For its, a train will be found that can be served by the same employee after the route. Then another train will be assigned and so on by the end of the day. Assigned trains are placed in the set of assigned trains. The bee algorithm will be used to determine the value of decision variables $x(p, z)$.

6 Summary and Conclusions

To summarise, the issues of railway traffic organization [30–34, 44] are a complex decision-making problem. Its complexity can be analysed by breaking down individual decision-making problems into individual elements. One of such problems is the assignment of conductor crews to perform tasks. In order to be enable that, it is first necessary to determine the tasks that employees have to perform. This can be done directly - with the a'priori method or by searching them with the use of optimization methods - with the aposteriori method. In this article it is assumed that a timetable is given, based on which it is necessary to develop the tasks performed by employees.

The problem of assigning conductor crews to tasks can be formulated using different types of mathematical apparatus. One of them is task assignment. In this article the mathematical model was developed with the use of that formula. This allowed for a precise mapping of the railway operator's needs in terms of train service, as well as the railway operator's capabilities in terms of the resources necessary to carry out scheduled tasks.

The developed method of assigning conductor crews to tasks in Polish conditions takes into account the specific Polish working time regulations according to the Labour Code. This is extremely important in the scheduling of work, as it ensures that all applicable standards in that area are met. In addition, it is highly likely that the problem of overwork of staff responsible for the safety of travellers on board can then be avoided. Worker fatigue could pose a risk to human life and health.

References

1. Abbink, E.: Solving large scale crew scheduling problems by using iterative partitioning. Econometric Institute Report EI2008-03, pp. 1–15 (2008)
2. Abbink, E., Albino, L., Dollevoet, T., Huisman, D., Roussado, J., Saldanha, R.L.: Solving large scale crew scheduling problems in practice. Public Transp. **3**(2), 149–164 (2011)
3. Abbink, E., Fischetti, M., Kroon, L., Timmer, G., Vromans, M.: Reinventing crew scheduling at Netherlands Railways. Interfaces **35**(5), 393–401 (2005)
4. Alefragis, P., Sanders, P., Takkula, T., Wedelin, D.: Parallel integer optimization for crew scheduling. Ann. Oper. Res. **99**(1–4), 141–166 (2000)
5. Bengtsson, L., Galia, R., Gustafsson, T., Hjorring, C., Kohl, N.: Railway crew pairing optimization. In: Algorithmic Methods for Railway Optimization, pp. 126–144. Springer, Heidelberg (2007)
6. Booler, J.: A method for solving crew scheduling problems. J. Oper. Res. Soc. **26**(1), 55–62 (1975)
7. Brucker, P., Qu, R., Burke, E.: Personnel scheduling: models and complexity. Eur. J. Oper. Res. **210**(3), 467–473 (2011)
8. Burdzik, R., Nowak, B., Rozmus, J., Słowiński, P., Pankiewicz, J.: Safety in the railway industry. Arch. Transp. **44**(4), 15–24 (2017)
9. Cai, Z., Yan, J.: Analysis of residents' travel characteristics along Beijing rail transit line based on binary choice model. Arch. Transp. **47**(3), 19–27 (2018)
10. Caprara, A., Fischetti, M., Guida, P., Toth, P., Vigo, D.: Solution of large-scale railway crew planning problems: the Italian experience. In: Computer-Aided Transit Scheduling, pp. 1–18. Springer, Heidelberg (1999)
11. Caprara, A., Fischetti, M., Toth, P., Vigo, D., Guida, P.: Algorithms for railway crew management. Math. Program. **79**(1–3), 125–141 (1997)
12. Caprara, A., Kroon, L., Toth, P.: Optimization problems in passenger railway systems. Wiley Encyclopedia of Operations Research and Management Science (2010)
13. Caprara, A., Kroon, L., Monaci, M., Peeters, M., Toth, P.: Passenger railway optimization. Handb. Oper. Res. Manag. Sci. **14**, 129–187 (2007)
14. Caprara, A., Monaci, M., Toth, P.: A global method for crew planning in railway applications. In: Computer-Aided Scheduling of Public Transport, pp. 17–36. Springer, Heidelberg (2001)
15. Caprara, A., Monaci, M., Toth, P.: Models and algorithms for a staff scheduling problem. Math. Program. **98**(1–3), 445–476 (2003)
16. Chu, S., Chan, E.: Crew scheduling of light rail transit in Hong Kong: from modeling to implementation. Comput. Oper. Res. **25**(11), 887–894 (1998)
17. Cordeau, J., Toth, P., Vigo, D.: A survey of optimization models for train routing and scheduling. Transp. Sci. **32**(4), 380–404 (1998)
18. Dyrekcja Generalna PKP: H-21 – Instrukcja pracy drużyn konduktorskich w Przedsiębiorstwie Państwowym "PKP". Warszawa (1998). http://kolej.krb.com.pl/h21/h21.htm

19. Ernst, A., Jiang, H., Krishnamoorthy, M., Sier, D.: Staff scheduling and rostering: a review of applications, methods and models. Eur. J. Oper. Res. **153**(1), 3–27 (2004)
20. Ernst, A., Jiang, H., Krishnamoorthy, M., Owens, B., Sier, D.: An annotated bibliography of personnel scheduling and rostering. Ann. Oper. Res. **127**(1–4), 21–144 (2004)
21. Ernst, A., Jiang, H., Krishnamoorthy, M., Nott, H., Sier, D.: Rail crew scheduling and rostering optimization algorithms. In: Computer-Aided Scheduling of Public Transport, pp. 53–71. Springer, Heidelberg (2001)
22. Freling, R., Lentink, R., Odijk, M.: Scheduling train crews: a case study for the Dutch railways. In: Computer-Aided Scheduling of Public Transport, pp. 153–165. Springer, Heidelberg (2001)
23. Gołębiowski, P., Jacyna, M.: Wybrane problemy planowania ruchu kolejowego. Prace Naukowe Politechniki Warszawskiej. Transport **97**, 123–133 (2013)
24. Gorman, M.F., Sarrafzadeh, M.: An application of dynamic programming to crew balancing at Burlington Northern Santa Fe Railway. Int. J. Serv. Technol. Manag. **1**(2–3), 174–187 (2000)
25. Guillermo, C., José, M.: Hybrid algorithm of tabu search and integer programming for the railway crew scheduling problem. In: 2009 Asia-Pacific Conference on Computational Intelligence and Industrial Applications (PACIIA), vol. 2, pp. 413–416. IEEE (2009)
26. Hanafi, R., Kozan, E.: A hybrid constructive heuristic and simulated annealing for railway crew scheduling. Comput. Ind. Eng. **70**, 11–19 (2014)
27. Hao, Z.: Research on optimization of crew scheduling. J. China Railway Soc. 4 (1998)
28. Hoffmann, K., Buscher, U., Neufeld, J.S., Tamke, F.: Solving practical railway crew scheduling problems with attendance rates. Bus. Inf. Syst. Eng. **59**(3), 147–159 (2017)
29. Huang, S., Yang, T., Wang, R.: Ant colony optimization for railway driver crew scheduling: from modeling to implementation. J. Chin. Inst. Ind. Eng. **28**(6), 437–449 (2011)
30. Jacyna, M., Gołębiowski, P., Krześniak, M.: Some aspects of heuristic algorithms and their application in decision support tools for freight railway traffic organization. Sci. J. Sil. Univ. Technol. Ser. Transp. **96**, 59–69 (2017)
31. Jacyna, M., Gołębiowski, P., Urbaniak, M.: Multi-option model of railway traffic organization including the energy recuperation. In: Communications in Computer and Information Science: Tools of Transport Telematics, pp. 199–210. Springer, Heidelberg (2016)
32. Jacyna, M., Wasiak, M., Lewczuk, K., Chamier-Gliszczyński, N., Dąbrowski, T.: Decision problems in developing proecological transport system. Annu. Set Env. Prot. **20**, 1007–1025 (2018)
33. Jacyna-Gołda, I., Żak, J., Gołębiowski, P.: Models of traffic flow distribution for various scenarios of the development of proecological transport system. Arch. Transp. **32**(4), 17–28 (2014)
34. Jacyna-Gołda, I., Gołębiowski, P., Izdebski, M., Kłodawski, M., Jachimowski, R., Szczepański, E.: The evaluation of the sustainable transport system development with the scenario analyses procedure. J. VibroEng. **19**(7), 5627–5638 (2017)
35. Jütte, S., Thonemann, U.: Divide-and-price: a decomposition algorithm for solving large railway crew scheduling problems. Eur. J. Oper. Res. **219**(2), 214–223 (2012)
36. Jütte, S., Thonemann, U.: A graph partitioning strategy for solving large-scale crew scheduling problems. OR Spectr. **37**(1), 137–170 (2015)
37. Jütte, S., Albers, M., Thonemann, U., Haase, K.: Optimizing railway crew scheduling at DB Schenker. Interfaces **41**(2), 109–122 (2011)
38. Kisielewski, P.: The system of IT support for logistics in the rail transport. Arch. Transp. **40**(4), 39–50 (2016)

39. Kozachenko, D., Vernigora, R., Kuznetsov, V., Lohvinova, N., Rustamov, R., Papahov, A.: Resource-saving technologies of railway transportation of grain freights for export. Arch. Transp. **45**(1), 63–74 (2018)
40. Kroon, L., Fischetti, M.: Crew scheduling for Netherlands railways "destination: customer". In: Computer-Aided Scheduling of Public Transport, pp. 181–201. Springer, Heidelberg (2001)
41. Kur, H.: Automatyzacja opracowywania rozkładów jazdy pociągów. Problemy Kolejnictwa **153**, 7–21 (2011)
42. Kwan, R.: Case studies of successful train crew scheduling optimisation. J. Sched. **14**(5), 423–434 (2011)
43. Kwaśnikowski, J., Gill, A., Gramza, G.: Szacowanie stopnia strat ponoszonych przez przewoźników kolejowych w wyniku zdarzeń niepożądanych w ruchu kolejowym. TTS Technika Transportu Szynowego **18**, 63–65 (2011)
44. Mikulski, J., Gorzelak, K.: Conception of modernization of a line section example in the context of a fast railway connect. Arch. Transp. **44**(4), 47–54 (2017)
45. Mingozzi, A., Boschetti, M., Ricciardelli, S., Bianco, L.: A set partitioning approach to the crew scheduling problem. Oper. Res. **47**(6), 873–888 (1999)
46. Nishi, T., Muroi, Y., Inuiguchi, M.: Column generation with dual inequalities for railway crew scheduling problems. Public Transp. **3**(1), 25–42 (2011)
47. PKP Intercity: Br-21 (H-21) Instrukcja dla zespołu drużyn konduktorskich w zakresie obsługi pociągów pasażerskich uruchamianych przez "PKP Intercity" spółka z o.o. Warszawa (2006). https://docplayer.pl/14029390-Pkp-intercity-spolka-z-o-o-br-21-h-21.html
48. PKP Przewozy Regionalne: Pr-1 (H-21) Instrukcja o technice i organizacji pracy drużyn konduktorskich w pociągach pasażerskich. Poznań (2008). https://docplayer.pl/27430867-Pr-1-h-21-instrukcja-o-technice-i-organizacji-pracy-drusyn-konduktorskich-w-pociagach-pasaserskich.html
49. Şahin, G., Yüceoğlu, B.: Tactical crew planning in railways. Transp. Res. Part E: Logist. Transp. Rev. **47**(6), 1221–1243 (2011)
50. Strupchanska, A., Penicka, M., Bjørner, D.: Railway staff rostering. In: FORMS2003: Symposium on Formal Methods for Railway Operation and Control Systems, pp. 15–16 (2003)
51. Sysyn, M., Gerber, U., Kovalchuk, V., Nabochenko, O.: The complex phenomenological model for prediction of inhomogeneous deformations of railway ballast layer after tamping works. Arch. Transp. **47**(3), 91–107 (2018)
52. Tian, Z., Songa, Q.: Modeling and algorithms of the crew scheduling problem on high-speed railway lines. Procedia Soc. Behav. Sci. **96**, 1443–1452 (2013)
53. Ustawa z dnia 26 czerwca 1974 r. Kodeks pracy. Dz.U. 1974 Nr 24 poz. 141
54. Wang, Y., Liu, J., Miao, J.: Modeling and solving the crew scheduling problem of passenger dedicated line. J. China Railway Soc. **1**(004) (2009)
55. Wiśnicki, B., Chybowski, L., Krukowski, D.: Analiza efektywności eksploatacyjnej taboru pasażerskiego. Systemy Wspomagania w Inżynierii Produkcji (2014)

Analysis of Lean Angle Influence on Three Wheeled Vehicle Steerability Characteristics

Witold Grzegożek, Bartosz Zagól, and Adam Kot(✉)

Cracow University of Technology, Al. Jana Pawla II 37, 31-864 Cracow, Poland
witek@mech.pk.edu.pl, bartoszzagol@gmail.com,
adam.kot@pk.edu.pl

Abstract. Despite the many opportunities offered by large cities, they have also some disadvantages. One of the main problems is the enormous pollution of the environment which is caused, in large extent by the number of vehicles. From this point of view the vehicles should be very economical: using three wheels instead of four can significantly reduce the weight of the vehicle, improving its energy consumption. Moreover, a tricycle vehicle is a viable alternative to four-wheeled and two-wheeled vehicles, combining positive aspects of the car, such as comfort and safety, as well as the dynamics and dimensions of the motorcycle. To prevent the vehicle from rollover, it should lean like a motorcycle. Motorized tilting vehicles have been studied and developed since the beginning of 1950s. This article presents the possibilities offered by the angle of lean during cornering. The example results of the initial tests of the own prototype tilt vehicle are presented. The conducted tests show clearly that a vehicle body lean angle influences significantly on the steering characteristics.

Keywords: Steerability characteristics · Three wheeled vehicle · Lean angle influence

1 Introduction

Due to a sharp increase of environmental pollution during last years, there is an increasing demand for producing energy efficient passenger cars. In cities, a two wheeled vehicle, generally seems to be a very good solution taking into an account its low weight, dynamic driving performance and dimensions. On the other hand, in this model the occupant is exposed to environmental conditions, and moreover the vehicle's safety factor is significantly lower. Furthermore, a motorcycle driving requires special skills due to its stability. Three wheeled vehicle is viable alternative to four and two wheelers for personal mobility. It combines positive aspects of a car, such as comfort and safety and motorcycle's, such as dynamics and dimensions. Since the three wheeled vehicle for urban transportation should be narrow, it has a higher probability of tip over during cornering. To prevent the rollover the vehicle should lean like a motorcycle [1]. Narrow tilting three wheeled (NTTW) vehicle with electric drive appears as highly suitable means of transportation in congested cities.

Motorized tilting vehicles have been studied and developed since the beginning of 1950s. The pioneering prototype was proposed by Ernst Neumann [10]. The F300 Life

M. Siergiejczyk and K. Krzykowska (Eds.): ISCT21 2019, AISC 1032, pp. 164–172, 2020.
https://doi.org/10.1007/978-3-030-27687-4_17

Jet [10], Clever [2, 11], Carver One [10], TTW one [12] and Toyota i-road [13] are the recent projects of tilting vehicles. There are two types of three wheeled vehicles namely - the tadpole and delta configuration, the former one has two wheels at the front axle and one at the rear whereas the later one has one wheel at the front and two at the rear. The F-300 Life Jet is characterized by two-wheeled front axle and a single rear axle. The front suspension mechanism allows the vehicle to lean up to 30° while maintaining the wheels almost parallel to the body. The Carver One (Fig. 1) and the Clever are characterized by a single front wheel that tilts with the main body and by a non-tilting two wheel rear axle. These vehicles have a conventional automotive steering line and are equipped with an active tilt control system. Due to a conventional steering line the active direct tilt control (DTC) is used in these vehicles, Recently narrow tilting vehicles are equipped with the steering tilt control (STC) together with DTC [3] what was suggested by Karnopp [4], [Toyota i-road, TTW one]. Steering tilt control (STC) uses steering to control the tilt as it is done in motorcycle. If STC driver indicates his desire to turn by steering the wheel, the control system converts the steering wheel angle into the desired lean angle and steers the wheels to lean the vehicle and to execute the turn corresponding to the driver's steering input [3]. Toyota i-road [13] (Fig. 2) steering into a corner uses the system, that incorporates rear wheel steering, a lean actuator motor and gearing mounted above the front suspension, automatically moves the front wheels up and down in opposing directions. The vehicles mentioned above are tilted by special hydraulic or electric actuators. On the other hand, the Yamaha Tricity [14] (Fig. 4) and Piaggio MP3 [15] (Fig. 3) are equipped with special kind of front suspension, which allows the wheels to lean together with the main body, up to 40°. The lean angle is controlled, in the same way as in a motorcycle, by the rider. The limited front wheel track allows these vehicles to be driven as a conventional motorcycle but the comfort and safety factors are low and advantages of these vehicle in city traffic are not prevailing. The Sway [16] tilting scooter has the front suspension with a larger track than the Piaggio MP3 or Yamaha Tricity to prevent falling at low speeds or stand still when the leaning lock is not used. The lean angle of scooter body is controlled by the operator's legs.

Fig. 1. Carver One

Fig. 2. Toyota i-road

Fig. 3. Piaggio MP3

Fig. 4. Yamaha Tricity

The stability of a narrow three wheeled vehicle should be matched with that of a standard four-wheeled vehicle [5]. It is clear that leaning of the vehicle in a turn under lateral acceleration should maintain a safe ride in a curve. However the maximum lean angle is limited by the vehicle design and due to this limitation the rollover stability in any driving condition is not ensured. Moreover, the lean angle of front wheels is very often the same as vehicle body. The vehicle steerability characteristics simultaneously depends on steer and lean angle. As mentioned above the vehicle lean angle is limited. It can be the reason of steerability characteristics sharp changes. The focus of this paper is to analyze the influence of lean angle on steerability characteristics of a three wheeled tilting vehicle.

2 Vehicles Description

The prototype was designed and built in the Institute of Automobiles and Internal Combustion Engines of the Cracow University of Technology. It is an electric three-wheeled tilting vehicle characterised by two-wheeled front axle and a single rear wheel (tadpole arrangement). The wishbone mechanism of front suspension allows the vehicle to lean up to 15° during maintaining the wheels almost parallel to the body. The elliptic leaf spring is adopted as the suspension resilient element. The leaf spring is connected to the body by a pivot so vehicle tilt motion is fully independent from the suspension. The prototype has a conventional automotive steering line on the front axle. The vehicle lean angle is controlled by the driver's body balance. The rear suspension is constructed as a strut and the rear wheel tilts with the main body in the same way as in a motorcycle. The vehicle is powered by 1 kW hub motor in the rear wheel. Figure 5 presents the CAD model of the main frame and the view of the prototype. The primary technical specification is presented in Table 1.

For the research purposes the vehicle tilt locking mechanism was assembled so that there is a possibility to determine different lean angles.

Fig. 5. Test vehicle – CAD model and test prototype

Table 1. Technical specification of test vehicle (values in brackets correspond to vehicle with driver)

Vehicle mass	104 (186) kg
Front axle load	64 (104) kg
Rear axle load	40 (82) kg
Wheelbase	1209 mm
Front axle track of wheels	976 mm
Wheel radius	230 mm
Maximal vehicle lean angle	15°
Distance between front axle and CG	500 (790) mm
Height of CG position	324 (499) mm
Moment of inertia about vertical axis	35 (43) kgm^2
Maximal vehicle speed	40 m/h

3 Measurement Apparatus

The angle of rotation of the steering wheel was measured using the Honeywell RTY270HVNAX sensor (number 2 on Fig. 6). The sensor uses an integrated Hall effect detector circuit to detect rotary motion in the range from 0 to 270°. Rotation changes the position of the magnet relative to the integrated circuit, which results in a change in magnetic flux density, transformed into a linear voltage change. The supply voltage is from 10 to 30 V and the output voltage from 0.5 to 4.5 V. The RTY series sensors have been tested in terms of EMI/EMC for automotive applications, they are shock resistant up to 50 g and vibrations resistant from 10 to 2000 Hz [17].

Using the Kubler D8.3A1 linear encoders (number 3 on Fig. 6), the side tilting of the body was measured with respect to the vertical Z axis. The encoders were powered with 12 V from the battery. When the vehicle body did not show lateral tilt in relation to the vertical axis Z, the output voltage of the encoders was 2.85 V and changed with the lateral tilt of the vehicle by approx. 0.12 V/° [18].

The RACELOGIC measuring device VBOX was used for the tests (number 1 on Fig. 6). The VBOX 3i has the ability to use data from the inertial unit and the IMAC (Inertial Measurement Unit) RACELOGIC and uses the Kalman filter to improve all parameters measured in real time. The IMU provides very accurate measurements of pitch (x), roll (y) and yaw rate (z) using three speed gyros, as well as acceleration using three accelerometers. The VBOX allows to record data at 100 samples per second enabling detailed recording of data received by the GPS receiver. A very low latency of 8.5 ± 1.5 ms ensures fast processing and saving of data in real time. Data acquired by the VBOX is saved on a flash card. The device can register in real time: longitudinal speed, longitudinal acceleration, lateral acceleration, yaw rate, transverse tilt speed, voltage values of electrical signals from external sensors. The angles defining the inclination, heel and deviation of the vehicle are measured with high accuracy (tilt/incline at 0.06° RMS, deviation at 0.5° RMS). The supply voltage is from 7 to 30 V with low current consumption [19].

Fig. 6. Vehicle with test equipment

4 Road Tests

Driving is one of the most important factors of active road safety. The steering characteristics of a road vehicle are basically the vehicle's response to the driver's commands, taking into account the environment in which the vehicle is surrounded, such as road conditions, etc. [6]. The most reliable method for determining the steering characteristic of a car is to measure the vehicle parameters during particular maneuver [7].

The basic maneuvers that allow to determine the steering characteristics of the vehicle are the fixed circle driving and the constant steer angle test [6, 8]. On the basis of these tests, it is possible to determine a parameter called "the understeer gradient" [9].

In a **constant radius test**, the vehicle moves in a circle with a constant radius at different speeds. The understeer coefficient can be determined while the vehicle is moving in a circle with a fixed radius, observing the steer angle relative to the lateral acceleration. The steady-state lateral acceleration can also be deduced from the vehicle's forward speed and the known turning radius. If the vehicle has neutral steer the steer angle required to maintain the vehicle on a constant radius path is the same for all forward speeds (i.e., the slope of the steer angle-lateral acceleration curve is zero) [7]. Performing the test, the vehicle is driven in a constant radius circle at a very low speed for which the lateral acceleration is small and the steering angle (Ackerman angle) required to keep the turn is recorded. The vehicle speed is then increased in equal steps, which causes an increase in lateral acceleration of about 0.1 g, paying attention to the steering angle at each speed [8].

The steering of the vehicle can also be measured at a **steady steering angle**, while changing the driving speed. The constant steer angle test is executed by locking the steering wheel at a constant steer angle value. The lateral acceleration at different driving speeds is measured.

It has to be noticed that the conducted road tests are similar to those performed for a typical four-wheeled car. In the considered test vehicle the body lean angle is an extremely significant parameter (besides steer angle) from the point of view the assessment of steering characteristic. This thesis is confirmed by test results presented in the next chapter. For this reason the application of a classical definition of "the understeer gradient" seems to be inappropriate for a tilt vehicle.

5 Tests Results

The example results of the initial tests of the prototype tilt vehicle are presented below. Figure 7 shows time courses of following quantities: velocity, steering wheel angle and yaw rate during quasi-static acceleration at a circle with a constant radius of 15 m and a maximal lean angle. The maximal vehicle velocity is determined by the electric motor limit revs and the driving wheel diameter. The steering wheel angle-against-time-profile indicates neutral steering characteristic for the considered lean angle.

The steering characteristics as steering wheel angle-against-lateral acceleration-profile for the different lean angles is shown in Fig. 8. It should be noted that the considered test vehicle has a non-linear steering characteristics. The prototype construction causes oversteer characteristic for non-tilt body position. The steering wheel angle is decreased with increasing lateral acceleration. This could be the effect of the rear suspension susceptibility. However, as it can be seen, this undesirable feature is extremely reduced by the vehicle tilt. For maximal lean angle the vehicle has almost neutral steering characteristics.

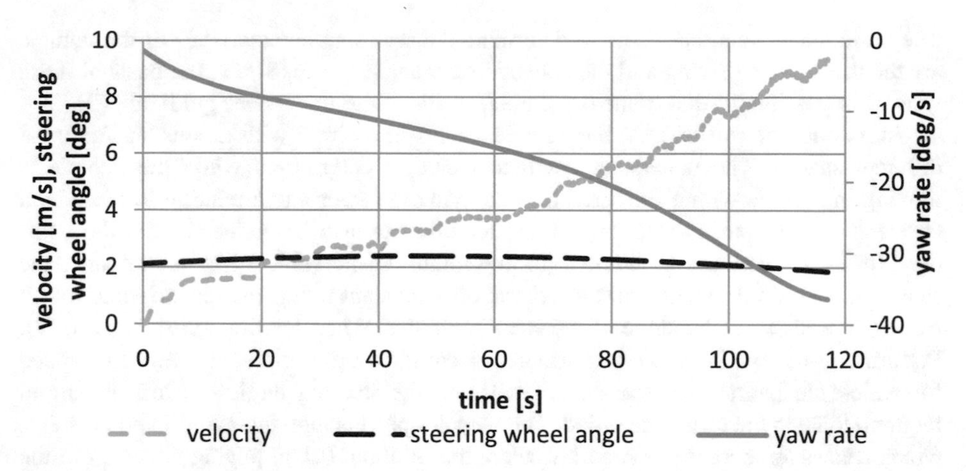

Fig. 7. The example time courses of velocity, steering wheel angle and yaw rate for quasi-static acceleration at constant radius test

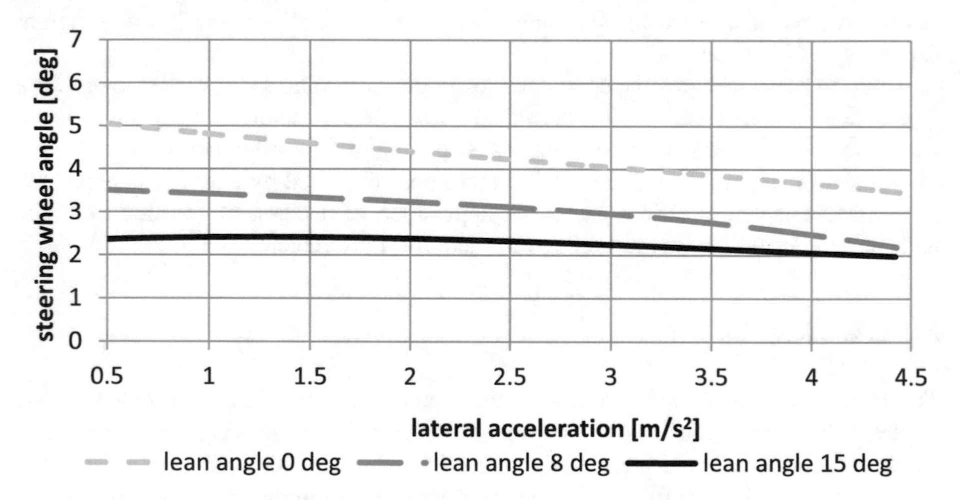

Fig. 8. The steering wheel angle against lateral acceleration for the different lean angles

Figure 9 presents the vehicle trajectory for fixed straight ahead steering wheel position and different lean angles. Based on conducted tests, it can be noticed that the vehicle's lean angle has a significant influence on the movement direction. For the 0° lean angle there is no the vehicle displacement in y direction. The larger lean angle, the greater deviation from the rectilinear path.

Considering the above results the vehicle trajectory radius was specified for each lean angle. In the next step the wheels equivalent steer angle was determined for the assumption of no lateral slip (the Ackermann range). Figure 10 shows the relationship between the wheels equivalent steer angle and the vehicle lean angle. For example, the maximal vehicle lean angle corresponds with the wheels equivalent steer angle even

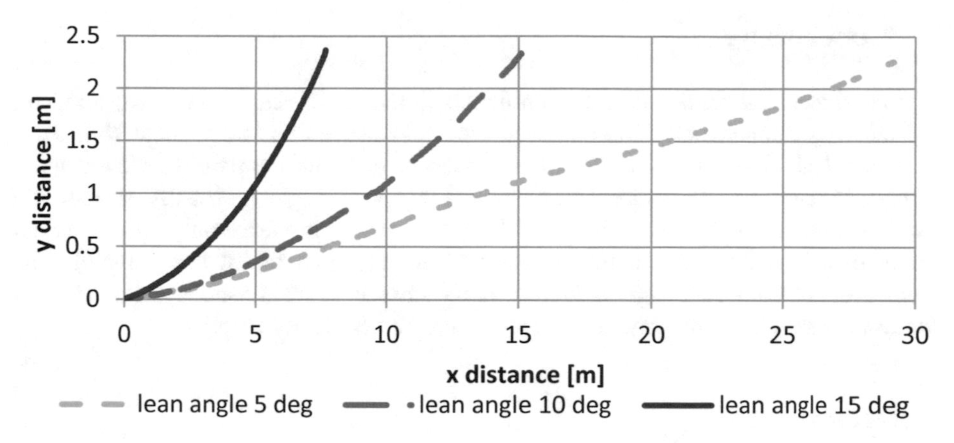

Fig. 9. The vehicle trajectory with the steady 0° steering wheel angle for the different lean angles

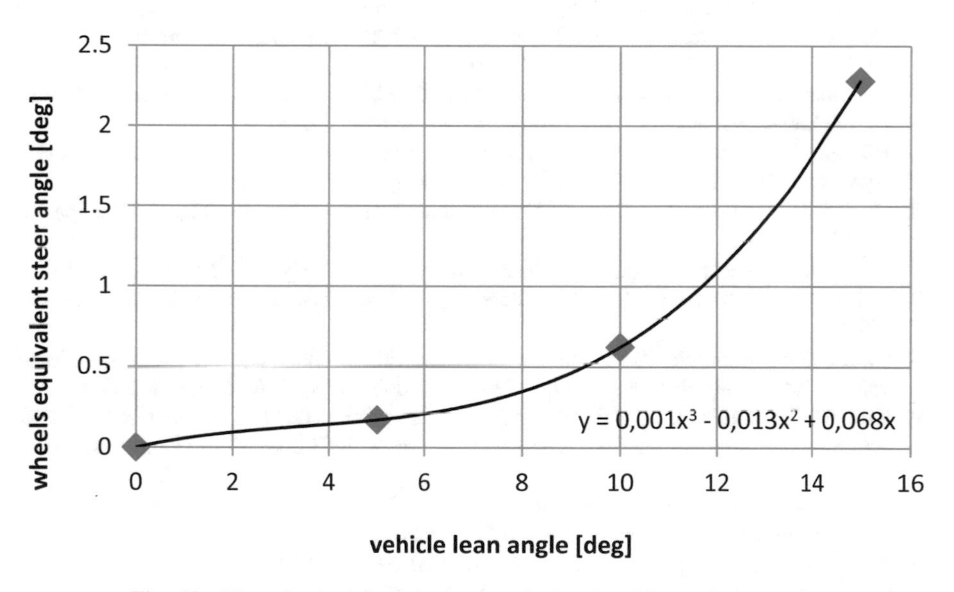

Fig. 10. The wheel equivalent steer angle for the different lean angles

2.3°. This value is similar to the average value of steering values of wheels during typical vehicle's motion.

The leaning of a three-wheeled vehicle plays an important role during cornering. In tilt vehicles, the allowance of the influence of the lean angle on the steering characteristics seems to be a necessity.

6 Conclusions

The conducted test results indicate significantly influence of vehicle body lean angle on the steering characteristic. The observed effects evidence that independent consideration of a lean angle and a steer angle is unacceptable from the point of view a mentioned steering characteristic. The vehicle body lean angle is usually structurally limited. This restriction causes the body lean angle has an influence only up to a certain value of lateral acceleration (about 2,8 m/s^2 for maximal tilt of test vehicle). The exhausting of body lean angle influence on the steering characteristic is not rapid. This feature constitutes significant advantage for driver who is not surprised.

References

1. Karnopp, D., Fang, C.: Simple model of steering-controlled banking vehicles. ASME Dyn. Syst. Control Div. (DSC) **44**, 15–28 (1992)
2. Hollmotz, L., Sohr, S., Johannsen, H.: Clever – a three wheel vehicle with a passive safety comparable to conventional cars. In: Proceedings 19th International Technical Conference on the Enhanced Safety of Vehicles (2005)
3. Robertson, J.W.J., Darling, J., Plummer, A.R.: Combined Steering and Direct Tilt Control for the Enhancement of Narrow Tilting Vehicle Stability (2014)
4. Hibbard, R., Karnopp, D.: Methods of controlling the lean angle of tilting vehicles. ASME Dyn. Syst. Control Div. (DSC) **52**, 311–320 (1993)
5. Huston, J.C., Graves, B.J., Johnson, D.B.: Three wheeled vehicle dynamics. SAE, Paper no. 820139 (1982)
6. Wong, J.Y.: Theory of Ground Vehicles, 3rd edn. Wiley, Hoboken (2001)
7. Hejtmánek, P., Blaťák, O., Suchý, J., Vančura, J.: Comparison of Vehicle Handling Models (2015)
8. Panke, D.A., Ambhore, N.H., Marathe, R.N.: Review on handling characteristics of road vehicle. Int. J. Eng. Res. Appl. **4**(7), 178–182 (2014). ISSN 2248-9622, (Version 4)
9. Kruger, H.P., Neukum, A., Schuller, J.: A Workload Approach to the Evaluation of Vehicle Handling Characteristics. SAE Technical Paper Series, SAE 2000 World Congress, Detroit (2000)
10. https://www.maxmatic.com/. Accessed 29 Apr 2019
11. https://www.bimmertoday.de/. Accessed 29 Apr 2019
12. https://alessandrobertin.wordpress.com/. Accessed 29 Apr 2019
13. https://www.toyota-europe.com/. Accessed 29 Apr 2019
14. https://www.yamaha-motor.eu/pl/pl/. Accessed 29 Apr 2019
15. https://www.piaggio.com/. Accessed 29 Apr 2019
16. http://www.swaymotorsports.com/. Accessed 29 Apr 2019
17. https://sensing.honeywell.com/. Accessed 29 Apr 2019
18. https://astat.pl/. Accessed 29 Apr 2019
19. https://www.vboxautomotive.co.uk/. Accessed 29 Apr 2019

Rolling Stability Control Based on Torque Vectoring for Narrow Vehicles

Witold Grzegożek and Krzysztof Weigel-Milleret(✉)

Cracow University of Technology, Al. Jana Pawla II 37, 31-864 Krakow, Poland
{witek,krzysztof.weigel-milleret}@mech.pk.edu.pl

Abstract. Narrow vehicles have many possible qualities then the normal track width cars. However, the main disadvantage of the narrow vehicles is their tendency to roll over. That is why the active roll mitigation system seems to be necessary. Torque vectoring has huge impact on vehicles steering characteristics. The paper contains a description of simulation models. The three degree of freedom model (3DOF) of a vehicle as a development of former bicycle model is presented as an appropriate model for the test subject. The rolling stability is described. The basic roll detection system is presented and its usefulness is discussed. Another approach based on the roll angle and roll rate is presented. The paper contains the description of a tested narrow car, results of preliminary tests and simulations results compared with road tests results of narrow vehicle. The usefulness of a 3DOF vehicle model to describe the movement of the vehicle was tested. The usefulness of roll angle calculation using 3DOF model in roll mitigation system at narrow car is discussed.

Keywords: Narrow vehicle · Torque vectoring · Active roll mitigation

Nomenclature

m	Mass of the vehicle [kg]
I_z	Moment of inertia around vertical axis [kg m^2]
I_x	Moment of inertia around longitudinal axis [kg m^2]
I_φ	Moment of inertia around rolling axis [kg m^2]
δ_1	Steering angle of the front wheels [rad]
δ_2	Steering angle of the rear wheels [rad]
l_1	Distance from front axle to centre of mass [m]
l_2	Distance from rear axle to centre of mass [m]
h	Height of a center of mass [m]
b	Track width [m]
Y_1	Horizontal force on the front axle [N]
Y_2	Horizontal force on the rear axle [N]
M_{ext}	Additional yaw moment [Nm]
F_{ext}	Additional horizontal force [N]
K_1	Cornering stiffness of the front axle [N/rad]
K_2	Cornering stiffness of the front axle [N/rad]
c	Rolling stiffness of a vehicle [Nm/rad]
k	Torsion damping factor [Nm/rad/s]

M. Siergiejczyk and K. Krzykowska (Eds.): ISCT21 2019, AISC 1032, pp. 173–183, 2020.
https://doi.org/10.1007/978-3-030-27687-4_18

F_{int} Force of inertia [N]
M_{int} Moment of inertia [Nm]

1 Introduction

Traffic jams, congestions and a lack of parking space, especially in urban centers are a serious problem in modern cities. In academic papers it is estimated that more than 5 billion hours are annually spent waiting on freeways in traffic congestions [1]. Efficient utilization of existing infrastructure (e.g. roads, parking places) is necessary, due to increasing costs of road construction [2]. Sharp increase of environmental pollution during last years is a significant problem. Some of this pollution comes from road vehicles and transportation powered by fossil fuels [2]. Numerous studies on occupancy rate indicate that passenger cars are not fully utilized in cities [3, 4]. An average number of passengers per vehicle in Europe at most frequent car trips is 1.7 persons/car. It varies, however, from a minimum of 1.4 in Denmark to a maximum of 2.7 in Romania [5, 6]. In U.S. are occupancy rates remains unchanged from 2009 to 2017 at about 1.54 [7]. The solution to these problems seems to be narrow cars, as the special kind of microcars. These are small-size vehicles designed to move short routes only by a driver or with one passenger with a small luggage. Due to their small dimensions, high maneuverability, dynamic driving performance and lower fuel consumption the narrow cars perform perfectly in the urban agglomeration spaces. Narrow vehicles are gaining interest not only for vehicle manufacturers but also research centers. They are increasingly the object of scientific research [8]. In European countries are special zones with limited or total ban of combustion engines cars (zero emission zone), where electric vehicles can freely move. Legal systems in many countries are favoring microcars in relation to full size vehicles; e.g. by lower taxes and lower insurance rates. That is why the electric motors are often used to drive narrow cars.

Vehicle rollover should be consider as an important safety problem. Accidents involving vehicle rollover often have fatal consequences [9]. Especially narrow cars are very susceptible to rollover, because they have a small track and a high center of gravity. Development of active roll mitigation systems have been described in the literature [9, 10]. The roll mitigation systems are commonly based on static or steady-state rollover models [11]. Dynamic models are rarely presented in literature [9, 12]. The roll mitigation systems for dynamics are based on continuous normal load force calculation. The lateral load transfer ratio during maneuvers is used [13]. It is difficult to measure the value of the vertical load [14]. In most cases, roll mitigation systems using lateral acceleration and roll angle are described in the literature [15]. The most common approach is to use vehicle brakes to create the deflection or tilting moment of the vehicle, both at the track stabilization and anti-rollover systems. However, each activation of the vehicle's brakes dissipates a portion of kinetic energy of the vehicle as heat. The vehicle received kinetic energy thanks to electric motors powered by an onboard battery pack. Therefore, each operation of the vehicle's stabilization system contributes to reducing the range of the electric vehicle. On the other hand, the distribution of driving forces on individual wheels of the driven axle can be used to prevent rolling over and to stabilize the vehicle motion. Such solution can be implemented only in vehicles with independent power supply for electric motors, which drive a single wheel. It causes the increase of the energy efficiency of the vehicle, which makes increasing the maximum range of the vehicle possible. Currently, active torque vectoring control systems are mainly used to limit the yaw rate and vehicle slip angle [12, 16, 17].

In this paper, three degree of freedom (3DOF) vehicle model is presented as a development of former bicycle model. The preliminary and road tests of a narrow car are described. The usefulness of roll angle calculation using the 3DOF model (based on a single line maneuver) in roll mitigation system at narrow car is discussed by comparison of simulation results with road tests results.

2 Tested Vehicle Characteristics

MIST car was chosen for tests. This narrow car was designed and developed in Institute of Automobiles and Internal Combustion Engines of the Cracow University of Technology. It is driven by two electric motors built in the rear wheel hubs. The vehicle lightweight composite body is shown in Fig. 1 and its basic technical data is presented in Table 1 [18]. The combination of trailing-arm and pushed-arm suspension causes the rolling axis of the vehicle body to be located in the plane of the road.

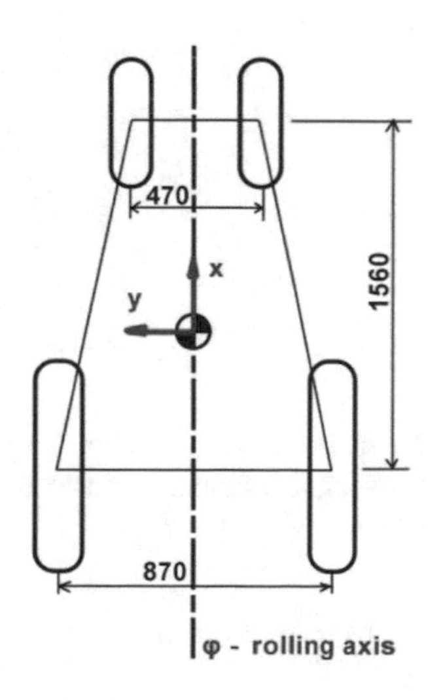

Fig. 1. MIST - tested narrow vehicle [19]

Table 1. MIST technical data

Frame	Steel, welded space frame
Front suspension	Dependent suspension with parallel wheel movement, doubled pushed arm
Rear suspension	Semi-independent twist beam with towed arms
Motor	2 × Brushless DC motor (3.5 kW peak power)
Brakes	Hydraulic, disc brakes
Vehicle mass	197 kg unloaded; 268 kg with driver
Wheelbase	1560 mm
Track width	470 mm front axle, 870 mm rear axle

3 Rollover Stability

The ability of the vehicle to resist overturning is called a rollover stability. To ensure rollover stability, the values of normal reactions acting on the inner wheels do not fall to zero. This case is showed in Fig. 2. It should be noticed that the vehicle is considered as a rigid body (impact of the suspension of the vehicle is not taken into account).

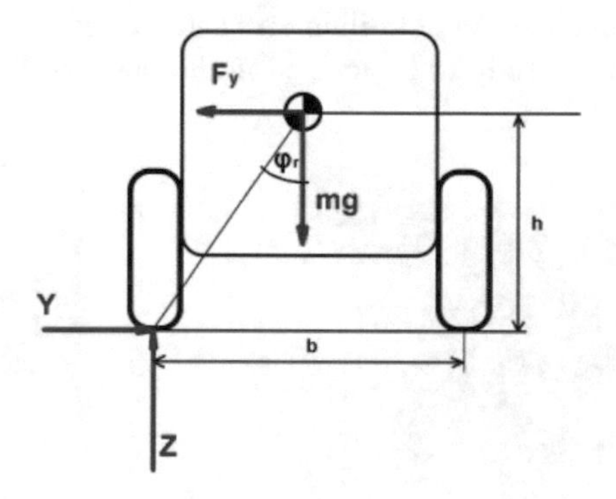

Fig. 2. Balance of forces acting on a car in a curve

The quotient of lateral force and vehicle's weight must be less than tangent of the line between the center of mass and rolling axis [20]:

$$\frac{F_y}{mg} \leq \tan \varphi_r \tag{1}$$

$$F_y \leq \frac{b}{2h}\,\mathrm{mg} \tag{2}$$

It is highly inadequate for narrow vehicles. The limitation of lateral acceleration reduces the possibility of cornering while traveling at high longitudinal velocity and significantly reduces maneuverability of the vehicle. For short maneuvers lateral acceleration may exceed the value described above, and do not cause the vehicle rollover. For the tested narrow vehicle described above, the lateral acceleration values depend on the longitudinal acceleration or deceleration (see Fig. 1) and are within the following limits (car loaded only with driver):

$a_{y\,acc} = 4.88\,\mathrm{m/s^2} \approx 0.5\,\mathrm{g}$ while accelerating
$a_{y\,vx=const} = 4.20\,\mathrm{m/s^2} \approx 0.43\,\mathrm{g}$ while driving at constant speed
$a_{y\,dec} = 3.81\,\mathrm{m/s^2} \approx 0.39\,\mathrm{g}$ while decelerating with full braking force

In another approach, the moment around the rolling axis, should have the opposite sign than the lateral inertia force.

$$\sum M_\varphi < 0 \tag{3}$$

4 Vehicle Modelling

In previous papers the single track vehicle model was presented [19, 21]. The single track model is sufficient for modeling the trajectory of the vehicle in advance. It is a flat model with three degrees of freedom. The vehicle itself is symmetrical. Axes have been reduced to individual wheels. The model assumes that the left and right wheels of the vehicle generate equal lateral forces and depend only on the slip angle of axles. This means that the description of the vehicle is identical to the description of a single-track vehicle. The appropriately modified single track model takes into account the impact of wheel torque distribution on the vehicle trajectory. However, narrow vehicles have a high tendency to roll over because of the short track and the high center of mass. Therefore, the approach based only on the prediction of lateral accelerations is insufficient. The short-term occurrence of high lateral acceleration values does not cause the vehicle overturning. A roll mitigation system supported by lateral acceleration measurements also does not give the desired effect in narrow vehicles. A different approach seems necessary. In order to develop the active roll mitigation system a model with three-degrees of freedom (3DOF) including yaw, roll (around the rollover axis) and lateral motion of vehicle is used. The model must consider the suspension stiffness and damping to predict vehicle movement. Because the longitudinal motion is at a constant speed (vx = const) the x-axis motion of the vehicle is not taken into account. Figure 3 shows the vehicle's motion in two perpendicular planes. The forces and moments acting on the vehicle can be calculated as:

$$\begin{aligned} \sum F_y &= m\ddot{y} \\ \sum M_z &= I_z\ddot{\psi} \\ \sum M_\varphi &= I_\varphi\ddot{\varphi} \end{aligned} \tag{4}$$

Where F_y, M_z and M_φ are lateral force and the moments around z and rolling axis; m, I_z and I_φ represent the mass and the moment of inertia around vertical and rolling axels. y, ψ, φ denote lateral position, yaw and roll angles. Suspension model is taken into account as presented in Fig. 3.

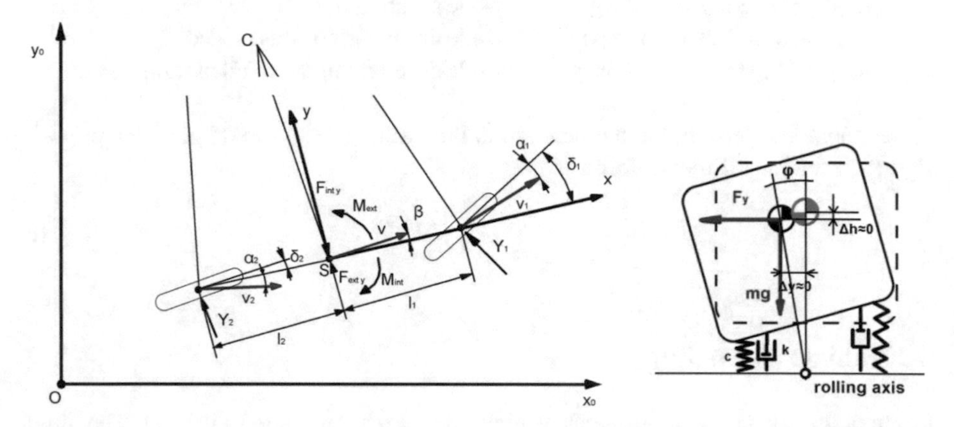

Fig. 3. 3DOF model used for simulations

The equations of forces and moments acting on the vehicle take the form:

$$-F_{int\,y} + F_{1_y}\cos\delta_1 + F_{2_y}\cos\delta_2 + F_{ext_y} = 0 \tag{5}$$

$$-M_{int\,z} + F_{1_y}\cos\delta_1 l_1 - F_{2_y}\cos\delta_2 l_2 + M_{ext\,z} = 0 \tag{6}$$

$$-M_{int\,\varphi} + F_y h\cos\varphi + mgh\,\sin\varphi - c\varphi - k\dot{\varphi} + M_{ext\,\varphi} = 0 \tag{7}$$

Moment of inertia around the rolling axis:

$$I_\varphi = I_x + mh^2 \tag{8}$$

The lateral forces of the front and rear axles can be determined as:

$$Y_1 = K_1\alpha_1 \tag{9}$$

$$Y_2 = K_2\alpha_2 \tag{10}$$

After taking into account some assumptions: $\delta_2 = 0$; $\cos\delta_1 \approx 1$; $\cos\beta \approx 1$ and $\beta = \frac{\dot{y}}{\dot{x}}$ the equations take form:

$$m\ddot{y} = K_1\delta_1 - \frac{K_1 + K_2}{v}\dot{y} - \frac{K_1 l_1 - K_2 l_2 + mv^2}{v}\dot{\psi} \tag{11}$$

$$I\ddot{\psi} = K_1 l_1 \delta_1 - \frac{K_1 l_1 - K_2 l_2}{v}\dot{y} - \frac{K_1 l_1^2 + K_2 l_2^2}{v}\dot{\psi} + M_{ext\,z} \quad (12)$$

$$I\ddot{\varphi} = F_y h \cos\varphi + mgh \sin\varphi - c\varphi - k\dot{\varphi} \quad (13)$$

5 Tests and Simulations

To perform the simulation it was necessary to obtain mass and geometric parameters of the vehicle. The following preliminary tests were carried out:

- measurement of the moment of inertia around the z axis on a three-link test stand
- measurement of the moment of inertia around the x axis on a vertical torsion pendulum test stand (presented in Fig. 4).
- the rolling stiffness of the vehicle around rolling axis
- damping coefficient.

Fig. 4. Measurement of the moment of inertia around the x axis

The measurement of the rolling stiffness allowed to determine the maximum safe range of rolling angle of the car body. After reaching the roll angle $\varphi = 11° \approx 0.2$ rad, the inner wheel of the vehicle is detached from the road surface.

The following test equipment was installed in the vehicle:

- Datron Correvit apparatus measuring the longitudinal and lateral velocities of the vehicle
- Crossbow measuring velocities and lateral, longitudinal and rolling accelerations
- Honeywell RTY 270HVNAX Hall-Effect rotary position sensor measuring the steering angle

- MPU-6050 integrated 6-axis MotionTracking device that combines a 3-axis gyroscope, 3-axis accelerometer, and a Digital Motion Processor™ (DMP) measuring lateral, longitudinal yaw and roll accelerations. The device is used to calculate the roll angle of the vehicle.

The test object equipped with apparatus is shown in Fig. 5.

Fig. 5. MIST narrow vehicle with test equipment

The tests and the simulations included a single turn and single lane change maneuver performed at different driving velocities and various drive torque distribution between the wheels. Three states of torque distribution were taken into account:

- Equal torque distribution between drive wheels
- Torque distributed only to the outer wheel in relation to the overcame arc
- Torque distributed only to the inner wheel in relation to the overcame arch

Matlab R2015b software was selected for the motion simulation. The geometric and mass parameters of the vehicle determined at preliminary tests and previously determined [21] cornerring stiffness for the front and rear axles were used in the 3DOF model. The input to the model was the steering angle as a function of time. Steering angle function obtained from real road tests. The result of the simulation is the lateral movement of a vehicle $y(t)$, $y'(t)$ in local CS, yaw angle $\psi(t)$, yaw rate $\psi'(t)$, roll angle $\varphi(t)$ and roll rate $\varphi'(t)$ as a function of time. The obtained data of the yaw rate, roll angle and roll rate was compared with the values measured during the tests. In order to determine the suitability of the 3DOF model to an active roll mitigation system, a correlation coefficient was determined (PCC - Pearson's Correlation Coefficient). Figures 6, 7 and 8 show simulation result examples.

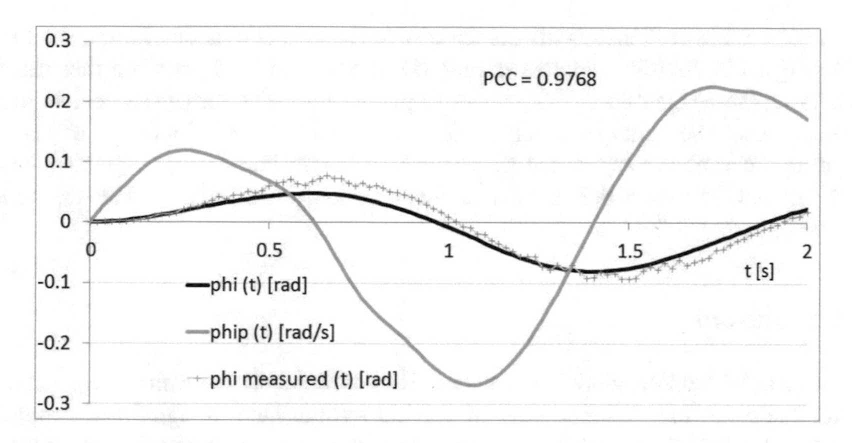

Fig. 6. Single lane change with only outer wheel driven. Additional yaw torque included

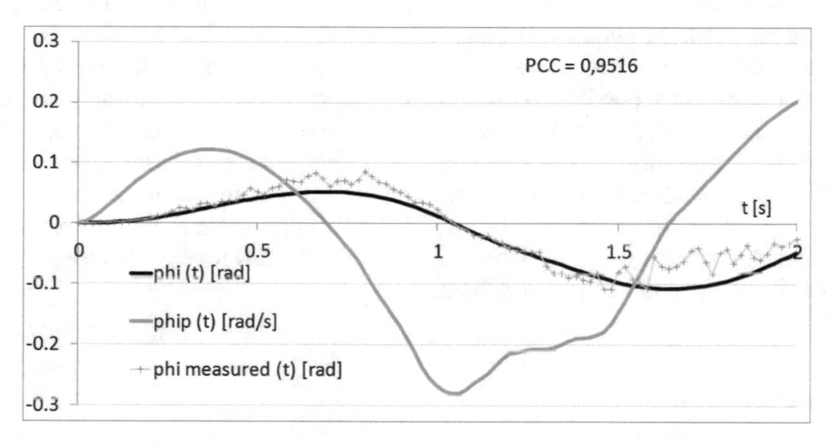

Fig. 7. Single lane change with an equal torque distribution

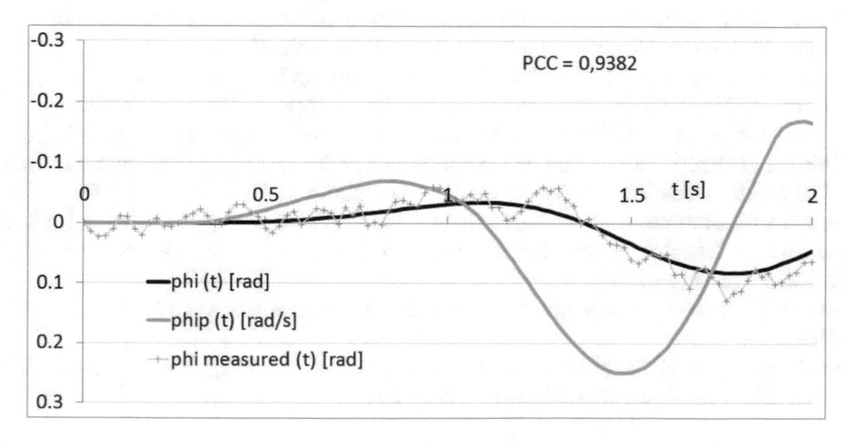

Fig. 8. Single lane change with only inner wheel driven. Additional yaw torque included.

The above simulations results coincide with the results obtained during road tests. The comparable vehicle response during simulation and road tests are the basis for considering the usage of a 3DOF model to predict the behavior of the vehicle during the movement. The anticipated roll angle and roll rate of the vehicle can be used for activating the vehicle's active safety systems. A single line change trajectory can be taken into account as a defensive maneuver taken by the driver when the sudden obstacle appears on the road.

6 Conclusion

A. The 3DOF vehicle model can be used to describe the movement of a narrow vehicle using the differentiating of driving torque. Due to significant simplifications, this vehicle model does not require much computing power. Predictions of vehicle behavior can be calculated in real time.
B. The roll angle and roll rate prediction can be used at an active roll mitigation systems based on torque vectoring.
C. The road test results presented an appropriate selection of the driving torque distribution makes it possible to obtain understeer characteristics corresponding to the safety range; torque vectoring restrictions in narrow vehicles are crucial because varying torque distribution can result in a change of understeer characteristics. Changing the characteristics in critical situations can prevent an accident.
D. The result of this work can be useful for defining optimal time of active roll mitigation system reaction. It is also the basis for developing the proper torque distribution at roll mitigation system.

References

1. Rajamani, R., Gohl, J., Alexander, L., Starr, P.: Dynamics of narrow tilting vehicles. Math. Comput. Model. Dyn. Syst. **9**(2), 209–231 (2003)
2. Michalek, J., Papalambros, P., Skerlos, S.: A study of fuel efficiency and emission policy impact on optimal vehicle design decisions. J. Mech. Des. **126**(6), 1062 (2004)
3. Ivanov, V., Augsburg, K., Savitski, D.: Torque vectoring for improving the mobility of all-terrain electric vehicles. In: Proceedings of the European Regional Conference of the International Society for Terrain-Vehicle-Systems Pretoria South Africa, pp. 1–8 (2012)
4. Chiou, J., Chen, C.: Modelling and verification of a diamond-shape narrow-tilting vehicle. IEEE/ASME Trans. Mechatron. **13**(6), 678–691 (2008)
5. Fiorello, D., Martino, A., Zani, L., Christidis, P., Navajas-Cawood, E.: Mobility Data across the EU 28 Member States: Results from an Extensive CAWI Survey Transportation Research Procedia (2016)
6. https://trimis.ec.europa.eu/project/occupancy-rate-vehicles#tab-results. Accessed 15 Apr 2019
7. 2017 report U.S. Department of Transportation, Federal Highway Administration, National Household Travel Survey. Accessed 22 May 2019

8. Ding, F., Huang, J., Wang, Y., Matsuno, T., Fukuda, T., Sekiyamahes, K.: Modeling and control of a novel narrow vehicle. In: 2010 IEEE International Conference on Robotics and Biomimetics (2010)
9. Yoon, J., Yi, K.: A rollover mitigation control scheme based on rollover index. In: Proceeding of the 2006 American Control Conference (2006)
10. Rajamani, R.: Vehicle Dynamics and Control. Springer, New York (2006)
11. Kiencke, U., Nielsen, L.: Drivetrain Control Automotive Control Systems - For Engine, Driveline and Vehicle. Springer, Heidelberg (2005)
12. De Novellis, L., Sorniotti, A., Gruber, P.: Wheel torque distribution criteria for electric vehicles with torque-vectoring differentials. IEEE Trans. Veh. Technol. **63**(4), 1593–1602 (2013)
13. Sawase, K., Inoue, K.: Maximum acceptable differential speed ratio of lateral torque-vectoring differentials for vehicles. J. Automobile Eng. **223**(8), 967–978 (2009)
14. Sawase, K., Ushiroda, Y., Miura, T.: Left-right torque vectoring technology as the core of super all wheel control (S-AWC). Mitsubishi Motors Techn. Rev. **18**, 16–23 (2006)
15. Wang, J., Wang, Q., Song, C., Chu, L., Wang, Y.: Coordinated control of differential drive assisted steering system with vehicle stability enhancement system. In: Proceedings of the Intelligent Vehicles Symposium Baden-Baden Germany, pp. 1148–1155 (2011)
16. Ahangarnjead, A.: Integrated Control of Active Vehicle Chassis Control Systems, Doctoral dissertation (2017)
17. Yang, D., Idegren, M., Jonasson, M.: Torque vectoring control. In: Advanced Vehicle Control Conference, AVEC 2018, Beijing (2018)
18. Grzegożek, W., Weigel-Milleret, K.: Torque vectoring for improving stability of small electric vehicles. In: IOP Conference Series: Materials Science and Engineering (2016)
19. Grzegożek, W., Weigel-Milleret, K.: Wheel torque distribution for narrow cars. In: 11th International Science and Technical Conference Automotive Safety, Automotive Safety 2018 (2018)
20. Reński, A.: Bezpieczeństwo czynne samochodu Oficyna Wydawnicza Politechniki Warszawskiej, pp. 247–268 (2011)
21. Grzegożek, W., Weigel-Milleret, K.: Modelling and simulation of narrow car dynamic. In: IOP Conference Series: Materials Science and Engineering (2018)

Driver's Reaction Time in the Context of an Accident in Road Traffic

Marek Guzek[✉]

Faculty of Transport, Warsaw University of Technology, Koszykowa St. 75, 00-662 Warsaw, Poland
mgu@wt.pw.edu.pl

Abstract. Road accidents are everyday in road traffic. The driver can influence the possibility of avoiding an accident in emerging situations. The basic defensive manoeuvre made by drivers in an emergency situation is braking before an obstacle, and one of the most important factors shaping the vehicle stopping distance, and consequently the possibility of collision and its potential effects at a given vehicle speed is the driver's reaction time. The article presents selected results of the assessment of the reaction time of a group of drivers during tests in the driving simulator and on the test track during the actual driving. Based on these results, using the Monte-Carlo method and using the braking model known from the basics of car motion mechanics, calculations of the stopping distance were made and were determined collision probability characteristics depending on vehicle speed and distance from the obstacle. Such characteristics, apart from the cognitive aspect, can also be used in practice, e.g. by court experts, who in cases involving road accidents often have to answer the court's question about the possibility of avoiding collisions.

Keywords: Driver's reaction time · Drivers' research · Stopping distance · The probability of collision

1 Introduction

The reaction time is one of the most important parameters characterizing the driver, especially in situations of traffic danger. It is also one of the basic input parameters used in the calculations carried out in the process of reconstruction/simulation of a traffic accident.

Using literature data, we can come across many sources in which the values of this parameter are given. Apart from general items such as [7, 12], in recent years there have been published in Poland several publications on the driver's reaction time, e.g. [4, 8–11]. There is necessary also to mention work [6]. The presented values of driver reaction time are different. Authors presenting results usually indicate the conditions in which they were obtained. Generally speaking, the differences result from different features of tests methodologies: test environment (laboratory, driving simulator, real vehicle, etc.), type of reaction assessed (reaction to simple or complex stimuli, type of element to which the examined driver exerts reaction), specific conditions of tests (e.g. day/night/lighting), characteristics of the tested group of drivers (e.g. group size, age,

M. Siergiejczyk and K. Krzykowska (Eds.): ISCT21 2019, AISC 1032, pp. 184–193, 2020.
https://doi.org/10.1007/978-3-030-27687-4_19

sex of examined drivers), others, see also e.g. [8–10]. Differences may also result from the definition of reaction time, as well as the manner of results presentation. Comparisons of reaction time for various research environments (test track, driving simulator, dedicated laboratory measurement stations are shown, for example, in [2, 3]. The second chapter of this paper presents the research results of reaction time in two environments - on a test track and in a car driving simulator.

Another issue is how the driver's reaction time may affect the risk of an accident (road accident) in the emerging emergency situation. Research on reaction time (also works mentioned above) indicates that this parameter is characterized by high dispersion. In this paper will be carried out an analysis showing how uncertainty about this parameter (but also other factors affecting the car motion) can shape the probability of a collision in traffic. To illustrate this effect, there was chosen a basic (according to the different research) defence manoeuvre performed by drivers in an emergency situation - the so-called emergency braking.

2 Driver Reaction Time – Results of Test

The study uses the results obtained as part of the research project N509 016 31/1251 "Development and updating of the database on reaction times of road vehicles drivers" conducted by teams from the Kielce University of Technology (leader), Warsaw University of Technology and Cracow University of Technology.

In this project, an assessment of reaction time on vehicle controls elements (brake pedal, steering wheel, and acceleration pedal) for the same group of drivers (around 100 male, aged 19–64) was made using a few different methods, including:

- tests in the autoPW car simulator, in an arranged accident situation (hereinafter referred to as "simulator");
- road tests in a real car (on research track) in an arranged accident situation analogous to that performed in the simulator (hereinafter referred to as "track").

The project took on the basic assumption that reaction times will be determined for a complex traffic situation, not for a simple stimulus. There were implemented, both using the actual vehicle on the test track and in the driving simulator (a simulator description can be found e.g. in [1]), tests for 3 selected accident situations scenarios. The common feature of the scenarios was the obstacle appearing on the road in a sudden way, additionally in conditions of limited visibility. The scenarios differed in the type of obstacle (passenger car, pedestrian, a set of tractor and semi-trailer), the way of the obstacle movement and other elements on the road, which affected the complexity of the situation. In both research environments, tests were carried out for various risk time values (the ratio of the distance from the obstacle to the speed of the car driven by the tested driver - see hereinafter). Detailed descriptions of these tests as well as their results can be found in the previously mentioned sources, e.g. [4, 5, 8, 11]. Here, the research results for the so-called Scenario 1 are used, in which the obstacle the tested driver was reacting to, was a passenger car (see Fig. 1).

Fig. 1. The sample illustration of tests in the driving simulator autoPW (left) and on the track (right).

In the further part we would be interested in the reaction time for the case of braking manoeuvre. Therefore, only the results for operating of drivers on the brake pedal are shown below.

One of the basic established assessment criteria was the so-called risk time (time to collision), i.e. "distance in time". It expresses the period until the potential collision, if the driver does not undertake preventive actions. Thus, it is the ratio of the distance of the vehicle from the obstacle to its speed at the moment the hazard arises. Figures 2 and 3 show the results obtained as a function of the risk time.

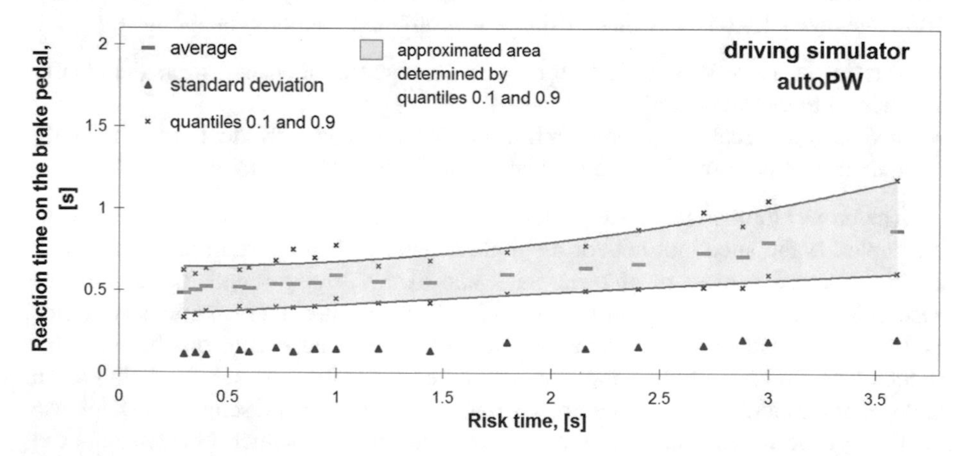

Fig. 2. Reaction time on the brake pedal for scenario I in the driving simulator.

The fact of the dependence of the reaction time on the risk time which describes the situation is an important factor that should be taken into account when analyzing the accident situations. It is also noticeable that in the most dangerous situations (for low risk time) this relationship is not visible (for risk times up to about 1.5 s). Therefore, this effect was not considered in the further part, and the results obtained by drivers for a situation with a risk time of less than 1.5 s were taken to analysis. Figure 4 shows the

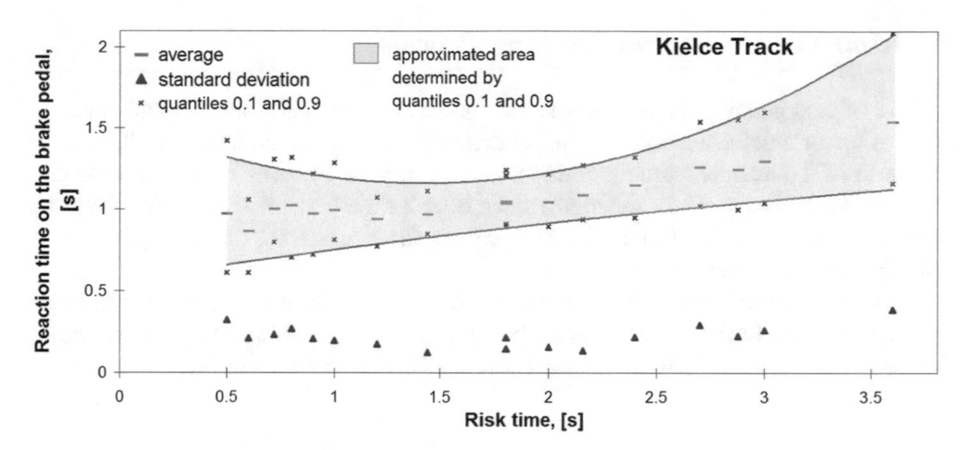

Fig. 3. Reaction time on the brake pedal for scenario I on the testing track.

histograms of reaction time on the brake pedal in both research environments only for situations with a risk time of less than 1.5 s. Table 1 presents the basic parameters of the obtained distributions. More precisely, the distributions are described e.g. in [2, 3].

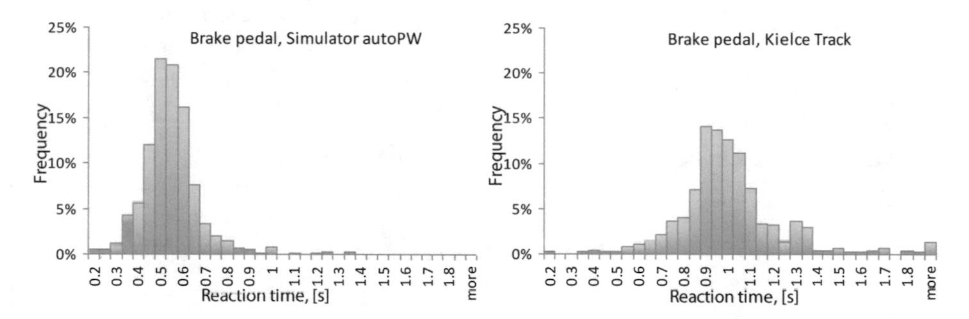

Fig. 4. Histograms of the reaction time obtained in the driving simulator autoPW (left) and on the track (right).

Table 1. Selected parameters of distributions of reaction time on brake pedal.

	Driving simulator autoPW	Kielce track
Mean [s]	0,531	0,994
Median [s]	0,525	0,960
Standard deviation [s]	0,129	0,275

Despite good qualitative accordance of the results obtained on the track and in the simulator, the values obtained in the second environment are clearly smaller. Therefore, the further analysis was limited to results obtained on the track, which was considered to be more realistic.

3 Model of Emergency Braking Process

A case of emergency braking was selected for the analysis. This kind of manoeuvre is a typical driver's defensive manoeuvre. In the project described earlier, regardless of the scenario of the accident situation, drivers in about 90% of cases took such a defence reaction to the threat. Only for high values of risk time (over 2 s) this percentage has been decreasing, but it still remains higher than, for example for the attempt to bypass the obstacle by means of a turning, [8].

To describe the straight-line braking process, it is convenient to use a simple model known from the basics of car motion theory (e.g. [7, 12]). Simplified time courses of brake pedal force and braking deceleration are assumed here - see Fig. 5.

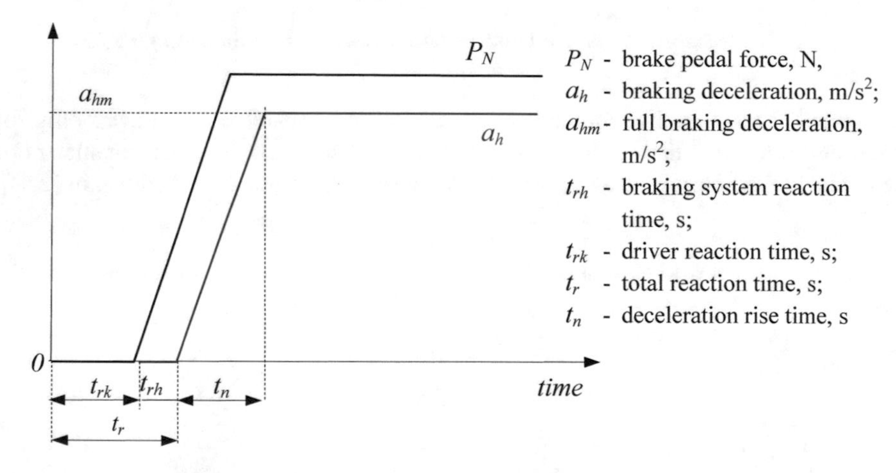

Fig. 5. Time history of car braking

Using this model, we can get an approximate dependence determining length of the stopping distance S_z from the initial car speed V_0 as the sum of the reaction distance S_r and the braking distance S_h:

$$S_z = S_r + S_h = V_0 \cdot \left(t_{rk} + t_{rh} + \frac{t_n}{2}\right) + \frac{V_0^2}{2 \cdot a_{hm}} \tag{1}$$

If we additionally assume that the vehicle brakes with the maximum deceleration due to road-tires friction coefficient, we can assume that:

$$a_{hm} = \mu \cdot g \tag{2}$$

where μ - coefficient of friction, g - gravitational acceleration.

4 Calculations of Stopping and Braking Distance

In the general, all the quantities appearing in formulas (1) and (2) have, in fact, some uncertainty. They can therefore be treated as random variables. In the case of the driver's reaction time, the distribution was taken according to the results obtained in the tests on the track (Fig. 4). For the others, the nominal values and uncertainties corresponding to the real braking of real vehicles on the dry asphalt road were assumed. Sets of values describing the parameters are summarized in Table 2. In the case of response time, the average value is given as the nominal value and the tripled standard deviation as uncertainty.

Table 2. Adopted braking parameters.

Parameter		Nominal/mean value	Uncertainty	Distribution
Driver reaction time	t_{rk}	(0,994 s)	(~0,825 s)	Empirical
Initial vehicle speed	V_0	*var*	±1 m/s	Uniform
Braking deceleration	a_{hm}	7,0 m/s^2	±0,5 m/s^2	Uniform
Braking system reaction time	t_{rh}	0,25 s	±0,05 s	Uniform
Deceleration rise time	t_n	0,3 s	±0,1 s	Normal

The calculations were made using the Monte Carlo method. The method works particularly well in computational problems, where random phenomena should be taken into account. Generally, it consists in repeatedly repeating the experiment, with randomly changing values of parameters within the scope determined by the type of experience and the phenomenon under research.

Figure 6 shows a generated distribution of pseudo-random numbers illustrating the driver's reaction time in the form of a histogram and functions of density and cumulative distribution. In Fig. 7 only the histograms of the remaining parameters are shown. Table 3 presents the values of selected measures of pseudo-distributions obtained (in the case of initial speed for a nominal value of 90 km/h). These values (as well as histograms) confirm the correctness of the generated distributions. Figure 8 shows the results for stopping distance S_z and braking distance S_h for initial speed of 90 km/h.

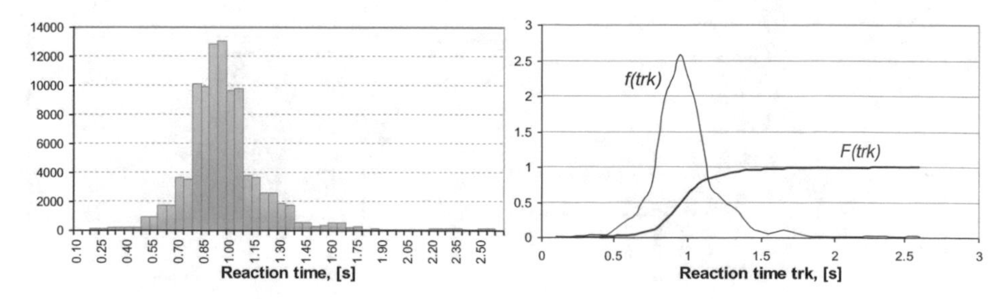

Fig. 6. Driver reaction time t_{rk} distribution by Monte Carlo method (histogram – left, probability density *f(trk)* and cumulative distribution *F(trk)* – right)

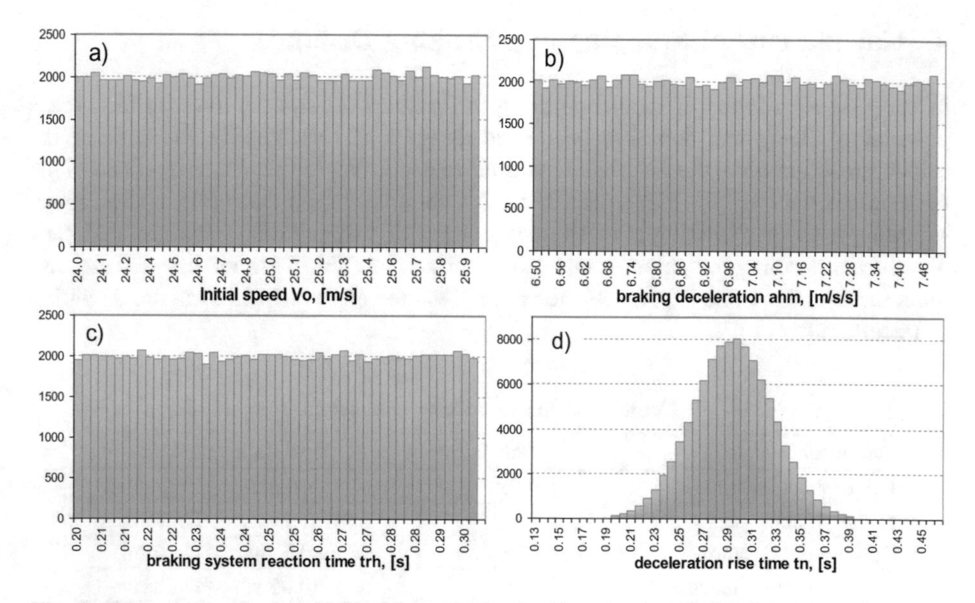

Fig. 7. Histograms of: a - initial velocity V_0, b - braking deceleration a_{hm}, c - braking system reaction time t_{rh}, d - deceleration rise time t_n.

Table 3. The obtained values of distribution parameters.

Parameter		Mean value	Standard deviation	Min/max
Driver reaction time t_{rk}	s	0,993	0,256	0,20/2,60
Initial veh. speed V_0 (90 km/h)	m/s	25,003	0,577	24,0/26,0
Braking deceleration a_{hm}	m/s^2	7,0	0,289	6,5/7,5
Braking system reaction time t_{rh}	s	0,2501	0,00289	0,2/0,3
Deceleration rise time t_n	s	0,2998	0,0328	0,2/0,4
Stopping distance S_z	m	79,6	7,32	53,7/128,5
Braking distance S_h	m	47,9	2,85	40,9/55,8

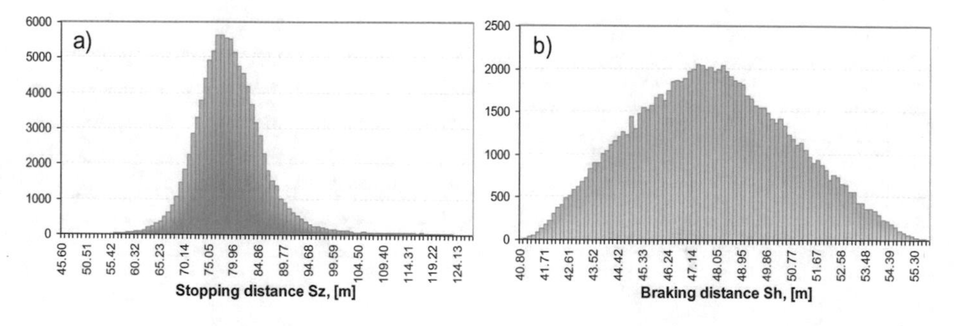

Fig. 8. Histograms of: a - stopping distance S_z, b - braking distance S_h.

5 The Probability of a Collision

Analogous calculations were made for different initial speeds from 30 to 140 km/h. Figure 9 shows the received cumulative distributions $F(S_z)$ and the probability density functions $f(S_z)$ of the stopping distance S_z. Cumulative distributions show directly the probability of avoiding a collision (stopping the vehicle ahead of the obstacle) at a given initial speed and a given distance from the obstacle at the moment the hazard arises (noticing the obstacle by the driver).

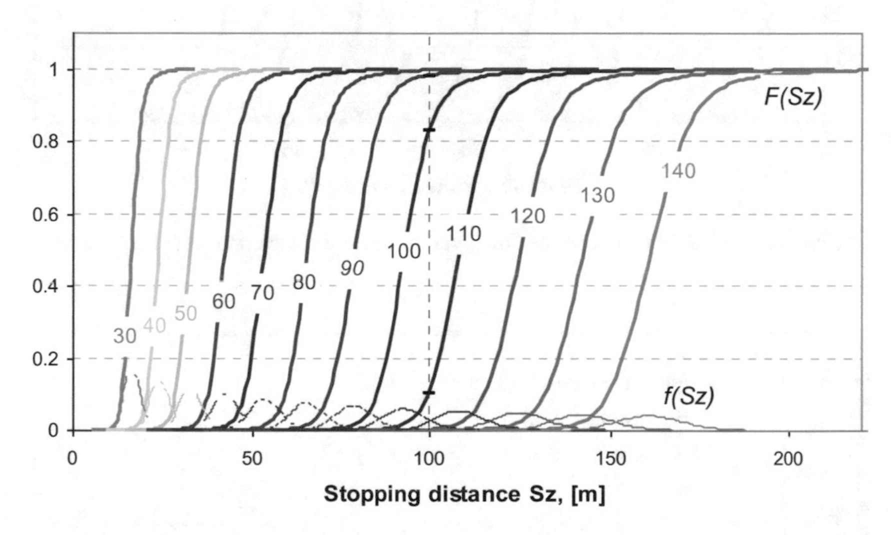

Fig. 9. Cumulative distribution $F(S_z)$ and probability density $f(S_z)$ of stopping distance S_z for different initial vehicle speed.

Using these cumulative distributions, we can present the probability of a collision (as an opposite event probability) - see Fig. 10. Another information on the possibility of collision occurrence or avoidance may be, for example, the distance from the obstacle at which it is possible to stop the vehicle ahead of it. In Fig. 10 the resulting quantiles 0.1, 0.5 (median) and 0.9 of the stopping distance of the car as a function of the initial velocity are therefore shown. For example: if the distance from the obstacle at the initial moment was S = 100 m, the driver driving at about 100 km/h has about 83% chance to avoid collision, but at a speed of about 110 km/h these chances would drop to about 10% (Figs. 9 or 10). At an initial speed of 90 km/h, the chances of avoiding an accident would be over 98%. If we set the initial speed of 90 km/h (let's assume that, the driver drives a car at the permitted speed on a given road), we can say that if the obstacle appeared more than 88 m, the chances of avoiding collisions are above 90% (quantile 0,9 = 87.7 m, Fig. 11). If the distance would be less than 71 m, the chance of avoiding the accident is less than 10% (quantile 0,1 = 71.7 m, Fig. 11).

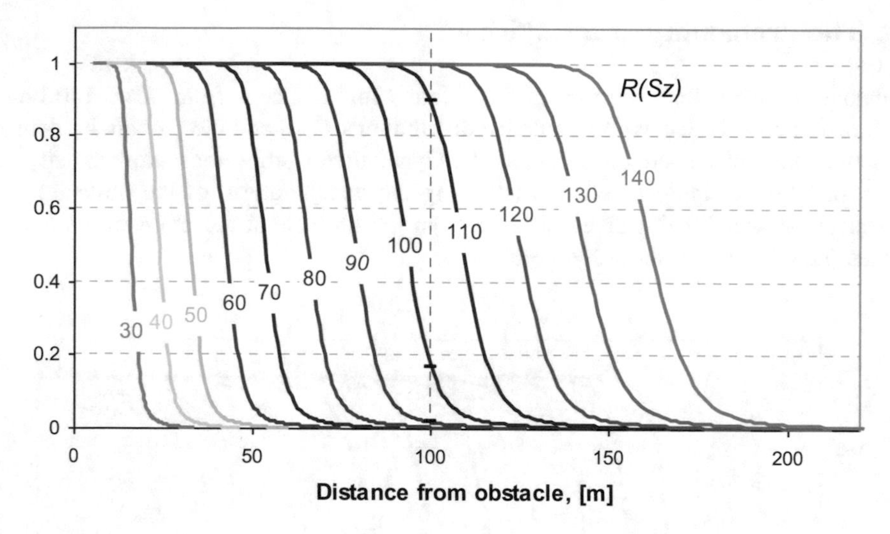

Fig. 10. Probability of collision for given obstacle distance and initial car speed

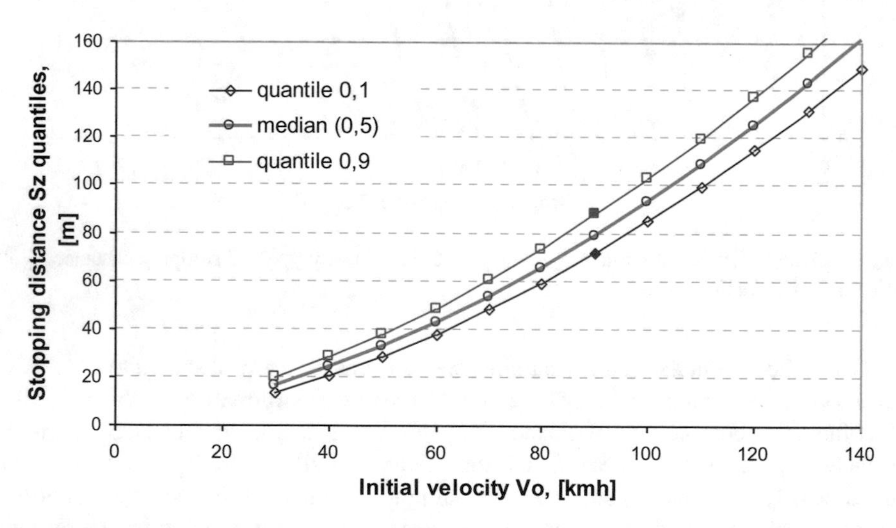

Fig. 11. Quantiles 0,1; 0,5 and 0,9 of stopping distance as a function of initial car speed

6 Conclusion

In the article there are presented the results of the drivers' reaction time research and the evaluation of the stopping distance of the car in the case of emergency braking. An important element of the study is the fact that the probabilistic character of the value of braking parameters is taken into account. The mentioned driver reaction time is one of the most important parameters affecting the stopping distance, due to its high uncertainty resulting from the features of the driver and the occurred road situation. The article uses the results of experimental research on a population of 100 drivers and the probability distribution of the reaction time obtained in these studies was used to calculate the stopping distance. Calculations made using the Monte-Carlo method allowed to obtain the probability distributions of the road stopping distance. Such distributions can be used to determine the possibility of occurrence or avoid traffic collisions with a given probability. In addition to their cognitive value, they may have practical use and help e.g. a court expert to give the court a more credible answer to the question about such a possibility.

The calculation results presented refer to a certain combination of values of braking parameters. However, there are no restrictions to make them for any combination. In the case of reaction time, additional factors (risk time, driver's age, type of accident situation, etc.) influencing the probability distribution of this parameter can be taken into account in further works. This may reduce uncertainty and increase the credibility of the assessment.

References

1. Chodnicki, P., Guzek, M., Lozia, Z., Mackiewicz, W., Stegienka, I.: autoPW driving simulator as a tool for driver-vehicle-surroundings system research. In: Weiss, Z. (ed.) Virtual Design and Automation, Sub-chapter in Chap. 5, pp. 331–338. Publishing House of Poznan University of Technology, Poznan (2008)
2. Guzek, M.: Uncertainty in the Analysis of Road Accidents, Monograph. Publishing House of Warsaw University of Technology, Warsaw (2016). (in Polish)
3. Guzek, M.: Driver reaction time in relation to evaluation method. Transport (112), 127–138 (2016). (in Polish)
4. Guzek, M., Lozia, Z., Zdanowicz, P., Jurecki, R., Stańczyk, T., Pieniążek, W.: Assessment of driver's reaction times in diversified research environments. Arch. Transp. **XXII**(2/2012), 149–164 (2012)
5. Jurecki, R.S., Jaśkiewicz, M., Guzek, M., Lozia, Z., Zdanowicz, P.: Driver's reaction time under emergency braking a car – research in a driving simulator. Maint. Reliab. **14**(4), 295–301 (2012)
6. Muttart, J.W.: Driver response in various environments estimated empirically. In: Proceedings of IX Conference "Problems of Reconstruction of Road Accidents", Zakopane (2004)
7. Prochowski, L.: Motor Vehicles. Mechanics. WKŁ, Warsaw (2008). (in Polish)
8. Stańczyk, T.: Driver's action in critical situations. Experimental research and modelling, Monographs, Studies, dissertations M43. Kielce University of Technology, Kielce (2013). (in Polish)
9. Stańczyk, T.L., Jurecki, R.: Precision in estimation time of driver re action in car accident reconstruction. In: Proceedings of XIV EVU Annual Meeting, Cracow, pp. 325–334 (2007)
10. Stańczyk, T.L., Jurecki, R., Lozia, Z., Pieniążek, W.: Influence of age and experiences of drivers on their reaction time value. Paragraf na drodze 10/2011, 339–353 (2013). (in Polish)
11. Stańczyk, T., Lozia, Z., Pieniążek, W., Jurecki, R.: Research on the reaction of drivers in simulated accident situations. In: Proceedings of the Institute of Vehicles, Warsaw, no. 1/77, pp. 27–52 (2010). (in Polish)
12. Wierciński, J., Reza, A. (eds.): Road Accidents. Vademecum of the Forensic Expert. ES Publishing House, Cracow (2011). (in Polish)

Developing a Method for Automated Creation of Interlocking Tables for Railway Traffic Control Systems

Mateusz Jurczak(✉) and Jakub Młyńczak

Faculty of Transport, Silesian University of Technology, Katowice, Poland
Jurczak.mateusz@azdpl.eu, Jakub.mlynczak@polsl.pl

Abstract. On account of the growing number of railway upgrading projects implemented in Poland, there are numerous problems emerging while preparing and performing contracts. These include adequate designing and conceptualisation of upgrading, the altering of which affects the initial assumptions made by contractors. While preparing detailed plans and specifications for railway traffic control equipment intended to secure traffic at a railway station, one must deal with numerous alterations as well as with a working mode referred to as phasing of railway traffic control works. On account of all such factors, different documentation versions are ultimately prepared, including a schematic plan and interlocking tables for railway traffic control systems. The problems caused by the interlocking tables being created incorrectly and unreliably may be solved by using a method which enables these documents to be prepared in a way which minimises the work effort required of designers and engineers. The very essence of this method is to create a model whose basic components are interdependent in such a manner that every alteration in the design part, as it is referred to, serves to generate completely new and automatically updated interlocking tables. The article provides a follow-up to the authors' previous studies in which they also addressed interlocking tables and the problems tackled while designing them. What has been proposed is to make a division of the relevant elements (objects) on the basis of which algorithms could be developed for the designing phase. Using such data, one can generate routes for the station being designed.

Keywords: Railway traffic control · Interlocking tables · Railway route

1 Introduction

1.1 Railway Route

A railway route should be understood as a set of ordered states that should be the states of elements of the railway traffic control equipment which set, secure and monitor a certain block [2].

There are two kinds of routes which must be taken into consideration in an interlocking table:

(a) Train route, characterising the route in which trains run [1];
(b) Shunting route – a route set for shunting vehicles [1].

M. Siergiejczyk and K. Krzykowska (Eds.): ISCT21 2019, AISC 1032, pp. 194–203, 2020.
https://doi.org/10.1007/978-3-030-27687-4_20

Such a set of routes along with their elements constitute a complete record of relationships in play at the given station. It defines the routes which may be covered simultaneously and the elements (railway traffic control devices) to be engaged for that purpose. Below is a railway route example (see Fig. 1).

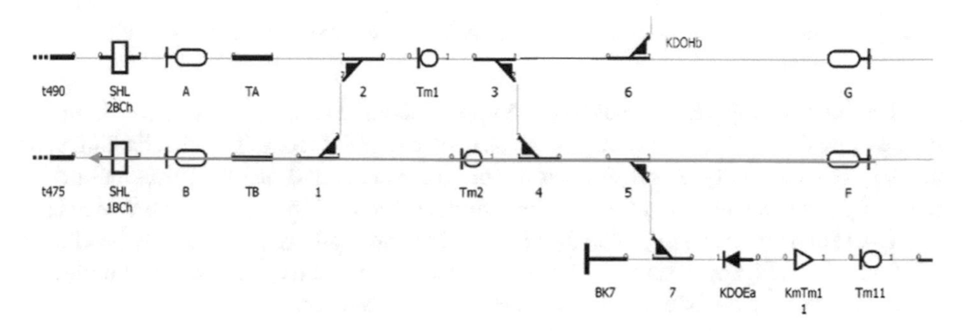

Fig. 1. Example of a train route starting at signal F [3].

The manner in which the route has been covered is described by the following equation:

$$R = P \cap D \cap S \cap TC \cap LB \cap LC \tag{1}$$

One should distinguish between individual factors whose specific states allow for the route (R) to be covered:

- Point,
- Derailer,
- Signal,
- Track circuit,
- Line block,
- Level crossing.

1.2 Interlocking Table

Interlocking tables are used to represent the location and state of the devices engaged in a train or shunting route in a legible manner. They are created with reference to a schematic plan of the given railway station (see Fig. 2). Interlocking tables are included in the internal documentation of railway traffic control systems, and they are intended for train dispatchers to safely manage traffic at the railway station from signal boxes.

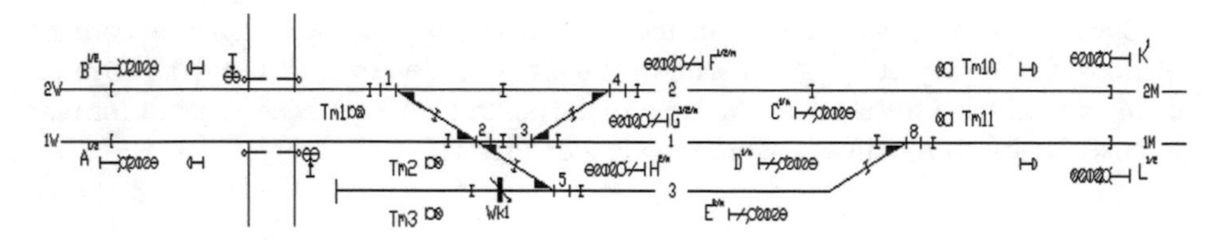

Fig. 2. Sample schematic plan of railway traffic control devices on a fictive station [4].

The interlocking table consists of the upper section, known as an interlocking table header, and the lower section, referred to as a table of point locks. The interlocking table header contains the type and number of internal control and block devices, while the table of point locks indicates the locks and mutual relationships between the devices [1].

The sample interlocking table (see Fig. 3) illustrates all organised routes handled at the Poznań PoD (proper name of the post) junction signal box, distinguished under the upgrading project for line 351 between Poznań and Szczecin.

L.P.	ROUTE	ROUTE DESCRIPTION From	To	ALTERNATIVES	ISOLATED TRACK AND POINT SECTIONS track sections	point sestions	sections in side protection
1	A	Poznań Główny, tor 2PG	tor 2a		A, 2a		
2							
3	B	Poznań Główny, tor 1PG	tor 1a		B, 1a		
4							
5	C	track 1a	Poznań Główny, tor 1PG		B		
6							
7	D	track 2a	Poznań Główny, tor 2PG		A		
8							
9	E	track 2a	Poznań Wola, tor 2PW		2b, K, 50, 50a	2/3, 6	
10	E	track 2a	Poznań Wola, tor 1PW	z	2b, J, 33, 39a	2/3, 6, 7	[illegible]
11	E	track 2a	Poznań Wola, tor 1PW	w	2b, J, 33, 39a	2/3, 4/5, 7	[illegible]
12	E	track 2a	Suchy Las, 2S	z	2b, H	2/3, 6, 7	[illegible]
13	E	track 2a	Suchy Las, 2S	w	2b, H	2/3, 4/5, 7	[illegible]
14	E	track 2a	Suchy Las, 1S		2b, G	2/3, 4/5	[illegible]
15							
16	F	track 1a	Poznań Wola, tor 2PW		1b, K, 50, 50a	1, 2/3, 6	[illegible]
17	F	track 1a	Poznań Wola, tor 1PW	z	1b, J, 33, 39a	1, 4/5, 7	[illegible]
18	F	track 1a	Poznań Wola, tor 1PW	w	1b, J, 33, 39a	1, 2/3, 6, 7	[illegible]
19	F	track 1a	Poznań Wola, tor 1PW	w	1b, J, 33, 39a	1, 2/3, 4/5, 7	[illegible]
20	F	track 1a	Suchy Las, 2S	z	1b, H	1, 4/5, 7	[illegible]
21	F	track 1a	Suchy Las, 2S	w	1b, H	1, 2/3, 4/5, 7	[illegible]
22	F	track 1a	Suchy Las, 2S	w	1b, H	1, 2/3, 6, 7	[illegible]
23	F	track 1a	Suchy Las, 1S	z	1b, G	1, 4/5	[illegible]
24	F	track 1a	Suchy Las, 1S	w	1b, G	1, 2/3, 4/5	[illegible]
25							
26	G	Suchy Las, 1S	track 1a	z	G, 1b, 1a	4/5, 1	[illegible]
27	G	Suchy Las, 1S	track 1a	w	G, 1b, 1a	4/5, 2/3, 1	[illegible]
28	G	Suchy Las, 1S	track 2a		G, 2b, 2a	4/5, 2/3	[illegible]
29							
30	H	Suchy Las, 2S	track 1a	z	H, 1b, 1a	7, 4/5, 1	[illegible]
31	H	Suchy Las, 2S	track 1a	w	H, 1b, 1a	7, 4/5, 2/3, 1	[illegible]
32	H	Suchy Las, 2S	track 1a	w	H, 1b, 1a	7, 6, 2/3, 1	[illegible]
33	H	Suchy Las, 2S	track 2a	z	H, 2b, 2a	7, 6, 2/3	[illegible]
34	H	Suchy Las, 2S	track 2a	w	H, 2b, 2a	7, 4/5, 2/3	[illegible]
35							
36	J	Poznań Wola, tor 1PW	track 1a	z	J, 1b, 1a	7, 4/5, 1	[illegible]
37	J	Poznań Wola, tor 1PW	track 1a	w	J, 1b, 1a	7, 6, 2/3, 1	[illegible]
38	J	Poznań Wola, tor 1PW	track 1a	w	J, 1b, 1a	7, 4/5, 2/3, 1	[illegible]
39	J	Poznań Wola, tor 1PW	track 2a	z	J, 2b, 2a	7, 6, 2/3	[illegible]
40	J	Poznań Wola, tor 1PW	track 2a	w	J, 2b, 2a	7, 4/5, 2/3	[illegible]
41							
42	K	Poznań Wola, tor 2PW	track 1a		K, 1b, 1a	6, 2/3, 1	[illegible]
43	K	Poznań Wola, tor 2PW	track 2a		K, 2b, 2a	6, 2/3	
44							

Fig. 3. Example of an interlocking table [5].

It should be stressed how much information of essential nature this document contains. The five leftmost columns pertain to routes, and they are successively:

- Item number;
- Route notation;
- Route beginning;

- Route end;
- Where more than one running movement has the same beginning and the same end, movement path variants are specified (alternatively, variants of what is referred to as an overlap);
- Another part of the table consists of columns providing data of all routes which, when combined with respective lines, form a square matrix (43 lines and columns) containing mutually exclusive routes;
- The columns marked as "Points" form the section known as the table of point locks. It contains data of all points and derailers along with designations of their locations within routes;
- The last part corresponds to all isolated track and point sections within the routes movement path and under flank protection.

The example provided is lacking such information as:

- Block signalling systems the state of which affects the execution of a station's outbound routes;
- Level crossings dependent on routes;
- Isolated sections within an overlap, which also affect the possibility of enabling a signal to proceed.

2 The Concept of the Model

2.1 Requirements Towards the Model

Studying the process in which interlocking tables are created on the basis of a plan of a station's track layout involves the following elements (see Fig. 4).

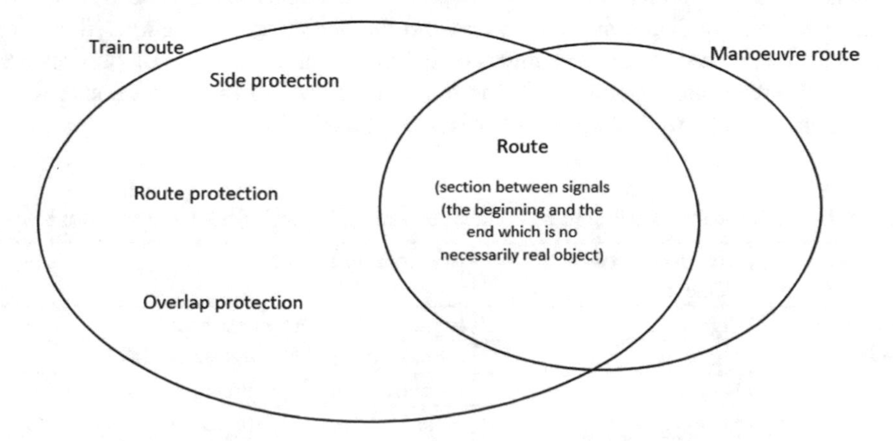

Fig. 4. Elements involved in the process of creating interlocking tables (authors' own study)

Modelling of the process of creating interlocking tables is understood as examining and describing relationships between elements of this process in order to determine the relevant requirements and the way to perform this process. The model comprises two fundamental elements:

- Design component,
- Relationship (generation) component.

2.2 Design Component

Table 1 provides a breakdown of the model's design component according to the occurrence of its elements in individual process elements. It should be noted that protective elements particularly characteristic of the train route (overlap, flank protection) are described separately on account of the additional types of relationships.

This is the reason why they are specified as separate elements in Tables 1 and 2.

Table 1. Breakdown of the design component according to route type (authors' own study).

Process element	Design component's element	Relationship type
Train route	Signal	Route beginning or end
	Level crossing	One-sided dependence in station equipment
	Block signalling system	Dependence on the block signalling system's state
	Fictive signal	Route end
Shunting route	Signal	Route beginning or end
	Shunting signal	Route beginning or end
	Fictive signal	Route end

It is characteristic that the design component's elements being parts of a specific route type typically define the route's end and the beginning of its enforced locking. This is the route boundary. The elements contained in Table 2 are the components used to determine the route's specific path and to secure the latter in relation to other routes (flank protection) as well as speed and distances (overlap).

Table 2. Breakdown of the design component according to block parts (authors' own study).

Process element	Design component's element	Relationship type
Movement path	Signal	Signal state in movement path
	Point	Element determining movement path
	Derailer	As above, only for shunting routes
	Level crossing	State in movement path depending on category
	Track or point circuit	State in movement path, particularly for train routes

(*continued*)

Table 2. (*continued*)

Process element	Design component's element	Relationship type
Overlap	Point	Element determining overlap
	Derailer	Special case of front protection
	Level crossing	State in overlap
	Track or point circuit	State in overlap, particularly for train routes
Flank protection	Point	Protective element, particularly train routes
	Derailer	Protective element, particularly train routes
	Track or point circuit	State in flank protection

The design component's elements summarised in Tables 1 and 2 are characterised by their own specific attributes which directly affect the notation in interlocking tables. As an example, one may refer to the most complex object, i.e. a point. It is a particularly important part of a route, as it determines the movement path but may also be involved in block protection. There may be serious consequences to how a point is individually planned over the track layout and noted in the interlocking table. Errors in this matter extremely rare at signal boxes handling relatively uncomplicated track layouts. Nevertheless, one must take special precautions when designing more complex stations. Below is an example of a tool bar in AutoCad (see Fig. 5).

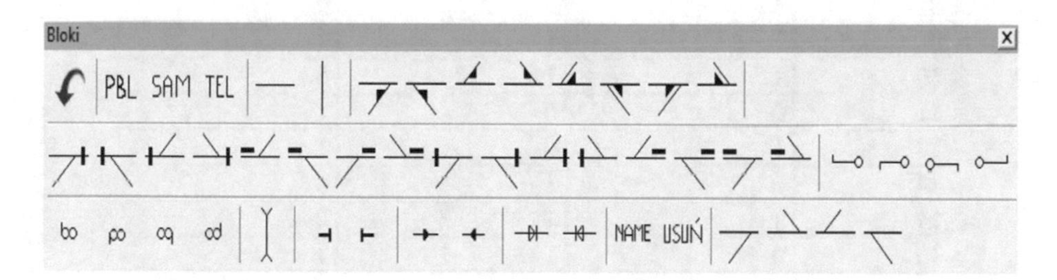

Fig. 5. Objects in the *Blocks* tool bar (authors' own study).

Figure 5 shows 8 different point types with 3 attributes defined in their names.

2.3 Attributes of Objects and an Example of Their Use

The most fundamental attributes – those that are subsequently to be used when creating routes – taken into account when generating positions and locks of points are as follows:

- X – turn in facing point movement,
- Y – direction of side movement,
- Z – point's normal position.

Specific attributes have been assigned to individual values of X, Y and X in the Table 3.

Table 3. Point attributes and their values (authors' own study).

Attributes/values	0	1
X	Right	Left
Y	Right	Left
Z	Straight	Sideways

With such a breakdown, one can investigate which of the attribute values exerts the greatest impact on specific routes at the station. Where this is the case, every single railway route should be represented as a function of three variables: x, y and z. For that reason, when considering a sample signal box (see Fig. 1), one obtains the following table containing values of individual attributes for points (see Table 4).

Table 4. Values of attributes for all fictive points at signal box [6].

Point number	Value of attribute X	Value of attribute Y	Value of attribute Z
1	0	1	0
2	1	1	0
3	0	0	1
4	1	0	1
5	0	1	0
6	0	1	1
7	0	1	1
8	1	1	0
9	1	1	1
10	1	1	0
11	0	0	1
12	1	0	1
13	0	1	0
14	1	1	0
15	0	1	1
16	1	1	0
17	1	1	0
18	0	1	0
19	1	1	0
20	0	0	1
21	1	0	1

Table 4 summarises data of all points handled at the signal box shown in Fig. 1. Each of them has specific attribute values and can be distinguished from the remaining ones. These values can naturally come in greater numbers, which depends on what object is analysed and what one intends to verify.

Another table (see Table 5) represents a chosen route beginning at signal A and extending between adjacent stations in direction K on track 1. Such functions of three variables may be represented as a sum of minimum polynomials, each multiplied by an appropriate coefficient equalling 0 or 1.

Table 5. Table of the function of route A to 1K

x	y	z	F = A to 1K	P
0	0	0	0	$P_{000} = \bar{x}\bar{y}\bar{z}$
0	0	1	0	$P_{001} = \bar{x}\bar{y}z$
0	1	0	1	$P_{010} = \bar{x}y\bar{z}$
0	1	1	1	$P_{011} = \bar{x}yz$
1	0	0	0	$P_{100} = x\bar{y}\bar{z}$
1	0	1	0	$P_{101} = x\bar{y}z$
1	1	0	1	$P_{110} = xy\bar{z}$
1	1	1	1	$P_{111} = xyz$

Function F may be noted as $F = f(x, y, z)$, where the variables are turn, direction and basic position. What was taken into account and used in the calculations was the properties of the canonical form of the sum characterising Boolean functions (2) and (3).

$$F = f(x, y, z) = (0 * P_{000}) + (0 * P_{001}) + (1 * P_{010}) + (1 * P_{011}) + (0 * P_{100}) + (0 * P_{101}) + (1 * P_{110}) + (1 * P_{111}), \tag{2}$$

therefore

$$F = \overleftrightarrow{\bar{x}y\bar{z} + \bar{x}yz} + \overleftrightarrow{xy\bar{z} + xyz}, \tag{3}$$

by putting the factor in front of the bracket and by following the law of contradiction for sum (4),

$$\bar{z} + z = 1 \tag{4}$$

one obtains the following function:

$$F = \overleftrightarrow{\bar{x}y + xy} \tag{5}$$

$$\bar{x} + x = 1 \tag{6}$$

$$1 + y = y \tag{7}$$

Once properties (6) and (7) have been applied, the function assumes the following final form:

$$F = y \tag{8}$$

By that means, one can determine the attribute exerting the greatest impact. However, other attributes of objects which affect the notation of routes in the interlocking table are also their coordinates. Since the simplified schematic diagram of a station has been developed in the AutoCad format, each element can be distinguished from others by considering such attributes as:

- Name,
- Coordinate X,
- Coordinate Y,
- Turns.

The example provided in Fig. 6 represents an algorithm for inserting a single signal object. On account of the complexity of the data, which should be used in the relationship component to generate interlocking tables, additional filters need to be applied, and they are special-purpose layers in this case.

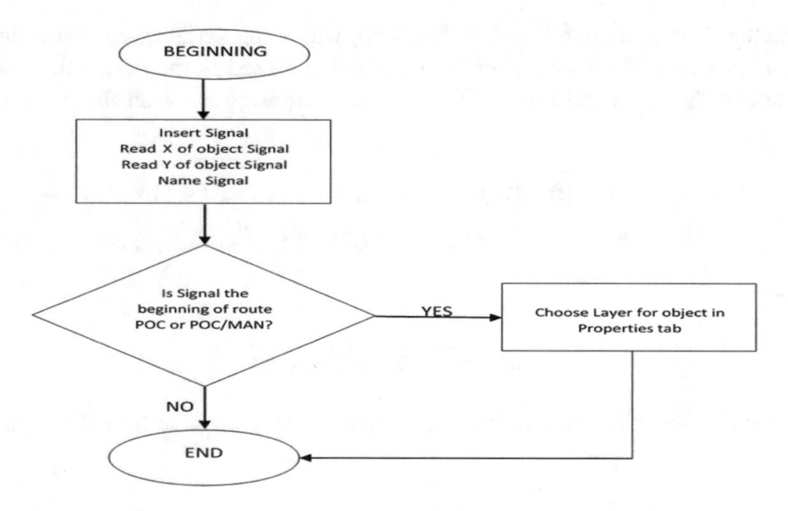

Fig. 6. Algorithm for inserting a single signal

What characterises the above transition graph is that the designer should be allowed to draw the station plan and to choose basic properties of objects in the simplest manner available. Designing the diagram and generating tables should be very smooth and as uncomplicated as possible. Furthermore, this is absolutely prerequisite when considering a signal box where the number of points handled is around one hundred or more.

3 Conclusions

Interlocking tables are necessary documents at every railway station in Poland. As a railway line is being upgraded, one is often forced to phase the works related to the railway traffic control equipment. Consequently, successive versions of the tables are created. It seems that the most popular computer program commonly used by design offices is AutoCad. For that reason, both the proposed solution and its follow-up comprising adequate representation and characterisation of the relationship component may significantly enhance all the design activities.

The process will be performed using an original tool created in commercial off-the-shelf software, Visual Studio Community, using a dedicated code developed for this purpose exclusively. At the first stage of the designing phase, all objects comprising the station setup will be extracted from the dxf file along with the variables describing their properties. All the objects will be divided according to the types they represent, e.g. points, signals, etc. In the next step, routes are generated along with their components (drive-on points, flank protected points, etc.).

References

1. Technical guidelines for designing railway traffic control equipment Ie-4 (WTB-E10). PKP Polskie Linie Kolejowe S.A. Warszawa (2017). (in Polish)
2. Dąbrowa-Bajon, M.: Podstawy sterowania ruchem kolejowym (Fundamentals of railway traffic control). Oficyna Wydawnicza Politechniki Warszawskiej, Warsaw (2007). (in Polish)
3. Jurczak, M.: Wpływ zmian układu stacyjnego na zależności (Effect of alterations in the station layout on relationships). Graduation dissertation for post-graduate studies, Katowice (2015). (in Polish)
4. Jurczak, M.: Structure of interlocking table. Arch. Transp. Syst. Telemat. **4**(2) (2011)
5. AŽD Praha S.R.O.: Zabudowa stacyjnych urządzeń srk typu ESA 44-PL na p. odg. Poznań PoD, Tablica zależności (Installation of type ESA 44-PL station railway traffic control equipment at the Poznań PoD junction signal box. Interlocking table), Prague (2019). (in Polish)
6. Jurczak, M.: Przebieg kolejowy jako funkcja trzech zmiennych (Railway route as a function of three variables), Szczyrk (2018). (in Polish)
7. Ostasz, M.: Metoda opisu formalnego układów torowych i charakterystyk techniczno-ruchowych stacji kolejowej oraz algorytmy wybranych problemów automatyzacji prac projektowych urządzeń sterowania ruchem kolejowym. rozprawa doktorska, Politechnik Warszawska, Instytut Transportu, Warszawa (1973). (in Polish)
8. Apuniewicz, S., Cegłowski, L.: Projektowanie urządzeń sterowaniem ruchem kolejowym. Wydawnictwo Politechniki Świętokrzyskiej, Kielce (1974). (in Polish)
9. Zając, L., Wdowiak, W., Mitura, G., Gryłka, A., Szafruga, J.: Tabzal- system wyznaczania tablic zależności i kart przebiegów dla urządzeń srk. Katowice (1993). (in Polish)
10. Sumiła, M.: Metoda tworzenia oprogramowania sterującego w systemach sterowania ruchem kolejowym. rozprawa doktorska, Politechnika Warszawska, Wydział Transportu, Warszawa (2007). (in Polish)

An Influence of Design Features of Tramway Vehicles on Kinematic Extortion from Geometry of a Track

Dariusz Kalinowski[1(✉)], Robert Konowrocki[2], and Tomasz Szolc[2]

[1] PESA Bydgoszcz SA, Zygmunta Augusta 11 Street, Bydgoszcz, Poland
dariusz.kalinowski@pesa.pl

[2] Institute of Fundamental Technological Research of the Polish Academy of Sciences, Pawińskiego 5B Street, Warsaw, Poland

Abstract. In the paper, simulation results of safety against derailment for a tramway vehicles with an arbitrary configuration of wagons and bogies is presented. The existing European standard EN 14363 covers all necessary tests for different railway vehicles, but it is inadequate for tramway vehicles, especially in safety against a derailment examination. Its operational conditions are much different. The described observations suggest that the methodology of safety against derailment testing described in the EN 14363 standard cannot be used without any modifications in the case of testing of tramway vehicles. On the basis of the computational results, a significant influence of different configurations of urban tramway vehicles on the wheel-rail contact forces was discussed, in particular on the Y/Q derailment factor.

Keywords: Tramway dynamics · Safety against derailment · Simpack

1 Introduction

Derailment has been a major concern for railway operations since the first day of wheels running on rails. Advances in technology have resulted in increasing speeds of railway vehicles. Increased speeds and wheel loads can result in greater consequences for casualty and property loss in the event of derailment.

The advances in computer simulation techniques and test measurement methods have resulted in a variety of derailment safety assessment methods being applied worldwide, and a corresponding wide array of regulations for the acceptance of new vehicles. Significant advances have occurred since the authors of paper [4] provided a review of the state-of-the-art methods for vehicle assessment.

The subject of a railway vehicle riding safety is still a topic taken up by researchers from around the world [1, 2, 5, 6, 8–14, 17, 18]. In most cases, the evaluation criteria based on the value of safety against derailment index *Y/Q*, wheel unloading and wheel lift D_z are used for riding safety analyses.

This paper is intended to an introduction to the study of light rail vehicles to classify the influence of their design and configuration on safety against derailment

M. Siergiejczyk and K. Krzykowska (Eds.): ISCT21 2019, AISC 1032, pp. 204–214, 2020.
https://doi.org/10.1007/978-3-030-27687-4_21

problems. Numerical vehicle dynamics simulations were carried out in order to determine safety against derailment index *Y/Q* and wheel lift D_z for different tram carbodies and bogies configuration.

2 Numerical Simulations of Tram Vehicle Models

2.1 Numerical Models

In order to determine theoretical values of forces in the wheel-rail contact zones, numerical models of three tram vehicles were created. These models are implemented into the computer simulation program Simpack in the form of a system of rigid bodies mutually connected by springs and dampers of linear characteristics. Such system of rigid bodies interacts with the railway track through the higher kinematic pairs of wheels and rails profiles which are inputted in the form of coordinates measured on real PST/Ri60N profiles shown in Fig. 4. This approach to modeling is called the multi-body simulation (MBS) method [6, 15].

To verify safety against derailment, simulation models of tram vehicles were prepared in various configurations shown in Fig. 1. Here, different kinds of carbody-carbody couplings were also considered.

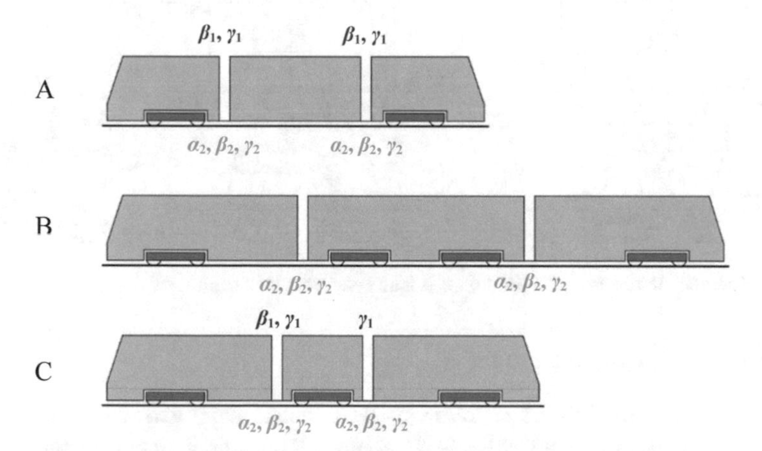

Fig. 1. Vehicle configurations for test of safety against derailment: A – 3-car articulated tramway with non-swiveling bogies; B – 3-car articulated tramway with swiveling bogies; C – 3-car articulated tramway with swiveling (outer) and non-swiveling (middle) bogies

Each of the tramway simulation models was divided into three basic elementary groups, which are wheelsets, bogie frames and carbodies. The wheelset was assumed as a rigid body with three degrees of freedom (DOF). The bogie frame is represented by a rigid body with five degrees of freedom, and the carbody model has five degrees of freedom, as well.

Motion of such structures of a tramway vehicle is mathematically described by means of second order ordinary differential equations. Assuming that oscillations of each particular tramway model relative to the reference system can be regarded as small, such motion can be described by means of the system of linear equations formulated here in the matrix form (1).

$$\left[\mathbf{M}\frac{d^2}{dt^2} + \mathbf{C}\frac{d}{dt} + \mathbf{K}\right] \cdot \mathbf{q} = \mathbf{F} \tag{1}$$

where: **q** denotes the vector of generalized coordinates, **M** is the symmetrical mass matrix, **C** describes the damping matrix, **K** is the stiffness matrix, **F** denotes the force vector and d/dt is the differential operator. A similar approaches and description of tramway models we can find in the papers [3, 16]. Due to the small paper volume, the authors limited themselves only to the shortened vector-matrix form of the equation of motion. The scheme of the physical model of the single tramway carbody with the bogie is shown in Fig. 2. Depending on the number of bogies and used carbodies, the presented model can be properly made complicated.

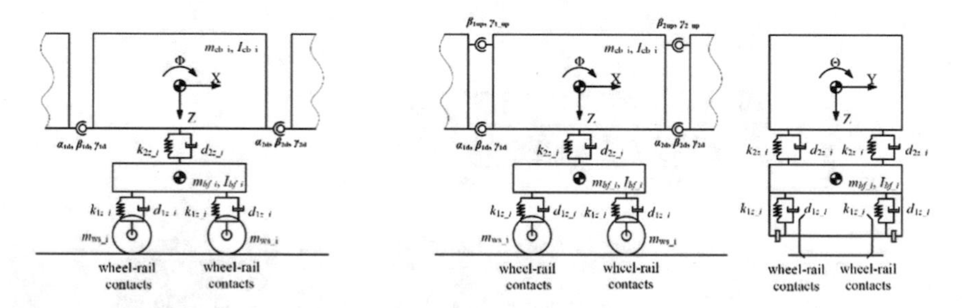

Fig. 2. Scheme of the vehicle model of a tramway at two variants of the carbody couplings

2.2 Wheel-Rail Contact

The mathematical models of a tramway have been integrated with the numerical procedures that described the wheel-rail contact. The model of wheel-rail contact is based on Kalker's simplified theory and the FASTSIM algorithm [10]. In order to calculate tangential contact forces the algorithm requires such input data as normal contact forces, coefficient of friction, length of the semi-axes of the contact ellipses (calculated using Hertz theory) [7], and creep values which are given in the form of relative rigid slip (2).

$$\begin{bmatrix} rs_x \\ rs_y \\ rs_z \end{bmatrix} = \frac{1}{vu_x}\begin{bmatrix} sv_x \\ sv_y \\ sv_y \cdot \sin(\alpha) + sv_z \cdot \cos(|\alpha|) \end{bmatrix}, \tag{2}$$

where rs_x, rs_y describe the creepages (relative rigid slip) in the longitudinal and lateral directions, r_{sz} denotes the spin, α is the contact angle, vu describes the speed of the moving reference frame (equal to the tramway speed); and sv is the slip velocity (3):

$$\begin{bmatrix} sv_x \\ sv_y \\ sv_z \end{bmatrix} = \begin{bmatrix} vu_x \\ 0 \\ 0 \end{bmatrix} + \begin{bmatrix} 0 \\ vr_y \\ 0 \end{bmatrix} + \bar{\omega}_w \times \bar{r}, \tag{3}$$

where ω_w is the relative angular velocity of the wheelset; r denotes the coordinates of the contact point in the reference frame connected to the wheelset mass centre; and vr describes the relative velocity of the wheelset mass centre.

The contact geometry parameters, that depend on the transverse displacement of a particular wheel, as input values to the FASTSIM procedure, i.e.: actual rolling radius, contact ellipse longitudinal half diameter and contact point coordinate on rail and on wheel have been shown in Fig. 3. The nominal PST profile shape [20] was assumed as a wheel profile and Ri60N profile shape was used for rails in the simulation models for testing of dynamic properties of the tramway vehicles. A contact geometry of the modelled tramway wheel/rail is presented in Fig. 4.

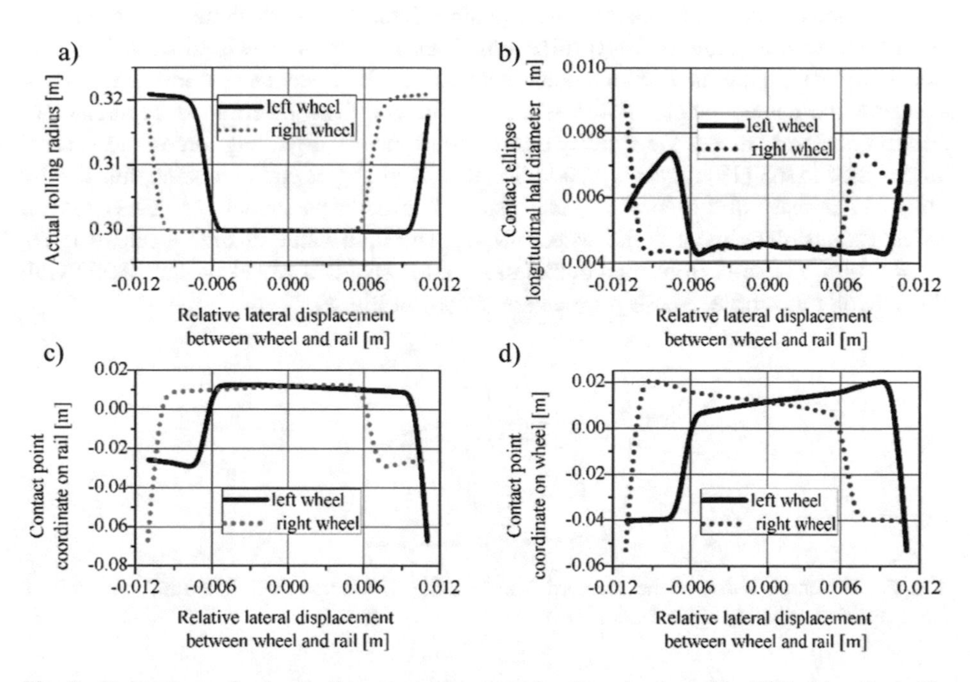

Fig. 3. Parameters of wheel-rail contact geometry for the wheel profile PST and rail profile Ri60N: (a) actual rolling radius, (b) contact ellipse longitudinal half diameter, (c) contact point coordinate on rail, (d) contact point coordinate on wheel as a function of wheelset lateral displacement y

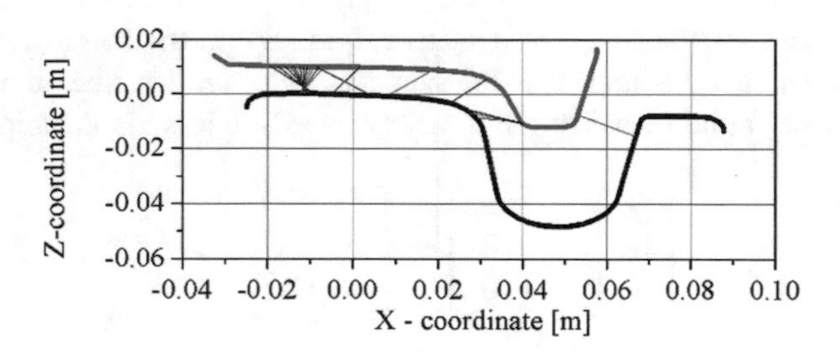

Fig. 4. Wheel rail contact position for the wheel profile PST and rail profile Ri60N

2.3 The Requirements of EN 14363 for Safety Against Derailment

Safety against derailment of railway vehicles is defined in EN 14363:2016 standard [19]. In order to confirm safety of railway vehicle negotiating curve with a small radius, safety against derailment index *Y/Q* is determined (a quotient of the lateral guiding force and the vertical force of the wheel on the rail) and a measurement of the lift of the guiding wheel D_z. The tests are performed on a track with a radius of R = 150 m with a twisted section of 3‰. In accordance with EN 14363, the twist is achieved by changing height of the outer rail. Here, a typical UIC60 profile is used. Then, EN 14363:2016 [19] recommends that simulation safety against derailment tests should be carried out on a track with a gauge of 1440–1465 mm. During tests on a twisted track it is also necessary to measure the friction coefficient between the wheel and the rail as well as to measure the contact angle of the wheel on the rail. The determined coefficient of friction is used to assess the validity of the measurement of guiding forces and vertical forces. EN 14363 [19] requires an additional twist of bogies and carbodies, fulfilled by shims in primary and secondary suspension. The main parameter for assessment of safety against derailment is the wheel lift D_z. The limit value of the wheel lift is $D_{z, lim}$ = 5 mm. It results from the combination of the standard wheel profile S1002 with the UIC60 rail profile. These profiles are shown in Fig. 5.

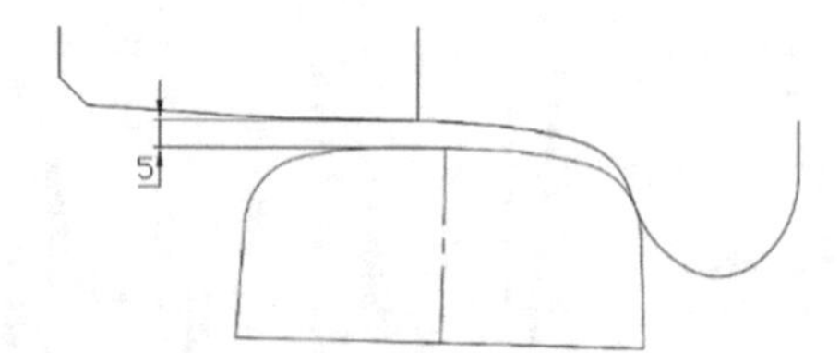

Fig. 5. The combination of the standard wheel profile S1002 and the UIC 60 rail profile for a limit position of safety against derailment

In Fig. 5 the wheel is shown in a limit position related to the rail head. The point of contact between the wheel and the rail is located at the end of the straight section of the flange. With a high probability further lifting of the wheel may lead to derailment.

2.4 Assumptions for Simulation Load-Cases

The methods presented in EN 14363:2016 [19] cannot be directly related to trams. These vehicles (especially low-floor ones) have much more complex kinematic systems than railway vehicles, where practically two configurations are used: "wagon" configurations (one carbody based on two bogies) or "articulated" configurations (adjacent carbodies based on common bogies). Trams usually run in the urban area and negotiate curves with much smaller radii than R = 150 m which is required in EN 14363. It does not have to be a critical curve for these vehicles. In view of the above, due to the lack of normative guidelines, the authors developed an original methodology for simulational testing of safety against derailment of trams on curves with different radii.

The document "Technical guidelines for the design and maintenance of tram tracks" [21] defines the radii of track curves used in Polish tram systems and it provides the minimum and recommended cant value for these curves. These guidelines assume an inclination ramp of 1:300. An 1:150 inclination was taken into consideration as a more critical case, which corresponds to the track twist of ca. 6.67‰. Radii of 18, 25, 50 and 150 m were selected as representative. The simulation analysis of safety against derailment was performed on each of the above curves. At the curve exit on the outer rail an additional twist of about 3,33‰ was added (a 20 mm rail height shift on a distance of 6 m), which resulted in a total twist of 10‰. Additional twist is a safety margin, that represents the local bad condition of the tram infrastructure. According to [21], the Ri60N grooved rail was used on curves with radii up to R = 150 m. Here, a nominal tram track width of 1435 mm was assumed. Trams are equipped with a significant number of rubber and metal elements, stiffness values of which depend on the tolerances and weather conditions (stiffening in low temperatures). It was taken into account by introducing the stiffening factor of rubber and metal elements equal to 2 for each of the simulation cases. Additional stiffening factors of bogies primary and secondary suspension elements were assumed to define their impact on the assessment parameters: safety against derailment factor *Y/Q* and wheel lift D_z. The combination of track and vehicle suspension parameters assumed for simulations is shown in Table 1.

Table 1. Combination of track and vehicle suspension parameters

Curve radius [m]	Velocity [m/s]	Stiffening factor of: (gradation: 0,1)	
		Primary suspension [−]	Secondary suspension [−]
18, 25, 50, 150	1 m/s	1,0–1,5	1,0–1,5

It is necessary to determine an assessment criterion of safety against derailment of the tram. The value of wheel lift is the most important parameter that defines the safety against derailment margin. In the case of trams, however, different profiles of wheels and rails are used than in railway vehicles. In Polish tram networks there are mainly two wheel profiles used, i.e. "T" and "PST" profiles. Their limit positions on the grooved rail Ri60N are shown in Fig. 6.

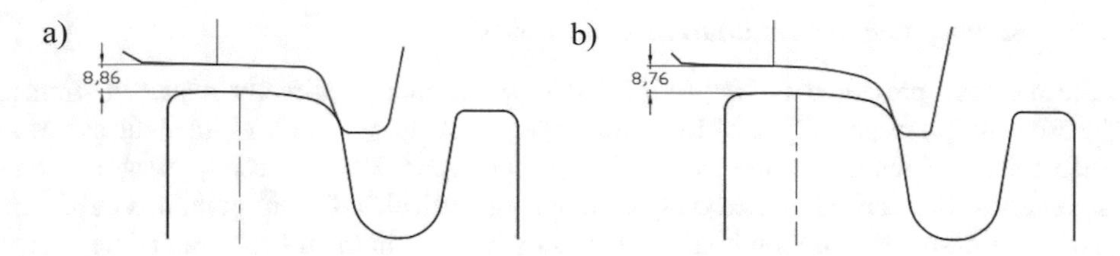

Fig. 6. The combination of wheel profiles T (a) or PST (b) and the Ri60N rail profile for a limit position of safety against derailment

The minimum wheel lift value, achieved for the combination of T/PST wheel profiles and the Ri60N rail profile, was D_z = 8.76 mm. According to the above, the value of $D_{z,lim}$ = 8 mm is proposed as the limit value for a determination of safety against a derailment of tramway vehicles with T or PST profiles.

3 Results of Numerical Simulations

In the paper there was presented the safety against derailment index and the wheel lift guiding (attacking) tram wheels in various configurations depending on the radius of the track curve and the vertical stiffness of primary and secondary suspension elements. As the original contribution of the work, a different track configuration has been assumed than in standard [19]. During these tests there was adopted the most unfavorable combination of the vehicle's operational configuration, i.e. the empty vehicle. During simulation tests the vertical wheel load forces Q, the lateral forces Y and the wheel raise D_z for guiding wheel were recorded. On the basis of the determined forces in the contact zone, safety against derailment indexes *Y/Q* were determined as the maximum value of the ratio of transverse force to the vertical force of the wheel on the rail. This criterion is commonly used to evaluate proneness to a derailment situation of a single wheel. The admissible value of the Y/Q index can be calculated using Nadal's formula (4):

$$\frac{|Y|}{|Q|} \leq \frac{tg\alpha - \mu}{1 + \mu \cdot tg\alpha} \tag{4}$$

where α is the angle made when the wheel flange is in contact with the rail edge, Y and Q refer to the lateral and vertical forces acting between the rail and wheel, and μ is the coefficient of friction between the wheel and the rail. For the wheel flange inclination angle α = 70° of the assumed the wheel profile T/PST and for the predefined friction coefficient μ = 0.36, the limit of safety against derailment index is $(Y/Q)_{lim}$ = 1.20. Exceeding this value may result in an increased probability of the derailment. Figure 7 presents the wheel lift of guiding wheel for tram configuration C as a function of the radius of the track curve and the stiffening factor of the primary and secondary suspension.

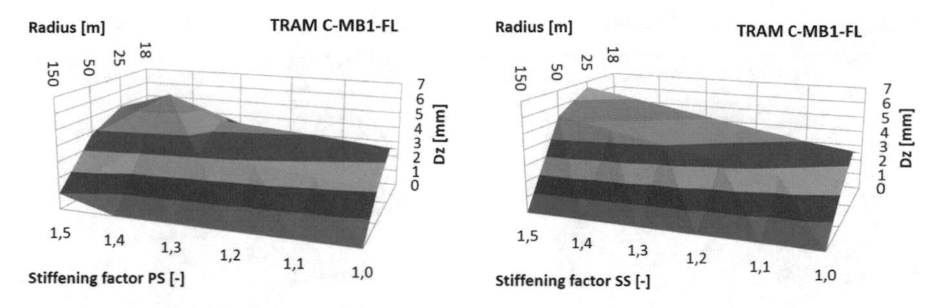

Fig. 7. Wheel lift D_z of guiding wheel for tram configuration C as a function of the radius of the track curve and the stiffening factor of the primary and secondary suspension.

For trams A and B the wheel lift did not occurred for any of assumed simulation conditions, so the graphs for these configurations were omitted. Indexes Y/Q for the guiding wheels of the analyzed tram configurations that depend on the radius of the curve and the stiffening factor of the primary and secondary suspension are shown in Figs. 8, 9 and 10. In these figures the results for guiding wheels of leading bogies are presented. In the following figures the configurations of trams have been marked from A to C according to Fig. 1.

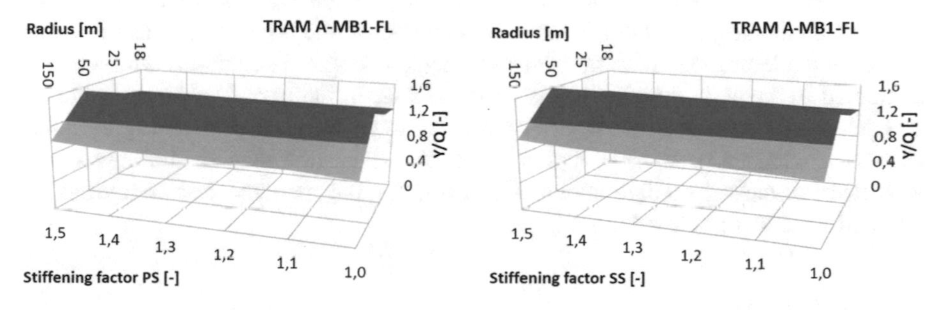

Fig. 8 Safety against derailment with index *Y/Q* as a function of the radius of the track curve and stiffening factor of the primary and secondary suspension for guiding wheel of tram A

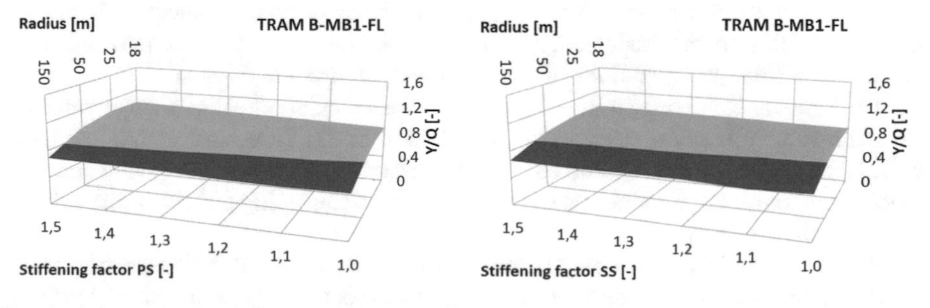

Fig. 9 Safety against derailment with index *Y/Q* as a function of the radius of the track curve and stiffening factor of the primary and secondary suspension for guiding wheel of tram B

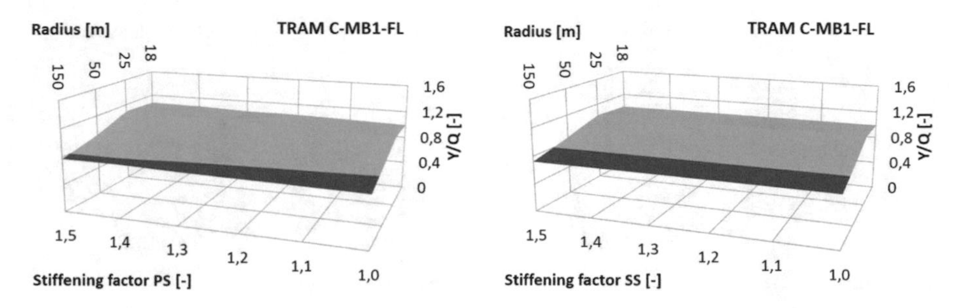

Fig. 10 Safety against derailment with index *Y/Q* as a function of the radius of the track curve and stiffening factor of the primary and secondary suspension for guiding wheel of tram C

For the first two configurations of trams, i.e. A and B, no wheel lift appeared in any of the simulational cases. Both types of trams provide a high level of safety against derailment for the assumed conditions. Despite the double stiffening of rubber-metal elements and adding additional stiffening in the range of 10–50%, the wheel lift of guiding wheel of the C-type tram did not exceed the assumed 8 mm limit. The maximum wheel lift for tram type C equal to 6.6 mm was achieved for the radius of R = 25 m and 50% stiffening of the primary and secondary suspension.

An explanation of a the large difference in the obtained values of the wheel lift for the all considered configurations should be seen in the way where the carbody is mounted on the bogie. In tram A the body is based on the bogie by means of flexicoil springs and a lemniscate with traction rods. Type B bogies is equipped with a bolster, connected to the carbody with a large-size bearing. In the tram C outer carbodies are connected to bolsters of motor bogies by pivots and based on transom slide plates. Such solution is characterized by a higher rotational resistance than in the case of two previously discussed configurations. However, for the assumed test conditions, the obtained results are below the limit value.

4 Final Remarks

In these investigations a comparison of an influence of some design features of tramway vehicles on a derailment possibility has been carried out. The investigation allows for a characterization of tramway dynamic behavior in the context of derailment safety according to the Nadal criteria. The simulations have been carried out using models of several configurations of track, carbody and bogie designs.

The results obtained from investigations indicated small effect of change values of the stiffness of primary and secondary suspension on safety against derailment index Y/Q. In most cases, the Y/Q ratio remained constant for a given radius value and for a specific carbodies configuration despite significant changes in stiffness in the primary and secondary suspension.

Based on the obtained computational results, an essential influence of various municipal tramway vehicle designs and configurations on wheel-rail contact forces, in particular, on the derailment factor Y/Q, is discussed. The simulation results confirm the assumption that the track conditions for safety against derailment described in standard [19] addressed to the railway vehicles are not suitable for investigations of the light municipal trams.

Further work in this topic will be focused on a determination of the impact of the traction drive and wheelsets with independent wheels on changes of the index value of safety against derailment *Y/Q*.

References

1. Bogacz, R., Konowrocki, R.: On new effects of wheel-rail interaction. Arch. Appl. Mech. **82**, 1313–1323 (2012)
2. Diana, G., Bruni, S., Di Gialleonardo, E., Corradi, R., Facchinetti, A.: A study of the factors affecting flange-climb derailment in railway vehicles. In: Proceedings of the Third International Conference on Railway Technology: Research, Development and Maintenance, Caligary, Italy, April 2016
3. Eken, E., Friberg, R.: Modelling of dynamic track forces generated by tram vehicles. Master thesis in Applied Mechanics, Chalmers University of Technology, Gothenburg, Sweden, p. 130 (2015)
4. Elkins, J.A., Carter, A.: Testing and analysis techniques for safety assessment of rail vehicles: the state of-the-art. In: Proceedings, presented at 13th IAVSD Symposium on the Dynamics of Vehicles on Roads and on Tracks, Chengdu, China, August 1993
5. Elkins, J., Wu, H.: New criteria for flange climb derailment. In: Proceedings of the ASME/IEEE Joint Railroad Conference, pp. 1–7 (2000)
6. de Jalon, J.G., Bayo, E.: Kinematic and Dynamic Simulation of Multibody Systems. Springer, Berlin (1994)
7. Hertz, H.: Über die berührung fester elastischer Körper (On the contact of rigid elastic solids). J. reine und angewandte Mathematik **92**, 156–171 (1882)
8. Kalinowski, D.: Current knowledge regarding safety against derailment (Aktualny stan wiedzy dotyczący zagadnienia bezpieczeństwa przed wykolejeniem, Pojazdy Szynowe). Railway Veh. **4**, 44–53 (2013). (in Polish)
9. Kalinowski, D., Konowrocki, R., Szolc, T.: The new simulation approach to tramway safety against derailment evaluation in term of vehicle dynamics. In: Proceedings of the 11th International Scientific Conference "Transbaltica: Transportation Science and Technology", Vilnius, Lithuania, May 2019
10. Kalker, J.J.: A fast algorithm for the simplified theory of rolling contact. Veh. Syst. Dyn. **11** (1), 1–13 (2007)
11. Konowrocki, R., Bajer, C.I.: Friction rolling with lateral slip in rail vehicles. J. Theor. Appl. Mech. **47**(2), 275–293 (2009)
12. Liang, L., Xin-Biao, X., Xue-Song, J.: Development of a simulation model for dynamic derailment analysis of high-speed trains. Acta. Mech. Sin. **30**(6), 860–875 (2014)
13. Opala, M.: Study of the derailment safety index Y/Q of the low-floor tram bogies with different types of guidance of independently rotating wheels. Arch. Transport **38**(2), 39–47 (2016)
14. Sato, Y., Matsumoto, A., Ohno, H., Oka, Y., Ogawa, H.: Wheel/rail contact analysis of tramways and LRVs against derailment. Wear **265**(9–10), 1460–1464 (2008)
15. Shabana, A.: Dynamics of Multibody Systems, 3rd edn. Cambridge University Press, Cambridge (2005)

16. Sowińska, M.: An analysis of running properties of light cars for railway vehicles on the example of bogie with independent rotating wheels. Doctoral thesis, Poznan University of Technology, Poland, p. 133 (2018). (in Polish)
17. Weinstock, H.: Wheel climb derailment criteria for evaluation of rail vehicle safety, Paper no. 84-WA/RT-1, 1984 ASME Winter Annual Meeting, New Orleans, LA, November, 1984
18. Wilson, N., Fries, R., Witte, M., et al.: Assessment of safety against derailment using simulations and vehicle acceptance tests: a worldwide comparison of state-of-the-art assessment methods. Veh. Syst. Dyn. **49**, 1113–1157 (2011)
19. EN 14363: 2016 – Railway applications – testing and simulation for the acceptance of running characteristics of railway vehicles – running behaviour and stationary tests (2016)
20. PN-K-92016: 1997 – Tram wheelsets, flexible – surfaced tires – requirements and testing (1997)
21. Technical guidelines for the design and maintenance of tram tracks, Ministry of Administration, Local Economy and Environmental Protection, Department of Public Transport and Roads, Warsaw (1983)

Mathematical Model of the Movement Authority in the ERTMS/ETCS System

Andrzej Kochan(✉) and Emilia Koper

Faculty of Transport, Warsaw University of Technology, Warsaw, Poland
{ako, eko}@wt.pw.edu.pl

Abstract. The increasing complexity of the track system and the number of trains running on it makes it more likely that dangerous situations will arise, which requires the development of modern railway traffic control systems. With subsequent technical solutions, new forms of transmission of information about the possibility of entering a given section of the railway line appeared, starting from the kind of tokens through shaped and light signals, to the movement authority sent in the ERTMS/ETCS system. The complexity of the effective determination of movement authority for an area of the railway network supervised by the ERTMS/ETCS system makes it necessary to study the phenomena associated with this process. The authors present in the article a proposal for a mathematical model of movement authority in the ERTMS/ETCS system. The model description shall be followed by a description of the essential characteristics of the ERTMS/ETCS system. Its different application levels give various possibilities to transmit the movement authority. Consideration of the model will be carried out with particular regard to level 2 and level 3 attributes. The proposed model is static and covers all actual elements of the movement authority such as sections, end (of movement authority), overlap, danger point. To the extent necessary, it also includes a simplified train model.

Keywords: Movement authority · ERTMS/ETCS · Train control

1 Introduction

In a certain context, the rail system consists of two basic elements: trackside infrastructure and rolling stock running on it. In further consideration, the track infrastructure will also be called a track layout. Rolling stock moving on a track layout may have different configurations, e.g. traction units, single locomotives, shunting trains. For the presented subject matter it is assumed that the rolling stock unit moving on the track system will always be called a train.

Control-command and signalling systems ensure the safety of railway traffic. This is only possible if a specific set of rules is respected. One of these is that the train starts or continues to run after it has received the movement authority.

Driving a train on the basis of a driver's direct observation of the route is in most cases impossible due to the length of the braking distance and the difficulty in visibility and recognition of the signals when the vehicle is not equipped with a cab signalling system. In the documents of the polish national infrastructure manager [9] it is assumed

M. Siergiejczyk and K. Krzykowska (Eds.): ISCT21 2019, AISC 1032, pp. 215–224, 2020.
https://doi.org/10.1007/978-3-030-27687-4_22

that at a speed of 100 kph the braking distance is min. 700 m, and at the speed of 160 kph it is min. 1300 m. These distances are the result of the physical characteristics of the rolling stock moving on the rail network, i.e. its mass, speed, braking performance and adhesion of wheels on steel rails. In position [8] you can read that such adhesion is eight times lower than that of a car moving on an asphalt road on rubber wheels. Thus, if a driver with even the best psychophysical characteristics notices an obstacle on the route, it is not possible to stop the train safely and securely against it. Therefore, for safety reasons, a vehicle may start running on a particular section of the track once it is certain that there are no obstacles on that section at the moment and that they do not occur until the end of the run. Please note that the available distance should be at least as long as the braking distance.

Moving on to the definition, it can be said that a movement authority for a train on a given section of a track is an authorisation for a train to enter that section after having first ensured that running can take place safely with the protection provided by the technical and organisational means of the signalling system.

2 ERTMS/ETCS

2.1 General Characteristics

The European Rail Traffic Management System is a system supported by the European Union which aims to harmonise the management and control system of rail traffic in order to ensure the interoperability of rail transport. The ERTMS system consists of the European Train Control System (ETCS) and GSM-R, a digital railway radio system based on the GSM standard.

For the ERTMS/ETCS system, levels and configurations are defined. The adoption levels of ERTMS/ETCS are defined by the different functionalities of ERTMS/ETCS. It is possible that a railway line and an ETCS vehicle operating on that line are equipped to operate at different levels of the ERTMS/ETCS system. Therefore, there is a need to distinguish between track-side and on-board [3]. There are three main levels of the ERTMS/ETCS (L1, L2, L3) and two additional (L0, NCT) [3, 5]. Only L1 and L2 are substantial for deliberations conducted by authors.

2.2 ERTMS/ETCS Level 1

The ERTMS/ETCS Level 1 (L1) is an overlay for existing station and line equipment. It shall ensure that the vehicle does not go beyond the point limiting the end of the movement authority and that it does not exceed the speed permitted at that particular section of the route. It is based on the transmission of information from track-side distributed control-command and signalling equipment. The transmitters of this information are usually balises, but Euroloops or radio equipment of the digital GSM-R system can also be used.

The L1 operation is based on the transmission by balises of the Movement Authority (MA), generated on the basis of information from trackside or interlocking devices. The Lineside Electronic Unit (LEU) encoder is an intermediate element that allows to read indications of signal and then to send these data to the balise. One of the possible variants of the application is that it is connected to the light circuits of railway signalling devices. The LEU encoder is connected via a cable (C interface) to a switchable balise, which transmits a movement authority to the ERTMS/ETCS on-board equipment, depending on the indication of the signal. Based on the information received, the ETCS on-board unit checks whether the driver is driving an ETCS vehicle according to the speed profile based on the current indication on the signal.

2.3 ERTMS/ETCS Level 2

The ERTMS/ETCS level 2 (L2) uses continuous digital bidirectional data transmission via the GSM-R communication system. At this level, balises are no longer used for the transmission of traffic data, but instead they are intended as location points. A line equipped with L2 devices must be equipped with a Radio Block Center (RBC).

Data transmission between the train and the Radio Control Centre is based on the GSM-R system. The RBC is responsible for generating movement authority for ETCS vehicles within the area controlled by RBC data based on information from the interlocking system on the occupation of track and switch sections and the position of switches.

ETCS vehicles are distinguished by their ETCS onboard identifiers. The interface between the RBC and the motion control layer is not standardised. A vehicle operating in L2 in addition to the equipment typical for L1 must be equipped with a digital GSM-R radio system for both voice and data communication.

Where there are only vehicles running on a given line equipped with active ERTMS/ETCS 2 on-board equipment (homogeneous traffic), the L2 implementation makes it possible to remove track-side signals because the functionality is taken over by a continuous digital transmission allowing information to be transmitted to the on-board equipment and consequently to the driver.

3 Movement Authority for ERTMS/ETCS

3.1 State of Art

Issues related to train control using ERTMS/ETCS L2 (ETCS-2) have been under consideration for several years. Books related to the system in general can be found in both Polish [12] and world [8] literature. They focus on the description of general characteristics of ETCS-2. Doctoral theses on this subject published in Poland mainly dealt with the issues of train location on the railway line [10] as well as with the appropriate balise distribution on the track system between stations [11]. However, these publications do not deal with mathematical models of the movement authority. The subject matter is still current, which can be proved by the published articles discussing the influence of the transmission system on the movement authority [15], the determination of the permit to drive focused on the vehicle [14] or new complete concepts of systems for determining the movement authority [13]. This article in Sect. 3 presents the results of the analysis of ETCS-2 specifications [3, 5] by the authors in search of relevant factors influencing the modelling of the movement authority determined by the RBC. Section 4 below presents selected elements of the author's mathematical description of the movement authority. Authors presented other problem related to the subject in previous works [1–3].

3.2 Definition

The term 'movement authority', like many others concerning ERTMS/ETCS is defined in the glossary of terms and abbreviations i.e. Subset 023 [6] and means "permission for a train to run to a specific location within the constraints of the infrastructure". Searching Polish railway bibliography, one can find that it is defined as a permission for a train to run to a specific location on a track layout on the route [7].

Data between train and track-side equipment is transmitted as packets [5]. The movement authority is transmitted as a packet 12 Level 1 Movement Authority - for L1 and as a packet 15 Level 2/3 Movement Authority - for L2. The characteristics of the ERTMS/ETCS movement authority and the packages above are described in the System Requirements Specification [5]. Particularly important is chapter 7 [5] describing so-called ETCS language, which describes the messages, packets and individual variables transmitted in the ERTMS/ETCS system.

3.3 Structure

Structurally the movement authority can be divided into sections, which may differ in terms of their properties (length, movement speed resulting from the static profile, etc.). An important element is the end of the movement authority, i.e. the point to which the train can move safely. In addition, a danger point is distinguished, the exceeding of which may cause a dangerous situation. There may be an overlap between the end of the movement authority and the non-safety point which, in an emergency, can be driven without leading to an unsafe situation.

3.4 Sections

A structurally defined movement authority is a track segment with the end at a specified location for which a maximum speed is set. The route of the movement authority is divided into sections. The last section of the movement authority is called the end section. At least the end section needs to be defined for the movement authority.

Each non-end section is described by:

- length,
- maximum speed,
- optional time-out period for a section, from the beginning of the section to the stopping point of the timer validity.

The end section is described by:

- length,
- maximum speed,

- validity period of the end section,
- release speed.

3.5 End

The end of the movement authority shall be clearly identified in the movement authority. This is the place where the conditions for driving are known. The maximum speed at which the vehicle can pass this point shall be determined for the end point.

As already mentioned, one of the parameters of the movement authority determines the maximum speed at the end of the movement authority, but the speed at which the movement authority approaches the end of the movement authority may also be specified. The value of this speed can be determined statically (L1) or calculated dynamically by the RBC (L2) and can also be calculated by the on-board equipment regardless of the level.

The end of the movement authority may be defined by the following characteristics/ parameters:

- time-out period of the end section,
- distance from the start point of the validity timer to the end of the section,
- critical point distance from the end of the movement authority,
- release speed assigned to a danger point,
- length of the overlap,
- time of validity of the overlap,
- distance from the timer start location to the end of the section,
- release speed to the end of the overlap.

3.6 Danger Point

The term 'movement authority' also refers to the term 'danger point'. This is the point on the road at which the front-end of a train can reach without leading to a dangerous situation.

3.7 Overlap

An optional element of the movement authority is the overlap. The overlap is the section of the track system behind the end of the movement authority that can be run over by the train after the end of the movement authority without introducing a hazardous situation.

4 Mathematical Model of the Movement Authority

4.1 Movement Authority

Movement authority zj is called tuple:

$$zj = (O, ok, pk, do, pn) \quad (1)$$

where:

- zj – movement authority,
- O – set of sections of a movement authority without the end section,
- ok – end section of movement authority,
- pk – end of movement authority,
- do – overlap of movement authority,
- pn – danger point.

4.2 Section

The route for the movement authority may be divided into sections. This division is due to the fact that the parameters defining the driving conditions on the individual sections can be differentiated. The O-set contains all sections except the end section.

$$O = \{o_1, \ldots, o_n\} \quad (2)$$

where:

- O – set of road sections $\boldsymbol{zj}$ with exclusion of the end section,
- n – number of sections other than the end section, n – is an integer greater or equal than 0.

The section o_i is described by the following tuple:

$$o_i = (d,\ l,\ v_{max},\ tw) \quad (3)$$

where:

- o_i – i-th section of zj,
- d – mileage of the beginning of o_i,
- l – length of o_i,
- v_{max} – maximum speed on o_i,
- tw – time-out for o_i.

This is the basic model of the section.

4.3 End Section

As mentioned above, the route of the movement authority is divided into sections. It consists of at least an end section. The end section is described by the following tuple:

$$ok = (d,\ l,\ v_{max},\ tw) \quad (4)$$

where:

- *ok* – end section,
- *d* – mileage of the beginning ok,
- *l* – length of ok,
- v_{max} – maximum speed on the section ok,
- *tw* – time-out for ok.

4.4 End

The end of the movement authority is described by the following tuple:

$$pk = (d, vk, vz) \tag{5}$$

where:

- *pk* – end of movement authority zj,
- *d* – mileage of pk,
- *vk* – maximum speed at pk,
- *vz* – maximum release speed to pk,
- *twvk* – time-out for vk.

4.5 Overlap

The overlap is modelled in thc form of tuple:

$$do = (d, l) \tag{6}$$

where:

- *do* – overlap,
- *d* – mileagc of the beginning of the overlap,
- *l* – length of overlap.

4.6 Danger Point

The danger point is modeled in the form of tuple:

$$pn = (d, l) \tag{7}$$

where:

- *pn* – danger point,
- *d* – mileage of pn,
- *l* – distance to pn.

5 Timing Nature of the Movement Authority Parameters

According to the specification [5], the parameters of the movement authority may have time limits. This reflects the fact that the movement authority is embedded in a dynamic operating process and cannot be treated statically in practice (fixed parameter values).

The first time-dependent parameter is the speed at the location of the end of the movement authority vk (5). Determined value *vk* is valid for a specified period of time *twvk*. In a particular case *twvk* may be equal to infinity. Time parameters may also be associated with movement authority component sections. The validity period can be defined for each section of the *O*-set. It means $(o_i(tw))$ the maximum time for which the section o_i is reserved within the scope of the *zj*.

6 Additional Information for the Movement Authority

The dependence of train speed on the distance travelled and time is called train profile *pj*.

$$pj(zj) = (v(zj)) \tag{8}$$

where:

pj – driving profile,
zj – movement of authority,
v(zj) – speed as a function of the movement authority

The maximum speed at which a vehicle can run on the sections of the movement authority in advance of the vehicle, i.e. the permissible driving profile for the movement authority, is calculated for the parameters of the movement authority.

$$pj_d(zj) = (v_d(zj)) \tag{9}$$

where:

pj_d – permissible driving profile,
zj – movement of authority,
$v_d(zj)$ – permissible speed as function of the movement authority

In addition to the parameters discussed so far, the determination of the driving profile requires additional data to describe the track infrastructure and the train. Additional data describing the track infrastructure is:

- gradients of the longitudinal profile of the route,
- wheel-rail adhesion factor,
- track conditions (e.g. low pantograph areas, non stopping areas, engineering objects locations)
- level crossings,
- mode profile,
- permitted braking distance.

Referring to the concept of sections, we assume that the above parameters are constant for a given section of o_i. Thus, formula (3) will be developed as follows:

$$o_i = (d, l, v_{max}, st, g, wps, wdj, pkd, tj, dh) \quad (10)$$

where:

- o_i – *i-th* section zj,
- d – mileage of the beginning o_i,
- l – length of o_i,
- vmax – maximum speed on the section o_i,
- tw – time-out for o_i,
- g – gradient of o_i.
- wps – wheel-rail adhesion factor,
- wdj - track conditions,
- pkd - level crossings,
- tj - mode profile,
- dh - permitted braking distance.

The exact description of train modelling parameters exceeds the scope of this Article. The minimum set of parameters that is relevant for modelling permission to run is the train length that affects the track layout area, which has to be reserved only for this train to avoid dangerous situations. So the train p for the purposes of this model will be described as tuple:

$$p = (l, pj_d(zj)) \quad (11)$$

where:

p – train running in accordance with the movement authority,
l – train length,
$pj_d(zj)$ – permissible driving profile for the given movement authority

7 Conclusion

The determination of the movement authority is a key activity of the train control system, which is the basis of railway traffic safety. As the complexity of railway traffic increases, the efficient determination of movement authority for a set of trains operating in a given area of the track system becomes increasingly important.

This task is even more important in considerations assuming the automation of train operation [1]. The presented mathematical model of the movement authority is a significant part of the model of train control command and signalling issues in the area. This model is the basis for other studies, including simulations, conducted by the WT PW Railway Traffic Control Team. This research includes formal methods and techniques of artificial intelligence in proving the safety of various solutions for railway traffic.

References

1. Kochan, A., Koper, E., Wontorski, P.: Automatyczne prowadzenie pociągu – analiza wymagań. Prace Naukowe Politechniki Warszawskiej. Transport **121**, 161–170 (2018)
2. Kochan, A.: System pokładowy w modelu warstwowym systemu kierowania i sterowania ruchem kolejowym. Zeszyty Naukowo-Techniczne Stowarzyszenia Inżynierów i Techników Komunikacji W Krakowie. Seria: Materiały Konferencyjne **107**(3), 79–87 (2015)
3. Kochan, A., Koper, E., Ilczuk, P., Gruba, Ł.: Tranzycje w systemie ERTMS/ETCS. Prace Naukowe Politechniki Warszawskiej. Transport **121**, 147–159 (2018)
4. Europejska, K.: (2016/919/UE) Rozporządzenie Komisji z dnia 27 maja 2016r. w sprawie technicznej specyfikacji interoperacyjności w zakresie podsystemów « Sterowanie » systemu kolei w Unii Europejskiej (Dz. U. L 158 z dnia 15.06.2016, str. 1 i n.) (2016)
5. Unisig: SUBSET-026 system requirements specification – issue 3.6.0
6. Unisig: SUBSET-023 glossary of terms and abbreviations
7. Bajon, M.: Podstawy sterowania ruchem kolejowym. Oficyna Wydawnicza Politechniki Warszawskiej (2014)
8. Theeg, G., Vlasenko, S.: Railway Signalling & Interlocking. DVV Media Group Eurail press, Hamburg (2009)
9. PKP Polskie Linie Kolejowe S.A.: Wytyczne techniczne budowy urządzeń sterowania ruchem kolejowym Ie-4 (WTB-E10), sierpień 2018
10. Toruń, A.: Metoda lokalizacji pociągu w procesie sterowania ruchem kolejowym. Rozprawa doktorska, Radom (2013)
11. Ilczuk, P.: Wpływ lokalizacji balis na parametry ruchowe linii wyposażonej w system ETCS. Rozprawa doktorska, Warszawa (2017)
12. Pawlik, M.: Europejski System Zarządzania Ruchem Kolejowym przegląd funkcji i rozwiązań technicznych – od idei do wdrożeń i eksploatacji, wyd, KOW, Warszawa (2015)
13. Yun, D.-G., Ahn, J., Yong, S.-w., Ko, J.-h.: Study on improvement of speed by design of movement authority. J. Korean Soc. Railway **22**, 150–157 (2019). https://doi.org/10.7782/JKSR.2019.22.2.150
14. Song, H., Shen, T.: A new movement authority based on vehicle-centric communication. Wirel. Commun. Mob. Comput. (2018). https://doi.org/10.1155/2018/7451361
15. Song, H., Liu, H., Schnieder, E.: A train-centric communication-based new movement authority proposal for ETCS-2. IEEE Trans. Intell. Transp. Syst. 1–11 (2018). https://doi.org/10.1109/TITS.2018.2868179

An Analysis of Electromechanical Interactions in the Railway Vehicle Traction Drive Systems Driven by AC Motors

Robert Konowrocki(✉) and Tomasz Szolc

Institute of Fundamental Technological Research of the Polish Academy of Sciences, Warsaw, Poland
{rkonow, tszolc}@ippt.pan.pl

Abstract. In the paper dynamic electromechanical interactions between the railway drive systems and their driving electric motors are investigated. These are drive systems of high-speed trains (HST) and locomotives driven by AC motors. In particular, there is considered an influence of negative electromagnetic damping generated by the asynchronous motor on a possibility of excitation of resonant torsional vibrations. The theoretical calculations have been performed by means of the advanced structural mechanical models. Conclusions drawn from the computational results can be very useful during a design phase of these objects as well as helpful for their users during a regular maintenance.

Keywords: Railway drive system · Electromechanical coupling · AC motors

1 Introduction

The knowledge about torsional vibrations of drive transmission systems of railway vehicles is of a great importance in the fields dynamics of mechanical systems [1–3]. In the modern electric traction drive systems particular attention is being paid to their reliability. The use of electrical drives in electric traction for high-speed trains has evolved from DC drives to AC drives [1, 2]. One of the main reasons, why AC drives are used, is due to the evolution of power electronics applied to these drives, for example, the functional improvement of inverters and the use of vector control techniques. Nowadays, the majority of high-speed trains (HST) and locomotives are driven by asynchronous and synchronous motors through elastic couplings and tooth gears [7, 8]. Thanks to their high efficiency and small sizes, many producers recently focused on using asynchronous motors for high-speed railcars. Their wheelsets and shafts are usually characterized by mass moments of inertia a many times greater than those of the driving motor rotors. As it follows from numerous observations, rotational motions of such devices are very sensitive to the electro-mechanical dynamic interaction that causes severe or even dangerous torsional vibrations, particularly in transient operating conditions [6, 11]. Such vibrations usually result in a significant motor rotor angular velocity fluctuation that causes more or less severe perturbations of the generated electromagnetic torque. Thus, these mechanical vibrations become very strongly coupled with electrical oscillations of electric currents in the motor windings [9, 10]. In

M. Siergiejczyk and K. Krzykowska (Eds.): ISCT21 2019, AISC 1032, pp. 225–235, 2020.
https://doi.org/10.1007/978-3-030-27687-4_23

such conditions the AC motors usually generate electromagnetic negative damping which can be a source of severe action instabilities that often lead to dangerous overloadings and even to damages of drive system components [9, 15]. These detrimental phenomena are observed in steady-state operating conditions as well as during start-ups and run-downs. For this reason, a mutual interdependence between various static and dynamic characteristics of the driving asynchronous motor and dynamic properties of the wheelset drive system is studied.

2 Rail System Description

In an electric multiple unit (EMU) several electric motors drive the train. Previously used DC motors and wire induction motors are now being replaced by squirrel cage motors in mature applications or asynchronous and asynchronous permanent magnet motors in new rail systems. Typically, one motor drives a one axle in each motorized carriage. Thus, there are two motors in one bogie [4, 14]. One motorized car usually has two bogies with motors. In this article dynamic properties of electromechanical drive system used in a high speed train (HST) were tested. A considered railway vehicle has two driven carbodies: at the front and at the end, respectively (Fig. 1). Each motorized car body has two bogies (Fig. 2). In these bogies each axle is driven by a one electric motor. Depending on the type of drive various locations of an electric motor connection to axles of the wheelsets are used. The main characteristics of the considered HST train are summarized in Table 1.

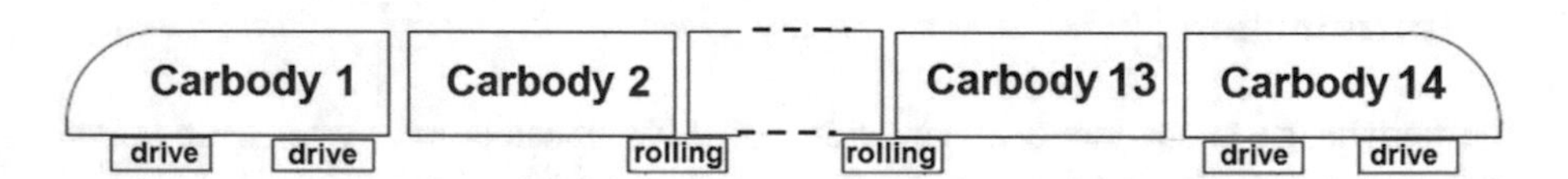

Fig. 1. Electric multiple unit of the high speed train

Table 1. Characteristics of example HST trains

Parameter	Value
Power	8000 kW
Voltage–frequency	25 kV–50 Hz
Number of motors	8
No load mass	322 t
Length	200 m
Number of axles	32

2.1 Structural Modelling of the Mechanical Drive Systems

In the paper a direct drive is analyzed. This drive is represented by the AC motor with the hollow shaft and with the special hollow joint shaft which surrounds an axle of the wheelset, see Fig. 2. This gearless direct wheelset drive in the variant with the AC motor was designed as an alternative solution of the drive for the high-speed electric-units [4]. In order to investigate the electromechanical coupling, a simple mechanical model of the drive system is applied. Here, the considered high speed train drive system is represented by a torsionally vibrating discrete model with two DOFs, as shown in Fig. 3, and a finite element model described in Sect. 4.

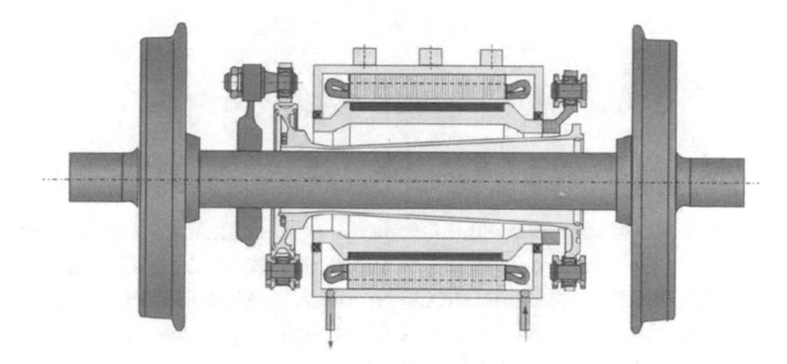

Fig. 2. The direct wheelset traction drive [5]

2.2 Torque Transmission System

For simulation purposes, the wheelset drive system of the HST was reduced into a two-mass discrete model which is presented in Fig. 3 [6]. The macro-slip phenomena were omitted according to assumption that superior adhesion control is implemented in this system. The small micro-slip phenomena were included in the viscoelastic element of the shaft line.

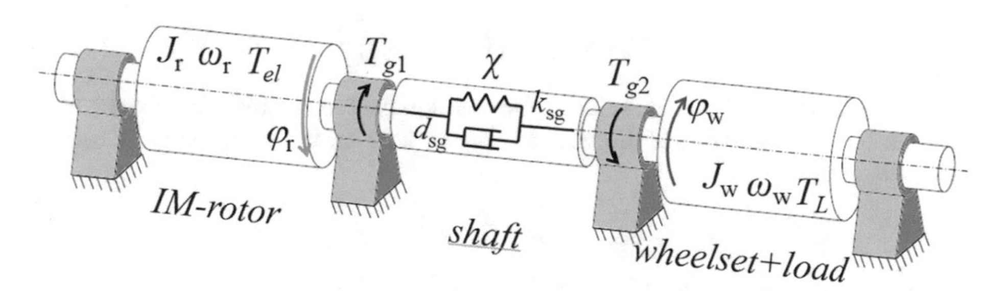

Fig. 3. Scheme of the torque transmission system

This simple model is described by the motion equations refered to the motor shaft

$$J_r \frac{d\omega_r}{dt} = T_{el} - T_{g1}, \tag{1}$$

$$J_w \frac{d\omega_w}{dt} = \chi T_{g2} - T_L, \tag{2}$$

$$\frac{d\varphi_r}{dt} = \omega_r \ and \ \frac{d\varphi_w}{dt} = \omega_w, \tag{3}$$

$$T_{g1} = k_{sg}(\varphi_r - \chi \cdot \varphi_w) + d_{sg}(\omega_r - \chi \cdot \omega_w), \tag{4}$$

$$|T_{g2}| = |\chi \cdot T_{g1}|, \tag{5}$$

where ω_r denotes the motor rotor angular speed, ω_w denotes the angular speed of the driven part, T_{el} is the motor torque, T_L is the load torque, T_{g1}, T_{g2}, are the torques transmitted trough the gear on input and output of the gear, respectively, k_{sg} denotes the stiffness, d_{sg} – the damping coefficient of 190 Nms/rad, J_r is the induction motor rotor mass moment of inertia, J_w denotes the mass moment of inertia of the driven part of 2900 kgm^2 and χ is the gear ratio of 1:1.

A misalignment of rotating drive system elements was modeled as an additional load torque component which appears in the motion equation. In the equation a viscous friction like additional load torque component was also assumed:

$$J_r \frac{d\omega_r}{dt} = T_{el} - T_{g1} - T_{w\omega} + T_{v-f}, \tag{6}$$

where $T_{w\omega}$ denotes the additional load torque with sinusoidal component and with frequency equal to the motor rotor angular velocity (7), and $T_{v\text{–}f}$ is the viscous friction load torque component (8).

$$T_{w\omega} = T_{mis} \cdot (1 + \sin(\varphi_r)), \tag{7}$$

where T_{mis} denotes the average value of the misalignment load torque component equal here to 720 Nm, and $T_{v\text{–}f}$ is the viscous friction load torque component defined by relationship (8).

$$T_{v-f} = F_v \cdot \omega_r, \tag{8}$$

where: F_v is the viscous friction coefficient equal 0.8 Ns/m.

The rolling stock of the coach moves due to an adhesion force between rail and driving wheel. An approximated relation between the adhesion force and the slip velocity is presented in Fig. 4.

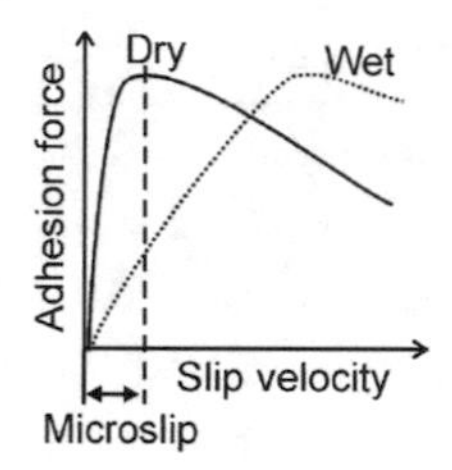

Fig. 4. Relation between the adhesion force and slip velocity of the coach [8]

Table 2. Parameters of the considered HST induction motor.

Parameter	Value
Power	1200 kW
Voltage	4000 V (Y)
Current	585 A
Torque	4308 Nm
Number of poles	6
Angular speed	278.6 rad/s
Frequency f_n	133 Hz
$\cos\,\varphi$	0.88
efficiency η_o	96%

3 The Induction Motor

For train propulsion a high-power induction motor is used. The power, voltage, nominal speed of the tested HST electric motor are respectively: 1.2 MW, 4 kV, 278 rad/s. All parameters of this asynchronous motor are presented in Table 2.

3.1 Modelling of the Electric Motor

In the case of a more advanced analysis of transient phenomena in the drivetrain a reliable circuit model of the electric motor is needed [9, 10]. Asynchronous motors are very commonly applied as driving sources of railway vehicles. From the viewpoint of electromechanical coupling investigation, as an introductory approach, a properly advanced circuit model of the electric motor seems to be particularly adequate, similarly as e.g. in [12, 13]. In the case of a symmetrical three-phase asynchronous motor, electric current oscillations in its windings are described by six circuit voltage equations. In order to simplify their form, they are transformed into the system of four Park's equations in the so called '$\alpha\beta$-dq' reference system

$$\begin{bmatrix} \sqrt{\frac{3}{2}}U\cos(\omega_e t) \\ \sqrt{\frac{3}{2}}U\sin(\omega_e t) \\ 0 \\ 0 \end{bmatrix} = \begin{bmatrix} L_s+\frac{1}{2}L_m & 0 & \frac{3}{2}L_m & 0 \\ 0 & L_s+\frac{1}{2}L_m & 0 & \frac{3}{2}L_m \\ \frac{3}{2}L_m & 0 & L'_r+\frac{1}{2}L_m & 0 \\ 0 & \frac{3}{2}L_m & 0 & L'_r+\frac{1}{2}L_m \end{bmatrix} \cdot \begin{bmatrix} \dot{i}^s_\alpha(t) \\ \dot{i}^s_\beta(t) \\ \dot{i}^r_d(t) \\ \dot{i}^r_q(t) \end{bmatrix}$$
$$+ \begin{bmatrix} R_s & 0 & 0 & 0 \\ 0 & R_s & 0 & 0 \\ 0 & \frac{3}{2}pM\dot{\varphi}_1(t) & R_r & p\dot{\varphi}_1(t)\left(L'_r+\frac{1}{2}L_m\right) \\ -\frac{3}{2}pL_m\dot{\varphi}_1(t) & 0 & -p\dot{\varphi}_1(t)\left(L'_r+\frac{1}{2}L_m\right) & R_r \end{bmatrix} \cdot \begin{bmatrix} i^s_\alpha(t) \\ i^s_\beta(t) \\ i^r_d(t) \\ i^r_q(t) \end{bmatrix}, \tag{9}$$

where U denotes the power supply voltage, ω_e is the supply voltage circular frequency, L_s, L'_r are the stator coil inductance and the equivalent rotor coil inductance, respectively, L_m denotes the relative rotor-to-stator coil inductance, R_s, R_r are the stator coil resistance and the equivalent rotor coil resistance, respectively, p is the number of pairs of the motor magnetic poles, $\dot{\varphi}_1(t)$ is the current rotor angular speed including the average and vibratory component and i^s_α, i^s_β are the electric currents in the stator windings reduced to the electric field equivalent axes α and β and i^r_d, i^r_q are the electric currents in the rotor windings reduced to the electric field equivalent axes d and q, [9]. Numerical values of parameters of the considered motor are presented in Table 3. Then, the total electromagnetic torque and its oscillatory part generated by such a motor can be expressed by the following formulae:

$$T_{el} = \frac{3}{2}pL_m\left(i^s_\beta \cdot i^r_d - i^s_\alpha \cdot i^r_q\right), \qquad T^{\mathrm{var}}_{el}(t) = S(\omega)\cdot\sin(\omega t) + T(\omega)\cdot\cos(\omega t). \tag{10}$$

By projecting the sine - $S(\omega)$ and cosine - $T(\omega)$ components of the electromagnetic torque oscillatory part as well as the analogous components of the rotor rotation angle, respectively, on the complex plane real and imaginary axes and using the proper definitions given e.g. in [9], the electromagnetic torsional stiffness $k_e(\omega)$ and the coefficient of damping $d_e(\omega)$ generated by the asynchronous motor are determined in the following form:

$$k_e(\omega) = -\frac{U\cdot S(\omega) + W\cdot T(\omega)}{U^2+W^2}, \qquad d_e(\omega) = -\frac{1}{\omega}\cdot\frac{U\cdot T(\omega) - W\cdot S(\omega)}{U^2+W^2}, \tag{11}$$

where:

$$W = \sum_{m=0}^{\infty} \frac{(X^S_m)^2 T(\omega)\cdot(\omega_m^2-\omega^2) - \left[(X^S_m)^2 S(\omega) - X^S_m X^R_m R\right]\cdot(\beta+\tau\omega_m^2)\omega}{\gamma_m^2\left[(\omega_m^2-\omega^2)^2 + (\beta+\tau\omega_m^2)^2\omega^2\right]},$$

$$U = \sum_{m=0}^{\infty} \frac{\left[(X^S_m)^2 S(\omega) - X^S_m X^R_m R\right]\cdot(\omega_m^2-\omega^2) + (X^S_m)^2 T(\omega)\cdot(\beta+\tau\omega_m^2)\omega}{\gamma_m^2\left[(\omega_m^2-\omega^2)^2 + (\beta+\tau\omega_m^2)^2\omega^2\right]}.$$

The above expressions derived in the framework of a qualitative analysis of properties of the asynchronous motor, that drives a linear model of the HST drive system, enable us to carry out thorough dynamic investigations of the coupled electromechanical system.

Table 3. Equivalent circuit parameters under rated conditions [7].

Parameter	Value	Description
R_s	0.0445 Ω	Stator resistance
R_r	0.0406 Ω	Rotor resistance
L_r'	0.00102 H	Stator inductance
L_s'	0.00082 H	Rotor inductance
L_m	0.03013 H	Mutual inductance

In Fig. 4a there are shown static characteristics of the considered asynchronous motor determined according to [5] for four variants of the starting torque values which depend on the rotor resistance $R_s'^{I} < R_s'^{II} < R_s'^{III} < R_s'^{IV}$: namely, the greater the resistance, the greater the starting torque. In Fig. 4b characteristics of the electromagnetic stiffness and coefficients of damping generated by the considered asynchronous motor within the torsional interaction frequency range of 0–200 Hz are plotted. In this figure the thin lines correspond to the stiffness characteristics and the thick lines, appropriately, to the damping ones. These characteristics have been determined by means of relationships (9). Using respectively the same colours, these plots correspond to the mentioned above four variants of rotor resistances and thus to four variants of motor static characteristics presented in Fig. 5a. It is to emphasize that all the electromagnetic damping characteristics indicate negative value zones. From Figs. 5a and b it follows that for the considered here four starting motor torque variants the breadths of these negative damping zones gradually decrease with the rise of the starting motor torque values.

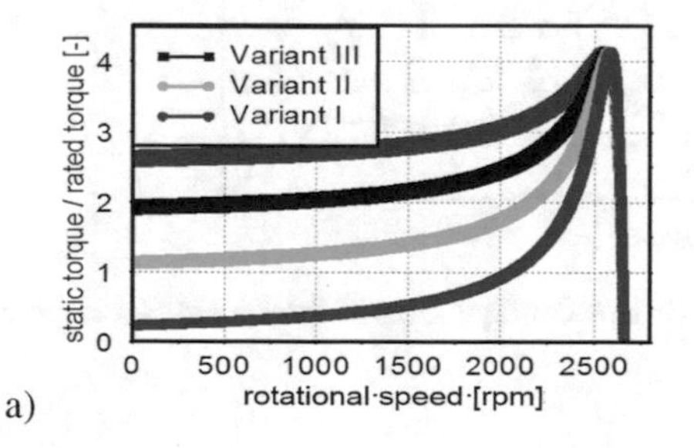

a)

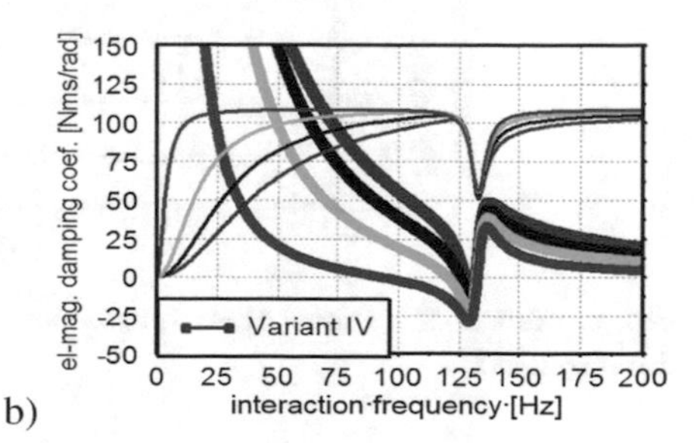

b)

Fig. 5. Static characteristics of the asynchronous motor (a) and dynamic characteristics of the rotor-to-stator electromagnetic damping and stiffness (b)

4 Analysis of Wheelset Torsional Vibration

Before analyzing the start-up of the railway drive described above, modal analyses for wheelset torsional vibrations were carried out. The modal properties of the wheelset were calculated from a finite element model created using the ABAQUS software[1]. This model includes the wheelsets with an electric motor connections in various locations on an axle. One of the location was in the middle of the wheelset axle (case I), while the second location was near the one wheel of the wheelset (case II). Depending on the type of drive various positions of drive torque imposition to axles of the wheelsets, several examples of such structures can be found in reference [14].

The results of analyses obtained by means of finite element method (FEM) are shown in Fig. 6. In this figure the wheel torsional vibration receptances within frequency range 0–1000 Hz are plotted. In order to analyze an influence of the electric motor connection location on wheelset torsional vibration properties (Case I and Case II), the flexible torsional eigenmode shapes of the wheelset are shown in Fig. 7. The first mode of the wheelset with electric motor connection in the middle of axle is characterized by natural frequency of 98 Hz and the second torsional eigenmode by the frequency of 350 Hz (Fig. 7a and c). Considering the location of the electric motor connection near wheel of the wheelset (Fig. 7b and d), the first and the second torsional natural frequency of this wheelset is equal to 109 Hz and 680 Hz, respectively. The first mode of the wheelset torsional vibration is that the wheel vibration direction is opposite to that of second wheel. The second mode of torsional vibration, however, is that the vibration directions of two wheels are coincident, and the direction of the AC motor rotor is opposite to that of the wheels. For an illustration, these features are marked in Figs. 7a and c by arrows. It is shown that an influence of the electromotor connection location with the axle on wheelset first torsional eigenvibration is small (a difference of a few Hz), but its influence on the second mode of torsional eigenvibration is much bigger.

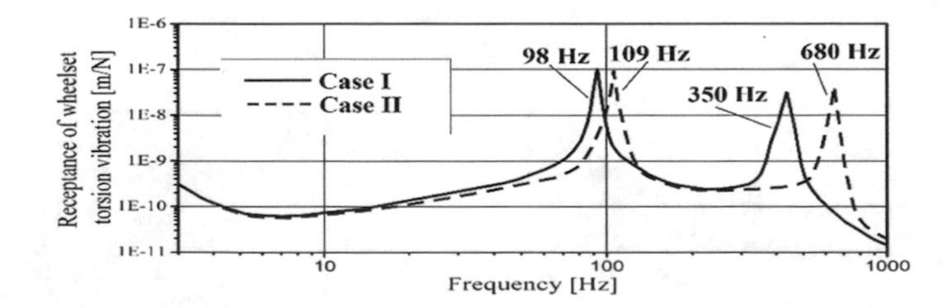

Fig. 6. Wheelset torsional vibration receptance for two configuration of the considered railway drive systems

[1] Academic license of ABAQUS obtained from Institute of Fundamental Technological Research.

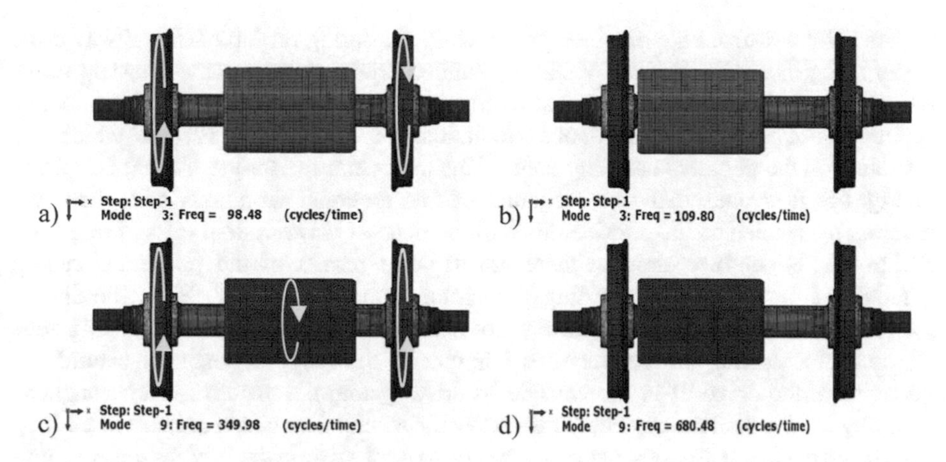

Fig. 7. Flexible torsional eigenmode shapes of the wheelset: the first mode - Case I (a); the first mode - Case II (b); the second mode - Case I (c); the second mode - Case II (d)

5 The Transient Dynamic Electromechanical Analyses

In order to demonstrate an influence of the electromagnetic negative damping on the electromechanical dynamic interaction, numerical simulations have been carried out for run-ups of the considered railway drive system. In the entire drive system there are assumed small retarding torques caused by rolling bearings and aerodynamic drag torques induced by the air gap between the rotor and stator of the driving motor. The macro-slip phenomena were omitted according to the assumption that superior adhesion control is implemented in this system. The small micro-slip phenomena were included in the viscoelastic element of the shaft line.

In transient operating conditions, e.g. during run-ups, the considered railway drive system can be affected by severe torsional vibrations excited by the electromagnetic torque generated by the AC motor and described by Eq. (5). To demonstrate a danger created by the negative electromagnetic damping generated by the asynchronous motor, run-ups of the considered drive to the rotational speed of 1870 rpm within 300 s have been performed. During start-ups the open-loop scalar control U/f = const of the asynchronous motor has been applied.

The parameters of torsional stiffness of the wheelset axle have been determined for two cases configuration of the system. In the first variant the location of the electric motor connection for the middle of the wheelset axle (Case I) was assumed, while in the second variant this connection of the electric motor was located at the end of the wheelset axle near wheel (Case II). The torsional stiffness of the axle has been determined for Case I and II as k_{sg} = 1.3e7 Nm/rad and k_{sg} = 6.8e6 Nm/rad, respectively. More parameters applied in this investigations are also given in the Table 4.

Table 4. Base mechanical parameters of model

χ	d_{sg}	J_r	J_w
1:1	190 Nms/rad	1.9 kgm^2	73 kgm^2

As it follows from the obtained plots in Fig. 8a, during run-up of the railway drive system, together with the rise of the transmitted driving torque some transient oscillations appear. Here, we can see that rotational speed 1547 rpm and 1700 rpm correspond respectively to synchronous excitations of 98 Hz and 108 Hz which are contained in the negative damping zone of the motor starting torque Variant I depicted in Fig. 5b. It is caused by an excitation of free torsional vibrations with their first natural frequencies of the considered configurations of drive system (Case I and Case II). In Fig. 8b in time domain there are depicted results of the performed run-up simulations for the greatest starting AC motor torque Variant IV. Here, the above-mentioned transient torque fluctuations not occur, because the negative damping zone of the motor starting torque Variant IV is outside the range of torsional natural frequencies of the drive. If in the considered drive system the negative electromagnetic damping cannot be naturally compensated by the mechanical one, a proper closed-loop control of the asynchronous motor is able to prevent such instability, as e.g. in reference [15]. However, due the limited size of manuscript, only the open-loop scalar control of the asynchronous motor are presented here.

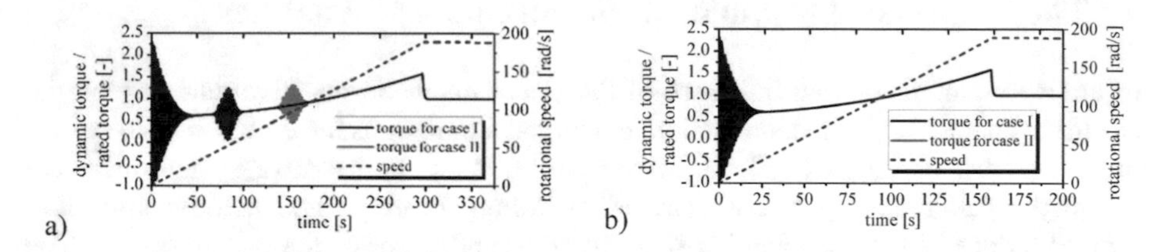

Fig. 8. Transient dynamic response due to the start-up of the railway drive system with the small starting AC motor torque-Variant I (a) and with the great starting AC motor torque-Variant IV (b) for both location of the electric motor of the axle (Case I, II)

6 Final Remarks

In the paper, a dynamic interaction between the torsionally vibrating HST drive system and the driving asynchronous motor was investigated. The results of wheelset torsional vibration analysis have shown that the first mode with torsional eigenfrequency of the considered wheelset with electric motor is equal to109 Hz and the second torsional eigenfrequency is equal to 680 Hz. The basic torsional eigenvibration property is that the first wheel vibration direction is opposite to that of second wheel. The second torsional eigenvibration property is that the vibration directions of two wheels are coincident and the direction of electric motor connection elements is opposite to the wheels. In the paper there was shown that static torque characteristics of the AC motor are correlated with its electromagnetic negative damping zones. This negative damping can be responsible for instability of the entire drive train. Concerning the transient and steady-state dynamic responses determined for the considered HST drive system, one can state that the studied phenomenon of electromechanical interactions should become an object of further investigations.

References

1. Steimel, A.: Electric Traction-Motive Power and Energy Supply. Oldenbourg Industrie verlag GmbH, Munich (2008)
2. Koseki, T.: Technical trends of railway traction in the world. In: Proceedings of the Power Electronics Conference (IPEC 2010), Sapporo, Japan, 21–24 June 2010, pp. 2836–2841 (2010)
3. Duda, S.: Numerical simulations of the wheel-rail traction forces using the electromechanical model of an electric locomotive. J. Theory. Appl. Mech. **52**(2), 395–404 (2014)
4. Kolář, J.: Design of a wheelset drive. Trans. Electr. Eng. **4**(1), 11–19 (2015)
5. Jöckel, A.: Getriebelose Drehstromantriebe für Schienenfahrzeuge. Elektrische Bahnen **101** (3), 113–119 (2003)
6. Matsui, K., Makin, T., Satoh, H.: Autocompensation of torque ripple of direct drive motor by torque observer. IEEE Trans. Ind. Appl. **29**(1), 187–194 (1993)
7. Torrent, M., Perat, J., Jiménez, J.: Permanent magnet synchronous motor with different rotor structures for traction motor in high speed trains. Energies **11**, 1549 (2018)
8. Ohnishi, K., Shibata, M., Murakami, T.: Motion control for advanced mechatronics. IEEE/ASME Trans. Mechatron. **1**(1), 56–67 (1996)
9. Szolc, T., Konowrocki, R., Michajłow, M., Pręgowska, A.: An investigation of the dynamic electromechanical coupling effects in machine drive systems driven by asynchronous motors. Mech. Syst. Sig. Process. **49**, 118–134 (2014)
10. Holopainen, T.P., Repo, A.-K., Järvinen, J.: Proceedings of the 8th IFToMM International Conference on Rotordynamics, KIST, Seoul, Korea, pp. 986–993 (2010)
11. Konowrocki, R., Szolc, T.: An analysis of the self-excited torsional vibrations of the electromechanical drive system. Vibr. Phys. Syst. **27**, 187–194 (2016)
12. Kumar, A.K.: Method and system of limiting the application of sand to a railroad rail, U.S. Patent 7,290,870B2, 6 November 2007 (2007)
13. Konowrocki, R., Szolc, T., Pochanke, A., Pręgowska, A.: An influence of the stepping motor control and friction models on precise positioning of the complex mechanical system. Mech. Syst. Sig. Process. **70–71**, 397–413 (2016)
14. Lata, M.: The modern wheelset drive system and possibilities of modelling the torsion dynamics. Transport **23**(2), 172–181 (2008)
15. Szolc, T., Konowrocki, R., Pisarski, D., Pochanke, A.: Influence of various control strategies on transient torsional vibrations of rotor-machines driven by asynchronous motors. In: 10 th IFTOMM, 23–27 January 2018, Rio de Janeiro, BR, vol. 4, pp. 205–220 (2018)

Maneuverability of Tracked Vehicle at the Border of Traction Between Tracks and Ground

Jarosław Kończak(✉)

Military Institute of Armoured & Automotive Technology, Sulejówek, Poland
jaroslaw.konczak@witpis.eu

Abstract. Heavy military vehicles are equipped with tracked tracks, because on a low-load bearing ground through a moving road - a track, you can distribute the pressure more evenly. One of the ways to protect the vehicle and the crew is to use the mobility of the vehicle combined with a continuous change of direction, i.e. permanent use of the maneuver against the targeting enemy. To assess the traction properties of the vehicle, in addition to the design features, the impact of the properties of the soil should be considered. The work reviews the characteristics related to the mobility of tracked vehicles. The issues of the load-bearing capacity of the soil, track adhesion to the ground, and turn resistance were taken into account. An original method was proposed to assess the maneuverability of vehicles moving on a deformable substrate. Trends in the development of suspensions and crawler chassis are presented. In the summary, further directions of analytical, computational and research works with the use of numerical techniques were specified. The results of the work can be useful in shaping the reliability of vehicles and issues related to the smoothness of the vehicle's movement (comfort of the crew's work, the possibility of precise aiming).

Keywords: Military vehicle · Rolling resistance · Grip of wheels · Maneuverability

1 Introduction

Military forces use various vehicles in order to provide forces and means with ability to move. The dominating type of vehicles are those, which transfer driving force to surface (ground) using wheels with pneumatic tyres. The remaining vehicles, typically heavy, are equipped with track traction because on a surface with small tonnage due to mobile road – a track, pressure can be distributed more evenly. Therefore, a tracked vehicle retains the ability to move, stop, and start driving.

In comparison to wheel traction, maintaining a track traction is by average 350–400% more expensive. However, vehicles providing their crews with armoured protection, characterised by weight exceeding 25 tonnes, and from which a high mobility is required are equipped with tracked chassis. The asymmetric military actions

M. Siergiejczyk and K. Krzykowska (Eds.): ISCT21 2019, AISC 1032, pp. 236–246, 2020.
https://doi.org/10.1007/978-3-030-27687-4_24

performed through the last 2 decades (the beginning of 21st century) have shown, that professional armies equipped with expensive heavy equipment lose it in consequence of using anti-tank guided missiles. Hits are executed after an anti-tank rocket is launched from fire position invisible for vehicle crews on an average distance between 700 and 2000 m. Vehicles are destroyed and soldiers are endangered to loss of health (burns, deafness, psychological injuries) or life.

1.1 Using Manoeuvre and Mobility in Vehicle Protection

2 trends are developed in order to provide vehicles with protection. The first one concerns increasingly more refined armours (multi-layered and reactive armours). The second trend – the construction of active protection systems focuses on the use of sensors, computer analysers, the purpose of which is to detect an incoming anti-tank rocket and counteract it, e.g. through a thermal trap or by shooting anti-rocket/grenade.

Another method to protect a vehicle and its crew is using the vehicle's mobility connected with a sudden change of movement direction, i.e. through a permanent use of a manoeuvre against aiming enemy.

Further lines of the article will be dedicated to increasing manoeuvrability of tracked vehicles.

Traction Requirements

Traction requirements set before military tracked vehicles cover the selected following features:

- V_{max} – maximum velocity of vehicle on a road with hard/hardened surface,
- T_{20mph} – acceleration time (on a hardened road) from a standing start up to the moment of acquiring velocity of 20 mph, i.e. around 32 km/h,
- R – value of vehicle's turning radius,
- Z – vehicle's range on hardened roads on one filling of gas tanks,
- α – ability to overcome hills with given angle,

However, these requirements do not determine other features that are significant for vehicle's protection and survival on battlefield, which should characterise a military vehicle. Those are:

- the ability to change the driving direction
- time of overstress exposure per crew and elements of vehicle caused by horizontal and vertical acceleration (resulting from the change of driving direction), which influences the combat ability and durability of track propeller or other vehicle components
- amplitude and frequency of angular vibrations of vehicle's hull, which influences the efficiency of kept fire.

2 Research Problem and Research Method

2.1 Comparative Research of Various Constructions

Traction research is performed according to the contents included in tactical and technical requirements and literature guidelines [1–4, 9].

In order to draw up an initial tracked vehicle manoeuvrability analysis, a characteristic of turn, which expresses the f_p driving force necessary during a turn on an overtaking track in the R turn radius function, is prepared. In the analytical method of drawing up a graphic turn characteristic are used the values of unit driving forces occurring on tracks: overtaking – f_{p2} and overtaken – f_{p1}, which determine dependencies:

$$f_{p1} = -\frac{1}{2}f + \frac{\mu L}{4B} \tag{1}$$

$$f_{p2} = \frac{1}{2}f + \frac{\mu L}{4B} \tag{2}$$

and from empirical dependency [2–4]:

$$\mu = \frac{\mu_{max}}{0,92 + 1,15\frac{R_g}{B}} \tag{3}$$

where:

f_{p1}, f_{p2}—unit forces (indicators of unit force) on the remaining and overtaking track and $f_{1,2}$ = F1.2/Gp (where F_{12}—driving forces on tracks),
G—weight of vehicle,
R—turning radius,
R_g—main calculation turning radius,
μ_s—turn resistance index,
μ_{max}—highest value of μ index at turn with R_g = B radius,
L—resistance length of a track in *m*,
B—track wheelbase,
f—rolling/movement resistance index.

The diagram of forces and torques [1] acting on a vehicle during a turn was adopted tor model considerations and the geometry of curvilinear movement of vehicle was determined (Fig. 1).

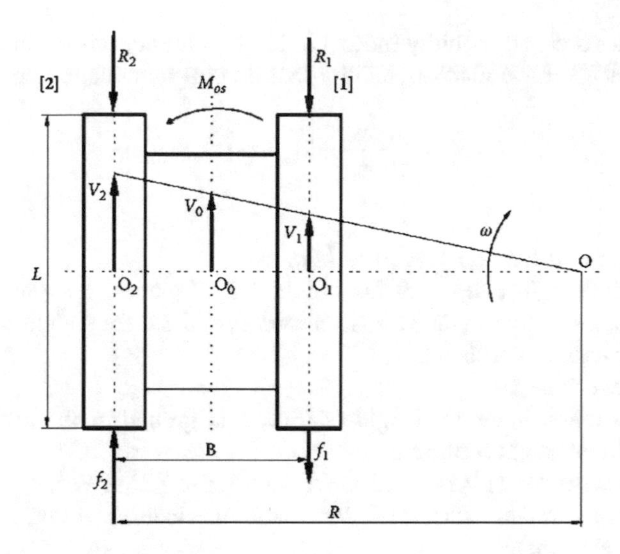

Fig. 1. Forces acting on the vehicle at the turn (R_i - resistance) forces on track belts; M_{OS} - the moment of turn resistance, source [1]

2.2 Influence of Surface Properties on Vehicle's Movement

Tracked vehicles move on various surfaces. The validity of their use [11] occurs on susceptible surfaces, e.g. clay soil, mud, or wet sand. The influence of surface properties should be considered in order to evaluate traction properties of a vehicle.

Index of track traction to deformable surface [4] in straight line movement determines the vehicle's ability to move and is determined for deformable surface using dependency:

$$\mu = \frac{10^6 b}{G} \cdot \int_0^L \tau_x \cdot dx \tag{4}$$

where:

μ — traction index,
G – vehicle weight in *kN*,
τ_x – shear loads in soft surface in *MPa*,
L – resistance length of a track in *m*,
b – width of a track in *m*.

The dependencies mentioned in works [1–4] take into account classic approach of mechanics of vehicle movement on deformable surface.

In another approach are included loads on ground and geometry of track system (L and B parameters) while taking into account the geometry of load bearing wheels.

Therefore a so-called mobility index [8, 12] was introduced for the evaluation of tracked (off-road) vehicle manoeuvrability that describes the dependency:

$$IM = \left(\frac{q_{śr} \cdot k_m}{k_t \cdot k_{gz}} + k_{kn} - k_p\right) \cdot k_s \cdot k_{un} \quad (5)$$

where:

$q_{śr}$ – average unit load on ground in *kPa*,
k_m – vehicle weight index (1.0 for weight < 22.7 tonnes, 1.2 - for weight 22.7–31.7 tonnes, 1.4 - for weight 31.7–45.3 tonnes, and 1.8 for weight > 45.3 tonnes),
k_{kn} – load bearing wheels index,
k_p – clearance index,
k_{gz} – construction index for height of track link ground hook (1.0 for height $\leq$ 38 mm, 1.1 for height > 38 mm),
k_s – unit force index (1.0 for 7.35 kW/t, 1.05 for < 7.35 kW/t),
k_{un} – driving system index (1.0 for hydro mechanical systems and 1.05 for mechanical systems),
k_t – track factor (track width in * 10^{-2}).

In work [12] were described numerical methods of determining the distribution of pressure under load bearing wheels. An important factor in determining the acting on the ground is a so-called belt of load bearing wheel (load bearing wheels) with an n amount of track belts, which influences the actual pressures on ground unlike average unit pressure, which is described [2, 9] by dependency (Fig. 2):

$$q_{śr} = \frac{G}{2 \cdot L \cdot b} \quad (6)$$

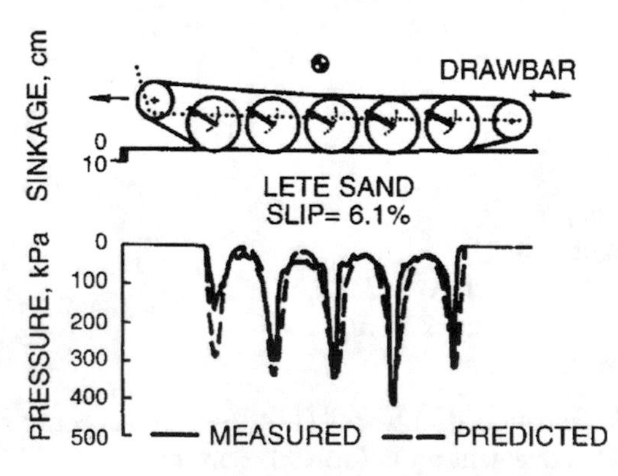

Fig. 2. Comparison of the measured normal pressure distribution under the track pad of vehicle (M 113) and the predicted one using a computer-aided method on sand, source [12]

In order to evaluate the ability to overcome deformable surface by vehicles, two substitutable calculation methods [8, 12], which are elaborated on the basis of experimental research: VCI (Vehicle Cone Index) and MMP (Mean Maximum Pressure) are used.

In the VCI method, to a vehicle is assigned a minimum load bearing capacity of the ground, which is necessary to overcome the ground without the vehicle getting stuck and which is compared with the actual RCI (Ratting Cone Index) load bearing capacity of the ground. The MMP index determines the level of mean maximum pressures in the ground under a track and road wheels (that decide about the depth of track marks) and in conditions of e.g. moderate climate. It should be stressed that the distribution of pressures under a single track link in a susceptible surface is not uniform and influences on the patchiness of tension concentrations in the ground, which causes a variable traction in longitudinal and transversal direction towards the section of track that is currently lying on a surface (moving vehicle).

Another construction parameters, aside from pressures created by a vehicle on the ground, are:

- height of the centre of mass of vehicle, which influences the ability to overcome terrain and steadiness in curvilinear movement. Recommended value [10, 11] for tracked vehicles is value $h < \frac{B}{2}$,
- possible to acquire unit force φ of tow capacity on P_u hook of a vehicle referenced to the G total weight acting on the surface, $\varphi = \frac{P_u}{G}$, which determines the ability to overcome terrain (off-road ability), Fig. 3.

The force of movement resistances of a tracked vehicle on susceptible surface that takes into account construction parameters and ground properties has been determined in [11] using dependency:

$$F_f = \left(\frac{1}{n+1}\right)\left(\frac{1}{k_e + bk_\varphi}\right)^{\frac{1}{n}}\left(\frac{G}{L}\right)^{\frac{n+1}{n}} \quad (7)$$

The depth of track mark caused by compressing a surface according to [11] is determined using formula:

$$z_0 = \left(\frac{G}{b \cdot k \cdot L}\right)^{\frac{1}{n}} \quad (8)$$

where:
k_e – ground vertical deformability index,
k_φ – ground concentration index,
n – index determining the structure and type of ground,
L – resistance length of a track in *m*,
b – width of a track in *m*,
G — weight of vehicle in *N*.

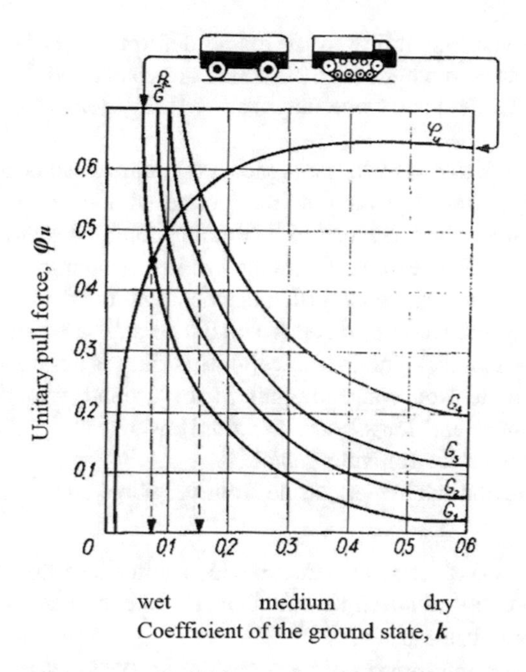

Fig. 3. Interdependence between unit pull force of a vehicle, different weight of trailers G_1, G_2, G_3, G_4 and rolling resistance coefficient *f* on a ground with different humidity, source [11]

Track traction to surface index determines the behaviour of high speed vehicle in extreme movement conditions, e.g. during braking, avoiding a driving obstacle with a permanent change of driving direction, or driving with velocity exceeding 50 km/h on the bend of a road.

Likewise, for wheeled vehicles it can be said that, unfortunately, the values of index of road wheels traction to surface, which can be found in literature, significantly differ from each other [7]. In literature are provided results of research on tyres of various constructions and generations and the research are conducted at various velocities that quite often are not provided. At the same time it should be stated that the results of research on traction of tyres to surface were acquired in various road conditions. Similar case occurs with tracked vehicles.

In literature [2–4], the values of indexes of longitudinal traction of tracks to surface are provided the most often. However, in curvilinear movement, aside from turn resistance index, a significant value is the index value of track traction in transversal direction.

This value is most often determined empirically, appropriately to a given type of surface.

On the next stage of manoeuvrability evaluation, author has proposed two tests, which should be virtually performed with the examined vehicle.

An original method of evaluating the ability of vehicle to change movement direction was adopted. In this method, the construction features of vehicle (e.g. values L and B) were used to create a driving or obstacle track. The method includes two tests.

The preparation of vehicle to tests consists of placing a measurement device with acceleration sensors, which operate in 3 directions (X, Y, and Z), in the centre of mass, as well as additional acceleration sensors in places potentially exposed to the occurrence of operating damage [6], e.g. at roll holding the upper branch of a track. The measured acceleration values in time function are subjected to numerical analysis in order to indicate critical values, which have impact on e.g. the quality of aiming or mechanical damages registered during operation.

Test 1. Driving on Eight-Shaped Track
The test covers driving a tracked vehicle on track presented in Fig. 4. The drive of examined vehicle should be carried out on various surfaces with various indexes of: traction, movement resistance, and load bearing capacity of the ground. The envisioned radius of movement track of vehicle's centre of mass should be equal to R = 3•B or 4•B. It is important not to exceed the critical velocity of vehicle during a turn [3], which is expressed with general formula:

$$V_{kr} = \sqrt{\mu_s \cdot g \cdot R} \tag{9}$$

and its specific case (in tests)

$$V_{kr} = \sqrt{\mu_s \cdot g \cdot 3B} \quad \text{or} \quad \sqrt{\mu_s \cdot g \cdot 4B} \tag{10, 11}$$

and the exceeding of which will result in a sudden skidding of the vehicle (a turn around the axle going through the centre of mass).

R = 3•B values are recommended for vehicles with group II (e.g. side clutch, planetary) turning mechanisms and R = 4•B values - for vehicles with group I (differential) turning mechanisms. This allows to test vehicles in the field with actual driving velocities in the upper range between 24 and 36 km/h.

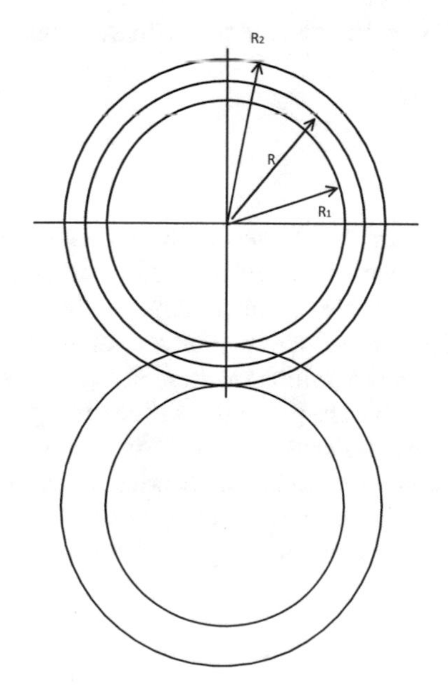

R_2 - turning radius of the vehicle, the center of mass,

R_1 - turning radius of the track belt remaining

R_2 - turning radius of the track belt overtaking

B – track base

$$B = R_2 - R_1 \tag{12}$$

$$R_1 = R - \tfrac{1}{2} B \tag{13}$$

$$R_2 = R + \tfrac{1}{2} B \tag{14}$$

Fig. 4. Test track of a high speed tracked vehicle – driving eight, own study

Test 2. Slalom Drive
The drive of examined vehicle should be carried out on various surfaces with various indexes of: traction, movement resistance, and load bearing capacity of the ground. The envisioned distances between road guards that indicate the outline of obstacles to be avoided were selected using vehicle's construction features, i.e. length L and B (Fig. 5).

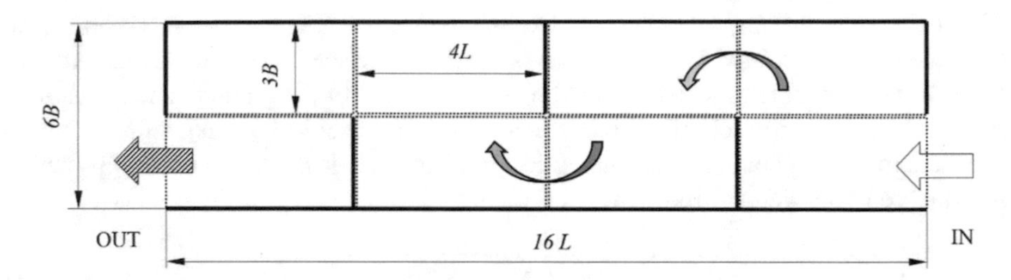

Fig. 5. Test track of a high speed tracked vehicle – slalom drive, own study

The time of speeding drive in a straight line drive of a vehicle is measured. A criterion of charging penalty points has been adopted: an additional 0.5 s is added for each case of touching/driving over a guard. The test corresponds to usual hazards for vehicles in the form of engineering barriers or driving in an urbanised area, e.g. urban fights in Syria, where tracked vehicle becomes an easy target.

The performance of two simple tests allows to compare the applied construction solutions: track mechanisms, turning mechanisms, features of driving system (unit force indicator, engine energy surplus), and the influence of placing the centre of mass on the manoeuvrability of a vehicle.

It is envisioned that average driving velocities on the test section with susceptible surface will amount to ab. 10–30 km/h.

3 New Construction Solutions

The operating practice shows that during vehicle's turn on an uneven surface occurs an uneven wear of double load bearing wheels (one wheel is burdened by 25%) or single load bearing wheels (increased wear of one side of a wheel by about 25%) at their co-operation with track belt and guiding combs. New construction solutions of tracked driving systems are aimed at the reduction of movement (rolling) resistances caused by uneven burden of load bearing wheels. Solutions with susceptible (tilting) mounting of wheel to suspension rocker arm (Fig. 6) [5] are encountered, especially in the developed rubber tracks. This facilitates the adjustment of a track and load bearing wheels to the shape of driven surface and influences the reduction of movement pores and causes less frequent damages in the driving system. On the basis of analysis of damages of classic suspensions [6], the application of innovative (tilting) systems may result in the extension of average run between suspension damages by 35%, which in turn will reduce the costs of maintaining a track traction.

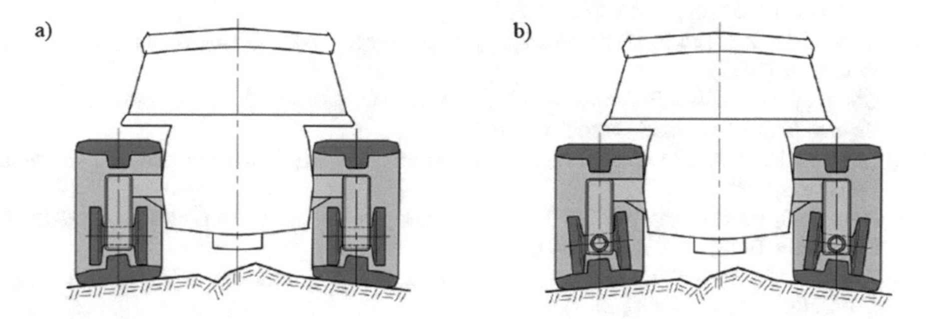

Fig. 6. Suspensions of tracked vehicles (a) classic, (b) with pivotable pivot wheels, source [5]

4 Summary

Manoeuvrability of tracked vehicles should be used to protect against effects of a hit. The identification of parameters of forced curvilinear movement on deformable surface is a contribution to:

1. evaluation of durability and reliability of track propellers (determination of wear for steel, rubber, or rubber-metal elements),
2. designation of fixes for on-board armament stabilisation system (influence of variable transversal accelerations of vehicle's hull),

Further research works will be focused on the construction of data bases with parameters of susceptible surfaces and on conversion of tool programs for comparative analyses of traction abilities of tracked vehicles into the Matlab environment in order to acquire the ability to envision the curvilinear movement of a vehicle on unknown surface in real time. In the construction of mathematical models of curvilinear movement of tracked vehicles should be considered:

1. the influence of increasing engine's power on the behaviour of a vehicle during a turn, which has a particular significance on surfaces with low load bearing capacity (turning resistances are decisive),
2. the variability of transversal traction index in the function of track turning (turning of a vehicle) in relation to a prior straight line direction of driving,
3. the estimations of turning resistance index for vehicle on a soft, deformable surface while taking into account the properties of the ground,
4. the empirical data from real object studies in order to perform changes in calculation algorithms, estimation of traction indexes, or approximation of numerical simulation results.

References

1. Borkowski, W., Rybak, P., Michałowski, B., Wysocki, J., Hryciów, Z.: Właściwości trakcyjne gąsienicowego wozu bojowego przy obniżonej mocy silnika napędowego. Biuletyn WAT **LIX**(1) (2010)
2. Burdziński, Z.: Teoria ruchu pojazdu gąsienicowego. Wyd. Komunikacji i Łączności, Warszawa (1972)
3. Chodkowski, A.W.: Badania modelowe pojazdów gąsienicowych i kołowych. Wyd. Komunikacji i Łączności, Warszawa (1982)
4. Dajniak, H.: Ciągniki Teoria ruchu i konstruowanie. Wyd. Komunikacji i Łączności, Warszawa (1982)
5. Dudziński, P., Chołodowski, J.: Energy efficiency of rubber tracked chassis. J. KONES Powertrain Transp. **23**(2), 97–104 (2016)
6. Kończak, J.: Reliability analysis of high-speed tracked vehicles in the polish army. J. KONBiN **42**, 143–163 (2017)
7. Luty, W.: Prawno-techniczne aspekty oceny przyczepności nawierzchni drogi w miejscu zdarzenia drogowego, Prace Politechniki Warszawskiej. Czasopismo Logistyka, Poznań (2010)
8. Łopatka, M., Dąbrowska, A.: Ocena zdolności pokonywania terenu o niskiej nośności przez pojazdy o dopuszczalnej masie całkowitej 14 ton. Logistyka **3**, Poznań, 916–921 (2015)
9. Simiński, P.: Zwrotność pojazdów. Belstudio, Warszawa (2012)
10. Simiński, P.: Wojskowe Pojazdy Kołowe. Belstudio, Suleówek-Warszawa (2015)
11. Sołtyński, A.: Mechanika układu pojazd – teren. Wyd. Ministerstwa Obrony Narodowej, Warszawa (1966)
12. Wong, J.Y.: Theory of ground vehicles. Willey, New York (2001)

Application of V2X Technology in Communication Between Vehicles and Infrastructure in Chosen Area

Daniel Kossakowski and Karolina Krzykowska(✉)

Warsaw University of Technology, Koszykowa 75, 00-662 Warsaw, Poland
kkrzykowska@wt.pw.edu.pl

Abstract. The Precision Agriculture (PA) concept is becoming more and more important in the modern world. Although it faces numerous hardships with technology the biggest problem is the lack of pure standard definition. Solution may be to get inspired from Vehicle to Everything (V2X) model. It is a widely spoken conception of the road system, where vehicle, infrastructure and other traffic elements would be able to communicate with others. However, it is not in the common use, it is a developed technology with tested and established assumptions for various of aspects like transmitting medium. Because such work has been done in that field, it should be considered to implement solutions from V2X to PA, due to the similarities between them.

Keywords: Precision Agriculture · Vehicle to Everything · Vehicle to Infrastructure · Vehicle to Vehicle

1 Introduction

Since the 1980s, there was a silent development of a precision agriculture as a fresh insight on a farming process. Precision Agriculture itself is a new concept of managing the implementation of new technologies in a agricultural field. Those technologies need to rise efficiency, profit and output values from farming. According to the 3-phases of innovation model [16], so called "green revolution" in 1950s/1960s could be recognized as a fluid phase, where we focus on product improvement. In this example, the efforts were put to develop better fertilizes, herbicides etc. Although, some efforts must have been done in the process development but when summarized, the disproportion between the process and the product can be easily recognized.

According to the model, agriculture is in the transitional phase, between product and process innovation. Slowly raising focus on the process innovation under the label of Precision Agriculture technology, prove the rightness of this statement. The only question is what exact standards need to be set. The variety of ideas and technologies must be unified to a certain level in order to achieve the base of common standard.

Surprisingly, the solution to this problem lays in the relative conception of Intelligent Transportation Systems. This technology focus on rising the efficiency of road system and road safety at the same time. Since some assumptions were made in the ITS

M. Siergiejczyk and K. Krzykowska (Eds.): ISCT21 2019, AISC 1032, pp. 247–256, 2020.
https://doi.org/10.1007/978-3-030-27687-4_25

and the similarities between mentioned conceptions, there should take place the discussion over implementing solutions from ITS to the PA.

2 Intelligent Transportation Systems

2.1 V2V

Vehicle to Vehicle (V2V) is a technology that allows establishing temporary link between two separate vehicles. It is also known as VANET (Vehicular ad hoc Networks), which describes the true concept behind the technology. While putting additional equipment, the vehicle can obtain the ability to read signals from its surroundings. The main reason for this technology to develop is to reduce the number of decisions that are needing to be made by a driver. In consequence, it is possible to rise the level of safeness just by filtering all information about road situation and to display them in a more ergonomic way.

In V2V the focus is put on the other vehicles using or attempting to use the road. The final system is a set of different applications that are to help driver in various situations. Examples:

- Parking Assistance;
- Collision Warning;
- Platooning;
- Do Not Pass Warning;
- Forward Collision Warning.

The simplest way to implement V2V in the vehicle is to use DSRC and GPS as a base for communication [1]. Dedicated Short Range Communication (DSRC) is a dedicated protocol based on IEEE 802.11p whose medium is based on microwaves at 5.9 GHz band. It is used as a medium for communication between two vehicles on a short distance to inform each other's drivers about their intentions and actions in advance.

The GPS is a second pillar for the system, informing about the current position of a unit and comparing it with others. The advantage of this element is its versatility due to the wide spectrum of potential activities. It can be a potential component not only in V2V but in Intelligent Transportation Systems (ITS) as well. Moreover, it is possible to use its components as well-established navigation systems. As mentioned before, the role of V2V would be the driver's assistance, not decision-making subject. System could possibly warn of obstacles on a road or other vehicles manoeuvres in advance, giving time for reaction. The low entry level in the terms of expenditures comparing gives it an advantage comparing to the full autonomous vehicle at early stage. However, the fact that V2V is limited from its assumptions, it may be seen as system not making full use of the capabilities of installed equipment.

2.2 V2I

Vehicle to Infrastructure (V2I) is a communication model that supports the information flow from the road system to vehicle and vice versa. When developed, this system could possibly increase the driving economy and emissions. Vehicles equipment with the V2I system have significantly better results compering to the standard one in terms of fuel consumption and vehicle expenditure. The drivers are driving in the smoother and steadier way due to the more given time for reaction.

V2I could use five main applications [7]:

- Informational that provide information in advance to a driver.
- Alert that is based on the external source of potential threats.
- Warning that is based on the vehicle parameters that are a threat in the current conditions.
- Control that may adjust the vehicle parameters in order to avoid threats.
- Data Exchange that is responsible for data exchange between vehicle and storage of service provider.

To make it possible for such results to happen, those systems demand various components installed alongside a road system. The test carried out in Bologna [5] has shown the problems with the signal usage in different environments such as urban or rural with various factors like trees or heavy vehicles considered. It also refers to the GPS signal which is reflected by ground objects. This can cause one or more secondary paths that can prolong the propagation time as well as phase or amplitude of waveform.

The other research has shown that all inspected parameters influence IEEE 802.11p V2I and must be considered while installing the system. It may require the landscape reshaping or adjusting the applying system to prevent the working errors. Although, the V2V uses DSRC medium, it has been presented [15] that the cellular networks preformed even better.

Apart from transmission, whether it would be the DSRC or cellular networks, vehicle must be equipped with additional sensors and elements in a similar way as in V2V. The need of GPS signal receiver installation shows again that ITS is a modular solution. Some of its elements can be universal and used for various purposes. The role of GPS itself, apart from being used for navigation and in V2V, here is to determine the location of vehicle in the intersection, between two elements of infrastructure.

2.3 V2X

Vehicle to Everything (V2X) is a communication model that allows vehicles to gather information from multiple sources [14]. Mentioned in the previous chapters V2I and V2V would be incorporated and enriched with systems enabling exchanging data between vehicle and network, pedestrian or grid (for electric vehicles). Alongside the system complexity, the profits should appear.

The motivation from the V2X would be like the V2I and V2V. The main reasons are the safety and human factor [17]. Both are strongly connected with each other and could easily be assisted by V2X system reducing economical costs and increasing road safety and traffic efficiency.

The original technology for V2X was WLAN, because it supported the establishing ad hoc network between two vehicles or between infrastructure and vehicle. As mentioned in 2.2, the drawback of this medium was its range.

A comparison done in this field [11] shows the superiority of cellular network over DSRC in some aspects or at least equality in other criteria. Challenges for V2X were:

- **high speed -** to support high speed vehicles, the changes in information coding were necessary. In DSRC the header and preamble are used as the pilots. This makes 4 pilot signals for every OFDM signal. On the contrary for cellular signal, the reference symbol is added per every two OFDM symbols. Orthogonal Frequency-Division Multiplexing (OFDM) is a multiplexing method in frequency domain that allows to transmit multiple data streams on different carrier frequencies. Its main advantage is the economy of bandwidth usage but with the drawbacks of vulnerability to the Doppler effect and synchronization with the carrier frequency;
- **long range communication -** for the long-range communication the upper hand has the cellular network as the combination of its certain features. Cellular V2X uses frequency multiplexing signal that compresses multiple communication bands allows the narrower signals to rise its spectral density to help its decoding. Moreover, the wave shape of cellular V2X needs less energy and in consequence the back off power is smaller than in DSRC;
- **decentralized multiple access -** because of the decentralized character of V2X there exists a threat of collision between many independent transmissions. In this field, DSRC has a natural advantage because of its multi access structure, similar to others from IEEE 802.11 family. The cellular solution was at a disadvantage because of the dedicated 5.9 GHz frequency. Although, using LTE and its band extremes to send information about circumstances of regular link availability. LTE technology provides a better feature to V2X technology than IEEE 802.11p standard [2].

3 Precision Agriculture Systems

3.1 Types of Precision Agriculture Systems

The process of developing the idea of Precision Agriculture in the last 20/30 years could be described as elaborating ways to use more precisely the products of so called “green revolution”. The main presumptions are to make farming more effective and profitably with lowering impact on the environment. Moreover, the UE [9, 17] is pointing towards a new technology like Big Data or IoT as a tool for agricultural improvement in terms of efficiency and environmental performance.

The process of transforming currently exploited cars into autonomous vehicles can be divided into main six stages [12]. Nowadays, the development reaches the stage three, where the vehicle can make basic safety decisions under the supervision of the driver. While the movement would be towards stage four and five, current state is enough to perform Precision Agriculture activities sufficiently, due to their auxiliary and assisting role. Different solutions were achieved separately and later combined for greater effect. Nowadays, it is possible to distinguish three main directions.

Yield Monitoring and Mapping (Fig. 1) is technology that main purpose is to manage fields in the most efficient way. In this way it can distinguish the productivity of soil over the certain area. It is a source of data that is essential in the farm management. Based on outcome, the specialty of production for different fields to maximize profits.

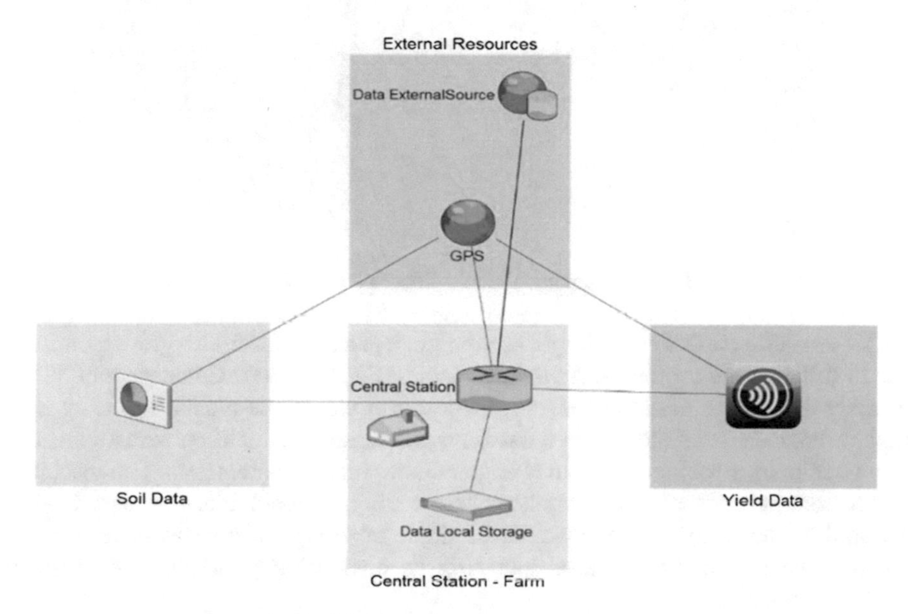

Fig. 1. Yield monitoring schematic [3, 8]

The method has been distinguished into two categories, based on the source of data [3]. The first one is a yield data that is collected by installing sensor on the harvester vehicles. By using algorithms, it is possible to precisely account the amount of grain from each area. GPS allows to apply collected data on a map and combine it with the previous years' outcomes. The second category is the soil data that comes directly from the soil examination process. By taking the samples from different areas with the flexible level of detail, estimated map of area can be created.

Vehicle System Guidance (Fig. 2) has been created to assist the operator of a field vehicle. By using precise GPS positioning systems, remote steering or auto steering machinery, the efficiency of a work can be increased on a two main levels. The first one is the rise of the precision of the activities by filtering the number of information and suggesting the optimal decision. The second method is the reduction of needed workforce by making the operating hour more efficient or even reducing the number of needed operators.

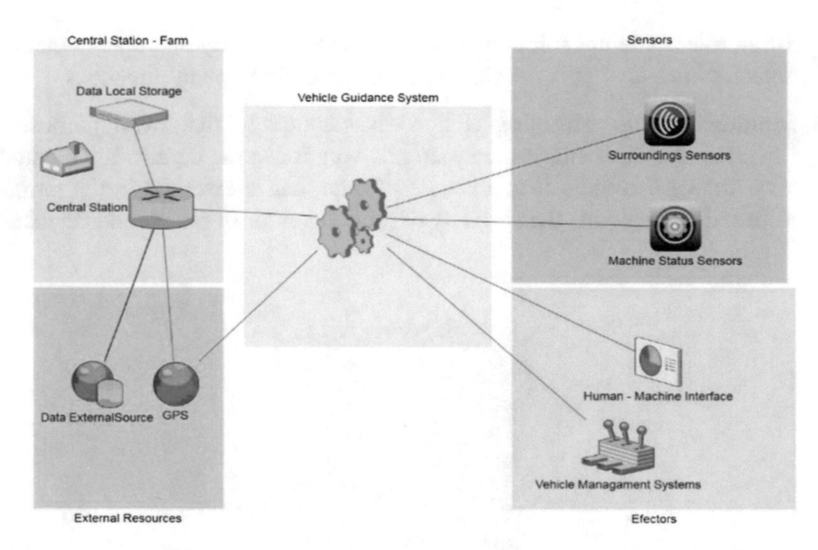

Fig. 2. Vehicle system guidance architecture [3]

VRT (Variable Rate Technology) has the great potential for future growth due to its role in fulfilling the ecological farming strategies (Fig. 3). The main concept of VRT is to inject the precise portion of chemicals, fertilizer etc. to the plants. By using information that have been storage in a database and sensors that add current weather and crop conditions machine is able to change ad-hoc working parameters. The machinery differs from each other and the precision of operating, although most of them represent standard of capability of manipulating of single or groups of nozzles or feeder. Processing data in real time, makes adjusting its own velocity or quantity of dosing substance possible.

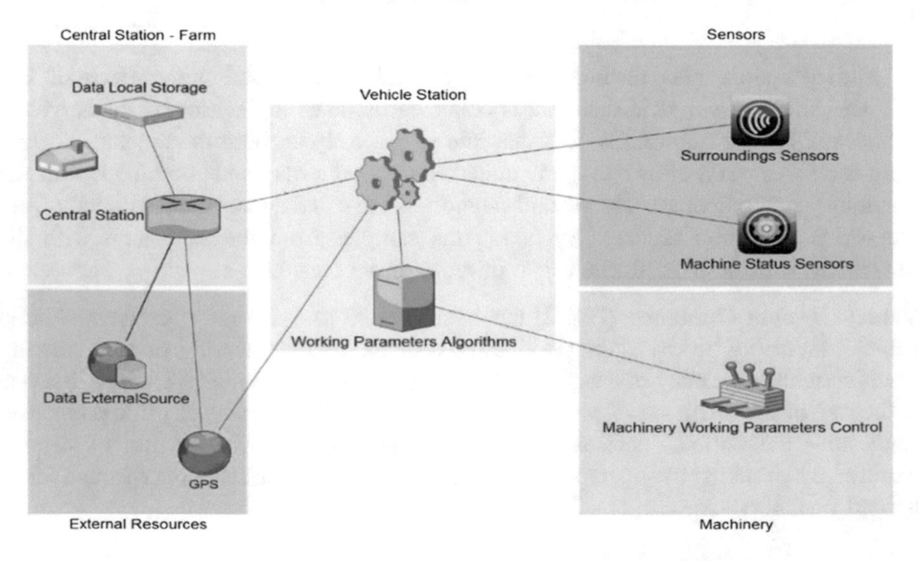

Fig. 3. Variable rate technology system architecture [3]

3.2 Profits

Because of different types agriculture business depending on the field varieties, business models etc., it is extremely hard to summary what are the real costs and savings from PA usage. While using different types of PA solutions, higher profit can be obtained by both reducing costs and increasing revenues. Example of system guidance shows great potential in reducing the labour costs. Some of the fields work can be done in shorter time than traditionally.

On the other hand, GPS and yield monitoring systems can make saving because of better use of available land. It is possible to map a farm area over years with full information about certain fields profitability. Because data was not averaged for bigger areas but shown in detail, it was easy to recognize high and low profitable sectors. In consequence, they can be managed more consciously, and many decisions can be made (such as changing low profit zones into bio buffer zones or lowering seeding rate).

Finally, VRT using results needs to be commented. It is hard to estimate exactly what are the profits but possibility of applying the exacts number of fertilities on a certain area is very promising. If not for sake of a pure profit, it would be researched for sake of environmental aspects.

Many researches have shown the truth of the thesis that methods above, adjusted to specified needs, could increase profit. An experiment [10] of compering two farms between 2008 and 2014 took place on neighbouring farms, testing different types of crops. One, that were using PA had a 54% higher output, 46% higher gross margin and in consequence, higher profit. It needs to be mention that the more vulnerable crops the more visible differences in profit were.

4 Concept of V2X Implementation in Precision Agriculture

4.1 Why Implementation of V2X into Precision Agriculture

Precision Agriculture has been described by the various systems. Because of the great variety of its purposes and different spheres of operating, PA is far from being a system of only one definition. From its very nature and presumptions, it has an open and modular structure to meet all kinds of expectations. Different needs and production specifications call for different solutions or their combinations. There are areas where close cooperatives of some methods would be essential (like VRT and mapping), nevertheless certain elements should have certain dose of independence.

Also, the very nature of V2X network is its multifunctionality. This system architecture maintains many vehicles, elements of infrastructure and other entities on a road. It supports the data flow from and to data bases, providing a real time assistance. The main advantage is the ability to link numbers of units from different service providers or producers. Although the system may not be effective enough to handle the traffic nowadays as full scale, the technology is promising.

Precision Agriculture is much less demanding environment than road traffic. The vehicle density is much lower, and the threat of collision is negligible. The main issue of the V2X technology are absent in farming like multi access connections, high speed vehicles or low level of threats.

Moreover, it would contain many exclusives networks, apart from data exchange with service delivers, farms that use PA would be independent and self sustainable units. This exclusiveness could cause the great diversity of the system modules mixes. Because, while creating V2X is supposed to differ depending on an environment and a country, it makes a perfect example and the source of solutions for the new coming Precision Agriculture.

4.2 How to Implement V2X into Precision Agriculture

In the Precision Agriculture, as mentioned before, the greatest challenge is to develop the end model for the system that will serve as the standard. Every single farm has different needs arranged in order from the most to the least desired [6]. This calls for the modular system arranged around the central station with different modules as plug ins. Those could have different origins, working parameters but must be able to communicate with at least central station. The whole PA cycle could be divided into four phases [4]:

- **Planning Phase** where based on a made decisions, maps are updated. Next, the conversion of data to the acceptable format for each destination has to be done.
- **Application Phase** where the data is used to perform farming activities. It refers to the regular and planed works but also to ad-hock actions. At the same time, during performances, some data is collected due to the additional sensors on a vehicle or from the post work report.
- **Results Phase** where the collected fresh data is storage and convert to the format for the central station. Some early phase mechanisms on data are done to help with evaluation.
- **Evaluation Phase** where the partially treated data can be evaluated by human or AI. Here the decision is made for the next cycle.

Depending on the conditions the cycle could be repeated annually or in the shorter basis. Triggered, it is also possible to finish it in the terms shorter than a day. The challenge is to ensure that all the cycles, which number definitely will be larger than one, could operate in a non-collisional way. The queueing based on the urgency of decision must be implemented. On the other hand, system of inter cycles conclusions exchange. Each cycle is partially autonomous in decision making in a sense that it bases on its own collected data and conclusions from other cycles. This would ensure the fluency of data processing and decision making.

The one of conditions for implementing solutions from V2X into PA would be to support one communication standard for all elements. The discussion about transmitting medium in ITS that took place, in this case is purposeless. The great range makes it very hard and presumably unprofitable to use DSRC instead of cellular network. Taking to account the landscape shape, various obstacles like trees or buildings shows that the easiest and cheapest way to cover large area with extreme low density of vehicles, would be to choose cellular network like LTE [13].

Another aspect that is worthy mentioning is a mistake margin that in agriculture is bigger than in a city traffic. Density of the vehicles around is pretty low and the threats of accidents are meaningless, what differ it from V2X's operating environment.

Moreover, the speed of farming vehicle is lower than average road traffic. In consequence, the high-speed vehicles communication problem is not taken into account in farming. Those aspects cause the lack of need for high class equipment such as receivers, antennas or data processing units.

Although, the farming operations must be done with precision in terms of location. Farming vehicles should equip with GPS equipment as an important element of PA system. Localization information must be combined with additional data to full support all aspects of PA technology: VRT, driver assistance and mapping. Moreover, it gives the system flexibility for the changes of working circumstances such as weather.

It is possible to imagine architecture where data is collected from moving vehicle, static sensors dislocated over farm and being supported form the external sources, both private (as a service) and public. Data could be processed on various levels, depending on their destination, purpose and urgency. The instructions could be received by number of effectors such as nozzle's driver, seeder's driver or operator assistance system.

5 Conclusions

The main problem with Precision Agriculture as a solution benefiting farm's production is the wide spectrum of types of agricultural activities. Unlike industry, farming is carried out on an annual basis and in consequence some tools are needed once or twice a year. By implementing V2X solutions and models, it is possible to make Precision Agriculture more flexible.

Moreover, the diversity of plant species influences the profitability of investment in PA as it states in the United States Department of Agriculture (Report Number 217). The use of some of the tools are only reasonable while enjoying benefits of the large scale, it does not mean that small and medium size farm will not benefit form adopting some of the solutions. Particularly when some of the V2X standards would be implemented, that could make available for the PA system to be adjusted to fixed conditions.

By making a reference to various tests on V2X, it can be justified to use cellular network as a main standard since it has not been well established in PA. It provides a good quality data flow in the conditions of farming environment. Moreover, the architecture of V2X relays on data exchange between vehicles, elements of infrastructure and central data stations. The whole system is supported by central stations that monitor the flow of data and make decisions automatically or by human. Although PA has slightly different needs and limitations (very low density and speed of vehicles) this architecture could be a point of reference for developing final standard of architecture for PA systems. V2X is a significant source of solutions for PA due to the huge similarities between the basic assumptions between those two concepts. Because of that, the project of Precision Agriculture system, enriched with the mentioned elements of V2X is an interesting concept for the future scientific work.

References

1. Andrews, S.: Vehicle-to-Vehicle (V2V) and Vehicle-to-Infrastructure (V2I) communications and cooperative driving. In: Handbook of Intelligent Vehicles, pp. 1121–1144 (2012). https://doi.org/10.1007/978-0-85729-085-4_46
2. Chen, S., Hu, J., Shi, Y., Zhao, L.: LTE-V: a TD-LTE-Based V2X solution for future vehicular network. IEEE Internet Things J. **3**(6), 997–1005 (2016). https://doi.org/10.1109/jiot.2016.2611605
3. David, S.: Farm profits and adoption of precision agriculture. United States Department of Agriculture. Economic Research Report Number 217; October 2016. http://purl.umn.edu/249773
4. Gebbers, R., Adamchuk, V.I.: Precision agriculture and food security. Science **327**(5967), 828–831 (2010). https://doi.org/10.1126/science.1183899
5. Gozalvez, J., Sepulcre, M., Bauza, R.: IEEE 802.11p vehicle to infrastructure communications in urban environments. IEEE Commun. Mag. **50**(5), 176–183 (2012). https://doi.org/10.1109/mcom.2012.6194400
6. Griffin, T.W., Shockley, J.M., Mark, T.B., Shannon, D.K., Clay, D.E., Kitchen, N.R.: Economics of precision farming. precision agriculture basics (2018) https://doi.org/10.2134/precisionagbasics.2016.0098
7. Kołodziejska, A., Krzykowska, A., Siergiejczyk, M.: Comparative analysys of V2V and A2A Technologies. J. KONBiN **45**(1), 345–364 (2018)
8. Lubkowski, P., Laskowski, D.: The end-to-end rate adaptation application for real-time video monitoring. Adv. Intell. Syst. Comput. **224**, 295–305 (2013). https://doi.org/10.1007/978-3-319-00945-2_26
9. Mihalis, K.: Precision agriculture in Europe. Legal, social and ethical considerations. EPRS Scientific Foresight Unit (STOA). PE 603.207, November 2017
10. Monzon, J.P., Calviño, P.A., Sadras, V.O., Zubiaurre, J.B., Andrade, F.H.: Precision agriculture based on crop physiological principles improves whole-farm yield and profit: a case study. Eur. J. Agron. **99**, 62–71 (2018). https://doi.org/10.1016/j.eja.2018.06.011
11. Nguyen, T.V., Shailesh, P., Sudhir, B., Kapil, G., Jiang, L., Wu, Z., Li, J.: A comparison of cellular vehicle-to-everything and dedicated short range communication. In: 2017 IEEE Vehicular Networking Conference (VNC). https://doi.org/10.1109/vnc.2017.8275618
12. Reddig, K., Dikunow, B., Krzykowska, K.: Proposal of big data route selection methods for autonomous vehicles. Internet Technology Letters, e36 (2018). https://doi.org/10.1002/itl2.36
13. Standards on agricultural electronics. ISO/TC23/SC19
14. Sumiła, M.: Pozyskiwanie informacji w systemach ITS (2013)
15. Ubiergo, G.A., Jin, W.-L.: Mobility and environment improvement of signalized networks through Vehicle-to-Infrastructure (V2I) communications. Transp. Res. Part C: Emerg. Technol. **68**, 70–82 (2016). https://doi.org/10.1016/j.trc.2016.03.010
16. Utterback, J.M.: Mastering the dynamics of innovation: how companies can seize opportunities in the face of technological change, SSRN (1994). https://ssrn.com/abstract=1496719
17. Weiß, C.: V2X communication in Europe – From research projects towards standardization and field testing of vehicle communication technology. Comput. Netw. **55**(14), 3103–3119 (2011). https://doi.org/10.1016/j.comnet.2011.03.016

Effect of Track Structure on Dynamical Responses of a Railway Vehicle

Jacek Kukulski(✉) and Ewa Kardas-Cinal

Faculty of Transport, Warsaw University of Technology,
Koszykowa 75, 00-662 Warsaw, Poland
jkukul@wt.pw.edu.pl

Abstract. A track structure with crashed stone composite designed at Faculty of Transport of Warsaw University of Technology is described. The article presents selected results of experimental measurements and simulation tests. Experimental studies included measurements of vertical track irregularities on the analyzed sections. The dynamical responses of railway vehicle for experimental track geometrical irregularities were determined in simulations for each analyzed track section.

Keywords: Railway vehicle · Track structure · Dynamical responses

1 Introduction

The development of the railway infrastructure at the turn of the 20th and 21st centuries and increase in passenger train speed to V_{max} = 300–350 km/h and freight train speed to V_{max} = 140–160 km/h are the results of railway vehicle design improvement and railway infrastructure optimisation. The purchases of new trains dedicated to high speed lines provide another stimulus for the improvement of track structures and its specifications. The question is whether using conventional ballasted track for high speeds is reasonable. Also economic aspects to the construction and maintenance of track structures can be important for the selection of a track structures design. Conventional track structures with track grid made from rails and ties embedded in a course of ballast and laid on subgrade work in an elastic and plastic state under the operating load. The weakest point in the design is ballast bed deformation: the main source of plastic strain. One of the solutions aimed at strengthening the existing track construction is the application of technological solutions in the form of geogrids, geotextiles or special resins used for gluing the top layers of ballast.

The deformation of track due to the traffic load leads to the track irregularities which perturb the motion of railway vehicles and thus decrease the running safety. In particular, these irregularities strongly affect the forces at the wheel/rail contact which may increase the risk of derailment. According to the original Nadal criterion [7], derailment due to wheel flange climbing on the rail occurs when the ratio of the lateral (Y) and vertical (Q) forces at the wheel-rail contact exceeds the limit value that depends on the flange angle and the coefficient of friction between the wheel and the rail. In current regulations the limit value $(Y/Q)_{\text{lim}}$ is applied to the running average of Y/Q obtained with the 2 m window to account for a finite duration of the derailment

M. Siergiejczyk and K. Krzykowska (Eds.): ISCT21 2019, AISC 1032, pp. 257–265, 2020.
https://doi.org/10.1007/978-3-030-27687-4_26

process. These regulations include the EN 14363 standard [2], in force in the European Union, and the UIC 518 code [9] published by the International Union of Railways. The limit value $(Y/Q)_{\lim} = 0.8$ adopted in the two standards corresponds to the coefficient of friction $\mu = 0.6$ and the maximum value of the flange angle $\gamma = 70^{\circ}$ for the S1002/UIC60 wheel/rail profiles while a typical value of $\mu = 0.36$ gives $(Y/Q)_{\lim} = 1.2$. The determination of the derailment coefficient Y/Q is also required in testing of railway vehicles by the relevant Polish regulations for certificates of admittance to operation for various types of railway vehicles, including locomotives, passenger cars, freight wagons as well as special and auxiliary vehicles [8]. The investigations of running safety based on the derailment criterion was also a core topic of previous works [4, 5] by one the co-authors.

2 Experimental Studies

2.1 Test Object

The objects of experimental of test solution has been developed by the Division of Transport Infrastructure of the Warsaw University of Technology Faculty of the Transport [1]. It is characterized by the mechanical and chemical resistance of the ballast to the phenomenon of deconsolidation.

The proposed crashed stone composite comprises a layer of crashed stone reinforced with a 2 geogrid and stabilized with a polyurethane resin (Fig. 1).

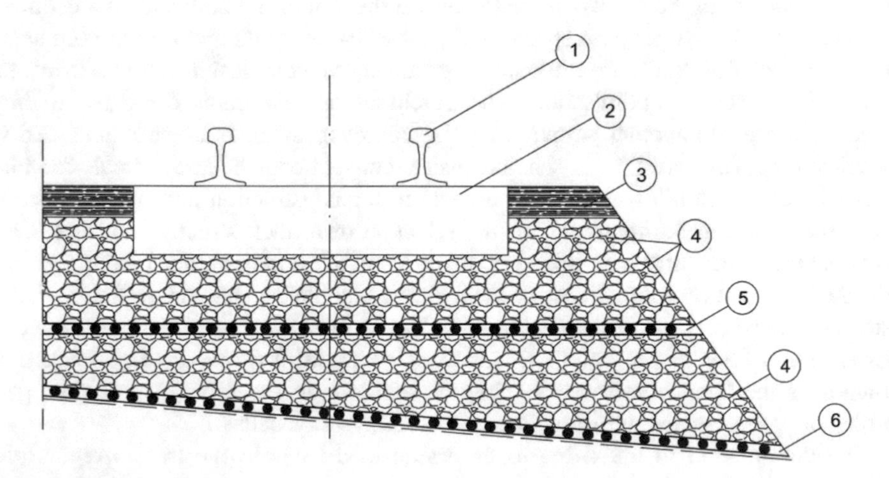

Fig. 1. Track structures with crashed stone composite [5] (1)-rail UIC 60 (60E1), (2)-sleeper, (3)-crashed stone layer with resin, (4)-ballast, (5)-top reinforcement (geogrid, geosynthetic), 6-bottom reinforcement (geogrid, geosynthetic)

The experimental track section with crashed stone composite was built on the Central Trunk Line in 2008. Given the local environment, the experimental section was built using an AHM machine enables laying and compacting the subgrade layer within, a single pass while the other pass builds and compacts the new ballast bed and lays crashed stone reinforcing geogrids. The experimental section was divided into four sectors and plots shown in Fig. 2.

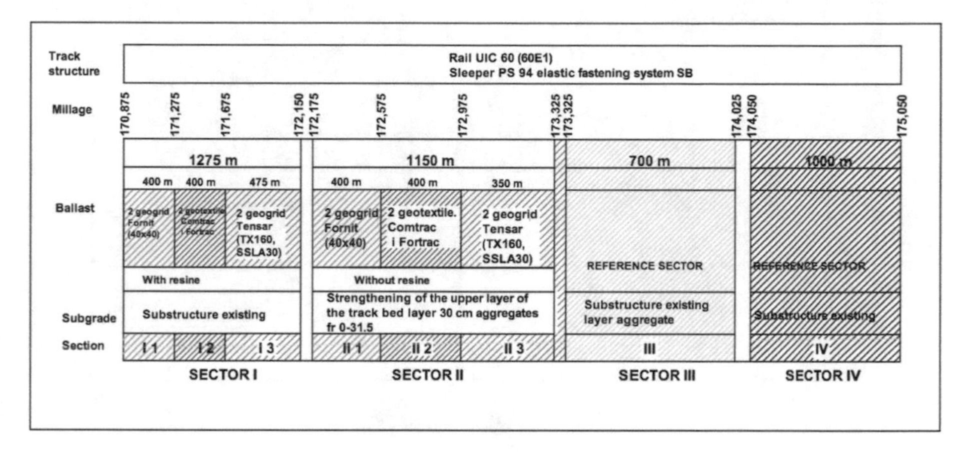

Fig. 2. Diagram of the experimental section of track with crashed stone composite on the central trunk line.

Figure 3 shows pictures of the experimental section of track structures with crashed stone composite.

Fig. 3. Experimental section of track with crashed stone composite (view of geogrids [1].

The analyzed test section was subjected to testing, on which the geometry deformability of the track was tested during operation. The main goal of the measurement of the geometric position of the track on the sections with the crashed stone composite was to assess its deformability during operation compared to the classic track (reference sector) on the adjacent section of the track. Results of the EM 120 (measuring motor car) measurements during 18 trips made between 2008 and 2014 were used in the evaluation of track geometry deformability. The traffic load from in that period was approx. 20 Tg. The selected results of averaged measurements values of standard deviation for vertical irregularities shows Fig. 4. The measurement cycles 11–13 were accepted for analysis.

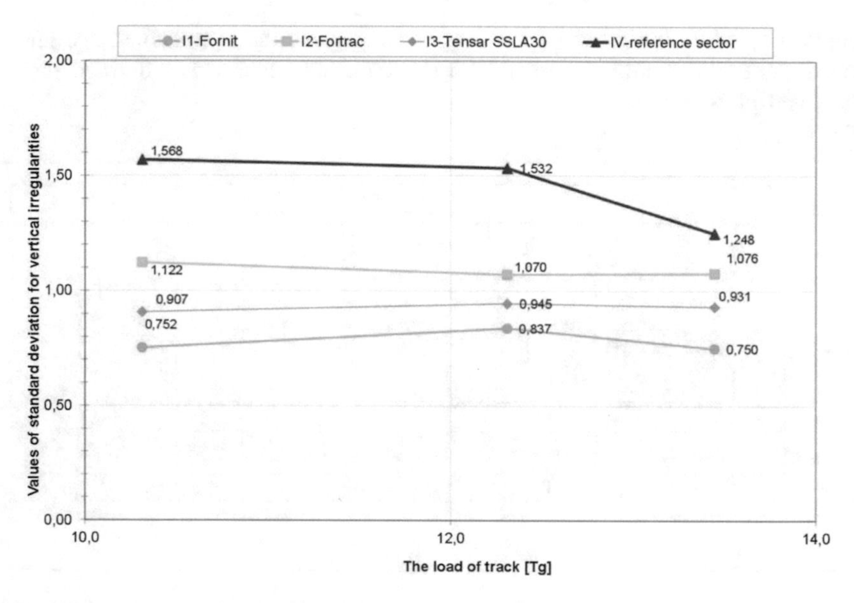

Fig. 4. Values of standard deviation for vertical irregularities for plots I1, I2 and I4 vs. reference sector IV.

Figures 5 and 6 show selected test results for vertical irregularities for reference sector IV and sector I1. The difference is that the reference sector has been prepared without the use of additional surface reinforcements (geogrids of type and crashed stone layer with resin). Such a solution is a standard solution performed in modernization works or the construction of new sections of the railway line.

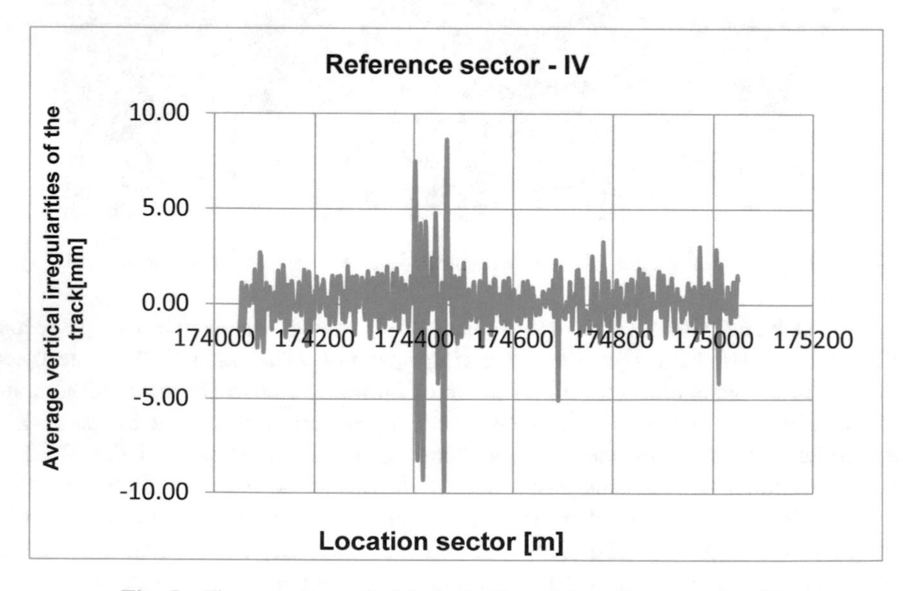

Fig. 5. The average vertical irregularities of the reference sector IV.

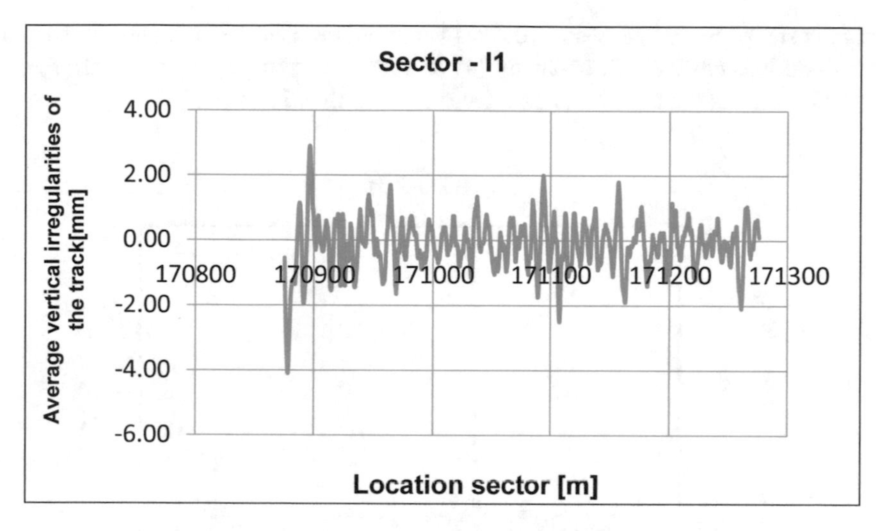

Fig. 6. The average vertical irregularities of the sector I1.

When analyzing the results of the averaged vertical irregularities of the track, it is possible to observe the differences between the reference sector and the reinforced sector. The reinforced standard track substructure (sector I1) causes 3-fold differences in the size of the track irregularities with the same transferred track load.

3 Simulation Studies

In a present study, the simulation were based on a non-linear dynamical model of a railway vehicle representing a passenger car that moves with constant velocity v along a nominally tangent track with geometrical irregularities. The model comprises 7 rigid bodies corresponding to the vehicle body, two bogies and four wheelsets, with 27 degrees of freedom and linear characteristics of primary and secondary suspensions [4]. The non-linear wheel (UIC 60) and rail (ORE S1002) profiles are used and the forces at the wheel/rail contact are calculated with Kalker's simplified nonlinear theory [3]. Dynamical responses that determine safety against derailment are lateral and vertical forces at the contact points between the wheels and rails. Following the UIC 518 code [9] and the EN 14363 standard [2], the running average $(Y/Q)_{2\mathrm{m}}$ was calculated at each track point x based on the ratio Y/Q obtained in simulations. Subsequently, the maximum value of $(Y/Q)_{2\mathrm{m}}$ is represented by the 99.85-percentile value of $(Y/Q)_{2\mathrm{m}}$ after rejecting 0.15% of its extreme values. The value of $(Y/Q)_{2\mathrm{m};0.9985}$ is found for each analyzed track sector and can be compared with the limit value $(Y/Q)_{\mathrm{lim}} = 0.8$ which corresponds to a immediate risk of derailment.

The simulations of the railway vehicle running at the speed of v = 160 km/h were performed for three track sectors: I1, I2, I3 with the crashed stone composite and one reference sector IV of the classic track. The variation of the derailment coefficient $(Y/Q)_{2\mathrm{m}}$ obtained for the leading wheelset of the front boogie in all analyzed track sectors is shown Figs. 7, 8, 9 and 10. This variation has characteristic short-lived peaks due to non-linear wheel/rail geometry [4] and the maximum peak values of $(Y/Q)_{2\mathrm{m}}$ are smallest in sector I1, around two times smaller than in sector IV.

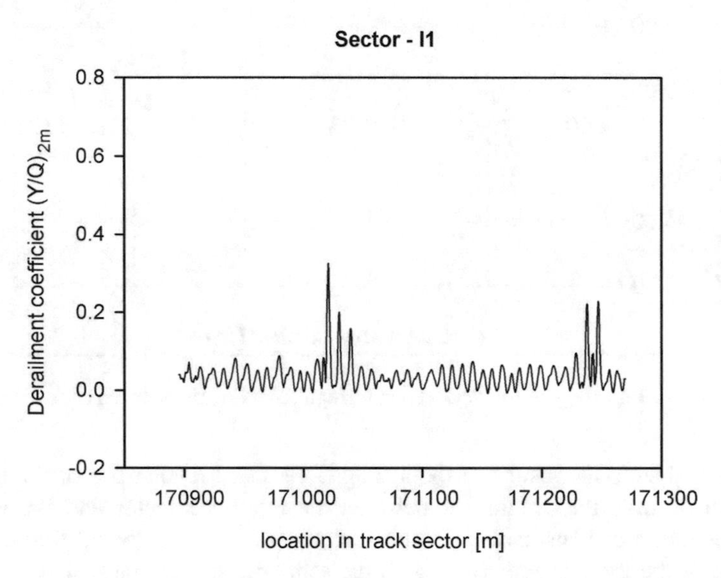

Fig. 7. Derailment coefficient: the running average $(Y/Q)_{2\mathrm{m}}$ for the leading wheelset of the front boogie in the sector I1.

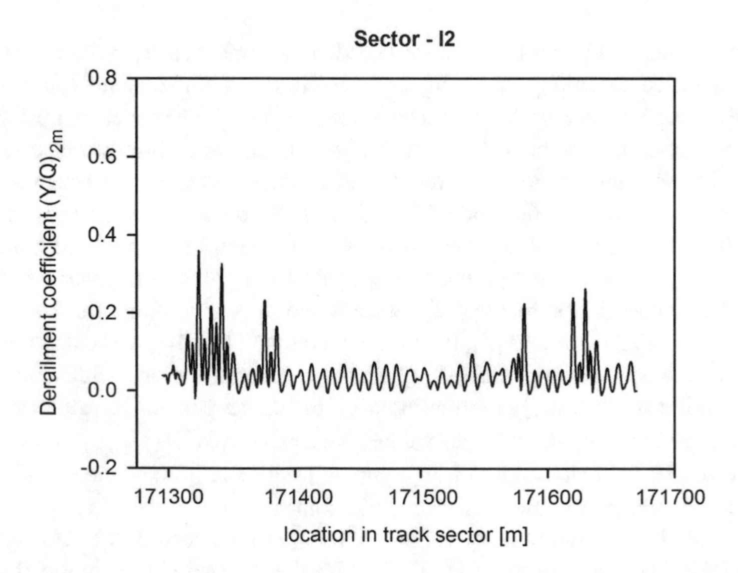

Fig. 8. Derailment coefficient: the running average $(Y/Q)_{2\mathrm{m}}$ for the leading wheelset of the front boogie in the sector I2.

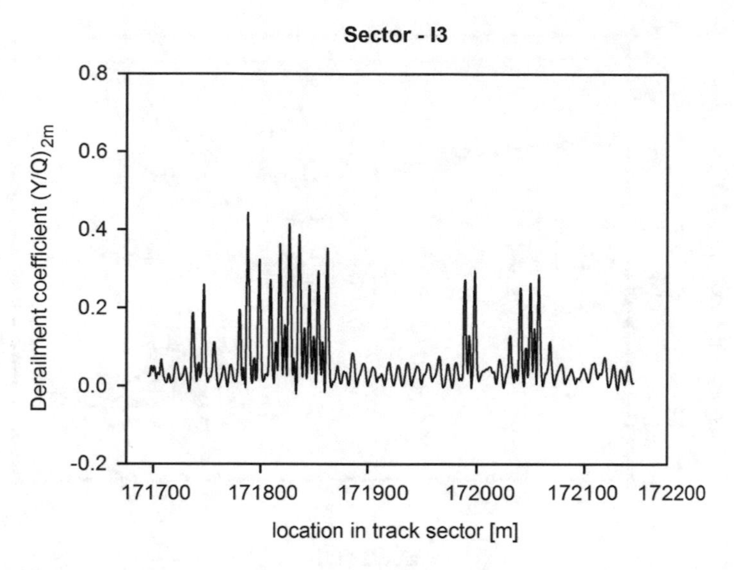

Fig. 9. Derailment coefficient: the running average $(Y/Q)_{2m}$ for the leading wheelset of the front boogie in the sector I3.

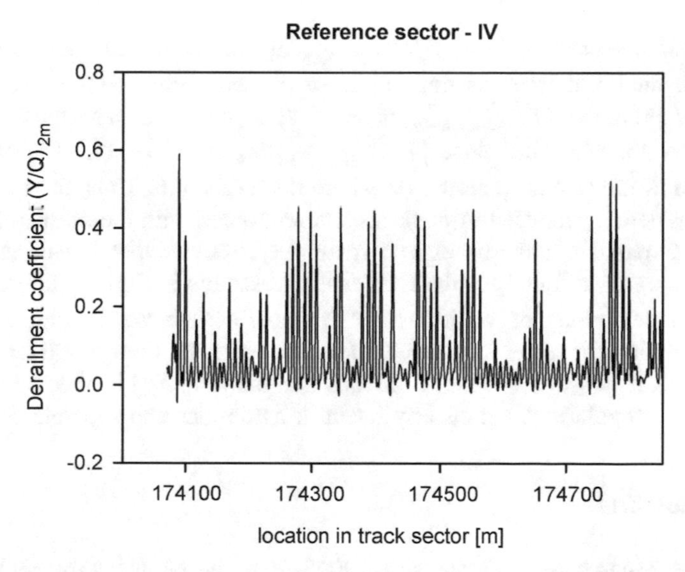

Fig. 10. Derailment coefficient: the running average $(Y/Q)_{2m}$ for the leading wheelset of the front boogie in the reference sector IV.

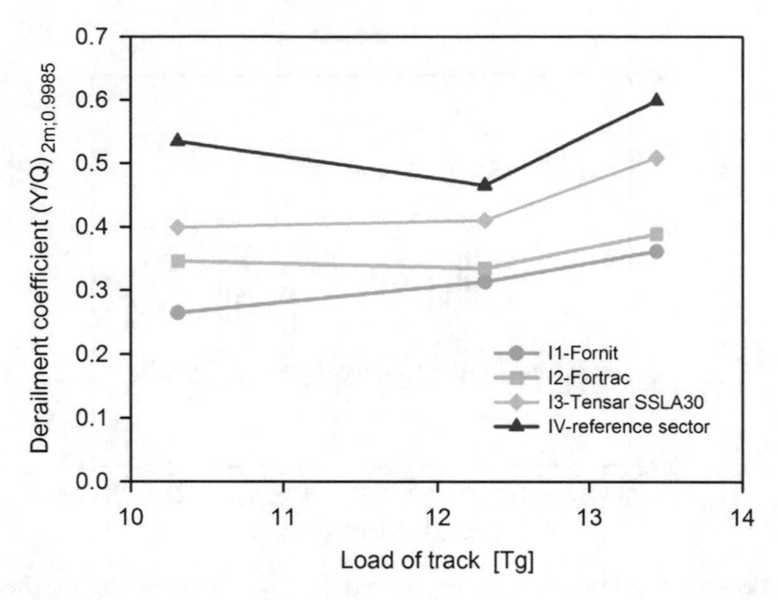

Fig. 11. Derailment coefficient: the 99.85-percentile values $(Y/Q)_{2m;0.9985}$ of the running average 2 m for sections I1, I2, I3 (sector I) vs. reference sector IV (see. Figure 2).

The 99.85-percentile values $(Y/Q)_{2m;0.9985}$ of the derailment coefficient obtained for the four analyzed track sections at three chosen traffic loads are compared in Fig. 11. It is found that $(Y/Q)_{2m;0.9985}$ in sector I2 is larger than in sector I1 and smaller than in sector I3, while the value $(Y/Q)_{2m;0.9985}$ in sectors I1, I2, I3 is considerably smaller than in the reference sector IV where it exceeds 0.6. Thus, the risk of derailment is considerably reduced in track sectors reinforced with the crashed stone composite in comparison with the classic track. Let us note that the ordering of the $(Y/Q)_{2m;0.9985}$ values found for the four analyzed sectors is slightly different than for the standard deviations of vertical track irregularities shown in Fig. 4 where the standard deviation for sector I3 is smaller than for sector I2, though still larger than for sector I1. This is possibly related to the fact, that the ratio *Y*/*Q* is affected not only by vertical track irregularities but also by lateral irregularities and superelevation.

4 Conclusions

Based on the research covering the period 2008–2014 during which the track supported the load of 18.6 Tg, it is evident that the track resistance to deformation at the sections with crashed stone composite is approx. 30% larger than on the track without composite. This conclusion is confirmed also by the results of vertical irregularities measurements. The use of simulation tools allowed for the assessment of the effect of track structure on the dynamic response of the railway vehicle. The simulation results have shown that the derailment coefficient Y/Q for a vehicle moving on track sections with crashed stone composite is even two times smaller than for the track without composite.

References

1. Basiewicz, T., Gołaszewski, A., Kukulski, J., Towpik, K.: Tests on a track structure with crashed stone composite on an experimental section of the CMK (Central Trunk Line). J. Civil Eng. Archit. Res. **3**(1), 1220–1227 (2016)
2. EN 14363: Railway applications - Testing for the acceptance of running characteristics of railway vehicles - Testing of running behaviour and stationary tests, European Committee For Standardization (2005)
3. Kalker, J.J.: A fast algorithm for the simplified theory of rolling contact. Veh. Syst. Dyn. **11**, 1–3 (1982)
4. Kardas-Cinal, E.: Spectral distribution of derailment coefficient in non-linear model of railway vehicle-track system with random track irregularities. J. Comput. Nonlinear Dyn. **8**, 031014 (2013)
5. Kardas-Cinal, E., Zboiński, K.: On investigation of safety of a railway vehicle in the presence of random track irregularities. In: Proceedings of the 10th Mini Conference on Vehicle System Dynamics, Identification and Anomalies, 2006, Zobory Istvan (ed.), ISBN 978-963-420-968-3, pp. 169–176 (2008)
6. Kukulski, J.: Experimental and simulation study of the superstructure and its components. In: Zboiński, K. (ed.) Railway research – Selected Topics on Development Safety and Technology, pp. 115–143. Intech Open Limited (2015)
7. Nadal, M.J.: Theorie de la Stabilite des Locomotives, Part 2. Movement de Lacet, Annales des Mines **10**, 232 (1896)
8. Rozporządzenie Ministra Transportu, Budownictwa i Gospodarki morskiej z dnia 7 sierpnia 2012 r. w sprawie zakresu badań koniecznych do uzyskania świadectwa dopuszczenia do eksploatacji typu budowli przeznaczonej do prowadzenia ruchu kolejowego, świadectwa dopuszczenia do eksploatacji typu urządzenia przeznaczonego do prowadzenia ruchu kolejowego oraz świadectwa dopuszczenia do eksploatacji typu pojazdu kolejowego, Dziennik Ustaw z 2012 r., poz. 918 (2012) (in Polish)
9. UIC Code 518 OR: Testing and approval of railway vehicles from the point of view of their dynamic behaviour - Safety – Track fatigue-Ride quality, International Union of Railways, 2 edn (2003)

Telematics as a New Method of Transport System Safety Verification

Andrzej Lewiński and Tomasz Perzyński(✉)

Faculty of Transport and Electrical Engineering,
Kazimierz Pulaski University of Technology and Humanities in Radom,
Malczewskiego 29, 26-600 Radom, Poland
{a.lewinski,t.perzynski}@uthrad.pl

Abstract. Telematics, as a set of technical solutions, has become an essential element of transport infrastructure and equipment of various means of transport. The introduction of new technical solutions in transport requires determining the impact of these solutions on the process of managing and controlling transport. With regard to transport safety, the research problem is to estimate the value of indicators characterizing safety. One of the research methods may be modeling, which allows to check various scenarios at the design and implementation stages. The article analyzes two examples of transport systems equipped with telematics systems. The analysis of the results shows that the use of additional telematics solutions improves the safety indicators. The proposed mathematical apparatus makes it possible to estimate the numerical values of safety.

Keywords: Telematics · Transports safety · Modelling

1 Introduction

Transport telematics systems allow for better transport management and control. The aim of introducing new systems in transport is to increase its functionality. Transport telematics can be verified by the main tasks, which include [1]:

- vehicle navigation,
- vehicle communication,
- vehicle management and control,
- traffic control and management,
- weather information
- navigation.

Implementation of the above mentioned tasks requires cooperation of solutions from the field of IT, telecommunication or electronics. Telematics systems are also designed to ensure an adequate level of safety in the transport process. The reliability of telematics systems can be achieved through effective management of the exploitation and diagnostic process [2].

M. Siergiejczyk and K. Krzykowska (Eds.): ISCT21 2019, AISC 1032, pp. 266–273, 2020.
https://doi.org/10.1007/978-3-030-27687-4_27

The number of dangerous road and rail accidents can be reduced by implementing new solutions not only in vehicles but also in infrastructure. Such solutions as V2V, I2V or V2I allow for better transport management and increased safety [3]. From an economic and social point of view, the following should be added, that the direct and indirect costs associated with accidents, especially car accidents, in the European Union account for up to a few percent of GDP (Gross Domestic Product) [4]. A solution to many problems may be the use of new technologies, including dedicated transport telematics systems. Telematics becomes a method for verifying the safety of transport systems.

In order to maintain an adequate level of safety in transport, means of transport must have defined safety indicators. In the case of traffic management and control systems, these systems must be constructed in accordance with the requirements and applicable standards, in which such requirements as the acceptable level of risk (THR - Tolerable Hazard Rate) or the method of coding the transmitted information have been defined [5]. Also for road transport there are requirements and guidelines for improving safety. European Parliament action focuses on legislation that can reduce fatalities and injuries on roads, 90% of which are caused by human error. It is worth mentioning, in 2018, there were around 25 100 fatalities in road accidents in EU [6]. Currently, new regulations approved by the European Parliament on 16 April 2019 will be submitted to the EU member states for approval. They predict that in a few years all new cars will have to be equipped with active safety systems (about 30 new systems are assumed) [7]. The use of new technologies, including telematics systems, will enable cars to better monitor the situation around them and, if necessary, automatically correct the driver's driving behaviour.

The paper presents current solutions that can significantly improve safety: collision avoidance system in cars and ETCS system in automatic train control. A mathematical apparatus in the form of Markov's processes was proposed for analysis. On the basis of the proposed mathematical apparatus it was shown that the safety of systems can be defined as the probability of a non-occurrence of a critical event.

2 Selected Solutions in Transport Systems

Telematics solutions are one of the tools to improve transport safety. One of the example of an active safety system for motor vehicles is the collision avoidance system. System operation is based on the analysis of information coming from sensors, cameras or radar systems [3]. An example of possible radar detection is shown in Fig. 1.

In the Fig. 1 we can distinguish:

- ACC - adaptive Cruise Control,
- CA - collision Avoidance,
- BS - blind-spot detection,
- CW - collision warning.

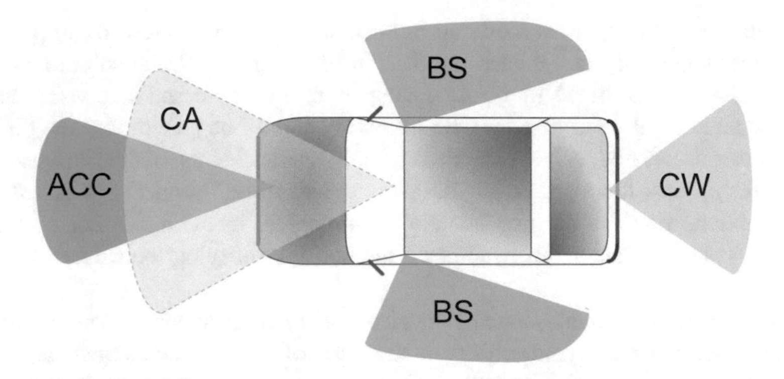

Fig. 1. Object detection via radar (own study based at [8])

It should be added that, with regard to braking systems, significant technological progress can be noted. Brake systems are no longer just mechanical solutions and an advanced system equipped with sensors, among others detecting obstacles or other vehicles moving in front of and around the vehicle. If the driver does not react properly to signals of a possible problem, the system will automatically use the braking system [9]. In Fig. 2 the example algorithm of the system is presented.

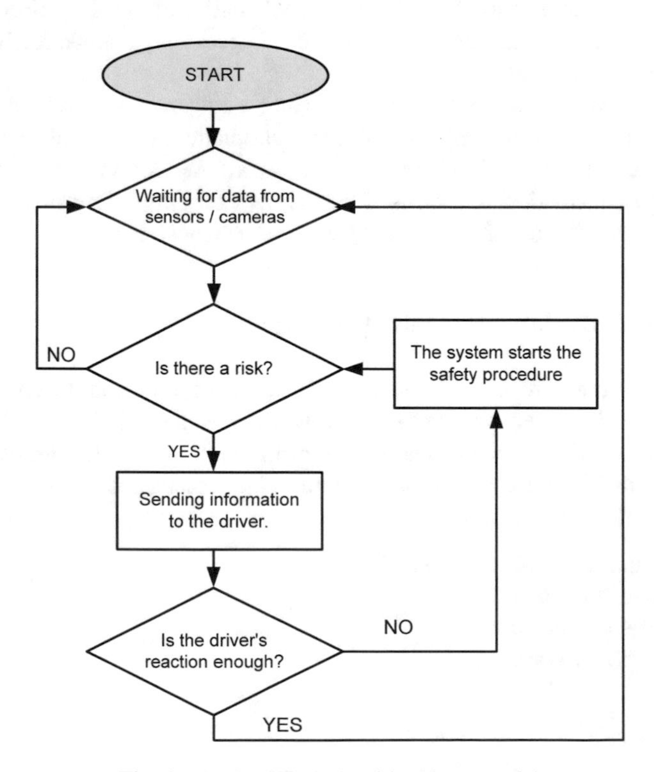

Fig. 2. A simplified algorithm (own study)

An example of the use of telematics solutions in rail transport is the European Rail Traffic Management System (ERTMS). In order to unify the infrastructure allowing for free movement of trains, ERTMS was introduced already in the 1980s. ERTMS is the basic means to implement interoperability. The solutions used in ERTMS allow for higher rolling stock speeds and higher throughput while maintaining a high level of safety. The ERTMS system consists of three main subsystems [10–12]:

- ETCS – European Train Control System,
- GSM-R – Global System for Mobile Communication – Railway,
- ETML – European Traffic Management Layer.

ETCS is a standardized and interoperable ATP/ATC system used in Europe. In the ETCS system, data transmission uses digital transmission through eurobalises, euro loops, digital radio transmission and specialized transmission modules. Based on data such as train weight, braking force, calculates fixed or variable speed profiles [13].

The ETCS system can be configured to work in one of the three levels:

- Level 1 - solution based on the transmission of permits for driving with balis, may also be used in conjunction with other national signalling devices such as track circuits,
- Level 2 - GSM-R is a transmission medium for the transmission of information, the track is additionally equipped with radio control centres (Radio Block Centre - RBC),
- Level 3 - it is a development of the second level and allows driving according to the changeable block distance. This level resigns from the axle counters and track circuits.

The use of new technologies, including automation in rail transport can brings significant benefits in terms of safety and performance and can offer better train capacity up to 8% only by eliminating the variability of manual driving [14].

3 Systems Modelling

Modelling is part of the verification of systems and allows for analyze the system assuming different scenarios of events [15]. The first of the proposed models presents an additional telematics system installed in the car, which informs the driver about approaching another vehicle, and if necessary, automatically starts the braking system. In the model shown in Fig. 3, we can distinguish:

- 0 - the state of correct work, no threats and information about approaching another vehicle,
- 1 - vehicle equipped with a telematics system, the system takes over the braking function,
- 2 - dangerous state, collision,
- 3 - vehicle not equipped with the system, the driver performs braking of the vehicle.

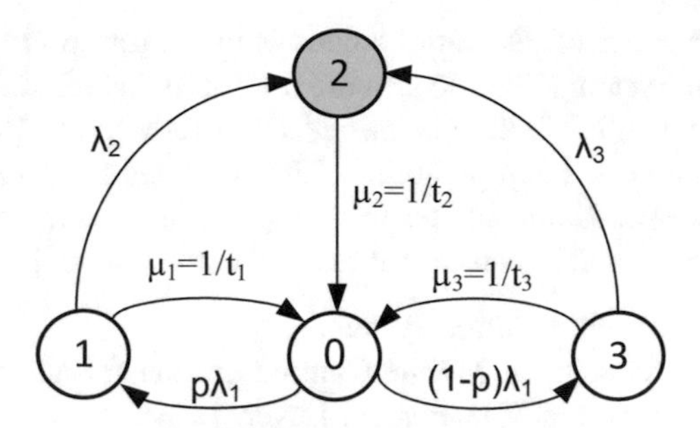

Fig. 3. Vehicle model with an additional telematics system (own study)

Description of transitions between states:

- λ - intensity of transitions,
- μ - the inverse of the time of returning to the state,
- p - the probability of equipment car in the additional telematics system.

For models in Fig. 3 we can write the equation in the form of operators:

$$\begin{cases} s \cdot \tilde{P}_0(s) - 1 = -p\lambda_1\tilde{P}_0(s) + \mu_1\tilde{P}_1(s) + \mu_2\tilde{P}_2(s) + \mu_3\tilde{P}_3(s) - (1-p)\lambda_1\tilde{P}_0(s) \\ s \cdot \tilde{P}_1(s) = -\lambda_2\tilde{P}_1(s) - \mu_1\tilde{P}_1(s) + p\lambda_1\tilde{P}_0(s) \\ s \cdot \tilde{P}_2(s) = \lambda_2\tilde{P}_1(s) - \mu_2\tilde{P}_2(s) + \lambda_3\tilde{P}_3(s) \\ s \cdot \tilde{P}_3(s) = (1-p)\lambda_1\tilde{P}_0(s) - \mu_3\tilde{P}_3(s) - \lambda_3\tilde{P}_3(s) \end{cases} \quad (1)$$

Using the properties of Laplace transform it was solved Eq. (1) and was calculated the critical probability $P_2(t)$ for $t \to \infty$ (state 2 in the model from Fig. 3 is the most undesirable):

$$\begin{aligned} P(t)_{t\to\infty} = {} & [\lambda_1((\mu_1+\lambda_2)\lambda_3 + p(\mu_3\lambda_2 - \mu_1\lambda_3))]/[\lambda_2(\lambda_1\lambda_3 + \mu_2(\mu_3+\lambda_1+\lambda_3)) \\ & + \mu_1(1-p)\lambda_1\lambda_3 + \mu_2(\mu_3+\lambda_1 - p\lambda_1 + \lambda_3) + p\lambda_1(\mu_3\lambda_2 + \mu_2(\mu_3\lambda_2+\lambda_3))] \end{aligned} \quad (2)$$

Assuming the values of indices:

- λ_1 – 15 h^{-1},
- λ_2 – damage to the electronic module at the level 0.00001 h^{-1},
- λ_3 – 0.001 h^{-1},
- μ_1 – after 1 s. returns to its normal state,
- μ_2 – after 0,5 h returns to its normal state,
- μ_3 – after 2 s. returns to its normal state.

The estimated values of safety with reference to probability of equip car with additional telematics system are shown below, in the Table 1:

Table 1. The calculation results for model from Fig. 3.

Probability of equip car with telematics system	Value of S(t) for $t \to \infty$ $S(t)_{t\to\infty} = 1 - P_2(t)_{t\to\infty}$
0,1	0,999985
0,3	0,999988
0,5	0,999992
0,7	0,999995
0,9	0,999998

The next model corresponds to the ERTMS system with the ATP subsystem.

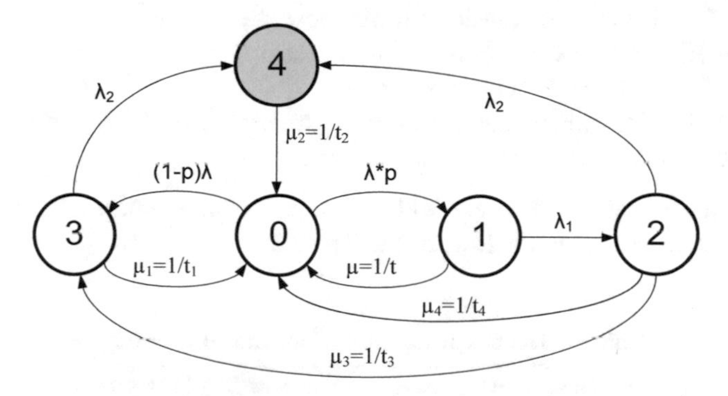

Fig. 4. Model for the ERTMS system with the ATP subsystem (own study)

We can distinguish in the model:

- 0 - normal operation status, waiting for data.
- 1 - implementation of commands by the telematics module, control of speed/ braking.
- 2 - condition of controlled damages (e.g. transmission loss, module failure),
- 3 - driving according to the track side signaling,
- 4 - condition of critical, uncontrolled damages.

For models in Fig. 4 we can write the equation in the form of operators:

$$\begin{cases} s \cdot \tilde{P}_0(s) - 1 = -p\lambda\tilde{P}_0(s) + \mu\tilde{P}_1(s) + \mu_4\tilde{P}_2(s) + \mu_1\tilde{P}_3(s) - (1-p)\lambda\tilde{P}_0(s) + \mu_2\tilde{P}_4(s) \\ s \cdot \tilde{P}_1(s) = p\lambda\tilde{P}_0(s) - \mu\tilde{P}_1(s) + \lambda_1\tilde{P}_1(s) \\ s \cdot \tilde{P}_2(s) = \lambda_1\tilde{P}_1(s) - \lambda_2\tilde{P}_2(s) - (\mu_4 + \mu_3)\tilde{P}_2(s) \\ s \cdot \tilde{P}_3(s) = -\lambda_2\tilde{P}_3(s) - \mu_1\tilde{P}_3(s) + \mu_3\tilde{P}_2(s) + (1-p)\lambda\tilde{P}_0(s) \\ s \cdot \tilde{P}_4(s) = \lambda_2\tilde{P}_3(s) + \lambda_2\tilde{P}_2(s) - \mu_2\tilde{P}_4(s) \end{cases} \tag{3}$$

Using the properties of Laplace transform it was solved Eq. (3) and was calculated the critical probability $P_4(t)$ for $t \to \infty$ (state 4 in the model from Fig. 4 is the most undesirable):

$$P_4(t)_{t\to\infty} = [\lambda_2((1-p)\mu(\mu_3+\mu_4+\lambda_2)+\lambda_1(\mu_3+p(\mu_1-\mu_4)+\mu_4+\lambda_2))] / [\lambda_1(\mu_3+\mu_4+\lambda_2)(\mu_2(\mu_1+\lambda) + (\mu_2+\lambda)\lambda_2)+\mu(\mu_3+\mu_4+\lambda_2)(\mu_2(\mu_1+\lambda-p\lambda)+(\mu_2+\lambda-p\lambda)\lambda_2+p\lambda(-\mu_2\mu_4\lambda_1+\mu_1\mu_2(\mu_3+\mu_4+\lambda_1) + (\mu_2(\mu_1+\mu_3+\mu_4)+(\mu_1-\mu_4)\lambda_1)\lambda_2+\mu_2\lambda_2^2)] \tag{4}$$

Assuming the values of indices:

- λ - 10 h^{-1},
- λ_1 - damage to the electronic module, signal loss, 0.00001 h^{-1},
- λ_2 - 0.00001 h^{-1},
- μ - after 5 s. follows the transition to the next state,
- μ_1 - after 10 s. follows the transition to the next state,
- μ_2 - after 15 h returns to its normal state,
- μ_3 - after 10 min. follows the transition to the next state,
- μ_4 - after 20 min. returns to its normal state.

The estimated values of safety with reference to probability of correct work of telematics system are shown below, in the Table 2:

Table 2. The calculation results for model from Fig. 4.

Probability of correct work of telematics system	Value of S(t) for $t \to \infty$ $S(t)_{t\to\infty} = 1 - P_4(t)_{t\to\infty}$
0,1	0,9999999837
0,3	0,9999999973
0,5	0,9999999909
0,7	0,9999999945
0,9	0,9999999981

4 Conclusions

Telematics is currently a technologically advanced tool supporting transport processes of various of transport means. Telematics, which is a collection of different technical solutions, is a tool to verify the safety of transport systems. The paper proposes mathematical models that reflect currently applied telematics solutions used in road and rail transport. In both examples, the telematics overlay does not reduce the level of safety related to Markov safety analysis with simultaneously improvement the functionality of the system. In the case of the railway ERTMS dedicated system, there is no doubt that the system is necessary to ensure interoperability. In the case of car-driving assistance systems, including collision avoidance systems, they are part of autonomous vehicles [16].

References

1. Perzyński, T.: Selected telematics systems in safety and management in land and inland transport. Publishing House Kazimierz Pulaski University of Technology and Humanities in Radom. Monographs Series, No. 201. Radom (2016)
2. Rosiński, A.: Modelowanie procesu eksploatacji systemów telematyki transport. Oficyna Wydawnicza Politechniki Warszawskiej, Warszawa (2016)
3. Perzyński, T., Lewiński, A.: The influence of new telematics solutions on the improvement the driving safety in road transport. In: Mikulski, J. (ed.) Management Perspective for Transport Telematics, TST 2018. Communications in Computer and Information Science, vol. 897. Springer, Cham (2018)
4. Wijnen, W., et al.: Crash cost estimates for European countries, deliverable 3.2 of the H2020 project SafetyCube. Loughborough: Loughborough University, SafetyCube (2017)
5. Lewiński, A., Perzyński, T.: The reliability and safety of railway control systems based on new information technologies. In: Transport Systems Telematics. Communications in Computer and Information Science, vol. 104. Springer (2010)
6. European Commission – Press release: Road safety. Data show improvements in 2018 but further concrete and swift actions are needed. Brussels, 4 April 2019
7. http://www.europarl.europa.eu/news/en/headlines/society/20190307STO30715/safer-roads-new-eu-measures-to-reduce-car-accidents. Accessed 01 May 2019
8. https://www.rs-online.com/designspark/lidar-radar-digital-cameras-the-eyes-of-autonomous-vehicles. Accessed 24 Apr 2019
9. https://www.euroncap.com/en/vehicle-safety/the-rewards-explained/autonomous-emergency-braking/. Accessed 01 May 2019
10. Młyńczak, J., Toruń, A., Bester, L.: European rail traffic management system (ERTMS). In: Sładkowski, A., Pamuła, W. (eds.) Intelligent Transportation Systems – Problems and Perspectives. Studies in Systems, Decision and Control, vol 32. Springer, Cham (2016)
11. UNIFE – The European Rail Industry International freight corridors equipped with ERTMS. ERTMS Factscheet, no. 15 (2014)
12. Winter, P., et al.: Compendium on ERTMS. UIC. 1st edn. (2009). ISBN 978-3-7771-0396-9
13. Lewiński, A., Toruń, A., Gradowski, P.: Modeling of ETCS with respect to functionality and safety including Polish railways conditions. In: Telematics – Support Of Transport. Communications In Computer And Information Science, vol. 471. Springer, Heidelberg (2014)
14. http://www.railway-technical.com/signalling/automatic-train-control.html. Accessed 01 May 2019
15. Leitner, B.: A general model for railway systems risk assessment with the use of railway accident scenarios analysis. Procedia Eng. **187**, 150–159 (2017). https://doi.org/10.1016/j.proeng.2017.04.361
16. Dąbrowski, T., Rosiński, A.: Review of car operation safety system elements. In: MATEC Web of Conferences, vol. 182, p. 01017 (2018). https://doi.org/10.1051/matecconf/201818201017

Critical Areas of the Autonomous Seagoing Vessel Concept Model - According to Selected Criteria

Zbigniew Łosiewicz[1(✉)] and Waldemar Mironiuk[2]

[1] West Pomeranian University of Technology, Szczecin, Poland
zbigniew.losiewicz@zut.edu.pl
[2] Polish Naval Academy, Gdynia, Poland

Abstract. The paper defines autonomous vessels as unmanned, remote-controlled vessels or vessels moving along a planned route, according to the program installed on the electronic control and control unit which selects the action, depending on the result of the analysis of current data in real time. Autonomy is also defined as the ability of a facility to perform design tasks, during which the safe exploitation of the facility requires the development of a decision. It was assumed that the consequences of these decisions should be consistent with the project assumptions, adequately to the conditions. In real operating conditions, the level of autonomy of vessels is dependent on the operating task specified in the design process. Operational tasks were identified by analysing the process of operating a sample merchant vessel. Since an autonomous craft may be devoid of technical solutions to ensure the safety of persons at work and to facilitate their operation on board, it must also be equipped with devices to replace persons.

The paper presents the types of navigation that can be performed by a ship, i.e. in open waters, in waters limited by approach lanes with specific parameters, in port areas with limited spaces and very complex line of quays, as well as in areas with high traffic intensity. It shows an example of a navigational infrastructure on land and a compatible ship system communicating with this infrastructure. The concept of a model of a sea autonomous vessel has been developed.

Keywords: Ship · Autonomous · Model

1 Introduction

Shipping is a very important area of the world economy. Maritime transport is a branch of transport which accounts for about 70% of the movement of goods and people and maritime vessels are the largest means of transport. At the same time, studies show that about 80% of accidents are caused by human error (human factor) [6, 8]. In order to eliminate human emotionality from the decision-making process, reduce the cost of hiring a crew and increasing difficulties with the manning of modern ships with highly qualified crew, intensive research is carried out on the creation of autonomous means of transport, including autonomous vessels [14, 15]. Initially this term was used for unmanned single-station vessels, remotely controlled by an operator from land, but

M. Siergiejczyk and K. Krzykowska (Eds.): ISCT21 2019, AISC 1032, pp. 274–283, 2020.
https://doi.org/10.1007/978-3-030-27687-4_28

during the research process the definition of autonomous seagoing vessel is subject to continuous modification.

In literature, autonomy is defined in a very wide range, depending on the field in which autonomy characterises the subject object. It seems justified to state that autonomy in technology is a contractual term, most often concerning a remotely controlled object or programmed in such a way that it can perform exploitation tasks without human intervention or with the lowest possible degree of such intervention [6].

Etymologically, according to [13]: The term "autonomy" is derived from the (old) Greek αυτονομία. Autonomy - the possibility of establishing norms for oneself, legal independence. Today it is used depending on the discipline (economy, law, politics) or context in the sense of sovereignty, independence, self-governance (total or partial). It concerns both the collective and the individual (independence in self-determination).

The autonomous seagoing vessels (hereinafter referred to as "ASV" are usually referred to as unmanned, remote-controlled or re-routed vessels operating along a planned route in accordance with a program installed on an electronic control and command unit (GSI), which selects an action based on the analysis of current data in real time.

In the work, autonomy is defined as the ability of the object to perform design tasks, during which the safe exploitation of the object requires the development of a decision. The consequences of these decisions should be consistent with the project assumptions, according to the conditions. By adopting the criteria for assessing autonomy, it seems appropriate at the same time to adopt the classification of the ASV according to different levels of autonomy. In real operating conditions, the level of autonomy of vessels is dependent on the operating task specified in the design process.

2 Genesis of the Autonomous Seagoing Vessel

Analysing the history of navigation, it can be seen that the rapid development of electronics, and thus of the computer industry (IT), has contributed to the growth of opportunities for the development of autonomous vessels. Radars, satellite telephones, navigation autopilots, electronic maps, electronic navigation systems and electronic maps (ARPA, GPS) [11] have appeared on the bridge, and in the marine engine room systems of control and control of engines and devices allowing for the use of watch-keepingless engine room (so-called class A-24), systems of control of conditions in cargo holds, tanks and mobile cargo spaces such as containers. However, the main burden of working out operational decisions on a conventional ship rests on the crew. The crew operates on-board equipment, performing such operations as opening and closing hold hatch covers, handling anchor winches and mooring winches, opening and closing ballast valves and cargo valves, maintenance of structural components and equipment. The presence of crew on board a ship is associated with specific technical solutions closely related to the functioning of human beings.

An example of a conventional vessel may be a merchant vessel. Its structural structure can be divided into three basic areas. The first one is the cargo area where the cargo is transported (cargo-unload, tanks or an adapted deck - e.g. on container ships). The second area is the power plant, where most of the equipment and machines are located and where the energy is processed. The third area is a superstructure with a navigation bridge and accommodation and social spaces.

A ship which is manned must have regulations in place to ensure safe and ergonomic work and rest (e.g. health and safety regulations). These regulations have an impact on human friendly technical solutions, including room dimensions, the construction of traffic routes including corridors and cages, equipment, lighting, gas atmosphere, noise levels and others [7]. They are all the more important because the human psyche (and thus also the consequences of exploitative decisions) is affected by the environmental conditions of work at sea, such as rocking [9], monotony of work with the same people, limitation of space, loneliness. The environmental protection regulations enforce the application of technical and logistic solutions eliminating the risk of environmental contamination with man-made waste (e.g. Annexes IV (domestic sewage) and V (waste, including plastics) of the Marpol 73/78 Convention) [3–5]. Ships must have tanks for domestic waste (from bathrooms and toilets), as well as for safety reasons for lifeboats and life rafts. Improving rest conditions requires the use of e.g. air conditioning, and storing food requires the use of cold stores. This is closely linked to the increase in energy consumption. A self-contained craft may be devoid of technical solutions to ensure the safety of persons at work and to facilitate their operation on board, but at the same time it must be equipped with devices to replace persons.

Existing systems of electronic maps included in the ARPA allow to identify the position of the ship on the water. Additionally, AIS (Automatic Identification System)

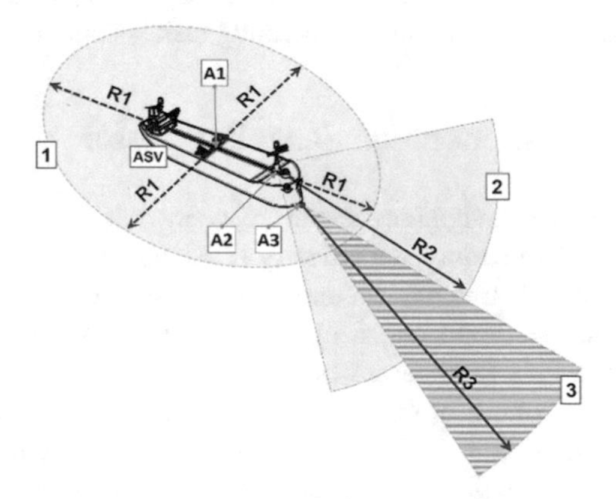

Fig. 1. General autonomous seagoing vessel (ASV) schema: where: ASV - autonomous seagoing vessel, SAT - Satellite, SAT_{ASV} - satellite communication with ASV (data transmission), A1 - conventional central location point of sensors for identification of the nearest surroundings ASV, R1 - radius of area of operation of sensors for identification of the nearest surroundings, 1 - area of operation of sensors for identification of the nearest surroundings, A2 - conventional central location point of sensors for identification of the area before ASV, R2 - radius of area of operation of area identification sensors in front of ASV, 2 - area of operation of space identification sensors in front of ASV, A3 - conventional central point of location of echo-sounder sensors - identification of the area of water under ASV, R3 - radius of area of operation of water area identification sensors under ASV, 3 - area of operation of water area identification sensors under ASV. [author's own study]

can be one of the sources of navigation information. By means of AIS, ships shall continuously and autonomously transmit and receive identification reports to and from other vessels equipped with AIS and located nearby (ship-to-ship mode) and to and from coastal AIS terminals (ship-to-shore mode). AIS can also operate in a designated area ("assigned mode"), e.g. an area under the control and monitoring of traffic by the coastal administration, e.g. VTS, and in polling mode, responding to a signal emitted by another station [12]. Figure 1 shows a general diagram of the basic equipment of the autonomous seagoing unit (ASV) with sensor systems used to identify the working area of the ASV.

3 Cooperation of the Autonomous Seagoing Vessel with the Land Infrastructure

The main task of a merchant vessel is to transport goods or people. This task consists of different stages in which the ship is in a specific operational state.

It is possible to distinguish basic operating states such as:

- vessel in port during cargo operations (loading and unloading). During unloading, the control of the mooring lines is carried out because the vessel is emerging. Since the ship usually comes under loading with ballast (e.g. in order not to get caught by land cargo equipment), during loading operations the ballast is removed. These activities are routinely performed by the crew,
- mooring manoeuvres (mooring and mooring to and from the berth or other berth) require precise positioning of the vessel in relation to the berth,
- a vessel during a voyage in a specific area. A cruise can consist of several stages. A vessel can navigate: in open waters (e.g. sea, bay, etc.), in waters limited by approach tracks with specified parameters of depth and width of track, marked with various types of point infrastructure of irrigation marking, in waters with point energy infrastructure (e.g. wind power plants), in harbour areas with limited space and a very complex line of quays as well as high intensity of traffic.

In the case of a self-contained unit, navigating in such waters requires the cooperation of the appropriate navigational infrastructure on land and a compatible ship's system communicating with it.

ASV may engage in coastal navigation, during which it may be within the range of the equipment of measuring coastal stations or, after leaving the range of coastal stations, engage in navigation in open waters in cooperation with the system of satellites (also in coastal areas devoid of dimensional coastal stations).

Figure 2 shows an illustrative diagram of autonomous seagoing vessel (ASV's) cooperation with navigation infrastructure.

Figure 3 shows the diagram of an exemplary arrangement of shore stations of the ASV control system of a specific range and degree of coverage of a coastal body of water of a complex shoreline or bay character.

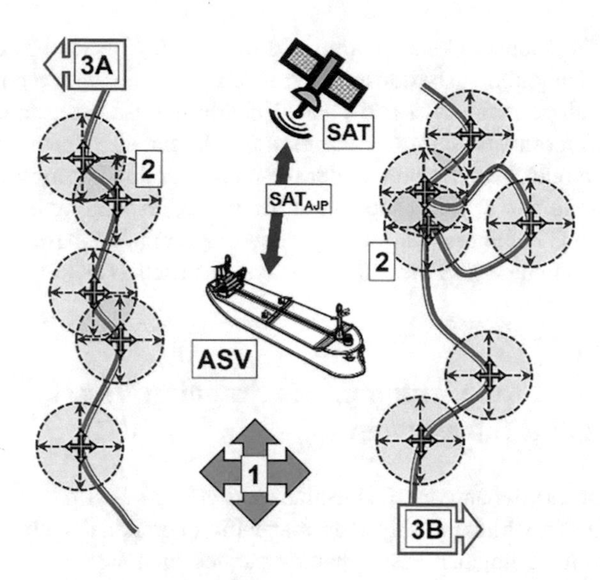

Fig. 2. Overview of autonomous seagoing vessel (ASV's) cooperation with navigation infrastructure: ASV - autonomous seagoing vessel, 1 - reservoir, 2 - coastal station field of operation, 3A - coastline (arrow indicates land side), 3B - coastline (arrow indicates land side), SAT - satellite, SAT_{ASV} - satellite communication between ASV and satellite. [author's own study]

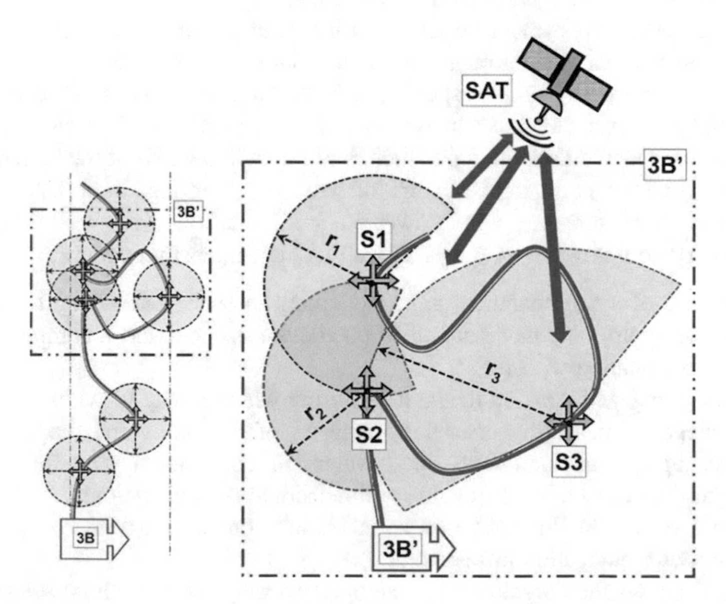

Fig. 3. Scheme of an exemplary arrangement of shore stations of the AJP control system with a specified range and degree of coverage of a coastal area with a complex shoreline or a bay character: where: S1 - shore station location within r_1 radius, S2 - shore station location within r_2 radius, S3 - shore station location within r_3 radius, 3B - shoreline (arrow indicates shore-side), 3B'; - selected bay area equipped with tracking stations cooperating with the ship and satellite, SAT - satellite. [author's own study]

4 The Measure of Autonomy of the ASV - Evaluation Criteria

Defining autonomy in the introduction to the work as the ability of the object to perform design tasks, during which the safe exploitation of the object requires the development of exploitation decisions consistent with the design assumptions, adequately to the conditions, the criteria for the assessment of autonomy should be adopted.

As mentioned in the introduction, in real operational conditions the level of autonomy of vessels is dependent on the operational task specified in the design process [1].

The criteria for autonomy were adopted:

- ability to perform tasks independently (the scope of the necessity of human intervention in the process of working out decisions adequate to the situation occurring in real time),
- time spent in the sea (energy reserve, exploitation media reserve),
- responsiveness to external conditions (e.g. adaptation of the course and speed of the AJP to wave and wind strength),
- navigational suitability for navigating large waters and manoeuvring on approach tracks to ports and mooring at quays,
- suitability to maintain the potential of the technical condition of the technical structure at a level allowing the performance of the exploitation task - reliability and durability of equipment and machinery,
- ability to switch systems and equipment in the event of failure of the elements of the technical structure of the facility according to a model that allows access to the place of refuge,
- ability to protect the physical space of the object,
- ability to protect the electronic-IT structure of control, control and protection of the object,
- degree (level) of independence from the operator or the facility crew,
- psychophysical resistance of the operator or crew.

At work, it's a term:

- "Capability" is defined as the potential to meet the design objectives of a device or system and the psycho-physical potential of a human being,
- "Fitness" is defined as the actual use of the capacity of a device or system and the psycho-physical potential of a human being in specific real operating conditions.

Figure 4 shows the diagram of cooperation of ASV with the navigation infrastructure of the land and with the navigation satellite system.

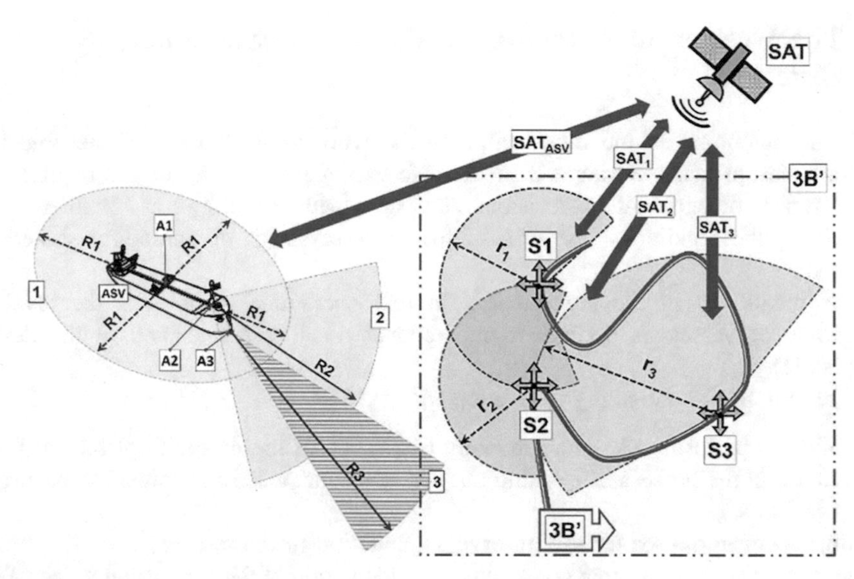

Fig. 4. Schematic diagram of ASV cooperation with land-based navigation infrastructure and navigation satellite system: ASV - autonomous seagoing vessel, SAT - satellite, SAT_{ASV} - satellite communication with ASV (data transmission), A1 - conventional central point of location of sensors for identification of the nearest surroundings ASV, R1 - radius of the area of operation of sensors for identification of the nearest surroundings, 1 - area of operation of space identification sensors in the immediate vicinity, A2 - conventional central location point of area identification sensors in front of ASV, R2 - radius of area of operation of area identification sensors in front of AJP, 2 - area of operation of space identification sensors in front of ASV, A3 - conventional central location point of the sonar sensors - identification of the body of water under ASV, R3 - radius of the body of water under ASV, 3 - area of operation of the body of water under ASV, S1 - location of the shore station with the range of operation under r_1, S2 - location of shore station with range of operation within r_2 radius, S3 - location of shore station with range of operation within r_3 radius, 3B - shoreline (arrow indicates the side of land), 3B'; - selected bay area equipped with tracking stations cooperating with the ship and satellite. SAT_{ASV}, SAT1, SAT2, SAT3, - satellite connectivity with ASV, shore stations S1, S2, S3 respectively [author's own elaboration].

5 Critical Areas of the Autonomous Seagoing Vessel Model

On the basis of the above considerations, a conceptual model for an autonomous craft has been developed, as shown in Fig. 5, consisting of the following elements: GSI - the main IT system, which is primarily responsible for identifying threats and developing operational decisions to ensure the safety of ASV in various operating states, PSI-1 - IT subsystem 1, which is responsible for e.g. identifying threats in the immediate vicinity of the AJP and sending to the GSI the developed data on the safety status in the immediate vicinity of the ASV with a defined R1 radius, as well as for other subsystems identifying risks associated with the AJP movement and communication with the operator or e.g. the ship owner, PSI-2 - IT subsystem 2, which is e.g. responsible for maintenance of ship engine room operation, and mainly main propulsion and identification of hazards related to proper operation of engine room, PSI-3 - IT subsystem 3, which is responsible, for example, for maintaining appropriate conditions for the transport of goods or people and identifying risks related to the quality of the transport conditions, other subsystems may be responsible for e.g. physical safety of units e.g. fire-fighting, anti-terrorist, protection against third party interference.

Systems and subsystems consisting of sensors, transducers, IT systems, equipment and mechanisms must be instrumented, connected by cable trays or radio equipment to the required safety level and be capable of being mirrored and protected, as appropriate to their structure of technical solutions.

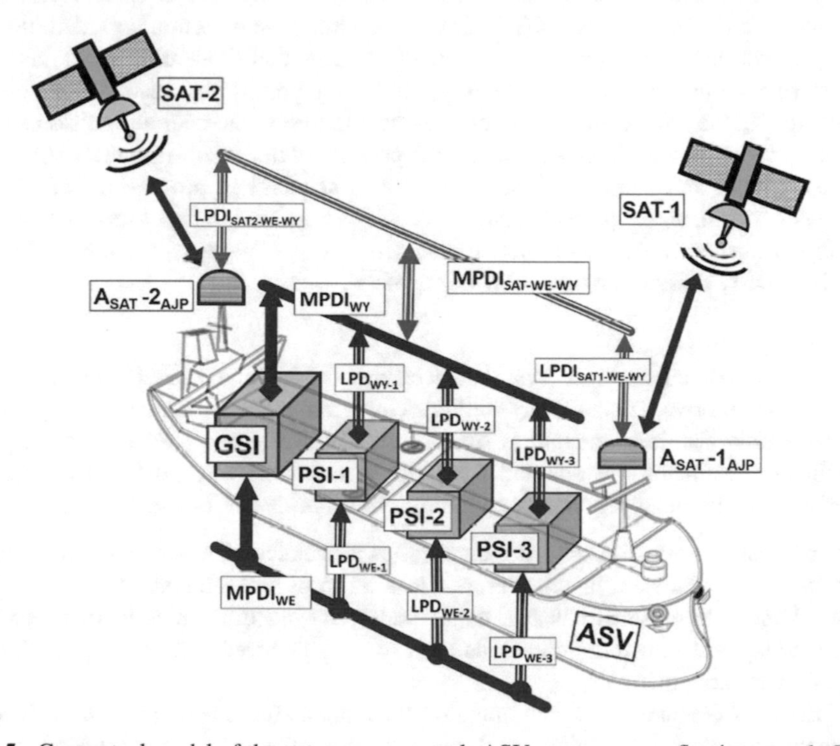

Fig. 5. Conceptual model of the autonomous vessel: ASV - autonomous floating vessel, GSI - main IT system, $MPDI_{WE}$ - main transmission bus for the transmission of IT input data, $MPDI_{WY}$ - main transmission bus for the transmission of IT output data, PSI-1 - IT subsystem 1, PSI-2 - IT subsystem 2, PSI-3 - IT subsystem 3, LPD_{WE}-1 - input data transfer line 1, LPD_{WE}-1 - input data transfer line 1, LPD_{WE}-2 - input data transfer line 1, LPD_{WE}-2 - input data transfer line 2, PSI-3 - IT subsystem 3, LPD_{WE}-1 - input data transfer line 1, LPD_{WE}-1 - input data transfer line 1, LPD_{WE}-2 - input data transfer line 1, LPD_{WE}-2 - input data transfer line 1, LPD_{WE}-2 - input data transfer line 2, PSI-3 - IT subsystem 3, LPDWE-1 - input data transfer line 1, LPDWE-2 - input data transfer line, LPD_{WE}-3 - input data line of subsystem 3, LPD_{WY}-1 - output data line of subsystem 1, LPDWY-2 - output data line of subsystem 2, $LPD_{WY\text{-}3}$ - output data line of subsystem 3, ASAT-1AJP - satellite dish 1 AJP, ASAT-2_{ASV} - satellite dish 2 $_{ASV}$, SAT-1, SAT-1 - satellite, $MPDI_{SAT}$-WE-WY - main transmission bus of input and output IT data transmitted via satellite, LPDISAT-WE-WY - transmission line of input and output IT data transmitted via satellite. [author's own study]

For the purposes of the paper, the definition of a critical area in the field of security engineering was adopted as the area of structure of a floating object, the damage to which may cause a potential threat of failure to perform the task or create a threat to the environment in which the object functions.

The critical areas of floating objects can be divided according to exemplary criteria: exploitation, covering activities related to the performance of exploitation tasks, which include, inter alia: navigation in open waters - the critical area is the loss of navigability (damage to ASV navigation equipment, damage to IT systems, damage to important sensors or power supply), navigation on designated approach lanes - manoeuvring during which excessive traffic may occur, especially in the case of small recreational craft which are difficult to identify and which do not have detection and identification devices, mooring manoeuvres, which may include collision-free, gentle berthing, ballast operations and mooring, energy, including processing, distribution, energy consumption, the critical area of which may be a failure of the main propulsion system or a failure of the ship's power plant and power failure, instrumentation structures, including input and output peripherals of a data acquisition, processing and visualisation subsystem, signal paths, the failure of which may lead to data loss or corruption, which may generate errors, the potential of the main IT system, the failure of which could seriously disrupt the operation of the ASV.

Summary

Design works on autonomous vessels are carried out by large global companies from aviation, automotive, IT and many military corporations. Many concepts of operation, and especially the instrumentation of sensors identifying the structure of the ASV working environment, are adapted from the aviation, railway [10] and automotive industry [2]. The main problems generating critical areas are factors such as

- harsh marine conditions causing continuous movement of a large amplitude object causing disturbance to measurements, loss of bases and reference points,
- the large size of ships causing large inertia causing difficulties in manoeuvring, especially when avoiding collisions with small high-speed craft, often provoked by poorly trained operators,
- if the ASV is conducted by an operator, the critical area is communication, delay in transmission of information and the problem of employing an operator with appropriate qualifications - this is an area in which research and development work is still in progress, and ultimately the population of the ASV should grow exponentially,
- Large dimensions and high inertia of the ASV force the use of high and very high power sensors and transmitters when identifying large areas. It is difficult to predict how they will affect each other and the environment.
- drastic sea conditions force the use of material solutions resistant to variable loads and chemical factors, which generates huge costs in the size of ships, and technical inspections and maintenance, especially of the electronic structure will require time and commitment of many specialists with the highest qualifications.

References

1. Abramowski, T.: Application of artificial neural networks to assessment of ship manoeuvrability qualities. Polish Marit. Res. **15**(2), 15–21 (2008)
2. Jacyna, M., Wasiak, M., Lewczuk, K., Kłodawski, M.: Simulation model of transport system of Poland as a tool for developing sustainable transport. Arch. Transp. **31**, 23–35 (2014)
3. Łosiewicz, Z.: Identifying the issue of reducing the emission of harmful compounds in the exhaust gas from marine main engines and description of the emission process of these compounds in probabilistic approach. J. Polish Marit. Res. **24**(2), 89–95 (2017)
4. Łosiewicz, Z.: Use of alternative fuels for hydrocarbon fuels for ships propulsion in the aspect of impact on the safety of navigation and environmental protection. In: 18th International Multidisciplinary Scientific Geoconference, SGEM 2018, Conference Proceedings Volume 18, Energy and Clean Technologies Issue 4.1, Renewable Energy Sources and Clean Technologies, Albena, 2–8 July 2018 (2018)
5. Łosiewicz, Z.: Effectiveness of ships propulsion operating on the alternative fuels to hydrocarbon fuels. In: 18th International Multidisciplinary Scientific Geoconference, SGEM 2018, Conference Proceedings Volume 18, Energy and clean technologies Issue 4.1, Renewable Energy Sources and Clean Technologies, Albena (2018)
6. Łosiewicz, Z., Mironiuk, W.: Elektroniczne systemy nadzoru stanu technicznego statku w aspekcie bezpieczeństwa żeglugi. Logistyka, Nr 6/2011, Poznań, pp. 2327–2333 (2011). ISSN 1231-5478
7. Łosiewicz, Z., Mironiuk, W.: Wpływ przepisów ochrony środowiska morskiego na konstrukcję wybranych typów statków w aspekcie bezpieczeństwa jednostki transportowej. Logistyka, nr 3/2012, Poznań, pp. 1401–1404 (2012)
8. Łosiewicz, Z., Mironiuk, W.: Ocena bezpieczeństwa statków handlowych różnych typów w warunkach morskich - wg przyjętych kryteriów. Technika Transportu Szynowego Nr 12/2015, Radom, pp. 2012–2015 (2015). ISSN 1232-3829
9. Mironiuk, W.: Model-based investigations on dynamic ship heels in relation to maritime transport safety. Arch. Transp. **33**, 69–80 (2015)
10. Siergiejczyk, M., Pas, J., Rosinski, A.: Issue of reliability–exploitation evaluation of electronic transport systems used in the railway environment with consideration of electromagnetic interference. IET Intell. Transp. Syst. **10**(9), 587–593 (2016)
11. Wawruch, R.: ARPA zasada działania i wykorzystania. WSM Gdynia, Gdynia (2002)
12. Zalewski, P.: Wyklad 11 - AIS.pdf, Instytut Inżynierii Ruchu Morskiego Akademia Morska w Szczecinie, Szczecin (2011)
13. http://www.Pl.wikipedia.org/wiki/Autonomia. Accessed 10 Apr 2018
14. https://www.gospodarkamorska.tv/relacje-tv/Statki-bezzalogowe-moga-zaczac-plywac-po-2030-r-11889.html. Accessed 10 Apr 2018
15. https://www.portalmorski.pl/stocznie-statki/36903-autonomiczny-statek-ze-szczecina-drony-z-trojmiasta. Accessed 10 Apr 2018

Method for Evaluation of the Actual Utilisation of the Train Maximum Speed

Andrzej Massel(✉)

Instytut Kolejnictwa, ul. Chlopickiego 50, 04-275 Warsaw, Poland
amassel@ikolej.pl

Abstract. The aim of the research is to establish method for evaluation of actual utilisation of the maximum line speed. Another goal is to identify factors influencing utilization of the maximum line speed in case of long-distance passenger trains on the Polish railway network. The extensive database covering the infrastructure data and rolling stock data for long-distance train services in Poland has been compiled. On infrastructure side main factor seems to be differentiation of the maximum speed along the line, which is taken into account with the harmonic weighted mean. As far as rolling stock is concerned main factors are type of train formation (loco-hauled, EMU) and power-to-weight ratio. Also impact of track works on speed utilisation has been studied. The best utilisation of maximum line speed is in the case of long sections passed without intermediate stops, at which the influence of acceleration and braking is relatively minor. The utilisation of maximum speed is negatively influenced by significant differentiation of the speed profile (frequent and large changes of speed along the line).

Keywords: Infrastructure · Maximum line speed · Commercial speed · Speed profile · Utilisation

1 Introduction

According to the Directive on the interoperability of the rail system within the European Union, the quality of rail services in the Union depends, inter alia, on excellent compatibility between the characteristics of the network (in the broadest sense, i.e. the fixed parts of all the subsystems concerned) and those of the vehicles (including the on-board components of all the subsystems concerned) [5]. Performance levels, safety, quality of service and cost depend upon that compatibility. Performance of the rail system can be evaluated according to numerous criteria and described through various parameters, like capacity, speed, accessibility.

In Poland, a speed qualification in the railway transport is based on the differentiation of two essential speed types [23]:

- maximum speeds, which are possible to be obtained depending on the structure and condition of railway lines, rail signalling systems, power supply systems and rolling stock used;

M. Siergiejczyk and K. Krzykowska (Eds.): ISCT21 2019, AISC 1032, pp. 284–294, 2020.
https://doi.org/10.1007/978-3-030-27687-4_29

- average speeds, which characterise a course of the operational processes on railways.

The most important characteristics of the rail infrastructure is the train maximum speed, being the key factor determining quality and competitiveness of the offer and demand for transport [20].

The relation between the maximum speed on particular railway line and the commercial speed of the trains has been investigated in the paper by Massel [11]. However that research was restricted to the high speed train services in selected European countries and to the services operated with high-speed rolling stock in Poland (EIP trains) [12].

The aim of the present research is to establish method for evaluation of actual utilisation of the maximum line speed. Another goal is to identify factors influencing utilization of the maximum line speed in case of long-distance passenger trains on the Polish railway network.

2 Literature Review

Good example of evaluation of overall performance of railway system can be found in the paper by Wróbel, which is related to the integration of timetables for passenger trains in Poland co-financed by competent authorities in the framework of Public Service Contracts (PSC) [22].

Several studies have presented methods of evaluating the robustness of simulated or real traffic performance, while some have considered other aspects of performance or an evaluation based around a more complete concept of quality of service (QoS). Lu et al. propose an independent evaluation framework based on the that concept, encompassing the priorities of all railway stakeholders, to assess the quantity and quality of the operational behaviour that results from a railway's use [10].

The issue of performance evaluation is valid in case of all transport systems, in particular it is necessary to develop methods supporting selection of optimum system for given local conditions. For example there numerous works comparing light rail systems (LRT) with bus rapid transit systems (BRT). One of the key features of particular transit system is its commercial speed. Therefore it is necessary to investigate factors influencing that speed. According to works by Kühn [9], in case of LRT systems, the direct relation exists between percentage of separated right-of-way and commercial speed.

The (sufficient) capacity is the key factor for the development of more sustainable European transport system with increased railway share of transport demand. When analysing the line capacity, the UIC states, that one first and foremost needs a definition of the infrastructure and timetable boundaries. The next step is to calculate the capacity utilisation, which is defined as "the utilisation of an infrastructure's physical attributes along a given section, measured over defined time period" [15].

In the paper by Rotoli et al. a synthetic methodology for the capacity and utilisation analysis of complex interconnected rail networks has been proposed [19]. The proposed methodology aims to evaluate capacity and utilisation rates of rail systems at different geographical scales and according to data availability.

Factors influencing commercial train speed and line capacity are discussed in the paper by Ramunas et al. [18]. The main parameter determining line capacity is the difference in train speeds. When the difference between the highest and the lowest train speed is increased, then the available line capacity decreases. Depending on operational conditions, the impact of the train load and the characteristics of the locomotive on the journey time, can be very significant, especially in case of heavy freight trains operated on line with long gradients (Fig. 1).

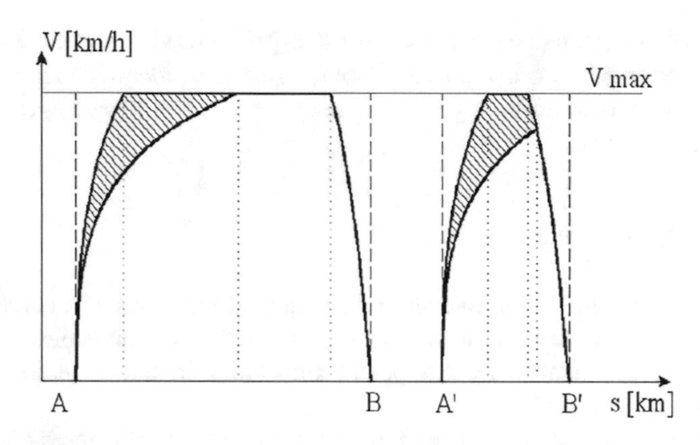

Fig. 1. A diagram showing movement of trains with different acceleration indicators in short and long districts [18]

Capacity can be also analysed in much broader sense. Boysen developed the general model of railway transportation capacity [4]. It models mass transportation capacity and volume transportation capacity per unit time. Application of the capacity model to cases of higher axle loads and higher speeds in freight transportation showed significant and mixed capacity effects that would not have been obvious otherwise.

Valuable in-depth assessment of railway transport system can be found not only in scientific papers but also in the reports of the European Court of Audit (ECA). A comprehensive performance audit on the long-term strategic planning of high-speed lines in the EU, on the cost-efficiency (assessing construction costs, delays, cost overruns and the use of high-speed lines which received EU co-funding), and on the sustainability and EU added value of EU co-funding was carried out in 2018 [1]. The analysis covered (inter alia) the speed on the audited high-speed lines and indicated that, on average along the course of a line, trains run at only around 45% of the line's design speed (in European conditions usually at the level of 300 km/h). Only two lines operate at average (commercial) speeds of more than 200 km/h, and no lines operate at an average speed above 250 km/h. The lowest speed yield on a completed high-speed line is on the Madrid-León high-speed line (39% of design speed). The cross-border Figures - Perpignan section also only operates at 36% of its design speed [1].

Two years earlier, in 2016, ECA evaluated rail freight transport across Europe [17]. The poor performance of rail freight transport in terms of volume and modal share in the EU is not helped by the average commercial speed of freight trains. On some international routes freight trains run at an average speed of only around 18 km/h. This is due to weak cooperation between the national infrastructure managers.

3 The Research Methodology

3.1 The Scope

The analysis covers the utilisation of the maximum line speed by the long-distance trains in Poland in the 2018/2019 timetable. Relatively large database has been prepared, covering 46 start-to-stop runs of Express Intercity Premium (EIP) trains and 448 runs of trains belonging to other categories: Express InterCity (EIC), InterCity (IC), Twoje Linie Kolejowe (TLK).

All involved trains are operated by PKP Intercity, to large extent in the framework of multiannual PSC contract, which has been concluded between the Ministry of Infrastructure of Poland and the train operating company on the basis of the National Transport Plan [8]. All EIP trains and a few EIC trains, however, are operated on purely commercial basis.

Practically all electrified railway lines in Poland, served with long-distance trains have been included in the analysis, apart from sections with seasonal traffic only. Due to different characteristics of diesel locomotives, this research does not cover the services operated on non-electrified routes.

3.2 Infrastructure Data

The official infrastructure data of PKP PLK were used to calculate exact distances and mean maximum speeds for all sections [2, 13, 14].

The maximum speeds are usually diversified on the train route, especially in the case of conventional railway lines, constructed to large extent in the XIX century as well in the beginning of the XX century. Main factor determining maximum speeds are the parameters of track geometry: radiuses of horizontal curves as well as the lengths of transitions. Therefore it is not possible to obtain desired uniform speed on the full length of the line. To characterise maximum train speeds it is necessary to adopt some statistical measures. The most important parameter seems to be the highest maximum speed on the section in question (V_{max}). This value defines the requirements for rolling stock making the full use from the infrastructure capabilities.

The differentiation of the maximum speed along the line has significant influence on journey time and, consequently, on line capacity. It is taken into account with the harmonic weighted mean $V_{0\ max}$, calculated according to the formula (1):

$$V_{o\,max} = \frac{\sum_{i=1}^{n} l_i}{\sum_{i=1}^{n} \frac{l_i}{V_{i\,max}}} \tag{1}$$

where $V_{i\ max}$ is maximum speed on section of track i, and l_i is the length of (sub)section i. Harmonic weighted mean is a quotient of the total length of the line (or the network) and the sum of theoretical journey times on particular (sub)sections with the constant speed. It is clear, that $V_{0\ max} \leq V_{max}$. For sections with uniform speed, $V_{0\ max} = V_{max}$. The example of real speed profile for line 131 section is presented in Fig. 2.

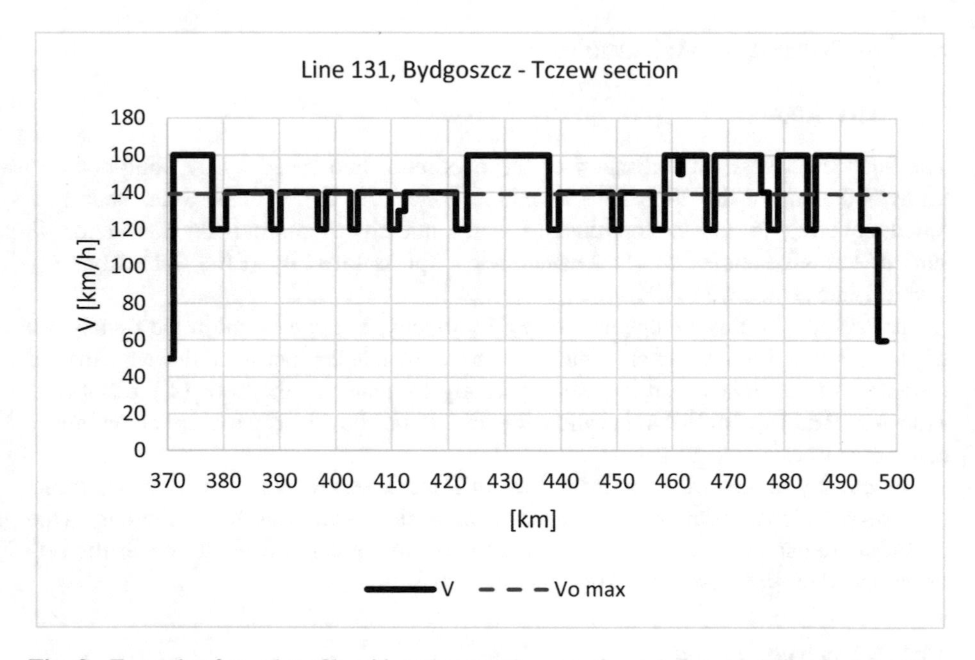

Fig. 2. Example of speed profile with numerous changes of speed. Source: Author's elaboration on the basis of official data of PKP PLK S.A [12, 13].

The speed of 140 or 160 km/h is generally in force on open sections (the sections between two consecutive stations), whereas the speed of 120 km/h is allowed through all the stations. In this case the weighted average maximum speed is $V_{0\ max} = 139.2$ km/h.

3.3 Train Data

The train characteristics have been extracted from the train composition plans, issued by PKP Intercity for 2018/2019 timetable [16]. It is noteworthy, that there are huge differences in the characteristics of passenger trains operated on the Polish railway network. The crucial parameter used in the research is the power-to-weight ratio (*P/m*). The highest power-to-weight ratios are in the case of some international trains operated with EU44 locomotives, manufactured by Siemens (more than 14 kW/t) and in the case of EIP trains served with ED250 Pendolino EMUs (more than 12 kW/t). On the other hand, there is a group of relatively heavy long-distance trains, hauled with EU07 or

EP07 locomotive (with constant power of only 2000 kW). In case of these trains power-to-weight ratio is just above 3 kW/t.

3.4 Operational Data

The journey times have been collected for all start-to-stop sections. They were extracted from official public timetables and from working timetables [21].

Practical measure for assessment of railway offer in passenger traffic is the value of commercial speed V_c, which is calculated according to the formula (2):

$$V_c = \frac{l}{\sum t_r + \sum t_s} \tag{2}$$

where l is the length of train route, t_r are the journey times on consecutive sections and t_s are the stopping times. It should be noticed, however, that for the analysis of utilization of the maximum line speed, the start-to-stop average speeds V_s are more convenient measure. Start-to-stop average speed was commonly used for presentation of the fastest train services in professional journals before World War II [3, 6]. It is noteworthy that the start-to-stop train speeds are the basis for train classification in the World Speed Survey, published bi-annually in Railway Gazette International until now [7].

The V_s values have been calculated for all sections of train run between consecutive stops. As the parameter characterising the utilisation of the maximum line speed, the speed utilisation ratio I_s has been defined (3):

$$I_s = \frac{V_s}{V_{0\,max}} \tag{3}$$

where V_s is the start-to-stop average speed of the train on particular section and $V_{0\,max}$ is the weighted average maximum line speed on the same section.

3.5 Compilation of Data for Analysis

Taking into account all infrastructure, rolling stock and operational characteristics, the following parameters have been listed for each start-to-stop section:

- Length of the section [14],
- Average maximum speed $V_{0\,max}$ calculated as harmonic mean according to [13],
- Maximum speed on the section V_{max},
- Journey time for the fastest train,
- Start-to-stop average speed V_s,
- Speed utilization ratio I_s,
- Weight of train,
- Power of the locomotive (or EMU) and power-to-weight ratio (*P/m*).

4 Results

4.1 Correlation-Regression Analysis

The results of analysis are presented for two major groups of trains: for EIC, IC and TLK trains and for EIP trains. In general, the values of start-to-stop average speeds show good correlation with respective average maximum speeds (Fig. 3).

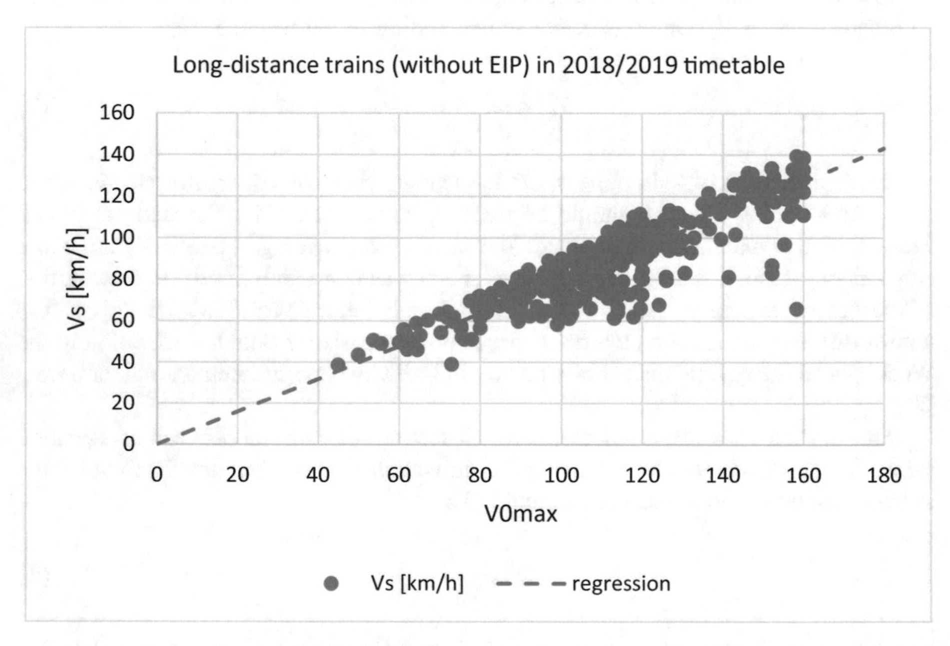

Fig. 3. Correlation between average maximum speeds and start-to-stop speeds for EIC, IC and TLK long-distance trains in the timetable valid from 9 December 2018

For 448 runs of EIC, IC and TLK trains the correlation coefficient r equals to 0.863. The regression equation can be formulated for this group of train services (4):

$$V_s = 0.795\, V_{0\,max} \tag{4}$$

It should be noted, that on several sections the average speeds of trains are affected by the extensive works performed in the framework of modernisation projects. Moreover, in some cases the speed restrictions, imposed on sections in bad condition of infrastructure, have also influence on train operation (Table 1).

Table 1. Correlation – regression analysis for EIC, IC and TLK trains

Parameter	All sections	Sections without works
Number of sections	448	394
Minimum value of I_s	0.416	0.543
Mean value of I_s	0.800	0.819
Maximum value of I_s	0.942	0.942
Average distance [km]	34.3	34.4
Correlation coefficient	0.863	0.926
Regression coefficient	0.795	0.816

When the correlation-regression analysis is limited to the sections without track works and without (severe) speed restrictions, the correlation coefficient is even better ($r = 0.926$).

The highest value of speed utilisation ratio is in the case of Starachowice Wschodnie – Ostrowiec Swietokrzyski section ($I_s = 0.942$). It can be attributed to rather low average maximum speed on that section ($V_{0\ max} = 53.7$ km/h).

Similarly to the above data, also the values of start-to-stop average speeds of EIP trains show very good correlation with (harmonic) mean maximum speeds. In case of 46 runs of these trains (served with ED250 EMUs) the correlation coefficient r equals to 0.942 (Table 2). The regression equation for this group of train services is as follows (5):

$$V_s = 0.818\ V_{0\,max} \tag{5}$$

It should be noted, that neglecting a few sections affected with track works results in even better correlation and in higher values of speed utilisation ratio. There are several sections with particularly favourable utilisation of maximum line speed, for example:

- Warszawa Wschodnia – Ilawa Glowna ($I_s = 0.900$),
- Lebork – Slupsk ($I_s = 0.903$).

For the start-to-stop runs of EIP trains, covering Central Trunk Line (at which 200 km/h is achieved) the speed utilisation ratio is also quite good:

- Warszawa Zachodnia – Krakow Glowny ($I_s = 0.858$),
- Krakow Glowny - Warszawa Zachodnia ($I_s = 0.859$),
- Warszawa Zachodnia – Zawiercie ($I_s = 0.851$).

Table 2. Correlation – regression analysis for EIP trains

Parameter	All sections	Sections without works
Number of sections	46	41
Minimum value of I_s	0.592	0.639
Mean value of I_s	0.813	0.829
Maximum value of I_s	0.903	0.903
Average distance [km]	90.4	90.4
Correlation coefficient	0.942	0.972
Regression coefficient	0.818	0.833

4.2 Impact of Power-to-Weight Ratio

The power of the locomotive or EMU has significant influence on the performance of the train, especially in case of acceleration (from stations and in the locations, where the maximum line speed changes). Therefore all train runs in the data base (494 runs) have been grouped into 3 subcategories, according the value of power-to-weight ratio (*P/m*). The first group (*P/m* > 8 kW/t) covers the runs of trains served with the modern rolling stock (new generation of EMUs and sets of cars hauled with EU44 electric locomotives). In the second group (4 kW/t < *P/m* $\leq$ 8 kW/t), the typical conventional trains of moderate weight are included (together with trains served with ED160 EMUs). The third subcategory groups the heaviest long-distance trains with conventional electric locomotives (usually EP07/EU07 type). The results are presented in Table 3.

Table 3. Correlation – regression for various groups of power-to-weight ratio

Parameter	All sections	*P/m* > 8 [kW/t]	4 < *P/m* $\leq$ 8 [kW/t]	*P/m* $\leq$ 4 [kW/t]
Number of sections	494	167	297	30
Minimum value of I_s	0.416	0.574	0.416	0.620
Mean value of I_s	0.801	0.810	0.798	0.778
Maximum value of I_s	0.942	0.942	0.935	0.884
Correlation coefficient	0.880	0.934	0.835	0.822
Regression coefficient	0.798	0.811	0.792	0.775

It is visible, that for higher values power-to-weight ratio, the speed utilisation ratio I_s is better than in case of underpowered trains (*P/m* < 4 kW/t).

5 Conclusions

The method for evaluation of actual utilisation of the maximum line speed has been developed. The proposed methodology has been verified for the comprehensive data base including the runs of long-distance trains in Poland (various train categories and types of rolling stock) in the 2018/2019 timetable. The analysis shows that typical value of speed utilisation ratio I_s for the long-distance passenger services in Poland is approximately 0.8 for traditional train compositions and 0.82–0.83 for the EIP trains operated with ED250 Pendolino EMU. The correlation between the average maximum line speed and the train commercial (start-to-stop) speed is rather good.

The differences between start-to-stop average speeds and average maximum speeds result from time losses due to train acceleration and braking at the beginning and the end of each run but also in all locations, where maximum speed changes. It should be remembered that timetable recovery margins are added to guarantee timekeeping.

The highest values of the speed utilisation ratio are observed for the longest sections passed without intermediate stops. On such sections the influence of acceleration and braking is relatively minor. I_s values at the level of 0.9 have been identified for some long runs of EIP trains, operated with ED250 trainsets.

The utilisation of maximum speed is negatively influenced by significant differentiation of the speed profile (frequent and large changes of speed along the line). The most effective utilisation of line capabilities is in the case of Electric Motor Units (EMUs) with distributed power and high power-to-weight ratio.

The results of the research show clearly, that the selection of passenger rolling stock for particular route can have significant impact on its day-to-day operation. Therefore it is very important to take the infrastructure characteristics (gradients, maximum speed profile) into account at the stage of drafting specifications for the rolling stock to be used on particular route. Similarly it is important to carry out in-depth analysis of rolling stock performance as a part of feasibility study for construction of the new railway infrastructure or modernisation of existing one. This is important field of applicability of the method and the results of the study as the tool to compare large number of various variants of the future infrastructure investment [11].

References

1. A European high-speed rail network: not a reality but an ineffective patchwork. European Court of Auditors, Special Report no 19 (2018)
2. Annual Report PKP Polskie Linie Kolejowe S.A., Warszawa (2018)
3. British Express Train Services in 1935 (by "Mercury"). The Railway Magazine, January, pp. 5–12 (1936)
4. Boysen, H.E.: General model of transportation capacity. Computers in railways XIII, pp. 335–347 (2012)
5. DIRECTIVE (EU) 2016/797 on the interoperability of the rail system within the European Union. Official Journal of the European Union 26.05.2016, L138, pp. 44–101 (2016)
6. European Express Trains in the Summer of 1938-II. Countries other than France and Great Britain (by "Mercury"). The Railway Magazine, November, pp. 361–367 (1938)
7. Hartill, J.: Italy joins the premier speed league. World Speed Surv. Railway Gaz. Int. **7**, 28–31 (2017)
8. Klemba, S., Wróbel, I.: Review of national transport plan and proposal for improvement. Problemy Kolejnictwa (Railway Rep.) **60**(172), 67–84 (2016)
9. Kühn, F.: Bus rapid or light rail transit for intermediate cities? In: Urban Mobility for All, Proceedings of the 10th International CODATU Conference, Lome, Togo, pp. 357–365 (2002)
10. Lu, M., Nicholson, G.L., Schmid, F., Dai, L., Chen, L., Roberts, C.: A framework for the evaluation of the performance of railway networks. Int. J. Railway Technol. **2**(2), 79–96 (2013)
11. Massel, A.: Infrastructure and operation – research on utilisation of the maximum train speed profile. In: Proceedings of 7th Transport Research Arena, Vienna, Austria, 16–19 April (2018)
12. Massel, A.: Wykorzystanie prędkości maksymalnej pociągów – przypadek pociągów EIP (Utilisation of the train maximum speed – EIP trains case). Prace Naukowe Politechniki Warszawskiej – Transport, vol. 111, pp. 371–378 (2016). ISBN 1230-9265
13. Network Statement 2018/2019. Annex 2.1(P). List of maximum speeds - wagon trains. PKP PLK, Warszawa (2018)

14. Network Statement 2018/2019. Annex 2.6. List of operating control points and forwarding points, Warszawa (2018)
15. Odolinski, K.: Railway line capacity utilisation and its impact on maintenance costs. J. Rail Transp. Plan. Manag. **9**, 22–33 (2019)
16. Plan zestawienia, obiegu i obslugi skladow pociagow pasazerskich krajowych (Plan of train composition and rosters for domestic passenger trains). PKP Intercity, Warszawa (2018)
17. Rail freight transport in the EU: still not on the right track. European Court of Auditors, Special Report no 08 (2016)
18. Ramunas, V., Gailiene, I., Podagelis, I.: Increment of railway line capacity. In: Environmental Engineering the 8th International Conference, Vilnius, 19–20 May (2011)
19. Rotoli, F., Malavasi, G., Ricci, S.: Complex railway systems: capacity and utilisation of interconnected networks. Eur. Transp. Res. Rev. **8**, 29 (2016)
20. Schumann, T.: Passenger demand for a high-speed network across Europe. Problemy Kolejnictwa (Railway Rep.) **57**(161), 67–86 (2013)
21. Sieciowy Rozklad Jazdy Pociagow (Network Train Timetable), valid from 9 December 2018, PKP PLK, Warszawa (2018)
22. Wróbel, I.: Evaluation of the integration of train timetables for public service trains. Problemy Kolejnictwa (Railway Rep.) **62**(178), 73–79 (2018)
23. Żurkowski, A. (ed.): High-Speed Rail in Poland. Advances and Perspectives. Taylor & Francis Group, London (2018)

Psychological Determinants of the Driver's Behaviour in the Context of Introducing Experimental Road Signs in Poland

Ewa Odachowska(✉), Monika Ucińska(✉), and Kamila Gąsiorek(✉)

Transport Telematics Center, Motor Transport Institute, Warsaw, Poland
{ewa.odachowska,monika.ucinska,kamila.gasiorek}@its.waw.pl

Abstract. Road signs as an important component of road infrastructure have an impact on the safety of the road users. Conducting research in this area is an important element of work in the field of improving road safety. In Poland, as in the world, there are analyzes conducted of the experimental signs and markings, which are unspecified in the applicable law, but meeting the technical conditions of conventional (standard) signs. The current researches point to the importance of psychological aspects regarding the effectiveness of marking. The presented analyzes are the result of a project implemented by the Motor Transport Institute as part of a project financed by the National Centre for Research and Development and the General Directorate of National Roads and Motorways (Agreement No. DZP/RID-I-36/5/NCBR/2016).

Keywords: Experimental marking · Traffic safety · Perception · Attention

1 Introduction

Road signs are an important component of the road infrastructure affecting the safety of road users. Intentionally, its impact affects both vehicle speed, traffic flow, and the drivers' behaviour. Providing appropriate guidance to the road users, warning them of the possible dangers, impediments present are one of the most important functions of the signs. Their task is to provide information to the road traffic participants in an abbreviated and/or encoded form, for which adequate knowledge is necessary to correctly understand them and make the right decisions resulting from the communication. The more clearly and unambiguous are the signs, the faster and more easily the driver of the vehicle is able to take appropriate action.

Testing a new type of signs called experimental is one of the elements that fit into a wide range of preventive activities for road safety. In this sense, it is also an element that fits in with the issues of transport psychology. The new signs are covered by the research program and are tested on the designated sections of public roads not only to examine the legibility of the description of the route, but also the influence on the drivers' behaviour.

M. Siergiejczyk and K. Krzykowska (Eds.): ISCT21 2019, AISC 1032, pp. 295–306, 2020.
https://doi.org/10.1007/978-3-030-27687-4_30

2 Psychological Aspects of the Road Environment Perception

Driving a vehicle is a complex activity that is influenced by many factors. It depends not only on the driver's skills or the ability to assess risk, but also on the environment, including the location of road equipment, e.g. road signs. Previous studies on the psychophysical features of drivers, which used modern measurement technology bring many insights and conclusions that may find implementation at the road design stage.

One of the most important cognitive functions enabling safe driving is an *attention*. The notion of attention refers to the determination of many cognitive processes responsible primarily for: keeping the body in a state of readiness for action, perception or other behaviour; maintaining the state of waiting for certain stimuli to appear; extracting essential elements from the perceptual field (e.g. visual, auditory) and suppressing (eliminating) non-essential elements (concentration of attention); properly organized course of a given mental activity, ensuring that side processes do not interfere with it; the ability to engage in the analysis of a given stimulus and the ability to detach from this stimulus and focus on another stimulus [6].

The mechanism of attention is understood here as an information processing system whose task is to select sensory information, select the right motor reactions and exercise control over their performance.

Another important psychological function involved in information processing are *perceptual processes*. The effectiveness of recognition of road signs is influenced by its spatial location in relation to the eyes of the person driving the vehicle. The road sign recognition takes place in several stages. The time of noticing and observation and the distance at which the mark is already noticed depends on the size of the sign and the speed at which the driver drives the vehicle. In the context of sign recognition, four stages can be distinguished:

1. Noticing an object that can be a road sign.
2. Recognizing the colour of the sign.
3. Recognizing the shape of the sign.
4. Interpreting the content of the sign and recognizing details - letters, numbers, etc. [4, 16].

The perception of the human environment is based on two types of vision: central vision and peripheral vision. The first aspect relates to the field of vision of the eye, and thus the section of the environment seen by the non-moving eye at the fixed position of the head. The total field of view of a human is about 120° in the vertical plane and 200° in the horizontal plane (with overlapping fields of vision of both eyes). The value of the angle of view (field of view expressed numerically) of the eye depends on the intensity of illumination, size and colour of the observed object. Peripheral vision, in turn, is responsible for spatial orientation in the environment and serves, among the others, to detect the danger.

For safety, the important thing is to see the contrast. Sensitivity to contrast gives information about the ability of the visual system to distinguish the object seen from the background [5]. When driving a vehicle, important elements of the field of vision are recognized by differences not only in contrast, but also in colour. With a daylight, where colour perception is possible, objects have sufficient contrast to be observed and recognized by the central system. In conditions of limited lighting, when colour perception is weak, contrast chiefly dominates. During the twilight driving, the information flowing into the sense of sight is reduced, the vision becomes worse, and the person perceives the world only in shades of blue and gray. During night vision, with a negligible amount of light, the driver sees a world completely devoid of colour, distinguishes only the degrees of ambient brightness. Under these conditions, the ability to recognize image details is significantly reduced [10].

Understanding information that is important for safety and anticipating the development of a traffic situation is associated with the functioning of working memory, where memory span is of particular importance. The capacity of short-term memory is limited. We can maintain in it seven simple information units at the same time. Based on Miller's research [7], a basic rule regarding memory capacity was created. It is 7 ± 2 elements. In difficult, complex or new situations, this number may decrease by 1–2 units, while in the known and predictable ones it may increase by the same value [7]. If the information transition to further processing phases is not disrupted by anything, it takes about 5 s. In multi-faceted situations, this time increases even twice to 10 s [9]. This information is of particular importance in the face of the development of messages presented on the sign, both in terms of their quantity and complexity.

Empirical data show that the driver can remember on average about 20% of signs he sees on the road [3]. Research confirms the relationship between the number of reminded signs and the distance the driver has passed and the knowledge of the road on which he travels [14]. Registration and memorization of the sign are related to the driver's age and experience. Better results are seen in younger drivers and drivers with less experience [8]. It is worth adding that the process of automating repeated activities plays a large role in the process of remembering and the process itself.

An important element on which the safety of road users depends, and which should be reflected in the aspect of the development of experimental signs is the speed and adequacy of response to stimuli. The reaction time characterizes the driver in terms of the possibility of taking preventive actions as early as possible in a pre-accident situation. The driver reaction time is influenced by factors such as: visual-motor coordination, nervous system performance, current psychophysical state, including fatigue, driver's personal characteristics (e.g. age), the specificity of a given road situation, e.g. the need to move the eyesight when perceiving an obstacle; night driving conditions (including the possibility of glare and contrast of the object in relation to the background) and biometeorological conditions [17]. Improperly selected road signs that are too complicated and not easily legible can lead to a longer reaction time of the driver and a significant increase in the risk of an accident.

The above-mentioned aspects show how many factors are involved in both information processing and perception.

3 Methodology of the Own Research

The study of functional features of experimental road signs took place in September 2017 in the Laboratory of Transport Psychology and Driving Simulators at the Motor Transport Institute. The test was conducted on an AS1200-6 high-end passenger car driving simulator, and the equipment used in psychological tests of drivers. During the development of the research methodology, the necessity of analyzing the processes involved in the processing of information from the external environment, including the way in which this information reaches the driver, were taken into account. For this purpose, both research devices (Vienna Test System - WST, Functional Vision Analyzer - FVA) and paper-pencil tests were used for the research.

In the studies a TUS (Attention And Perceptiveness Tests) [2] was used, intended for the analysis of the speed of perceptual work and attention. In terms of cognitive functions analysis, the tools included in the Vienna Test System (WST) were also used: Kognitron COG [18], checking the general predisposition regarding attention efficiency, peripheral perception test PP [12] and CORSI [1] – enabling examination of both visual - spatial fresh memory as well as visual and spatial learning. In addition, the Adaptive Tachistoscopic Traffic Perception Test ATAVT [13] (WST) was used to assess the speed of perception, at the same time referring to the ability of the visual identification of objects and patterns. The motor efficiency in the form of reaction time was analyzed using the Reaction Time Test (RT) [11] in the same set (WST). For a wider view of processes affecting perception of periphery elements, analysis of visual functions was conducted using the F.A.C.T. (Functional Acuity Contrast Test), which is part of the Advanced Functional Vision Analyzer - FVA, aimed at assessing sensitivity to contrast [5].

Differences in terms of temperament were checked using the FCZ-KT temperament questionnaire [19], which is a tool for the diagnosis of basic, biologically determined personality dimensions, describing formal aspects of behaviour (Briskness, Perseverance, Sensory sensitivity, Emotional reactivity, Endurance, Activity). It was created based on the Regulatory Temperament Theory by Strelau [15]. According to it, temperament refers to the formal characteristics of behaviour, and the results of research indicate that the temperament traits so understood play an important role in the regulation of human relations with the world. The regulatory function involves here primarily modifying the stimulus and temporal value of behaviours and reactions, as well as the situation in which the individual is located, but it manifests itself primarily in circumstances described as difficult or extreme for the individual. Thus understood, it affects every sphere of human life, including driving a vehicle.

The experimental signs tests were carried out while the participants were driving the simulator in three research scenarios. During the test run, measurements of parameters for the evaluation of the experimental signs were carried out. The length of a single test run for each person was between 10–20 min. The trips took place on the A or G class roads in the extra-urban area. The subject's task was to follow specific signs. Routes were clearly marked using both experimental and standard signs. There were also signs on the route that indicated other destinations than the participants' destination. The scenario was secured against going the wrong way. The participant was informed that he should drive the entire route without excessive haste (causing, for example, dangerous behaviour on the road, exceeding the speed limit, etc.), however in the shortest possible time. This was to "optimize" the selected route. The indication of the necessity of driving at the highest possible speed was also supposed to favour the evaluation of the experimental signs patterns used to reduce speed and maintain a safe distance between vehicles.

All experimental signs studies were carried out in respect to the reference variant. The reference sections were fragments of the similar scenario (or identical) to the research fragment, however, marked in a standard manner. The evaluation of the experimental marking involved comparing the motion parameters registered in both variants - reference and research (using OE). The following signs were used in the test: Non-standard boards, Warning sign prototype, Optical measures reduction, Standard warning sign.

During the test, normal traffic of other vehicles occurred, except for the situations specified in the assumptions of the test situation for individual sections.

Psychological variables were analyzed both in the case of velocity in individual experimental conditions and in the peripheral environment perception in the studies conducted in a simulated environment.

4 Results

76 people participated in the study, including: 31 women, 45 men. Targeted choice was used, and the condition for participation in the study was to have a category B driving license and at least half a year of driving experience. The average age of the subjects was M = 40.07 years old. The majority of the participants (47.4%) had higher education, followed by secondary education (38.2%) and the fewest (14.5%) with incomplete higher education. There were no people with basic, vocational or secondary education among the respondents.

The subjects were active drivers, regularly driving the vehicle under their driving privileges. Most of the drivers tested (73.7%) assessed their driving skills as good, 11 people (14.5%) as very good and only 9 people (11.8%) consider their skills as average. None of the respondents rated their skills as poor or very poor.

In the context of driving frequency, 37 respondents (48.7%) declared that they were everyday road users as drivers, 30.3% drive a vehicle several times a week, almost 20% (N = 15) drive vehicle rarely (several times a month), and only one person declared that he/she is involved in road traffic as a driver very rarely (several times a year). Over half of them (52.6%) travel between 250 and 850 km a month, 21% (N = 16) of the respondents travel between 860 and 1700 km, 14.5% (N = 11) up to 250 km, and only 11.8% (N = 9) of drivers drive over 1800 km a month. In order to trace the relationship between speeds at control points for individual measurement conditions (throbbing lines, standard signs, reinforcements in the form of pictograms on the roadway) and psychological variables, a correlation analyzes were performed (tables below).

The correlation analyzes using r-Pearson method showed statistically significant relationships between the speed at which participants drove and temperament (FCZ-KT). Correlations were observed mainly in the case of *Sensory sensitivity* and *Briskness*. In both cases, they were correlations with a negative value. Regardless of the sign used (throbbing lines, standard signs, pictograms), this component of temperament correlated with the speed achieved. In this case, an interesting observation is the fact that the main relationships of temperamental variables (two components) relate to speed in free driving conditions, and therefore, along with the increase in *Sensory sensitivity* and *Briskness* the speed that the subjects obtained in laboratory conditions decreased.

Then, the psychological variables relating to the properties of perception and observing were analyzed. In order to check the relationships between traffic behaviour and these properties, a correlation analysis was performed. The results are presented in the table below (Table 1).

Table 1. A rho-Spearman correlation matrix for velocities and psychological variables, N = 76

	Psychological variables					
	The angle of stereopsy	Visual acuity	Contrast	Peripheral vision field	Left vision field	Right vision field
Throbbing lines						
Free driving	−0,13	**0,24***	0,08	**0,25***	0,16	**0,27***
At the signs	−0,08	**0,26***	−0,07	**0,40****	**0,28***	**0,42****
Road curve beginning	−0,07	0,04	0,13	**0,33****	**0,26***	**0,30***
Road curve the end	−0,15	0,08	0,00	**0,38****	**0,32****	**0,34****
Free driving behind the curve	−0,11	0,16	0,03	0,22	0,12	**0,26***
Standard signs						
Free driving	**−0,24***	**0,35****	−0,03	**0,31****	**0,24***	**0,33****
At the signs	−0,18	**0,27***	−0,03	**0,44****	**0,33****	**0,47****
Road curve beginning	**−0,28***	**0,28***	0,00	**0,42****	**0,30****	**0,47****
Road curve the end	−0,17	0,21	−0,05	**0,25***	0,16	**0,30***
Free driving behind the curve	−0,14	0,22	0,04	**0,27***	0,19	**0,28***
Pictograms						
Free driving	−0,10	**0,34****	−0,11	**0,38****	**0,28***	**0,41****
At the signs	−0,13	**0,23***	0,07	**0,33****	**0,26***	**0,36****
Road curve beginning	−0,11	**0,35****	−0,05	**0,44****	**0,34****	**0,43****
Road curve the end	−0,18	0,19	0,03	**0,37****	**0,26***	**0,40****
Free driving behind the curve	**−0,08**	**0,26***	**−0,03**	**0,40****	**0,27***	**0,41****

Selected (bold type) correlations are statistically significant at the level: *p < 0.05; **p < 0.01; ***p < 0.001.

Source: own research.

Measurements were made during free driving condition, at the signs, road curve beginning, road curve the end, free driving behind the curve. The correlation analyzes using rho-Speraman method showed the relationships observed in the field of psychological variables in the form of perceptiveness, perception field and other visual properties and the speed obtained by the respondents in the individual road conditions (throbbing lines, standard signs, pictograms). The Table 2 shows the correlation analyzes using r-Pearson method.

Table 2. The r-Pearson correlation matrix for velocities and psychological variables, N = 76

	ATAVT		CORSI		RT		COG		
	Correct answers	Time to complete	Correct reactions	Time to complete	Average reaction time	Correct reactions	Time of correctly rejected	Sum of correctly accepted	Time to complete
Throbbing lines									
Free driving	0,04	−0,26*	0,25*	0,20	−0,22	−0,07	−0,25*	0,20	−0,27*
At the signs	0,22	−0,31**	0,43**	0,30**	−0,16	−0,05	−0,27*	0,24*	−0,25*
Road curve beginning	0,15	−0,12	0,31**	0,19	−0,11	−0,12	−0,24*	0,19	−0,22
Road curve the end	0,12	−0,11	0,35**	0,29*	−0,11	−0,17	−0,14	0,22	−0,09
Free driving behind the curve	−0,04	−0,27*	0,29*	0,19	−0,18	−0,05	−0,27*	0,16	−0,27*
Standard signs									
Free driving	−0,04	−0,33**	0,35**	0,26*	−0,13	−0,14	−0,31**	0,21	−0,31**
At the signs	0,10	−0,19	0,45**	0,37**	−0,21	−0,02	0,28*	0,25*	−0,27*
Road curve beginning	0,09	−0,17	0,33**	0,27*	−0,24*	−0,10	−0,27*	0,18	−0,30**
Road curve the end	0,17	−0,24*	0,27*	0,23*	−0,14	−0,12	−0,21	0,22	−0,21
Free driving behind the curve	−0,06	−0,31**	0,27*	0,19	−0,20	−0,11	−0,29**	0,20	−0,30**
Pictograms									
Free driving	0,05	−0,51**	0,25*	0,17	−0,17	−0,10	−0,39**	0,19	−0,40**
At the signs	0,10	−0,33**	0,36**	0,24*	−0,24*	−0,02	−0,37**	0,28*	−0,35**

(*continued*)

Table 2. (*continued*)

	ATAVT		CORSI		RT		COG		
	Correct answers	Time to complete	Correct reactions	Time to complete	Average reaction time	Correct reactions	Time of correctly rejected	Sum of correctly accepted	Time to complete
Road curve beginning	0,16	**−0,42****	**0,41****	0,22	**−0,33****	−0,06	**−0,45****	**0,26***	**−0,43****
Road curve the end	0,12	−0,21	**0,29***	**0,28***	−0,17	−0,14	**−0,30***	0,11	**−0,29***
Free driving behind the curve	**0,08**	**−0,33****	**0,24***	**0,20**	**−0,19**	**−0,05**	**−0,28***	**0,17**	**−0,29***

*p < 0.05; **p < 0.01; ***p < 0.001.
Source: own research.

The tables above show that most relationships were observed in peripheral vision (PP) and memory spans (CORSI). Positive relationships indicate that the high results obtained in these properties are accompanied by a faster car driving.

In the course of the analyzes, the evaluation was also made of the relationships between the analyzed psychological variables (temperament, attention and perceptiveness, memory span, perception field and other visual properties) and the velocities obtained by the respondents at the control points (3 curves). In order to trace the relationship between speeds at control points for individual measurement conditions (non-standard boards, warning sign prototype, optical reduction measures, standard warning sign), correlation analyzes were performed, the results of which are presented in the tables below (Tables 3 and 4).

Table 3. Rho-Spearman correlation matrix for velocities and psychological variables for non-standard boards and warning sign prototype, N = 76

Curve	Non-standard boards			Warning sign prototype		
	1	2	3	1	2	3
FVA – Functional Vision Analyzer						
Stereopsy angle	−0,19	−0,09	**−0,24***	−0,20	−0,17	−0,16
Visual acuity score	0,20	**0,25***	**0,23***	**0,25***	0,19	0,20
Eye contrast	−0,08	−0,06	−0,13	−0,10	−0,08	−0,04
Peripheral perception (PP)						
Peripheral field of view in degrees	**0,32****	**0,38****	**0,31****	**0,24***	**0,29***	**0,37****
Left field of vision in degrees	**0,29***	**0,33****	**0,29***	0,21	**0,27***	**0,33****
Right field of vision in degrees	**0,36****	**0,41****	**0,32****	**0,28***	**0,29***	**0,38****

(*continued*)

Table 3. (*continued*)

Curve	Non-standard boards			Warning sign prototype		
	1	2	3	1	2	3
Adaptive Tachistoscopic Traffic Perception Test (ATAVT)						
Correct answers	0,14	0,15	0,07	0,00	0,04	0,08
Working time	**−0,26***	**−0,26***	**−0,26***	−0,16	−0,22	−0,22
The memory range (CORSI)						
Correct reactions	0,20	**0,23***	0,18	0,22	0,17	**0,23***
Working time in seconds	0,08	0,14	0,07	0,14	0,10	0,16
Reaction Time (RT)						
Average reaction time	0,00	−0,07	−0,09	−0,02	−0,10	−0,04
Correct reactions	−0,05	−0,13	−0,09	−0,11	−0,13	−0,03
Attention and Concentration Test (COG)						
Time of the correctly rejected	−0,13	−0,22	−0,18	−0,11	−0,14	−0,08
Sum of the correctly accepted	**0,26***	**0,34****	**0,24***	**0,26***	**0,25***	**0,29***
Working time	**−0,13**	**−0,23***	**−0,20**	**−0,12**	**−0,15**	**−0,08**

*p < 0.05; **p < 0.01; ***p < 0.001.
Source: own research.

Table 4. Rho-Spearman correlation matrix for velocities and psychological variables for optical reduction measures and standard warning sign, N = 76

Curve	Optical reduction measures			Standard warning sign		
	1	2	3	1	2	3
FVA – Functional Vision Analyzer						
Stereopsy angle	−0,17	−0,11	−0,13	−0,11	−0,15	−0,17
Visual acuity score	**0,33****	**0,23***	0,19	0,11	0,20	**0,24***
Eye contrast	−0,16	−0,02	−0,04	0,02	−0,09	−0,15
Peripheral perception (PP)						
Peripheral field of view in degrees	**0,40****	**0,47****	**0,28***	**0,28***	**0,30***	**0,35****
Left field of vision in degrees	**0,35****	**0,40****	**0,27***	**0,27***	**0,27***	**0,34****
Right field of vision in degrees	**0,44****	**0,47****	**0,30****	**0,28***	**0,31****	**0,34****
Adaptive Tachistoscopic Traffic Perception Test (ATAVT)						
Correct answers	0,09	0,09	0,06	0,10	0,06	0,11
Working time	**−0,29***	**−0,29***	−0,19	−0,17	**−0,27***	−0,22

(*continued*)

Table 4. (*continued*)

Curve	Optical reduction measures			Standard warning sign		
	1	2	3	1	2	3
The memory range (CORSI)						
Correct reactions	**0,30****	**0,32****	0,19	0,16	0,19	0,16
Working time in seconds	0,21	0,22	0,10	0,12	0,13	0,09
Reaction Time (RT)						
Average reaction time	−0,04	−0,16	−0,06	−0,10	−0,01	−0,03
Correct reactions	−0,11	−0,06	−0,13	0,04	−0,14	−0,12
Attention and Concentration Test (COG)						
Time of the correctly rejected	−0,18	**−0,23***	−0,13	−0,06	−0,17	−0,14
Sum of the correctly accepted	**0,32****	**0,32****	**0,23***	**0,29***	0,22	**0,24***
Working time	**−0,19**	**−0,23***	**−0,14**	**−0,07**	**−0,18**	**−0,15**

*p < 0.05; **p < 0.01; ***p < 0.001.
Source: own research.

Many significant relationships were observed in the correlation between psychological variables responsible for perception and speed in individual road conditions. The most pronounced relationships were recorded in the field of peripheral vision. It turned out that both the general result and that of the properties of the right or left eye are significantly positively related to speed. The respondents who obtained better results in peripheral vision at the same time drove faster in each condition, regardless of the sign used. Such positive correlations were also observed in the eyes of visual acuity, attention and concentration (sum of correct answers). Negative relationships were noted in turn in the case of the results of the Traffic Orientation Test (ATAVT). The time of task performance in the test was negatively related to speed, especially for non-standard road signs. The analysis also showed statistically significant relationships between the speed at which participants drove and temperament (FCZ-KT). Correlations were observed mainly in the case of *Sensory sensitivity* and *Briskness*. The *Briskness* was negatively related to the speed at all signs, so regardless of the sign used, this component of temperament correlated with the speed achieved. The higher the results for *Briskness*, the lower the speed in the road conditions. Apart from the standard sign, in the remaining conditions it was observed that the higher the *Sensory sensitivity* the lower the speed on the curves of individual signs.

5 Conclusion

Driving a vehicle requires the driver to have proper situational awareness, i.e. the ability to perceive and understand important information for safety, as well as the ability to predict further development of the traffic situation. Perception process can be disrupted by over-stimulating by markings (in terms of location, format, colour), while understanding and prediction related to the functioning of working memory can be disturbed by the content of the sign. The studies of functional features of experimental road signs conducted at the Motor Transport Institute, showed that in perception of the peripheral reality, the characteristics of the individual, such as: attention, perception, memory as well as individual differences in the form of temperament traits, are of importance.

Analyzes showed the existence of a correlation between speed and temperament. The main relationships concerned Briskness and Sensory sensitivity. The Briskness describing the rate of reaction and performance as well as the ease of adjusting the behaviour to the changing situation was negatively related to the speed at all signs, so regardless of the sign used, this component of temperament correlated with the speed achieved. Similar relationships were noted for other features and skills. For example, respondents who achieved better results in peripheral vision at the same time drove faster in each condition, regardless of the sign used.

The psychological aspects discussed in the presented analyzes form the basis for developing a methodology for researching the impact of experimental signs on the behaviour of drivers and their safety, and as such should help to improve the work on the implementation of new solutions. The issue of testing experimental signs in terms of their relationship with the behaviour of the drivers and traffic conditions in the driving simulator environment as well as in the real traffic is the subject of research for several years. Nevertheless it is still a very young field, and the implementations are very few. The main problem occurring in Poland is the noticeable lack of indications for the use of experimental markings. The absence of the recommended methods of selection, implementation and verification of the impact of this marking described in the legal documents means that these measures are implemented very rarely and only based on the indications resulting from application in other countries. This is insufficient, and the expected results may not correspond to those assumed due to cultural differences or factors of communicativeness and acceptability of the sign.

References

1. Baddley, A.D.: Working Memory. Oxford University Press, Oxford (1986)
2. Ciechanowicz, A., Stańczak, J.: Testy uwagi i spostrzegawczości TUS Podręcznik. PTP, Warszawa (2006)
3. Costa, M., Dondi, G., Bucchi, A., Simone, A., Vignali, V., Lantieri, C.: Looking behavior for vertical road signs. Transp. Res. Part F: Traffic Psychol. Behav. **23**, 147–155 (2014)
4. Gruszczyński, J.: Oznakowanie dróg, a bezpieczeństwo ruchu drogowego [w:] Projektowanie i zarządzanie drogami – zasady, dobre praktyki, efektywność. Materiały Konferencyjne, nr 89 (zeszyt 149), pp. 57–76. Stowarzyszenie Inżynierów i Techników komunikacji Oddział w Krakowie, Kraków (2009)

5. Jędrzejczak-Młodziejewska, J., Krawczyk, A., Szaflik, J.P.: Badanie wrażliwości na kontrast testerem wzroku Functional Vision Analyzer. Okulistyka **3**(II)/2010, 60–64 (2010)
6. Maruszewski, T.: Psychologia poznania: Sposoby rozumienia siebie i świata. Gdańskie Wydawnictwo Psychologiczne, Gdańsk (2002)
7. Miller, G.A.: The magical number seven, plus or minus two: some limits on our capacity for processing information. Psycholog. Rev. **63**, 71–97 (1956)
8. Milošević, S., Gajic, R.: Presentation factors and driver characteristics affecting road-sign registration. Ergonomics **29**(6), 807–815 (1986)
9. Newell, A., Simon, H.: Human Problem Solving. Prentice Hall, Englewood Cliffs (1972)
10. Pas-Wyroślak, A., Siedlecka, J., Wyroślak, D., Bortkiewicz, A.: Znaczenie stanu narządu wzroku dla kierowcy. Łódź: Medycyna Pracy **64**(3), 419–425 (2013)
11. Prieler, J.: Reaction Test. RT Test Manual Version 31.00. Schuhfried GmbH, Vienna (2008)
12. Prieler, J.: Peripheral Perception. PP Test Manual Version 24. Schuhfried GmbH, Vienna (2009)
13. Schuhfried, G.: Adaptive Tachistoscopic Traffic Perception Test, ATAVT Test Manual version 22. Schuhfried GmbH, Vienna (2009)
14. Sanderson, J.E.: Driver recall of roadside signs. Traffic Research Report 1, Ministry of Transport. Wellington, New Zeland (1974)
15. Strelau, J.: Temperament, osobowość, działanie. PWN, Warszawa (1985)
16. Targosiński, T.: Koncepcja pomiarów eksploatacyjnych oznakowania dróg. Prace Instytutu elektrotechniki Zeszyt 237 (2008)
17. Ucińska, M.: Metodyka psychologicznych badań w zakresie psychologii transportu. ITS, Warszawa (2015)
18. Wagner, M., Karner, T.: Cognitrone. Manual Version 38. Schuhfried GmbH, Vienna (2008)
19. Zawadzki, B., Strelau, J.: Formalna Charakterystyka Zachowania – Kwestionariusz Temperamentu (FCZ-KT). Podręcznik. PTP, Warszawa (2010)

Operational System Modelling in a Focused Fire Alarm System with an Open and Signal Detection Circuit Supervising Railway Station Premises

Jacek Paś[1](✉) and Tomasz Klimczak[2]

[1] Faculty of Electronics, Military University of Technology, 00-908 Warsaw, Poland
jacek.pas@wat.edu.pl
[2] The Main School of Fire Service, 01-629 Warsaw, Poland

Abstract. Transport facilities commonly operate safety systems, fire alarm systems (FAS) in particular. According to a Regulation of the Minister of Interior and Administration of 7/6/2010 (Dz.U. No. 109, item 719), the use of FAS in Poland is required at railway stations and ports, designed for the presence of more than 500 people. Currently, designers and supervisors of operation of FAS located in railway areas do not include reliability indices, as no such requirements are present in the applicable standards and acts. The authors recommend taking selected safety indices into account already at the system design and operation stages, based on the methodology using Markov chains continuous over time and on the use of computer simulation in the Reliasoft Blocksim/Reno software. The example and calculations presented in the paper prove that this is possible. The developed FAS analysis method using the Markov process enables decision-making already at the design stage.

Keywords: Operation · Model · Fire alarm system

1 Introduction

The safety of transport facilities is equally important to the safety in transport. Transport facilities commonly use safety systems, especially fire alarm systems (FAS), acoustic warning systems, smoke venting systems and fixed extinguishing devices both aqueous and gaseous. The aforementioned safety systems supervise railway signal boxes, electrical switchgear, ICT containers on railway crossing, railway stations and platforms. In the age of technological progress involving railway facilities, each investment includes modern fire safety systems, mainly FAS integrated with numerous technical and fire-protection systems, as well as the Building Management System (BMS) and the intrusion detection system (IDS) [1, 4–6, 18].

Safety can be defined as a state providing a sense of confidence, lack of threat and lack of concerns regarding its loss in the future. It is one of the most important needs for humans, as well as for systems operated by the country and various organizations. There are various domestic requirements regarding system safety, such as requirements

M. Siergiejczyk and K. Krzykowska (Eds.): ISCT21 2019, AISC 1032, pp. 307–316, 2020.
https://doi.org/10.1007/978-3-030-27687-4_31

for the safety of electrical or fire safety systems [7, 8, 10, 11, 18]. The systems in railway facilities, which are responsible for fire safety are comprised of fire protection technical measures – Fig. 1 [15, 20, 21].

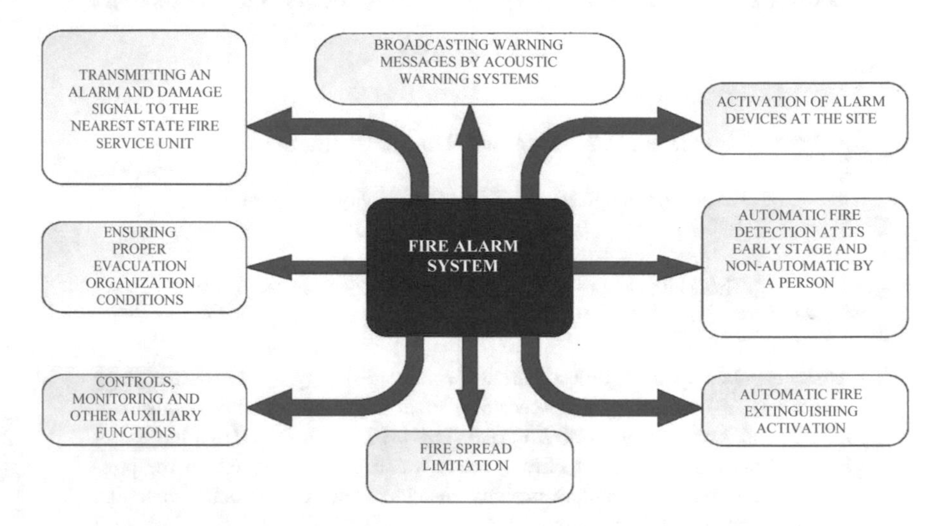

Fig. 1. Basic tasks of a fire alarm system within a transport infrastructure.

1.1 Focused Fire Alarm System with Open Detection Circuits

Several FAS types are distinguished, depending on the design, configuration and type of used linear elements [3, 5, 11, 18, 19, 22]. The application of a given FAS type depends on the legal requirements for such systems, a fire scenario, which must be implemented, legal requirements for a given transport facility subject to protection, the adopted protection scope and the functional-utility requirements, which must be satisfied by the system. The type of an FAS installed at a facility impacts its division into so-called detection zones. Detection zones should be determined also based on the fire risk within them, which requires a detailed analysis by the designer. According to the requirements, single, open detection circuits, called radial circuits, non-addressable or addressable, can supervise a fire zone area of up to 1600 m^2. The maximum number of rooms to be protected by a single open circuit is 10. It is allowed for 32 fire detector or up to 10 manual call points (MCP) to be installed within an open detection circuit. A fire alarm system with open, non-addressable detection circuits is used at small facilities, such as, e.g. small train stations or workplaces. Radial (open) circuits are marked as type B and are conventional – Fig. 2. In the case of an open detection circuit, a single break causes the elimination of linear elements, which are between the break and the end of the circuit. In an extreme case, i.e., with a break directly by the fire alarm control unit (FACU), all elements are eliminated [12, 18–21].

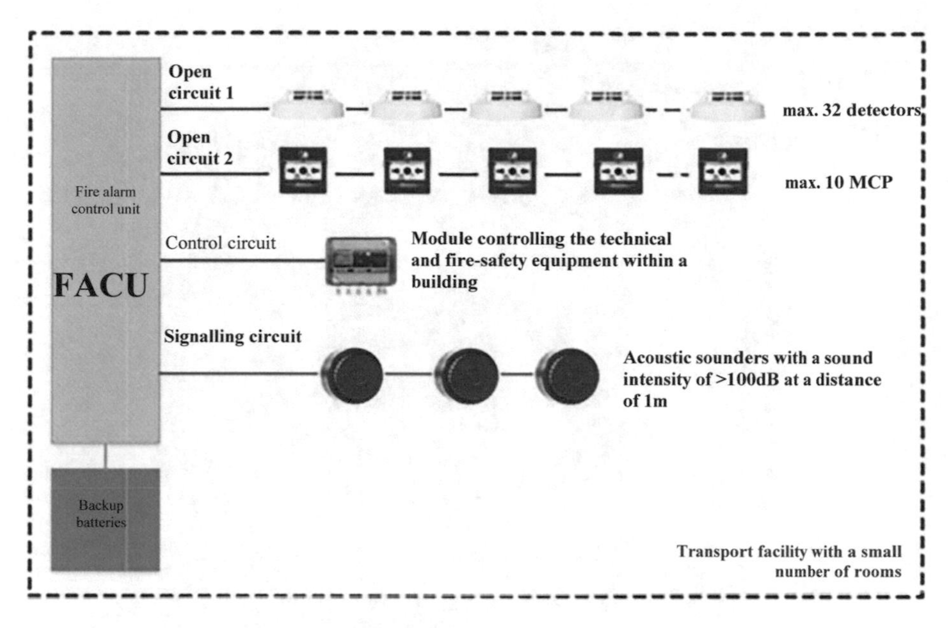

Fig. 2. Focused FAS, open detection circuits, no connection to the State Fire Service (SFS) notification system

2 System Operating Process Module

Operating a FAS is a set of deliberately taken organizational, technical and economic actions, occurring between the system and the facility it monitors. These relationships involve system elements from the moment of start-up (commissioning, acceptance) of a FAS, throughout the period of its operation as per its purpose, until it is decommissioned (liquidation) [2, 6, 8, 10, 13, 16, 22]. The reliability analysis of a fire alarm system focused with open detection circuits not connected to an SFS notification system is shown in Fig. 3 [12, 14, 17, 21]. The system shown in Fig. 3 can be described by the following Chapman–Kolmogorov equations:

$$\begin{gathered}
R_0'(t) = -\lambda_1 \cdot R_0(t) - \lambda_{11} \cdot R_0(t) - \lambda_{S1} \cdot R_0(t) - \lambda_{SP1} \cdot R_0(t) - \lambda_{CSP} \cdot R_0(t) + \mu_1 \cdot Q_{ZB1}(t) \\
+ \mu_{11} \cdot Q_{ZB2}(t) + \mu_{S1} \cdot Q_{ZB3}(t) + \mu_{SP1} \cdot Q_{ZBP1}(t) + \mu_{CSP} \cdot Q_B(t) \\
Q_{ZB1}'(t) = -\lambda_2 \cdot Q_{ZB1}(t) - \mu_1 \cdot Q_{ZB1}(t) + \lambda_1 \cdot R_0(t) + \mu_2 \cdot Q_B(t) \\
Q_{ZB2}'(t) = -\lambda_{22} \cdot Q_{ZB2}(t) - \mu_{11} \cdot Q_{ZB2}(t) + \lambda_{11} \cdot R_0(t) + \mu_{22} \cdot Q_B(t) \\
Q_{ZB3}'(t) = -\lambda_{S2} \cdot Q_{ZB3}(t) - \mu_{S1} \cdot Q_{ZB3}(t) + \lambda_{S1} \cdot R_0(t) + \mu_{S2} \cdot Q_B(t) \\
Q_{ZBP1}'(t) = -\lambda_{SP2} \cdot Q_{ZBP1}(t) - \mu_{SP1} \cdot Q_{ZBP1}(t) + \lambda_{SP1} \cdot R_0(t) + \mu_{SP2} \cdot Q_{ZBP2}(t) \\
Q_{ZBP2}'(t) = -\lambda_{SP3} \cdot Q_{ZBP2}(t) - \mu_{SP2} \cdot Q_{ZBP2}(t) + \lambda_{SP2} \cdot Q_{ZBP1}(t) + \mu_{SP3} \cdot Q_B(t) \\
Q_B'(t) = -\mu_2 \cdot Q_B(t) - \mu_{22} \cdot Q_B(t) - \mu_{S2} \cdot Q_B(t) - \mu_{SP3} \cdot Q_B(t) - \mu_{CSP} \cdot Q_B(t) \\
+ \lambda_2 \cdot Q_{ZB1}(t) + \lambda_{22} \cdot Q_{ZB2}(t) + \lambda_{S2} \cdot Q_{ZB3}(t) + \lambda_{SP3} \cdot Q_{ZBP2}(t) + \lambda_{CSP} \cdot R_0(t)
\end{gathered} \tag{1}$$

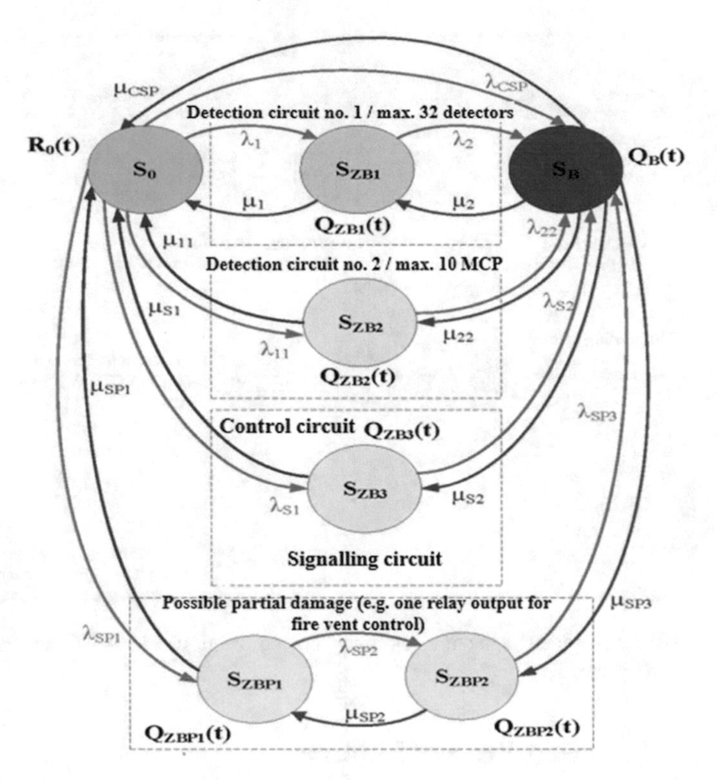

Fig. 3. Relations occurring within a focused system with a fire alarm central unit with connected open circuits with optical smoke detectors, an MCP open circuit, control circuit with a control module and a signalling circuits with sounders

Adopting the baseline conditions:

$$\begin{gathered} R_0(t) = 1 \\ Q_{ZBP1}(0) = Q_{ZBP2}(0) = Q_{ZB3}(0) = Q_{ZB2}(0) = Q_{ZB1}(0) = Q_B(0) = 0 \end{gathered} \tag{2}$$

where:

$R_0(t)$ – probability function of the system staying in the state of full fitness S_{PZ};
$Q_{ZB1}(t), Q'_{ZB1}(t), Q''_{ZB2}(t), Q''_{ZBP1}(t), Q''_{ZBP2}(t)$ – probability function of the system staying in individual safety hazard states;
$Q_B(t)$ – probability function of the system staying in the state of safety unreliability S_B;
Q_{ZB1} – the probability function of a system staying in a state of partial fitness I $\mathbf{S_{ZB1}}$ (damage to the detectors);
Q_{ZB2} – the probability function of a system staying in a state of partial fitness II $\mathbf{S_{ZB2}}$ (damage to manual call points);
Q_{ZB3} – the probability function of a system staying in a state of partial fitness III $\mathbf{S_{ZB3}}$ (damage to the control module);

Q_{ZBP1}, Q_{ZBP2} – the probability function of a system staying in a state of partial fitness IV $\mathbf{S_{ZBP1}}$, $\mathbf{S_{ZBP2}}$ (damage to acousto-optical sirens);
λ_{CSP} – intensity of transitions from the state of full fitness S_{PZ} to the state of safety unreliability S_B for a fire alarm central unit;
μ_{CSP} – intensity of transitions from the state of safety unreliability S_B to the state of full fitness S_{PZ} for a fire alarm central unit;
$\lambda_1, \lambda_{11}, \ldots$ – intensity of transitions from the state of full fitness S_{PZ} or safety hazard S_{ZB} to the state of full fitness, safety hazard or safety unreliability S_{ZB} – according to the designation on the graph;
$\mu_1, \mu_{11}, \ldots$ – intensity of transitions from the state of safety hazard S_{ZB} to the state of full fitness S_{PZ}, from the state of safety unreliability to the state of safety hazard or the state of full fitness to the state of safety hazards – according to the designation on the graph.

Applying the Laplace transform, the following system of linear equations is obtained:

$$\begin{gathered}
s \cdot R_0^*(s) - 1 = -\lambda_1 \cdot R_0^*(s) - \lambda_{11} \cdot R_0^*(s) - \lambda_{S1} \cdot R_0^*(s) - \lambda_{SP1} \cdot R_0^*(s) - \lambda_{CSP} \cdot R_0^*(s) \\
+ \mu_1 \cdot Q_{ZB1}^*(s) + \mu_{11} \cdot Q_{ZB2}^*(s) + \mu_{S1} \cdot Q_{ZB3}^*(s) + \mu_{SP1} \cdot Q_{ZBP1}^*(s) + \mu_{CSP} \cdot Q_B^*(s) \\
s \cdot Q_{ZB1}^*(s) = -\lambda_2 \cdot Q_{ZB1}^*(s) - \mu_1 \cdot Q_{ZB1}^*(s) + \lambda_1 \cdot R_0^*(s) + \mu_2 \cdot Q_B^*(s) \\
s \cdot Q_{ZB2}^*(s) = -\lambda_{22} \cdot Q_{ZB2}^*(s) - \mu_{11} \cdot Q_{ZB2}^*(s) + \lambda_{11} \cdot R_0^*(s) + \mu_{22} \cdot Q_B^*(s) \\
s \cdot Q_{ZB3}^*(s) = -\lambda_{S2} \cdot Q_{ZB3}^*(s) - \mu_{S1} \cdot Q_{ZB3}^*(s) + \lambda_{S1} \cdot R_0^*(s) + \mu_{S2} \cdot Q_B^*(s) \\
s \cdot Q_{ZBP1}^*(s) = -\lambda_{SP2} \cdot Q_{ZBP1}^*(s) - \mu_{SP1} \cdot Q_{ZBP1}^*(s) + \lambda_{SP1} \cdot R_0^*(s) \\
+ \mu_{SP2} \cdot Q_{ZBP2}^*(s) \\
s \cdot Q_{ZBP2}^*(s) = -\lambda_{SP3} \cdot Q_{ZBP2}^*(s) - \mu_{SP2} \cdot Q_{ZBP2}^*(s) + \lambda_{SP2} \cdot Q_{ZBP1}^*(s) \\
+ \mu_{SP3} \cdot Q_B^*(s) \\
s \cdot Q_B^*(s) = -\mu_2 \cdot Q_B^*(s) - \mu_{22} \cdot Q_B^*(s) - \mu_{S2} \cdot Q_B^*(s) - \mu_{SP3} \cdot Q_B^*(s) \\
- \mu_{CSP} \cdot Q_B^*(s) + \lambda_2 \cdot Q_{ZB1}^*(s) + \lambda_{22} \cdot Q_{ZB2}^*(s) + \lambda_{S2} \cdot Q_{ZB3}^*(s) \\
+ \lambda_{SP3} \cdot Q_{ZBP2}^*(s) + \lambda_{CSP} \cdot R_0^*(s)
\end{gathered} \tag{3}$$

Continuing the transformations gives:

$$\begin{gathered}
(s + \lambda_1 + \lambda_{11} + \lambda_{S1} + \lambda_{SP1} + \lambda_{CSP}) \cdot R_0^*(s) - \mu_1 \cdot Q_{ZB1}^*(s) - \mu_{11} \cdot Q_{ZB2}^*(s) - \mu_{S1} \cdot Q_{ZB3}^*(s) \\
- \mu_{SP1} \cdot Q_{ZBP1}^*(s) - \mu_{CSP} \cdot Q_B^*(s) = 1 \\
(s + \lambda_2 + \mu_1) \cdot Q_{ZB1}^*(s) - \lambda_1 \cdot R_0^*(s) - \mu_2 \cdot Q_B^*(s) = 0 \\
(s + \lambda_{22} + \mu_{11}) \cdot Q_{ZB1}^*(s) - \lambda_{11} \cdot R_0^*(s) - \mu_{S2} \cdot Q_B^*(s) = 0 \\
(s + \lambda_{S2} + \mu_{S1}) \cdot Q_{ZB3}^*(s) - \lambda_{S1} \cdot R_0^*(s) - \mu_{S2} \cdot Q_B^*(s) = 0 \\
(s + \lambda_{SP2} + \mu_{SP1}) \cdot Q_{ZBP1}^*(s) - \lambda_{SP1} \cdot R_0^*(s) - \mu_{SP2} \cdot Q_{ZBP2}^*(s) = 0 \\
(s + \lambda_{SP3} + \mu_{SP2}) \cdot Q_{ZBP2}^*(s) - \lambda_{SP2} \cdot Q_{ZBP1}^*(s) - \mu_{SP3} \cdot Q_B^*(s) = 0 \\
(s + \mu_2 + \mu_{22} + \mu_{S2} + \mu_{S3} + \mu_{CSP}) \cdot Q_B^*(s) - \lambda_2 \cdot Q_{ZB1}^*(s) - \lambda_{22} \cdot Q_{ZB2}^*(s) \\
- \lambda_{S2} \cdot Q_{ZB3}^*(s) - \lambda_{SP3} \cdot Q_{ZBP2}^*(s) - \lambda_{CSP} \cdot R_0^*(s) = 0
\end{gathered} \tag{4}$$

3 Operational Process Study

The study involved FAS operated in transport facilities, and Tables 1 and 2 show the types of damage and repairs regarding the representative systems, over one operating year [7, 12, 19, 21].

Table 1. Studying the operational process of FASs used in transport facilities

No.	Damage type	Failure time	Failure removal time	Repair duration	Repair type
1	Failure of detector 3/57	19/1/2018 08:10	19/1/2018 15:50	7h 40 min.	Central unit reset
2	Ground fault of loop no. 1	02/05/2018 13:30	02/05/2018 19:00	5h 30 min.	Central unit reset
n-1	....	....	....	....	....
n	Circuit no. 2 interference	11/03/2018 15:00	11/03/2018 16:30	1h 30 min.	Central unit reset

Table 1 shows representative types of damage for selected FAS. The data were compiled based on a set of damage for n = 20 systems, operated over a vast transport area. A maximum repair time was assumed for a given type of FAS damage. The repair time does not include the service personnel travel time (in the case of such FAS, such personnel is on site). Table 2 groups FAS repair types characterized by the same message on the fire unit control panel.

Table 2. Repair types with a maximum repair time T_{max} annualized

No.	Repair for a given damage type	Failure time	Failure removal time	Maximum repair time [T_{max}]
Detection loop damage				
1.	Circuit no. 3 interference	03/01/2018 14:32	03/01/2018 18:10	3h 38 min.
2.	Circuit no. 2 interference	11/03/2018 15:00	11/03/2018 16:30	1h 30 min.
....				
Detector damage				
1.	Failure of detector 2/11	05/09/2018 11:00	05/09/2018 16:10	5h 10 min.
2.	Failure of detector 3/51	11/10/2018 05:30	11/10/2018 09:30	4h

4 Validation of the Operational Process Model in the Reliasoft Blocksim/Reno Software

An FAS operational process chain, which is discrete in terms of states and continuous over time can be described using Markov equations through graphs of transitions between distinguished system states. The transitions between the states are defined by (constant) transition intensities [3, 4, 9, 11, 15]. It is common to use time-continuous Markov chains when analysing the issues of system reliability or availability. The BlockSim software was used to conduct the calculations of the availability and reliability for an FAS staying in individual states, using a model as per the assumed operation time of 8760 h and adopted repair and restoration time values resulting from Tables 1, 2 and Figs. 4, 5.

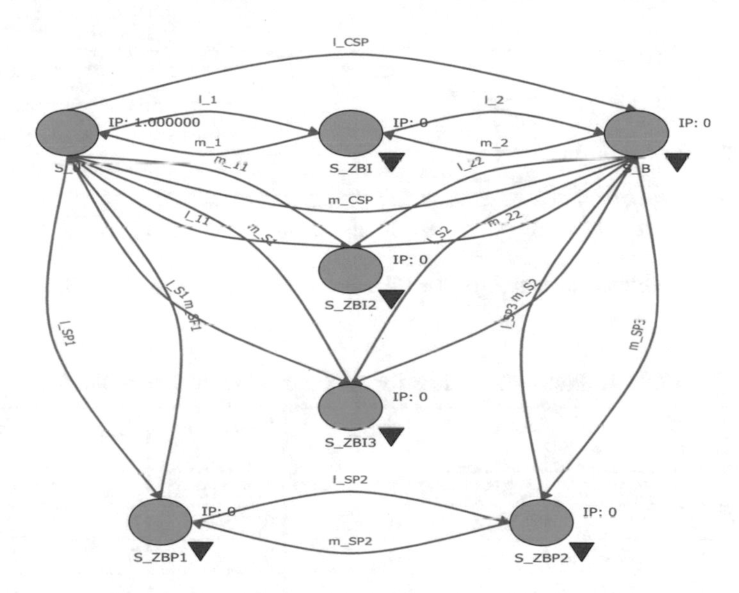

Fig. 4. Transition matrix for the adopted FAS model

Table 3. Calculation values for individual FAS states

Results for a time of 8760 h					
State	Initial probability	Average probability	Availability for time t	Reliability for time t	Time in state
$S_{\_0}$	1	0,99999344	0,999993439	0,991489928	8759,94257
$S_{\_B}$	0	2,245E−07	2,24528E−07	0,001526641	0,001966621
$S_{\_ZBI}$	0	3,75408E−06	3,75731E−06	0,003920964	0,032885763
$S_{\_ZBI2}$	0	8,25865E−07	8,26355E−07	0,001033117	0,00723458
$S_{\_ZBI3}$	0	7,10979E−07	7,11386E−07	0,000996234	0,006228174
$S_{\_ZBP1}$	0	8,16121E−07	8,16726E−07	0,001033117	0,007149221
$S_{\_ZBP2}$	0	2,24374E−07	2,24529E−07	0	0,001965516

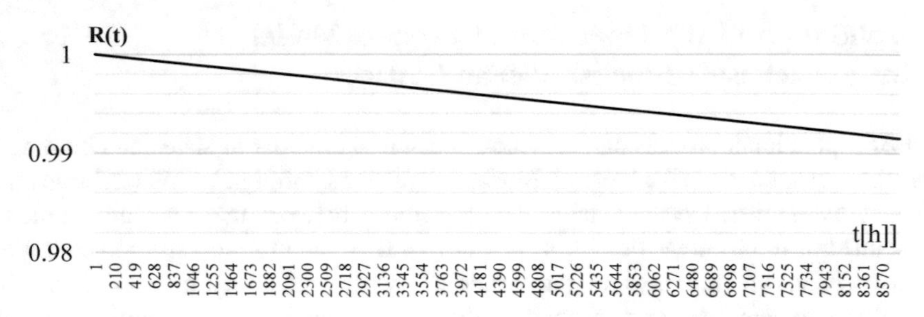

Fig. 5. Reliability R(t) for a focused FAS

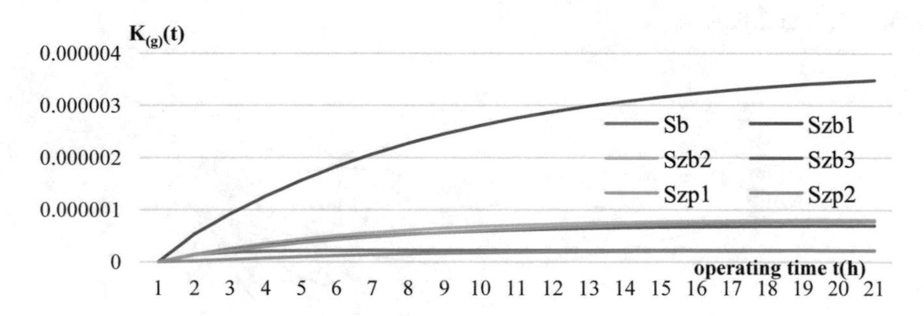

Fig. 6. Availability coefficient $K_g(t)$ for FAS states: S_B, S_{ZB1}, S_{ZB2}, S_{ZB3}, S_{ZP1}, S_{ZP2} SSP

Table 4. Probability values for FAS individual state transitions.

From to	S_O	S_B	S_{ZBI}	S_{ZBI2}	S_{ZBI3}	S_{ZBP1}	S_{ZBP2}
S_O	–	1,750E−07	4,4951E−07	1,184E−07	1,1421E−07	1,184E−07	0
S_B	0,0759	–	0,1818	0,1968	0,125	0	0,2
S_{ZBI}	0,1305	2,5296E−07	–	0	0	0	0
S_{ZBI2}	0,1968	5,7091E−08	0	–	0	0	0
S_{ZBI3}	0,2	1,4161E−08	0	0	–	0	0
S_{ZBP1}	0,2	0	0	0	0	–	1,18E−07
S_{ZBP2}	0	1,4161E−08	0	0	0	0,2	–

The simulation conducted in the Reliasoft BlockSim/Reno software involving a selected FAS enables determining the values of average probabilities (P_{avg}) for the system staying in given states – Table 3 (e.g. P_{avg} = 0,99999344 for the state S_O – full system fitness). The P_{avg} values for other FAS states are at a very low level, in the order of $10^{(-6 \div -7)}$. In the case of the considered FAS operation time (1 year), the FAS stayed in the state of full fitness to execute an operating task for a period of 8759, 94257 (h). This means that the adopted FAS operational process variant was correct [9, 12, 14, 20]. Figure 6 shows the growth process of availability coefficients for individual FAS states at the initial operational process stage. The dominating state during the initial operation time is S_{Zb1}. Table 4 shows the transition probability values for individual FAS states. The maximum probability values were achieved for the following transitions: $S_{ZBI3} \rightarrow S_0$, $S_{ZBP1} \rightarrow S_0$, $S_{ZBP2} \rightarrow S_{ZBP1}$ – and is 0,2. In the case of times of operation above t = 30 (h), the $K_g(t)$ for the distinguished states achieve constant values, - at a level from 0,00004 – to 0,000065 – Fig. 7.

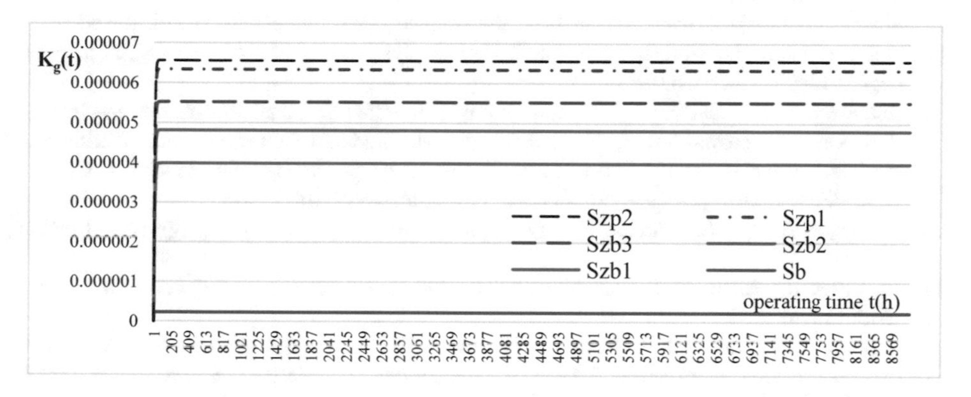

Fig. 7. Availability coefficient $K_g(t)$ for FAS states: S_B, S_{ZB1}, S_{ZB2}, S_{ZB3}, S_{ZP1}, S_{ZP2}

5 Conclusions

The basic operational state of transport FAS should be the S_{PZ} state of full fitness. In this state, the FAS executes all operating tasks arising from the developed fire scenario. Conducted simulations confirmed the fact that the system stayed within the distinguished state for practically 100% of the time of examination. The availability coefficients $K_g(t)$ for the distinguished FAS states are at a very low level – in the order of $(4 - 6{,}5) \cdot 10^{-6}$. The FAS achieves maximum value of the probability for a transition from the distinguished, various states – e.g. $S_{ZBI3} \rightarrow S_0$, $S_{ZBP1} \rightarrow S_0$. The availability coefficient $K_g(t)$ for a period of t (8760 h) and a considered FAS achieved the maximum value for state S_0 (full fitness), $K_g(t)$ = 0,999993439.

References

1. Dyduch, J., Paś, J., Rosiński, A.: The Basic of the Exploitation of Transport Electronic Systems. Publishing House of Radom University of Technology, Radom (2011)
2. Garmabaki, A.H.S., Ahmadi, A., Mahmood, Y.A., Barabadi, A.: Reliability modelling of multiple repairable units. Qual. Reliab. Eng. Int. **32**(7), 2329–2343 (2016)
3. Łubkowski, P., Laskowski, D.: Selected issues of reliable identification of object in transport systems using video monitoring services. In: Communication in Computer and Information Science, vol. 471, pp. 59–68. Springer, Heidelberg (2015)
4. Paś, J.: Operation of Electronic Transportation Systems. Publishing House University of Technology and Humanities, Radom (2015)

5. Rosiński, A.: Modelling the Maintenance Process of Transport Telematics Systems. Publishing House Warsaw University of Technology, Warsaw (2015)
6. Warczek, J., Młyńczak, J., Celiński, I.: Simulation studies of a shock absorber model proposed under conditions of different kinematic input functions. Vibroeng. Procedia **6**, 248–253 (2015)
7. Kaniewski, P., Gil, R., Konatowski, S.: Estimation of UAV position with use of smoothing algorithms. Metrol. Measur. Syst. **24**(1), 127–142 (2017)
8. Siergiejczyk, M., Paś, J., Rosiński, A.: Issue of reliability–exploitation evaluation of electronic transport systems used in the railway environment with consideration of electromagnetic interference. IET Intell. Transp. Syst. **10**(9), 587–593 (2016). https://doi.org/10.1049/iet-its.2015.0183
9. Stawowy, M.: Model for information quality determination of teleinformation systems of transport. In: Nowakowski, T., Młyńczak, M., Jodejko-Pietruczuk, A., Werbińska–Wojciechowska, S. (eds.) Proceedings of the European Safety and Reliability Conference ESREL 2014, pp. 1909–1914. CRC Press/Balkema (2015)
10. Verma, A.K., Ajit, S., Karanki, D.R.: Reliability and Safety Engineering. Springer, London (2010)
11. Billinton, R., Allan, R.N.: Reliability Evaluation of Power Systems. Plenum Press, New York (1996)
12. Zajkowski, K., Rusica, I., Palkova, Z.: The use of CPC theory for energy description of two nonlinear receivers. In: MATEC Web of Conferences, IManE&E 2018, vol. 178, p. 09008 (2018). 10.1051/matecconf/201817809008
13. Dziula, P., Paś, J.: The impact of electromagnetic interferences on transport security system of certain reliability structure. In: 12th International Conference on Marine Navigation and Safety of Sea Transportation TransNav, Gdynia, Poland, pp. 185–191 (2017)
14. Chen, S., Ho, T., Mao, B.: Maintenance schedule optimisation for a railway power supply system. Int. J. Prod. Res. **51**(16), 4896–4910 (2013)
15. Paś, J., Rosiński, A., Wiśnios, M., Majda-Zdancewicz, E., Łukasiak, J.: Electronic Security Systems. Introduction to the Laboratory. Military University of Technology, Warsaw (2018)
16. Jin, T.: Reliability Engineering and Service. Wiley, New York (2019)
17. Siergiejczyk, M., Paś, J., Dudek, E.: Reliability analysis of aerodrome's electronic security systems taking into account electromagnetic interferences. In: Safety and Reliability – Theory and Applications, 27th European Safety and Reliability Conference ESREL 2017, Portoraź, Słowenia, pp. 2285–2292 (2017)
18. Paś, J., Rosinski, A.: Selected issues regarding the reliability-operational assessment of electronic transport systems with regard to electromagnetic interference. Eksploatacja i Niezawodnosc – Maint. Reliab. **19**(3), 375–381 (2017)
19. Krzykowski, M., Paś, J., Rosiński A.: Assessment of the level of reliability of power supplies of the objects of critical infrastructure. In: IOP Conference Series: Earth and Environmental Science, vol. 214, pp. 1–9, 012018 (2019). https://doi.org/10.1088/1755-1315/214/1/012018
20. Lewiński, A., Perzyński, T., Toruń, A.: The analysis of open transmission standards in railway control and management. In: Communications in Computer and Information Science, vol. 329, pp. 10–17. Springer, Heidelberg (2012)
21. Klimczak, T., Paś, J.: Analysis of reliability structures for fire signaling systems in the field of fire safety and hardware requirements. J. KONBIN **46**, 191–214 (2018)
22. Duer, S., Scaticailov, S., Paś, J., Duer, R., Bernatowicz, D.: Taking decisions in the diagnostic intelligent systems on the basis information from an artificial neural network. In: 22nd International Conference on Innovative Manufacturing Engineering and Energy - IManE&E 2018. MATEC Web of Conferences, vol. 178, pp. 1–6, 07003 (2018). https://doi.org/10.1051/matecconf/201817807003

Rail Transport Systems Safety, Security and Cybersecurity Functional Integrity Levels

Marek Pawlik(✉)

Railway Research Institute, 50 Chłopickiego str., Warsaw, Poland
mpawlik@ikolej.pl

Abstract. Observed growing changes in the character and severity of the risks in rail traffic safety and rail transport security are associated with present development of utilized technical solutions. New hazards are coming out, besides known ones, including hazards associated with cyber-crime. As a result it is fully justified to undertake works dedicated to collect and settle all risks associated with technical solutions using modern technologies for acquisition, computing and transfer of the data, which are vital from the rail traffic safety and rail transport security point of view. Article defines rail transport systems safety, security and cybersecurity functional integrity levels thanks to knock-out and differentiating questions regarding identified key safety related functionalities. Proposed methodology was used for safety and security verification of a chosen homogenous rail transport system separated from the overall Polish railway system. Results have shown discrepancies in utilized protection measures. Proposed methodology can be used for assessment of existing systems as well as for specifying scopes of investments both for infrastructure and rolling stock modernizations. Applicability range covers railway transport, light rail services, metro, urban rail transport systems as well as rail based transport systems using autonomous vehicles.

Keywords: Rail traffic safety · Rail transport security · Railway cybersecurity

1 Introduction

1.1 Technical Safety and Personal Security

It is unambiguous, that all technical solutions, which are utilized by railway transport, are being defined, constructed and maintained having safety in mind. It is not so easy to declare, that railway system with all its solutions is safe, as incidents and accidents happen since time to time even if the amount of them is relatively low. Severity of the railway accidents is frequently high and therefore it is a real challenge to define how safe the railway system has to be declared as a safe one.

From the very beginning of the railway history safety was based on technical solutions and procedures. Technical solutions were, and still are, constructed in a way ensuring safe operation in case of errors, faults, failures, which may cause system malfunctioning. As degraded operational circumstances do appear during long lifecycles of the railway solutions it was, and still is, required to apply fail-safe principle. It means, that neither faults or failures nor errors or extreme external conditions, e.g.

M. Siergiejczyk and K. Krzykowska (Eds.): ISCT21 2019, AISC 1032, pp. 317–329, 2020.
https://doi.org/10.1007/978-3-030-27687-4_32

temperatures or loads, can lead to dangerous situations understood mainly as giving the train permission to run too far or too quick. As a result faults, failures, errors and extreme external conditions thanks to own inherent characteristics of the solutions lead to shifting responsibility from technical systems to staff and procedures in degraded operational circumstances.

Applying fail-safe principle was, and still is, not appropriate for some technical solutions – for ensuring appropriate endurance of the tracks, embankments, bridges, as well as vehicles' bodies, running gears and auxiliary systems. In that respect safety since the beginning was, and still is, based on widely accepted codes of practice e.g. UIC leaflets, EN standards, OTIF specifications.

Moreover applying fail-safe principle is not appropriate for technical solutions which are utilizing electronic, programmable systems and modules for which huge catalogue of possible faults, failures and errors forefend verifications of the appropriateness and completeness of the fail-safe principle. That mainly applies to railway control command and signaling equipment.

It was obvious from the beginning, that technical safety is a must, but insufficient requirement. Railways had, and still have, to ensure personal safety on stations and in trains. They had, and sill have, to ensure safety of the cargo as well as minimized undesirable influence on environment. Up to the beginning of the twenty first century that was ensured by procedures and dedicated staff – by railway police. Presently they are more and more supported by electronic equipment, and therefore relationships between technical safety and personal & goods security starts to be blotted.

1.2 Railway Traffic Safety and Rail Transport Security

Presently technical experts responsible for railway infrastructure – for tracks, switches and crossings as well as underlying embankments, bridges and viaducts as well as accompanying platforms, pathways and station buildings and their accoutrements apply nearly only technical solutions, which are fully in accordance with abovementioned codes of practice e.g. Vignole railway rails fully compliant with EN 13674 standard series. To large extend approach based on codes of practice apply also for traction power supply and rolling stock as adequate technical documents directly define requirements for materials and constructions as well as for verifications of the final solutions.

Codes of practice also exist for control command and signaling equipment installed both trackside and onboard. However in that respect technical documents define functionalities and safety verification rules generally omitting materials and the way how to construct technical solutions. This is largely linked to relatively quick development of technical solutions utilized for control command and signaling. As a result fail-safe principle and codes of practice are not sufficient for ensuring technical safety. Since twenty years in that respect railway signaling solutions respect so called Safety Integrity Levels defined by EN standards dedicated to reliability, availability, maintainability and safety – RAMS standards [3 ÷ 7]. Control command and signaling systems and devices inherent safety is assessed in case of electronic, digital and programmable solutions against **Safety Integrity Level SIL-4** defined by tolerable hazard rates $10^{-8} \div 10^{-9}$ for hazardous events per hour, which may be caused by random failures and external factors and recommended technics for minimization of hazards associated with human mistakes. The SIL based approach is required for control command under interoperability directive [2], which is applicable to new technical solutions. At the same time risk based approach is required under railway safety directive [1]. Both however omit security challenges, which the railways are facing all the time.

Railways are using presently technical solutions based on electronic, digital, programmable components not only for control command and signaling but also for supporting security staff – for generating emergency alerts, video monitoring, communication, supporting rescue and evacuation. In that respect SIL-4 based approach is not required. As a result it is reasonable to ask whether safety and security are fully ensured and whether they are ensured similarly well in relation to different risks which are presently common. Safety and security impact reference model was defined to answer those questions.

2 Safety and Security Impact Reference Model SSIRM

Safety and security impact reference model SSIRM is based on identification of functions which are supported by electronic, digital and programmable solutions. Safety is treated not only as an overall requirement which has to be ensured in normal operation but also as inherent characteristics of the technical solutions which ensure safety in degraded modes of operation. Key functionalities, understood as groups of individual functions, which influence railway traffic safety are shown at Fig. 1.

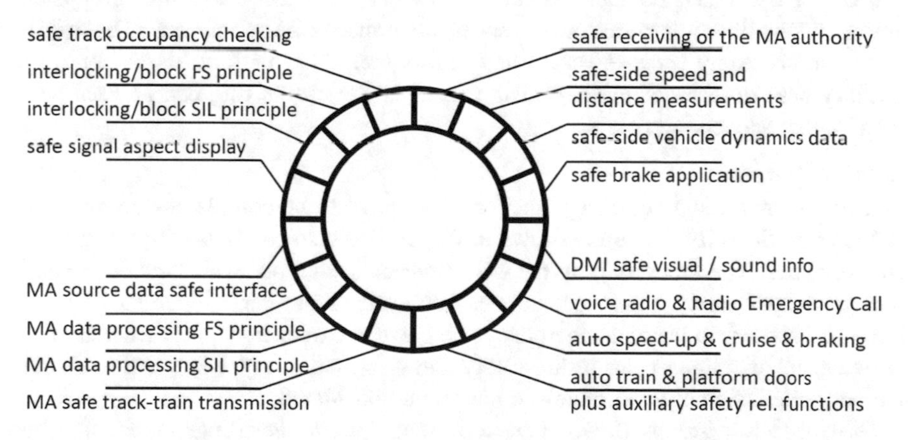

Fig. 1. Safety and security impact reference model SSIRM, safety related functionalities. FS – fail-safe principle, SIL – safety integrity level SIL-4 principle, MA – digital movement authority, DMI – driver machine interface/own elaboration/

Safety Domain

Sixteen functionalities can be subdivided into five groups. The first group (upper left) represents classic signaling equipment installed trackside. The interlocking and block systems are due to respect fail-safe principle and ensure Safety Integrity Level SIL-4. This would however not ensure safety if track occupancy checking or displaying signal aspects do not respect safety rules. Applying fail-safe and SIL-4 principles are required for individual solutions, however overall verification has to be performed taking into account dependences between elements working together.

The second group (lower left) represent control command equipment installed trackside. Its role is to prepare an electronic movement authority for the train on the basis of data taken from signaling equipment. The way how the data is taken cannot change the data in the wrong side neither on signaling nor on control command side. Data processing has to respect both fail-safe and SIL-4 principles. Prepared movement authority has to be sent in a safe way. Data acquisition, data processing and data transmission have to technically safe in all operational circumstances.

The third group (upper right) represents control command equipment installed onboard of the traction vehicles. Movement authorities have to be received, verified and respected. Obtaining digital movement authority is worth only if onboard equipment ensures reliable and safe information about relationship between location and speed of the train in relation to the braking curves imposed by authorities taking into account distance and speed measurements as well as considering vehicle dynamics. From safety point of view it is necessary to ensure also safe application of the brakes taking into account both full service brake and emergency brake.

Digital movement authority available onboard can be utilized by manually operated trains and by automatically operated vehicles. Therefore the four remaining functionalities are subdivided into two groups (lower right). Group fourth representing cab signaling by visual and audible information as well as radio communication between train driver and trackside staff together with emergency calls. The fifth group representing automatic driving – automatic speed-up, cruise and braking as well as functions which are necessary for safe access and egress including platform doors control and auxiliary auto train functionalities. The sixteen fields on the ring can be used to represent safety aspects by colors.

Security Domain

Control command and signaling functionalities have to be complemented by security related ones. Security domain is shown at Fig. 2. It is also composed by five groups of functionalities. The first group (upper left) represents solutions which are due to ensure basic passenger safety. Passenger information systems are very important both trackside and onboard in case of emergency, to prevent panic, to support evacuation etc. Railways are also due to ensure fire safety and electrical safety. More and more that is also supported digitally and therefore has to be considered.

The following groups (lower left) represent systems ensuring protection against crime and vandalism as well as enhanced protection for passenger health. The second group covers solutions enabling passing alarms to dedicated staff and video monitoring systems. The third group covers emergency call installations as well as medical equipment like e.g. automatic external defibrillators AED.

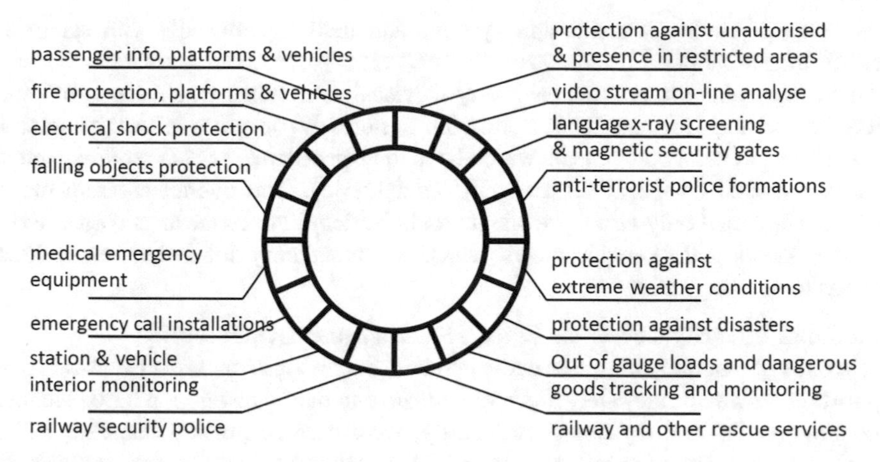

Fig. 2. Safety and security impact reference model e-SSIRM, security related functionalities/ own elaboration/

The fourth group (upper right) is representing enhanced technical means against crime, vandalism and terrorism. It covers protection against unauthorized access and presence based on simple solutions as well as video stream analyzers, which are able to detect persons entering restricted areas e.g. passing from one platform to another over tracks, unattended luggage, fake crowd, running persons, etc. As a result appropriate information is automatically identified and communicated to security staff immediately providing chance to react in due time and not only to document hazardous events for further investigations. Additionally stations and platforms can be protected by language screening and magnetic gates. That is already in use in some cases, however it is questionable especially in case of traffic in agglomerations.

The fifth group (lower right) represents technical protection means against natural and construction disasters. Also such technical solutions are already in use in some places e.g. in case of high speed lines going through seismic areas, in case of long railway tunnels and so on.

3 Enhanced Impact Reference Model e-SSIRM and Safety, Security and Cybersecurity Functional Integrity Levels FIL

Safety of the Data Transmission

The simple SSIRM model does not fully take into account data transmissions. As a result of considerations regarding cybersecurity it was therefore enhanced by adding data transmissions in a form of connections between different fields on the safety ring and between different fields on the security ring. Adding connections representing data transmission systems on one side enables showing safety by colors similarly to the fields. On the other side adding connections provides flexibility which ensures easy way for showing different transmission media arrangements covering individual transmission systems and complex transmission systems serving different functionalities as well as wired and wireless transmission systems and their relationships with safety and security functionalities.

Such representation enables easy way to visualize systems which have to respect safety related requirements for transmission systems [7] together with their relationships with equipment components which have to respect safety related requirements for hardware modules [6] and for software modules [5]. The overall assessments are however supported only visually while it would be helpful to create an add on to Safety Integrity Levels SIL-4 requirements which are mandatory for individual technical systems supporting safety.

Functional Integrity Levels for Safety, Security and Cybersecurity
Ten groups of functionalities, five dedicated to data transmission based technical means supporting railway traffic safety and five dedicated to data transmission based technical means supporting railway transport security, were used to prepare sets of questions containing two types of questions – knock-out questions and differentiating questions. The knock-out question can receive value "0" and value "1". The differentiating question can receive values "1" and value "2". Additional set of questions was prepared for transmission systems to reflect resistance against internal malfunctioning, extreme external conditions and cyber-crime. Also in that respect knock-out and differentiating questions were defined.

Table 1. Examples of the knock-out questions – knock-out questions for safety.

1	Whether all tracks with their full length are covered by track occupancy systems?
2	Whether all track occupancy systems fully apply fail-safe rules?
3	Whether all track occupancy systems utilizing electronic, programmable or simply digital solutions have safety cases elaborated by producers and verified by independent safety assessors proving SIL-4?
4	Whether all station interlockings, line block systems and level crossing protection systems fully apply fail-safe rules?
5	Whether all station interlockings, line block systems and level crossing protection systems have safety cases elaborated by producers and verified by independent safety assessors proving SIL-4?
6	Whether all track sections are protected by visible track-side signals or visualised in the cabs on-board in all trains permitted to run?
7	Whether all signal types are included in safety cases elaborated by producers and verified by independent safety assessors proving SIL-4?
8	Whether all technical solutions for data acquisition from interlockings, line block systems and level crossing protection systems (used for digital movement authorities) do not influence, under any foreseeable circumstances, safety of the interlockings, block systems and level crossing protection?
9	Whether all technical solutions used for data acquisition from interlockings, line block systems and level crossing protection systems fully apply fail-safe rules?
10	Whether all technical solutions for data acquisition from interlockings, line block systems and level crossing protection systems have safety cases elaborated by producers and verified by independent safety assessors proving SIL-4?

(*continued*)

Table 1. (*continued*)

11	Whether all technical solutions for data processing for electronic movement authorities have safety cases elaborated by producers and verified by independent safety assessors proving SIL-4?
12	Whether all messages, especially those containing movement authorities, contain data enabling sender authentication?
13	Whether all messages, especially those containing movement authorities, contain data enabling verification of validity?
14	Whether on-board control command equipment verifies sender authentications and message validities of all received messages, especially those containing movement authorities?
15	Whether on-board control command equipment properly estimates and takes into account errors of the distance and speed measurements?
16	Whether on-board control command equipment properly estimates and takes into account train braking dynamics?
17	Whether all on-board control command elements are fully taken into account in safety cases elaborated by producers and verified by independent safety assessors proving SIL-4?
18	Whether on-board control command equipment provides drivers with visualisation of the electronic movement authorities? or Whether on-board equipment automatically changes speed using on-board control command elements?
19	Whether voice radio communication is provided for drivers and signalmen? or Whether on-board equipment automatically controls and commands on-board and trackside equipment (e.g. pantograph, main circuit-breaker, train doors and platform doors)?
20	Each knock-out question may have positive answer (YES = 1) or negative answer (NO = 0). The overall value is a product of all of them. Even a single negative answer is a knock-out for safety of a whole solution

Table 2. Differentiating questions for safety, security, cybersecurity.

Safety:	
1	Whether control command messages contain data which are used by on-board control command equipment for verification of completeness and coherency of all received messages?
2	Whether on-board control command equipment verifies cryptographic protection of all received messages?
3	Whether drivers are informed by control command about latest places for starting braking and warned before equipment interventions?
4	Whether automatic braking interventions are using more than one braking mode (full service braking and emergency braking)?
5	Whether receiving emergency signal automatically initiates braking which ensures stopping in a place appropriate for evacuation and for security and rescue staff interventions?

(*continued*)

Table 2. (*continued*)

Security:	
1	Whether emergency medical equipment, especially automated external defibrillators, are available in all stations in areas accessible for passengers and provides with appropriate signs and instructions?
2	Whether video-monitoring system used for providing security is equipped with video-stream analyser ensuring immediate generation of security warnings?
3	Whether luggage scanning is provided?
4	Whether protection against possible natural disasters is provided?
5	Whether tracking of dangerous goods is provided?
Cybersecurity:	
1	Whether in case of detecting loss of communication for signalling automatic reconfiguration of communication system takes place or automatic switch on of the backup communication system takes place to ensure traffic control by technical means (and not only procedures)?
2	Whether safety related personnel is equipped with backup communication means?
3	Whether in case of detecting loss of communication for control command automatic reconfiguration of communication system takes place or automatic switch on of the backup communication system takes place to ensure train running supervision?
4	Whether in case when control command system is out of order trains can be driven on the basis of the signal aspects displayed on the track-side signals?
5	Whether technical systems and devices supporting security, especially video-monitoring systems are provided with backup power supply?

All questions are defined and described in a dedicated monograph [8].

Knock-out questions for safety are shown in Table 1, as an example. Differentiating questions for safety, security and cybersecurity are shown in Table 2. Each knock-out question may have positive answer (YES = 1) or negative answer (NO = 0). The overall value is a product of all of them. Even a single negative answer is a knock-out for safety of a whole solution. Each differentiating question also may have positive or negative answer, however in this case YES = 2 while NO = 1. Safety, security and cybersecurity are therefore represented by a vector.

$$[safety, security, cybersecurity] \tag{1}$$

in other notation

$$[SF, SC, CS] \tag{2}$$

where:

SF – product of all answers regarding safety,
SC – product of all answers regarding security,
CS – product of all answers regarding cybersecurity.

The Functional Integrity Level for safety, security and cybersecurity, FIL level, is defined as a sinus of an angle between vector and reference geometrical plane, for which maximum vector is perpendicular.

$$FIL_{SF,SC,CS} = sin\!<\!\left(\begin{matrix} 32 & 0 & 0 \\ 0 & 32 & 0 \\ 0 & 0 & 32 \end{matrix}, [SF,\ SC,\ CS] \right) \quad \left| \begin{matrix} SF \neq 0 \\ SC \neq 0 \\ CS \neq 0 \end{matrix} \right. \tag{3}$$

where:

$FIL_{SF,\ SC,\ CS}$ is a safety, security and cybersecurity functional integrity level.

An angle between vector and geometrical plane (represented by matrix) may only be right (=90°) or acute (<90°). FIL is defined only for non-zero values of the *SF*, *SC* and *CS*. Maximum FIL value equals "1" (as a sinus of 90°) when products of the answers regarding safety, security and cybersecurity are equal to each other. Growing discrepancies between products of the answers causes dropping of the FIL keeping it > zero.

4 Application for a Chosen Part of the Polish Railway System

Answering knock-out and differentiating questions is not credible in case of national and other large scale systems due to differentiation of the solutions utilized on different railway lines and by different vehicle types. As a result proposed approach is perfect for assessing functionally isolated homogeneous rail transport systems – for metro, for APMs - Advanced Peoples Movers where automated vehicles are running on a predefined routes and stopping with vehicle doors synchronized with platform doors and for other non-railway guided transport overall solutions like the airport one shown at Fig. 3.

Fig. 3. Airport guided transport system – view on infrastructure from an automatic vehicle/own photo/

Monograph [8] describes however application of the SSIRM and e-SSIRM models as well as calculation of the Functional Integrity Level for safety, security and cybersecurity for a chosen part of the Polish railway system. It describes the steps which are applied to obtain virtually functionally isolated system for assessment purposes together with pointing how individual assumptions influence applicability of the assessment results.

Chosen part of the Polish railway system is formed by part of the railway network, which was modernized recently and new multiple units used for the purpose of the passenger services. Application of the SSIRM model results are shown for the safety domain at Fig. 4 and for the security domain at Fig. 5.

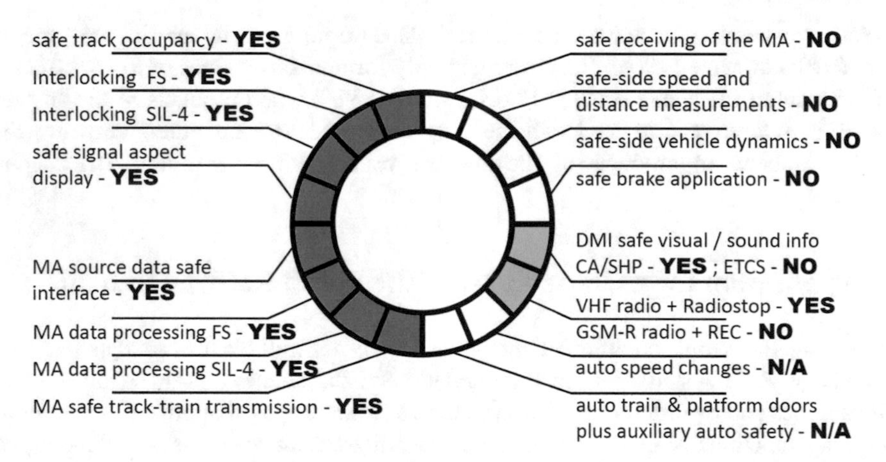

Fig. 4. Application for a chosen part of the Polish railway system. Safety and security impact reference model SSIRM, safety related functionalities/own elaboration/

The N/A (not applicable in the casa of the railway system under analysis) show, that traction units are operated by train drivers – there are no automatically operated driverless vehicles and platform doors. From the safety point of view in relation to calculating FIS level this is perfectly OK.

It is also visible, that railway line as well as running trains are using analogue radio although GSM-R system is already installed. The ongoing debate on Radiostop 2.0, which can be introduced together with GSM-R in Poland to achieve further availability of the automatic brake initiations in case of emergency is not reflected directly on SSIRM model, but it is in the FIS level questions. This means that use of the Radiostop 2.0 gives an answer YES = 2 instead of NO = 1 for one of the differentiating questions for determining safety for the purpose of the vector representing safety, security and cybersecurity.

Figure 4 in its upper right clearly shows, that the traction vehicles which are running on the infrastructure equipped with track-train transmission system, namely ETCS are not equipped adequately. The European Train Control System, which was implemented trackside is therefore not supporting traffic safety at all. Plentiful funds, which were spent for ETCS trackside installation will only be worth if traction vehicles running on the infrastructure are ETCS equipped ones or mainly ETCS equipped. Lack of onboard ETCS equipment results with NO = 0 for four knock-out questions.

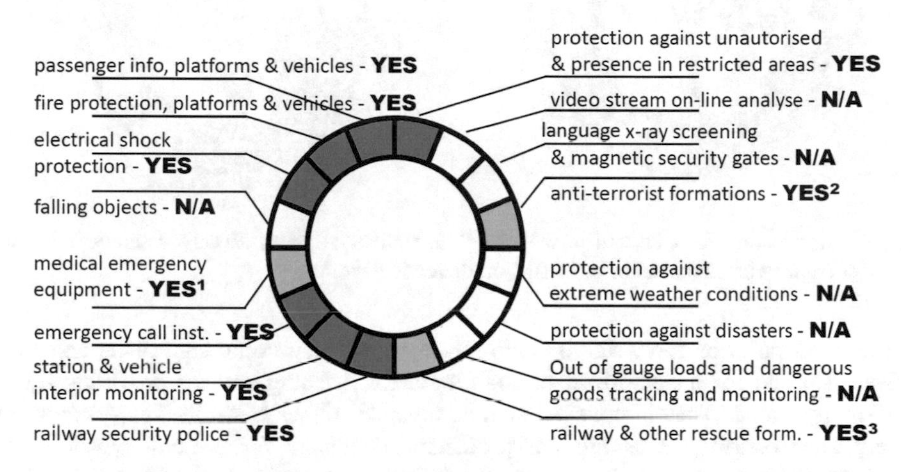

Fig. 5. Application for a chosen part of the Polish railway system. Safety and security impact reference model e-SSIRM, security related functionalities/own elaboration/

In security domain the N/A (not applicable in the casa of the railway system under analysis) show, that infrastructure does not require protection against falling objects. Such protections are mainly used in case of stations in high mountains. It also shows, that video monitoring is not supported by video stream analyses, which means that it is strongly supporting investigations but only partly security staff responsible for continuous monitoring. Stations are not equipped with x-ray screening and magnetic gates as it is dedicated for intense local traffic. There are no systems supporting out of gauge loads and dangerous goods tracking and monitoring as the infrastructure and services are dedicated for passenger traffic only. Protection against extreme weather conditions and disasters are ensured by staff and procedures without use of the dedicated data transmission based systems. That is perfectly OK from the point of view of the FIS level calculations.

However it is also visible, that in three cases the YES answers are only partly true. Access and egress ([1]) is not available for persons with reduced mobility. The only support for anti-terrorist formations ([2]) and rescue formations ([3]) are limited to setting predefined calls on the traffic controller communication panel. There are no means supporting interventions e.g. communication means and firefighting equipment.

Data transmission systems which are utilized by safety and security technical means are shown at Fig. 6 in accordance with e-SSIRM model. It is shown, that railway system as a whole uses many different transmission means. Only transmission systems 1 to 5 for the safety purpose are fully protected from the cybersecurity point of view. Control command track-train transmission systems responsible for passing movement authorities are proven from sender point of view, but were not proven from

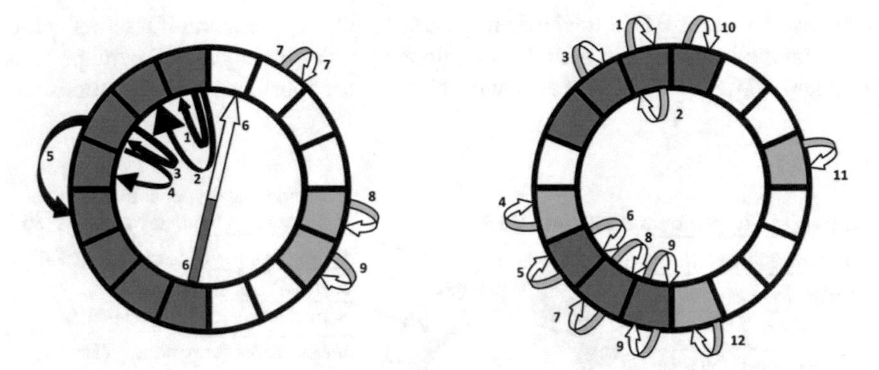

Fig. 6. Application for a chosen part of the Polish railway system. Received safety (left) and security (right) visualizations in e-SSIRM model/own elaboration/

the receiver point of view. Radio communication between traffic controllers and train drivers is open for unauthorized persons, both for listening and for providing information by voice. That is considered unacceptable. Other transmission systems for safety and security means are not protected adequately for present cybersecurity challenges. Low amount of cyber-attacks is achieved partly thanks to non-standard technical solutions, which are kept secret by signaling companies. They are however presently under standardization.

5 Conclusions

Presently technical means based on data acquisition, processing, transmission and storage are widely used for supporting railway traffic safety as well as rail transport security. Therefore safety, security and cybersecurity should be seen as complementary topics, which have to be provided on similarly high level. It is not reasonable to provide high safety for some functionalities and no safety for the others. The cyber-attacks which have already happened did not affected control command or signaling equipment but passenger information systems and timetabling.

A SSIRM and e-SSIRM models applicability covers not only railway and rail systems but also other guided transport systems. Application undertaken for functionally isolated homogenous railway system distincted from the Polish railway system has shown that all three safety, security and cybersecurity are provided only partly. The zero values do not allow calculation of the FIS level as a sinus of an angle between vector representing safety security and cybersecurity and a reference plane. It would be reasonable to undertake works aimed to achieve vector [4, 4, 4] having movement authorities completeness and coherency verification as well as automatic braking initiation in case of emergency for safety purpose, automated external defibrillators as well as video-stream analyzer for security purpose, and backup communication systems for degraded operational circumstances and backup power supply for all safety and security technical means for cybersecurity purpose. That would result with vector [4, 4, 4] and FIS level equal “1”.

References

1. Directive (EU) 2016/798 of the European Parliament and of the Council of 11 May 2016 on railway safety (EU OJ L 138/102)
2. Directive (EU) 2016/797 of the European Parliament and of the Council of 11 May 2016 on the interoperability of the rail system within the European Union (EU OJ L 138/44)
3. European Standard EN 50126-1:2017, Railway applications—The specification and demonstration of reliability, availability, maintainability and safety (RAMS) – Part 1: Generic RAMS Process
4. European Standard EN 50126-2:2017, Railway Applications - The specification and demonstration of reliability, availability, maintainability and safety (RAMS) – Part 2: Systems Approach to Safety
5. European Standard EN 50128:2011, Railway applications—Communication, signalling and processing systems—Software for railway control and protection systems
6. European Standard EN 50129:2003/AC:2010, Railway applications - Communication, signalling and processing systems - Safety related electronic systems for signalling
7. European Standard EN 50159:2010, Railway applications - Communication, signalling and processing systems - Safety-related communication in transmission systems
8. Pawlik, M.: Railway safety and security functional reference model built on data transmission based systems (Referencyjny model funkcjonalny wspierania bezpieczeństwa i ochrony transportu kolejowego przez systemy z transmisją danych). Warsaw University of Technology Publishing House (Oficyna Wydawnicza Politechniki Warszawskiej), Warsaw (2019). ISBN 978-83-7814-908-8

Comparison of Vehicle Exhaust Emissions with Gasoline Engines of Different Ecological Classes in Road Tests

Jacek Pielecha(✉), Jerzy Merkisz, and Karolina Kurtyka

Poznan University of Technology, Piotrowo Street 3, 60-965 Poznan, Poland
jacek.pielecha@put.poznan.pl

Abstract. The introduction of changes in the vehicle type approval testing procedures is a way to reduce the issue of exceeding the limits set in the standards. Previous laboratory tests of vehicle emissions did not contain information on the actual vehicle and engine operating conditions. Therefore, particular attention was paid to comparative tests of gasoline engines in the context of the particles number and nitrogen oxides road emission. The reason for this was a significant discrepancy in the ecological classification of engines (Euro 6 and Euro 6d-Temp) for vehicles of similar weight. The article presents the assessment of ecological vehicle indicators in real traffic conditions in accordance with the latest RDE (Real Driving Emissions) test proposals. It has been shown that despite the compliance of vehicles with emission limits of nitrogen oxides and the number of particles in tests carried out in the chassis dynamometer test, they are varied in real operation conditions. This proves that special attention should be paid to gasoline engines, which emit much more solid particles than the latest emission class compression-ignition (Euro 6d-Temp) engines. At the same time, the development of the automotive industry, new engine designs fueled by alternative fuels and the use of hybrid drives should constantly include the latest emission testing procedures.

Keywords: Exhaust emissions · Real driving emissions · Gasoline engines

1 Introduction

The dynamic development of road transport and the resulting increase in the number of vehicles has led to the fact that the car has become the most popular means of transport in the world. Despite the continuous improvement of vehicles, the most popular drive is still an internal combustion engine, used as a source of power for over a hundred years. Increasing the number of cars around the world will cause a rise in carbon dioxide emissions. This is a reason for more and more emphasis on sustainable development. This means that the expectations of car users are taken into account while at the same time being met by the producers. The challenges modern vehicle designers must meet apply to both the economic and ecological spheres. Currently, the main focus is on the environmental performance of drive units and the lowest fuel consumption. Earlier the most important issue was the performance of the engine; nowadays, it is the emission of toxic substances to the atmosphere, noise, and minimization of carbon dioxide

M. Siergiejczyk and K. Krzykowska (Eds.): ISCT21 2019, AISC 1032, pp. 330–340, 2020.
https://doi.org/10.1007/978-3-030-27687-4_33

emissions. A significant change in the development of internal combustion engines was the introduction of exhaust emission standards. These standards have radically reduced exhaust emissions in both gasoline and diesel engines. The Euro 6 emission standard [5] is currently in force, which must be met by all car manufacturers. To meet the requirements of the standard, vehicle constructors strive to achieve the lowest possible weight, while maintaining low rolling resistance and the best possible aerodynamics of the vehicle. Fulfilling the requirements of low emissions and fuel consumption has led to the introduction of downsizing technology. It means reducing the size of the combustion engine (mainly the displacement volume) while maintaining the operational parameters. Contemporary downsizing forces the use of engine supercharged various types and the elements of hybrid systems in such vehicles. Global trends related to the development of motor vehicles strive to reduce the main dimensions of the engine, which reduces its mass. In addition to downsizing, there is also downspeeding, which involves reducing the maximum engine speed, causing a decrease of fuel consumption and carbon dioxide emissions to the atmosphere. Starting from 2019, the process of approving a new type of passenger cars in the European Union includes a procedure for measuring exhaust emissions in real driving conditions. The European Union Regulations (715/2007/UE and 692/2008 [2]) concerning RDE tests [3, 4] is a response to the results of research on the increased nitrogen oxides emission from passenger cars equipped with compression ignition engines [7, 8], despite the fact that these vehicles met the permissible standards in laboratory conditions [13]. The results of the authors' research [9–12] indicate an increased emission of nitrogen oxides and solid particles in relation to compression-ignition engines. Some researchers [1, 6] consider that the newest gasoline with direct fuel injection engines emit more solid particles than the diesel engines (comparison assuming that both they are equipped with particulate filters).

2 Research Aim

The aim of the research was to compare the road emission of two vehicles with the Euro 6 and Euro 6d-Temp emission class in the RDE test, specified in the latest emission regulations. The scope of work included the comparison of exhaust emissions and fuel consumption in individual parts of the road test, and additionally the operating conditions of both engines. This approach allowed for the interpretation of the obtained results and drawing the appropriate conclusions.

3 Research Methodology

3.1 Research Objects

The vehicles were equipped with four-cylinder gasoline with direct fuel injection engines and were characterized by the same maximum power, and similar maximal torque value. The precise technical data of the tested vehicles are listed in Table 1.

Table 1. The characteristics of research objects.

Parameter	Vehicle A	Vehicle B
Model year	2016	2018
Emission standard	Euro 6	Euro 6d-Temp
Transmission	7-speed automatic	9-speed automatic
Cylinder number, arrangement	4, in series	4, in series + starter-generator 48 V
Displacement [cm^3]	1991	1497
Three-way catalyst	Yes	Yes
Gasoline Particle Filter	No	Yes
Max. power [kW] at [rpm]	135/5500	135/5800–6100
Max. torque [Nm] at [rpm]	300/1200–4000	280/3000–4000
Vehicle curb weight [kg]	1570	1635
Mileage [km]	60,000	30,000

3.2 Test Apparatus

The Semtech DS analyzer manufactured by Sensors was used to measure the concentration of harmful compounds in the exhaust gas. It allowed measurement of harmful compounds concentration including – carbon monoxide, hydrocarbons, nitrogen oxides and carbon dioxide. The analyzer processing unit received data directly from the engine diagnostic system. The analyzer consists of the measurement modules:

- FID (Flame Ionization Detector) used to determine the total hydrocarbon concentration in exhaust gases,
- NDUV (Non-Dispersive Ultraviolet), designed to measure the concentration of nitrogen oxide and nitrogen dioxide,
- NIDR (Non-Dispersive Infrared) infrared radiation, designed to measure the concentration of carbon monoxide and carbon dioxide,
- electrochemical for determining the oxygen content in the exhaust gas.

The TSI 3090 EPSS™ analyzer (Engine Exhaust Particle Sizer™ Spectrometer) was used to study the particle size distribution of particulate matter. It enabled the measurement of a discrete range of particle diameters (in the range from 5.6 nm to 560 nm) emitted in the exhaust gases based on their different velocity. Due to the device's data acquisition frequency of up to 10 Hz, the analyzer can be used for the study of particulate emissions in transient engine states.

3.3 Research Route

The test route has been chosen so that it meets the requirements of the European Commission as described in the Regulations [3, 4], with particular attention paid to its topography. Table 2 shows the characteristics of the research route in terms of terrain. The analysis of route topography confirms its compliance with the test requirement regarding the difference in the altitude of the test start and end which does not exceed 100 m.

Table 2. Characteristics of the research route in terms of terrain.

Parameter	Value	Route map
Distance of the test:		
Urban [km]	avg. 27.0	
Rural [km]	avg. 27.0	
Motorway [km]	avg. 24.0	
Altitude:		
minimum [m]	52	
average [m]	82	
maximum [m]	109	
Maximum road grade:		
increase [%]	4.1	
decrease [%]	–3.7	

4 Research Results

Despite the differences in the speed profiles for the test drives of both vehicles at the same test route, the parameters of compliance with the RDE test in each part have been retained (Fig. 1, Table 3).

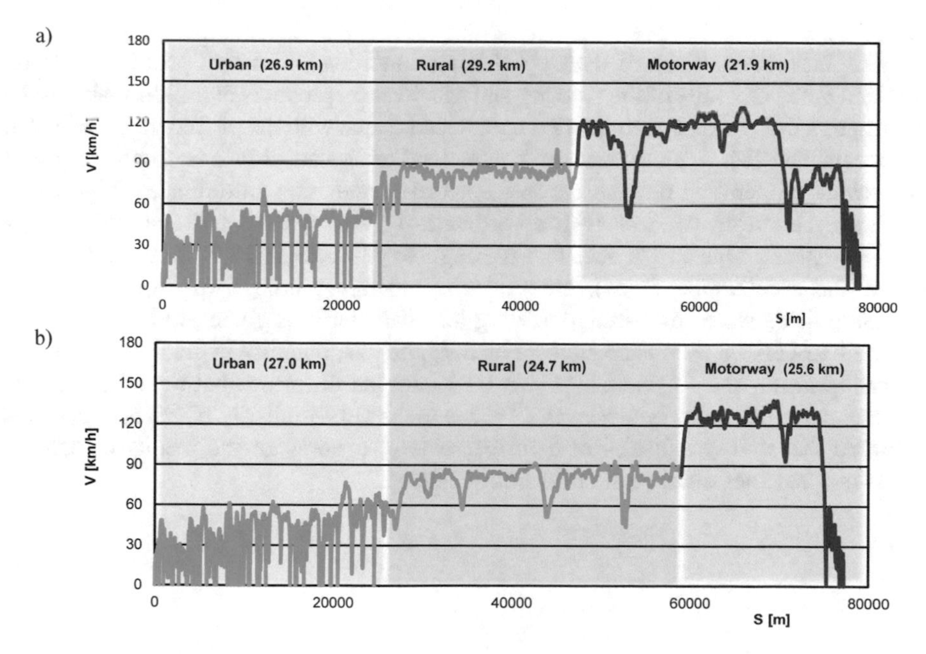

Fig. 1. The speed profile of vehicles test drives: (a) Euro 6 emission class (2.0 L), (b) Euro 6d-Temp emission class (1.5 L)

Table 3. Validation of trip characteristics in accordance with EU RDE requirements.

Trip characteristics	Vehicle A	Vehicle B	Requirement vehicle A/B	Valid?
Urban distance [km]	27.0	26.9	>16	OK
Rural distance [km]	24.7	29.2	>16	OK
Motorway distance [km]	25.6	21.9	>16	OK
Total trip distance [km]	77.3	78.0	>48	OK
Urban distance share [%]	34.9	34.5	29–44	OK
Rural distance share [%]	32.0	37.5	33 ± 10	OK
Motorway distance share [%]	33.1	28.0	33 ± 10	OK
Urban stop time [%]	25.4	27.8	6–30	OK
Motorway speed above 100 km/h [min]	11.5	10.4	>5	OK
Motorway speed max [km/h]	138.4	132.0	<160	OK
Motorway speed above 145 km/h [%]	0.0	0.0	<3	OK
Total trip duration [min]	101.88	102.7	90–120	OK
Urban 95th percentile $V \cdot a_+$ [m^2/s^3]	10.4	10.3	<17.7/<18.2	OK
Rural 95th percentile $V \cdot a_+$ [m^2/s^3]	14.1	13.4	<24.9/<24.2	OK
Motorway 95th percentile $V \cdot a_+$ [m^2/s^3]	18.7	20.4	<27.1/<27.2	OK
Urban RPA [m/s^2]	0.22	0.19	>0.137/>0.131	OK
Rural RPA [m/s^2]	0.09	0.09	>0.047/>0.061	OK
Motorway RPA [m/s^2]	0.08	0.09	>0.025/>0.025	OK

Graphical comparison of dynamic driving conditions – relative positive acceleration and 95th percentile of the velocity and acceleration product (Fig. 2) also allows the assumption that the tests were carried out in accordance with the procedure described in the Regulation [5]. A more accurate comparison of the test drives is a comparison of the vehicle's operating time shares in the speed-acceleration co-ordinates (Fig. 3).

Each measuring window with co-ordinates (ΔV, Δa) has been assigned successively numbers. Due to the fact that the number of fields is 14×9 (the number of speed and acceleration fields), the values of operating time share from the corresponding fields were compared, obtaining the characteristics given in Fig. 4. It shows that both test drives were characterized by a significant similarity in the share of vehicle operating time – the R^2 coefficient was 0.93, and the difference between the accurate representation of results (y = x) was 7%. Due to the very similarity of test drives, it was assumed that it is possible – in a further part – to compare the results of exhaust emissions and fuel consumption for both vehicles.

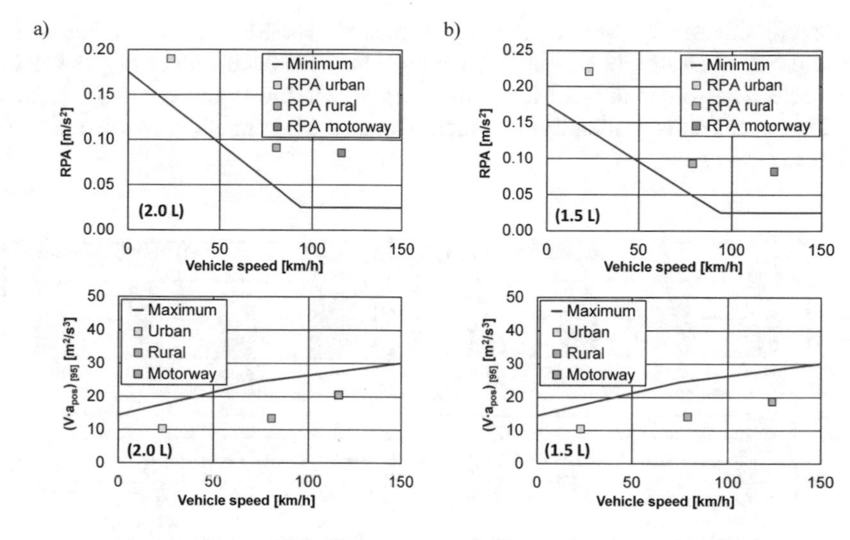

Fig. 2. Comparison of dynamic conditions in the test for vehicles with emission class: (a) Euro 6 (2.0 L), (b) Euro 6d-Temp (1.5 L).

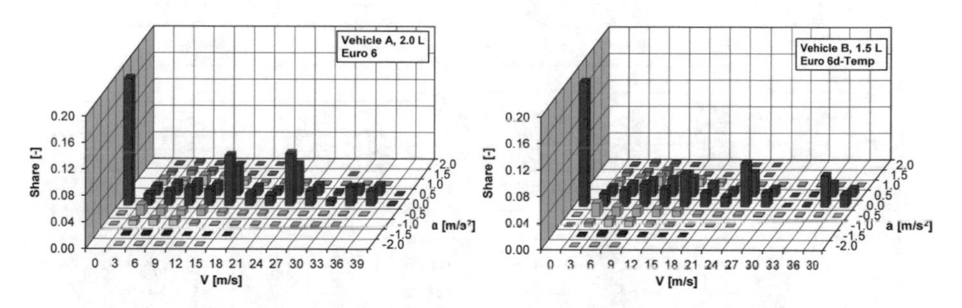

Fig. 3. Comparison of vehicle operating time share in the velocity-acceleration co-ordinates.

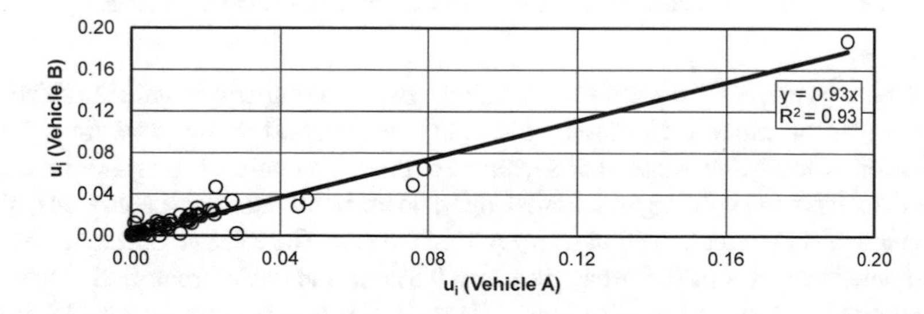

Fig. 4. Comparison of operation shares for vehicle A and B (based on Fig. 4) during RDE tests.

On the basis of the exhaust compounds concentration measurements, knowledge of the exhaust flow rate and the distance traveled in each part of the test, the values of road emissions of individual harmful compounds were determined (Fig. 5). The procedure was used for this MAW (Moving Average Window). The Moving Averaging Window method provides an insight on the real-driving emissions occurring during the test at a given scale. The test is divided into sub-sections (windows) and the subsequent statistical treatment aims at identifying which windows are suitable to assess the vehicle RDE performance.

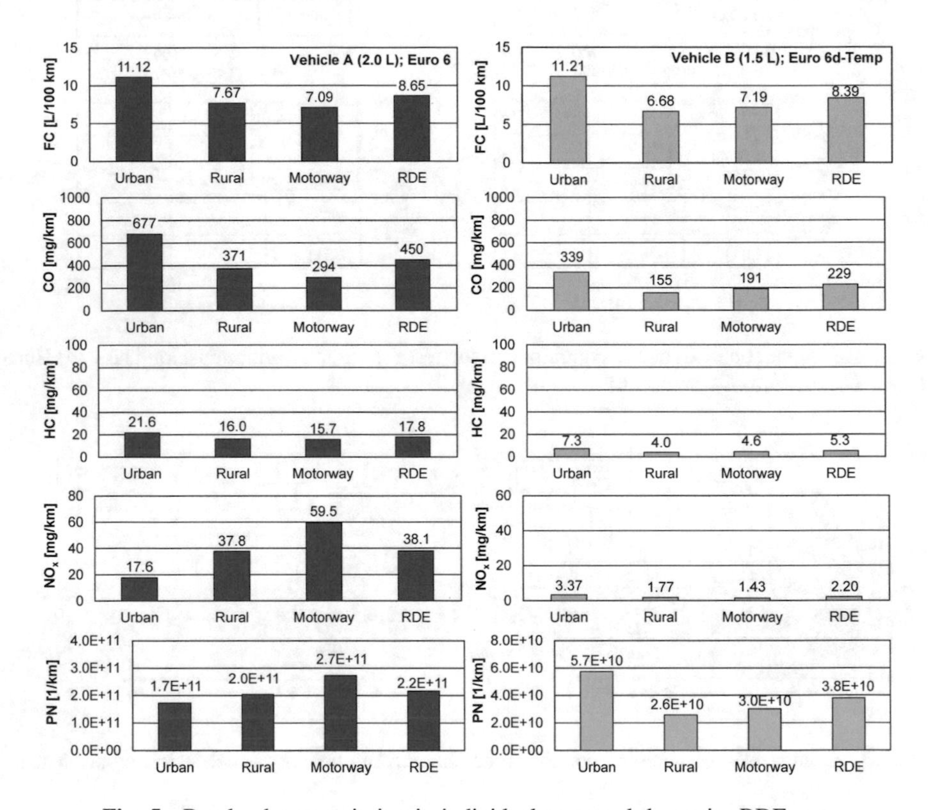

Fig. 5. Road exhaust emission in individual parts and the entire RDE test.

The comparison of the determined values reveals comparable values of fuel consumption (carbon dioxide emissions) for both vehicles, and at the same time different behavior of exhaust after-treatment systems for other compounds. Despite the unequal engine displacement volume and similar operational parameters of the engines, similar fuel consumption was achieved in every part of the test. The relative difference between the compared vehicles in the urban part was 0.8%, in the extra-urban part it was 13%, and on the motorway – 1.4%. The total difference in fuel consumption was 3% in favor of a vehicle with the 1.5 L engine meeting the Euro 6d-Temp standard.

The situation was different for harmful compounds: carbon monoxide road emissions were about 50% lower for the vehicle with smaller displacement volume and Euro 6d-Temp emission class compared to the vehicle with a larger drive unit. Similar relations were observed for hydrocarbon road emissions: a 3-fold smaller road emission of this compound occurs for the engine in the Euro 6d-Temp emission class. In the case of road emissions of nitrogen oxides – the distinction is even more significant (5–10 times higher for the engine with the lower emission class). Large differences were also obtained for the number of particles: the newer vehicle (Euro 6d-Temp) was fitted with a particulate filter, hence the emission of the number of particles is significantly lower compared to the vehicle without a particulate filter (about 5–10 times depending on the RDE test stage).

A detailed analysis of fuel consumption supported by determination of engine operation area in individual parts of the RDE test for vehicles A and B, reveals differences that may affect the values obtained (Fig. 6).

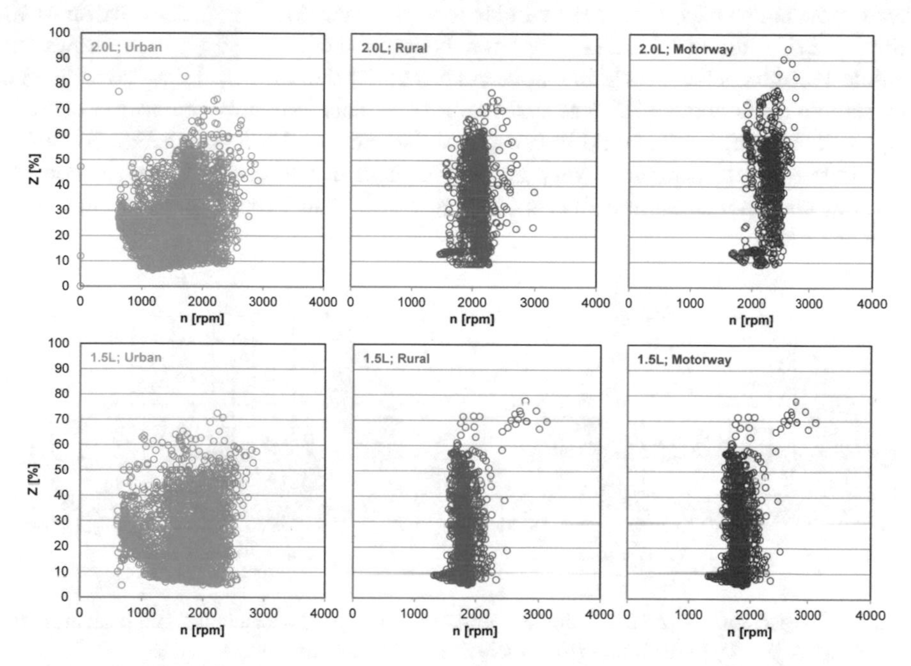

Fig. 6. Comparison of the speed and load values for individual parts of the RDE test for the vehicle A (2.0 L, Euro 6) and vehicle B (1.5 L, Euro 6d-Temp).

In the urban part of the RDE test, the average speed value is not close to each other – 1321 rpm and 1585 rpm, respectively for vehicles A and B, while the relative loads are also different, 26% and 23% respectively. It follows that higher speed and lower relative loads were the domain of the vehicle with the 1.6 L engine meeting the Euro 6d-Temp standard. It was mainly supported by the use of an automatic 9-speed gearbox (vehicle A was equipped with an automatic 7-speed gearbox). The design of the drive system was also reflected in the subsequent stages of the test: in the extra-urban part the average engine speed was respectively, for vehicle A and B, 2060 rpm and 1825 rpm, with a relative load of 37% and 27% respectively. A slightly different situation was observed during the motorway section, where more power of the drive system was required. While the engine speed was lower for a vehicle with a higher number of ratios (1990 rpm in relation to 2283 rpm), this required a greater engine load – 53% compared to 42%.

The presented conditions resulted in the diversification of fuel consumption – only in the part of the extra-urban RDE test for the vehicle with the smaller engine volume, significantly lower fuel consumption was noted (reduction by 13% compared to a vehicle with a 2.0 L engine displacement). In the other parts – where the rotational speed and load were not unambiguous - no such positive effects were noted.

Comparing the values of carbon dioxide emissions intensity (identified with fuel consumption) for vehicle A and B, there is a significant similarity in the obtained results. These values in individual fields defined by speed-acceleration co-ordinates (corresponding to each other) are similar to each other (Fig. 7). A comparison of the relevant fields allowed finding a relation between them (Fig. 8), which shows that vehicle B (with a smaller engine displacement) under the same operating conditions of the vehicle emits about 17% less carbon dioxide. This is mainly due to the obtained slope of the curve, but it should be noted that the value of the determination coefficient $R^2 = 0.45$ did not guarantee a very good agreement of the results. Thus, the comparison may contain a measurement error, which should be reduced by performing a larger number of tests.

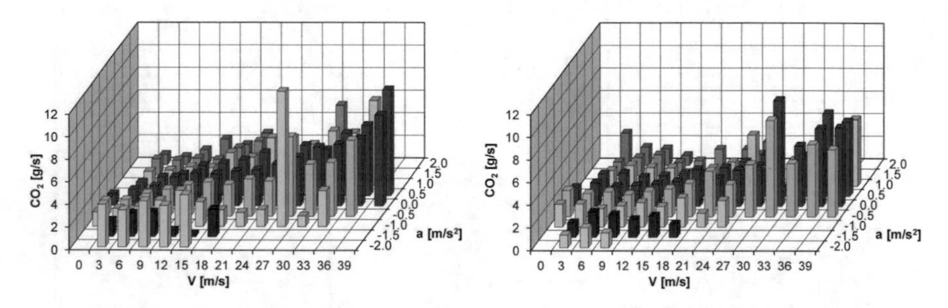

Fig. 7. Comparison of carbon dioxide emission in speed-acceleration co-ordinates for: (a) vehicle A (2.0 L, Euro 6) and (b) vehicle B (1.5 L, Euro 6d-Temp).

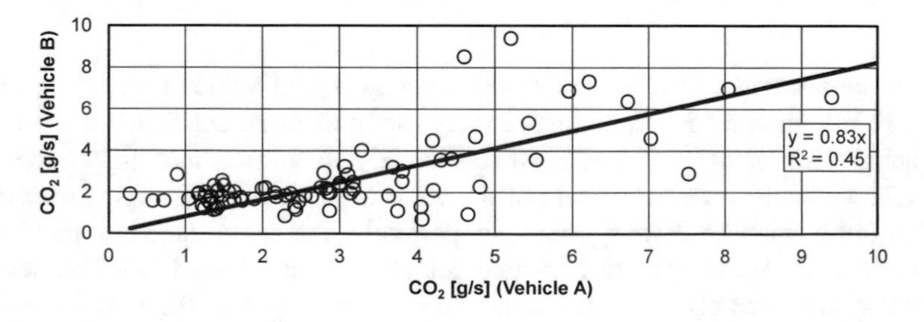

Fig. 8. Comparison of the carbon dioxide emission intensity for vehicle A and B (based on Fig. 8) during RDE tests.

5 Summary

New homologation testing procedures force vehicle tests in road emission tests. In this light, two vehicles with similar weights and drive units (in terms of operational parameters) are presented. They have been differed mainly in emission class and engine displacement: vehicle with Euro 6 emission class (2016) and displacement of 2.0 L, and class-vehicle Euro 6d-Temp (2018) and displacement of 1.5 L. Based on the carried out tests, it was found that despite significant reduction of road emissions of all harmful compounds (in the range of 50–90%), the most unfavorable value remains the value of mileage fuel consumption. Because when the engine displacement volume is reduced by 25%, the mileage fuel consumption is reduced by only 17%. The studies carried out do not exhaust the issue of ecological assessment of such vehicles, but only show technological progress, which reduces the environmental nuisance of means of transport.

References

1. Busch, S., Zellbeck, H.: Particle emission of the direct-injection gasoline engine under real driving emissions conditions. MTZ Worldw. **1**, 66–71 (2019)
2. Commission Regulation (EU) 692/2008 of 18 July 2008 implementing and amending Regulation (EC) 715/2007 of the European Parliament and of the Council on type-approval of motor vehicles with respect to emissions from light passenger and commercial vehicles (Euro 5 and Euro 6) and on access to vehicle repair and maintenance information, European Commission (EC), Official J. European Union, L 199 (2008)
3. Commission Regulation 2016/427 of 10 March 2016 amending Regulation No 692/2008 as regards emissions from light passenger and commercial vehicles (Euro 6) (2016)
4. Commission Regulation 2016/646 of 20 April 2016 amending Regulation No 692/2008 as regards emissions from light passenger and commercial vehicles (Euro 6) (2016)
5. Commission Regulation 2017/1154 of 7 June 2017 amending Regulation 2017/1151 supplementing Regulation No 715/2007 of the European Parliament and of the Council on type-approval of motor vehicles with respect to emissions from light passenger and commercial vehicles (Euro 5 and Euro 6) (2017)
6. Giechaskiel, B.: Real Driving Emissions (RDE) Particle Number (PN) Portable Measurement Systems (PEMS) Calibration. Technical Report. European Commission (2018)
7. Giechaskiel, B., Suarez-Bertoa, R., Lahde, T., Clairotte, M., Carriero, M., Bonnel, P., Maggiore, M.: Evaluation of NOx emissions of a retrofitted Euro 5 passenger car for the Horizon prize "Engine retrofit". Environ. Res. **166**, 298–309 (2018)
8. Kwon, S., Park, Y., Park, J., Kim, J., Choi, K., Cha, J.: Characteristics of on-road NOx emissions from Euro 6 light-duty diesel vehicles using a portable emissions measurement system. Sci. Total Environ. **576**, 70–77 (2017)
9. Merkisz, J., Fuc, P., Lijewski, P., Pielecha, J.: Actual emissions from urban buses powered with diesel and gas engines. Transp. Res. Procedia **14**, 3070–3078 (2016)
10. Merkisz, J., Pielecha, J., Lijewski, P., Merkisz-Guranowska, A., Nowak, M.: Exhaust emissions from vehicles in real traffic conditions in the Poznan agglomeration. In: 21st International Conference on Modelling, Monitoring and Management of Air Pollution (Siena), AIR POLLUTION XXI. WIT Transactions on Ecology and the Environment, vol. 174, pp. 27–38 (2013)

11. Pielecha, J., Merkisz, J., Markowski, J., Jasinski, R.: Analysis of passenger car emission factors in RDE tests. In: E3S Web of Conferences, vol. 10, p. 00073 (2016)
12. Stelmasiak, Z., Larisch, J., Pielecha, J., Pietras, D.: Particulate matter emission from dual fuel diesel engine fuelled with natural gas. Pol. Marit. Res. **24**(2), 96–104 (2017)
13. Varella, R.A., Giechaskiel, B., Sousa, L., Duarte, G.: Comparison of portable emissions measurement systems (PEMS) with laboratory grade equipment. Appl. Sci. **8**(9), 1633 (2018)

Differences in 2D and 3D Simulation's Impact on the Simulation Sickness

Małgorzata Pędzierska(✉), Mikołaj Kruszewski, Kamila Gąsiorek, and Arkadiusz Matysiak

Motor Transport Institute, 80 Jagiellonska Str., 03-301 Warsaw, Poland
malgorzata.pedzierska@its.waw.pl

Abstract. In recent years, a shortage of professional drivers has been observed in many countries. It is estimated that currently in Poland 100,000 only active drivers are missing [14]. It results from e.g. too high costs of training, low availability of training centers, rising employee requirements and an negative image of the driver's profession. At the same time, there are many defavorised groups in the labor market (NEETs, people over 50, immigrants). These people could fill the shortage of professional drivers on the market as soon as they complete their training. In order to decrease the training costs, an alternative to the conventionally conducted training is the use of virtual reality. Unfortunately, the occurrence of simulator sickness may become an important problem hindering its wide implementation. This can also be an important factor affecting the possibility of driver training and its effectiveness. The tests were carried out as a part of ICT-INEX project. The simulations were carried out using HTC VIVE PRO and Oculus Rift VR headsets as well as a computer screen, aiming to determine the possibility of using these tools in professional driver training. The research group included a group of 15 men aged 20–29 (NEETs) and 15 men over 50 years old. In order to determine how the particular simulations influenced the occurrence of simulation sickness symptoms, an analysis was carried out using the RSSQ questionnaires [8, 9]. The impact analysis of particular display types on the occurrence of simulation sickness was carried out.

Keywords: Virtual reality · Drivers training · Simulation sickness

1 Introduction

1.1 Virtual Reality

Virtual Reality is a term describing three-dimensional environment generated by computer. Person who explores VR becomes a part of virtual world. User can interact and manipulate with selected objects in artificially created, virtual world [7].

In VR the stimuli can be presented using a device called HMD - Head-Mounted Display, as well as special goggles which are placed on the subject's head. The glasses consist of headphones and two displays that allow 3D spatial vision and sound reception.

Navigating in the virtual environment is enabled by the "tracking system". The system consists of an antenna and sensors placed on the HMD. The sensors can also be

M. Siergiejczyk and K. Krzykowska (Eds.): ISCT21 2019, AISC 1032, pp. 341–350, 2020.
https://doi.org/10.1007/978-3-030-27687-4_34

used in other places, such as on the upper, lower limbs or a body. Sensors mounted on the hands or legs enable mapping their movements and interacting with virtual objects [4, 11].

Youngblut et al. [18] determined that the binocular field of human view reaches 180° in horizontal and 120° in vertical plane. However, during the head rotation, the field of view (FoV) in the horizontal plane can reach as much as 270°. Youngblut also stated that a 90–110° FoV would be enough to create an immersive virtual environment. Research on virtual reality technologies by Anthes et al. [1] shows that state-of-the-art systems comply to the abovementioned limits and even exceed them up to 200°.

However, a wide FoV can affect the severity of simulation sickness symptoms. Until today many studies about simulation sickness symptoms and its causes in 2D and 3D were conducted. Weidner et al. [17] compared the effect of 2D, stereoscopic 3D and VR simulations, regarding the simulation sickness symptoms. A total of 94 participants took part in the research. Their task was to change the road lane while driving in a simulator. Analyzes indicated a much higher simulation sickness in the VR-HMD-based projection than in a stereoscopic 3D image.

In a study conducted in 2017, the role of visually evoked sense of one's movement (vection), optokinetic nystagmus (OKN) and uncoordinated head movements in the formation of simulation sickness caused by rotation of the visual environment were examined. The results showed an increase in the symptoms of the sickness along with the increase in strength and intensity of the visually induced sense of one's movement (vection strength). There were no significant relationships between indicators related to head and eye movements and the occurrence of the sickness [12].

As part of the European ICT-INEX project, research was carried out using 2D and 3D simulations to determine the possibility of using them in professional driver training. The project was co-funded in scope of Key Action 2 of Erasmus + Programme. The publication presents partial results of tests carried out at the Motor Transport Institute.

2 Research Methodology

2.1 Drivers

The research group included 15 men aged 20–29 (NEET – Not in Education, Employment or Training) and 15 men over 50 years old. The requirement to participate in the study was to have a driving license for at least 3 years and to drive a vehicle for a minimum of 3,000 km per year.

2.2 Used Devices and Experimental Procedure

The most popular devices on the market are Samsung Gear VR, Oculus Rift, HTC Vive, HTC Vive PRO and Sony PlayStation VR. Despite very small sizes, they allow to completely indulge in the virtual world. Devices consist of goggles, a camera used to track the user's position and headphones.

The research was carried out using the HTC VIVE PRO VR HMD, Oculus Rift and a computer screen. The HTC VIVE PRO device is equipped with a dual-OLED display with a resolution of 2880 × 1600. The Oculus is equipped with two OLED displays with a resolution of 1080 × 1200, which gives a total of 2160 × 1200. For both devices the refresh rate was 90 Hz and the angular size of the generated image was 110° [5, 6]. The detailed technical specification of these HMDs used was described in [3].

The test stand consisted of a Playseat Forza, a Logitech steering system consisting of a steering wheel, a gearshift and a steering wheel equipped with indicator handles. A 40 in. TV set has been placed in front of the seat. During the 2D projection it was used to display the simulation. In the 3D mode, it displayed a view seen by subjects' in HMD.

The tests included 3 rides with the same research situation containing sections of motorway, two lane single-carriageway, two lane dual-carriageway, roundabout and crossroads with traffic lights, but were taken with the use of different devices. Every simulation lasted, according to driver's speed, from 8 to 10 min. Order of the device use was randomized. Before starting the test, drivers were familiarized with the procedure and purpose of the study. After they were asked to fill the questionnaires regarding their experience with computer games and virtual reality. The next step was to perform an adaptation ride during which the subjects could get used to the driving in the simulated conditions as well as the simulation environment itself. Additionally, the adaptation allowed to exclude people experiencing simulation sickness symptoms from the further stages of the research. After each run, the respondents were asked to complete the Revised Simulator Sickness Questionnaire RSSQ, immersion and the usability assessment of the equipment questionnaires.

3 Simulation Sickness

The acquired data has been analyzed in terms of immersion, device usability and simulation symptoms. The article describes the results of the conducted simulation sickness analyzes.

Simulation sickness is sometimes referred to as, visually brought about, motion sickness [16]. This is a phenomenon occurring either during or after the exposure to a virtual environment. The subject experiences unpleasant symptoms such as headaches or dizziness, nausea, and even vomiting [10]. It is estimated that the symptoms of this sickness vary in intensity from 30% to as much as 80% of the population [13].

The influence of the duration of the study on the severity of the sickness symptoms was checked by Stanney et al. 1,000 people took part in the study, and it lasted 15, 30, 45 or 60 min in the independent measurement scheme. 80% of participants experienced nausea, oculomotor disorders and disorientation. The feeling of disorientation lasted up to 24 h after the test. 12.9% of people did not complete the study due to feeling poorly, in 9.2% of them there was an emetic reflex, and in 1.2% vomiting. It can be assumed that exposure to a virtual environment should not exceed 2 h. It is also recommended to use breaks between simulation sessions. One session should not last longer than 1 h [2, 15].

3.1 Simulator Sickness Questionnaire (SSQ)

The key of subjective method used also in the estimation of simulation sickness is the Simulator Sickness Questionnaire (SSQ) by Kennedy [8]. By means of a questionnaire, the researcher himself, on the basis of his own feelings, assesses the intensity of the occurrence of given symptoms. The questionnaire lists 16 symptoms characteristic of simulation and simulation disease. The participant on a 4-point scale determines the weight of a particular symptoms. The project used a method using the RSSQ (Revised Simulator Sickness Questionnaire) questionnaire, which includes 28 symptoms belonging to the following categories: nausea (N), oculomotor (O) and disorientation (D). The following is the assignment of symptoms to particular categories:

- General discomfort (N, O),
- Fatigue (O),
- Boredom,
- Drowsiness,
- Headache (O),
- Eyestrain (O),
- Difficulty focusing (O, D),
- Increased salivation (N),
- Dry mouth,
- Sweating (N),
- Nausea (N, D),
- Difficulty concentrating (N, O),
- Depression,
- Confusion (D),
- Blurred vision (O, D),
- Dizzy (eyes open) (D),
- Dizzy (eyes closed) (D),
- Dizziness (D),
- Memory flashbacks,
- General weakness,
- The need to take a breath,
- Gastric problems (N),
- Loss of appetite,
- Increased appetite,
- The need for a bowel movement,
- Feeling lost,
- Eructation (N),
- Vomiting.

Symptoms not assigned to any of the categories are ignored when evaluating.

Each symptom is assessed on a four-level scale. The method of assigning numerical values to the subjective assessment is presented in the Table 1.

Table 1. The method of assigning numerical values to the subjective assessment [8]

Subjective assessment	Weight
None	0
Slightly	1
Mildly	2
Badly	3

The next step in the assessment is to sum up the results for each category and multiply this sum by the appropriate weighting factor. The following weighting factors are assumed:

- 9.54 for the Nausea category,
- 7.58 for the category of oculomotor symptoms,
- 13.92 for the Disorientation category.

The numerical result calculated in this way may be used to determine the degree of impact of a simulation disease on the examined person. Another way is to calculate the total score by multiplying the sum of all factors by 3.74 [8].

4 Analysis of Simulator Sickness Occurrence

The results of all 30 study participants were used in the statistical analyses. The participants uses different equipment in different order, which allowed to perform analyses of comparisons of average's and search for correlations between RSSQ's results and other dependent variables.

4.1 Comparison of SS Scores in Scope of Different Simulation Device

At first, the RSSQ results were compared in pairs for particular simulation devices: HTC Vive PRO device (hereafter HTC), Oculus Rift device (hereafter Oculus) and classic display on a computer screen (hereafter PC). Results were analysed both in scope of the overall score and the results in individual categories of symptoms. The results of the comparison of the overall impact on the occurrence of SS symptoms are presented in Fig. 1 and the results of the comparison in pairs in Table 2.

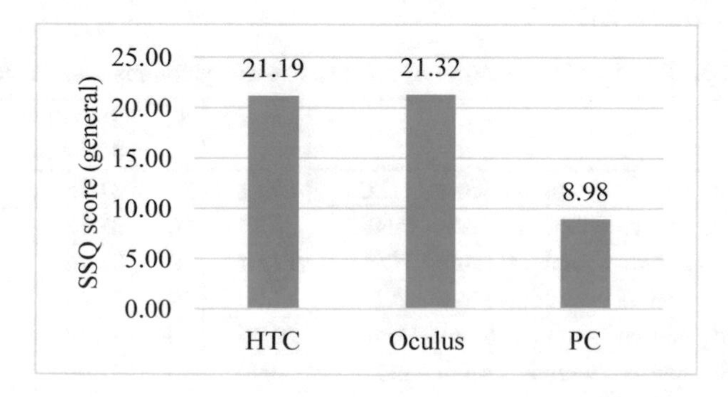

Fig. 1. Comparison of overall SS symptoms occurence for different simulation devices (source: own study)

Table 2. Results of comparison in pairs of overall score of RSSQ

		Mean	N	Standard deviation	Standard mean error
Pair 1	RSSQ overall score – HTC	21,19	30	34,77	6,35
	RSSQ overall score – Oculus	21,32	30	32,37	5,91
Pair 2	RSSQ overall score – HTC	21,19	30	34,77	6,35
	RSSQ overall score – PC	8,98	30	16,30	2,98
Pair 3	RSSQ overall score – Oculus	21,32	30	32,37	5,91
	RSSQ overall score – PC	8,98	30	16,30	2,98

Source: own study.

On the basis of the statistical analysis of the results, there were indicated statistically significant differences between Oculus and PC ($p = 0.027$). The comparison between HTC and PC showed a statistical trend ($p = 0.055$). Comparison of HTC with Oculus did not show statistically significant differences between the devices.

As the next step, a pairwise comparison was performed due to RSSQ results for particular symptoms categories (symptoms: nausea, oculomotor and disorientation). The results of RSSQ for particular symptoms and for all systems are shown in Fig. 2. The results of pairwise comparisons are presented in Tables 3, 4 and 5.

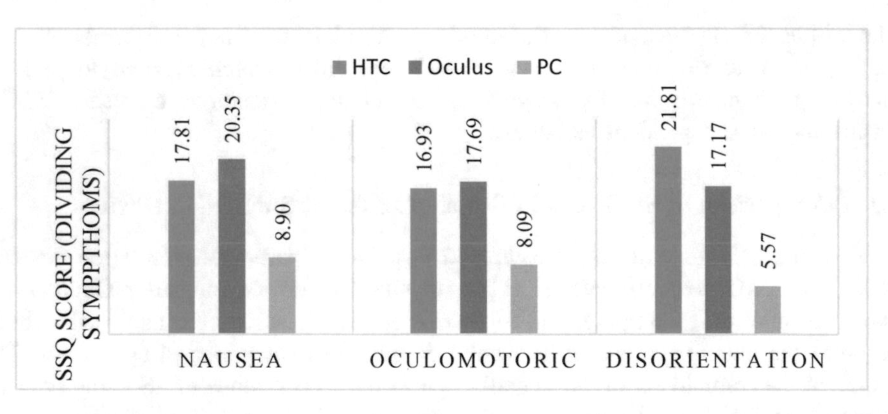

Fig. 2. Comparison of averages' of particular symptoms occurence for different simulation devices (source: own study)

Table 3. Results of comparison in pairs of nausea symptoms score of RSSQ

		Mean	N	Standard deviation	Standard mean error
Pair 1	RSSQ nausea symptoms score - HTC	17,81	30	31,76	5,80
	RSSQ nausea symptoms score - Oculus	20,35	30	33,12	6,05
Pair 2	RSSQ nausea symptoms score - HTC	17,81	30	31,76	5,80
	RSSQ nausea symptoms score - PC	8,90	30	21,69	3,96
Pair 3	RSSQ nausea symptoms score - Oculus	20,35	30	33,12	6,05
	RSSQ nausea symptoms score - PC	8,90	30	21,69	3,96

Source: own study.

Significantly more symptoms of nausea (p = 0.041) occurred after simulation in Oculus goggles (M = 20.4) than after projection on a PC screen (M = 8.9). There were no statistically significant differences between the different types of goggles (Oculus and HTC) or between HTC and PC.

Table 4. Results of comparison in pairs of oculomotoric sympthoms score of RSSQ

		Mean	N	Standard deviation	Standard mean error
Pair 1	RSSQ oculomotor symptoms score - HTC	16,93	30	25,70	4,69
	RSSQ oculomotor symptoms score - Oculus	17,69	30	24,08	4,40
Pair 2	RSSQ oculomotor symptoms score - HTC	16,93	30	25,70	4,69
	RSSQ oculomotor symptoms score - PC	8,09	30	12,26	2,24
Pair 3	RSSQ oculomotor symptoms score - Oculus	17,69	30	24,08	4,40
	RSSQ oculomotor symptoms score - PC	8,09	30	12,26	2,24

Source: own study.

For the PC simulation (M = 8.1), there were also significantly fewer oculomotor symptoms (p = 0.016) than for the Oculus (M = 17.7). On the level of statistical tendency (p = 0.060) there were found differences between PC simulation (M = 8.1) and HTC (M = 16.9). There were no statistically significant differences between the HTC and Oculus goggles.

Table 5. Results of comparison in pairs of disorientation symptoms score of RSSQ

		Mean	N	Standard deviation	Standard mean error
Pair 1	RSSQ disorientation symptoms score - HTC	21,81	30	40,16	7,33
	RSSQ disorientation symptoms score - Oculus	17,17	30	32,63	5,96
Pair 2	RSSQ disorientation symptoms score - HTC	21,81	30	40,16	7,33
	RSSQ disorientation symptoms score - PC	5,57	30	10,08	1,84
Pair 3	RSSQ disorientation symptoms score - Oculus	17,17	30	32,63	5,96
	RSSQ disorientation symptoms score - PC	5,57	30	10,08	1,84

Source: own study.

However, statistically significant more disorientation symptoms (p = 0.027) were indicated in comparison between HTC (M = 21.8) and PC (M = 5.6). The statistical trend (p = 0.051) was also shown by the Oculus (M = 17.2) with PC (M = 5.6) comparison. Slightly lower statistical trend (p = 0.057) of differences in results was shown by the comparison of HTC goggles with Oculus.

4.2 Comparison of SS Scores in Scope of Age Group

Due to the lack of compliance with the assumptions, the equality of the averages was verified using non-parametric tests. The Mann-Whitney Test was used in the context of non-parametric comparison of means for two independent groups.

There were significant differences between groups in the level of symptoms of simulation sickness caused by PC simulation. Group of participants between 20–29 y. o. had significantly less symptoms after simulation (M = 1.752000) than participants after 50 y.o. (M = 16.206667); p = 0.01.

A similar relationship was found in the case of a nausea symptoms. There were significant differences in the level of symptoms caused by PC simulation between groups. Group of participants between 20–29 y.o. had significantly less symptoms after simulation (M = 1,272,000) than participants after 50 y.o. (M = 16.536000); p = 0.016.

There were also significant differences in the level of oculomotor symptoms between groups after PC simulation and after using Oculus. After PC simulation, a group of participants between 20–29 y.o. had significantly less symptoms after simulation (M = 2.021333) than participants after 50 y.o. (M = 14.1493333); p = 0.01. After using Oculus, a group of participants between 20–29 y.o. had significantly less disease symptoms (M = 8.590667) than participants after 50 y.o. (M = 26.782667); p = 0.023.

There were significant differences in the level of disorientation symptoms between groups after PC simulation. Group of participants between 20–29 y.o. had significantly less symptoms after simulation (M = 0.928000) than participants after 50 y.o. (M = 10,208,000); p = 0.026.

4.3 Analysis of Correlations Between Time of Simulation and SS Score

An analysis of the correlation between particular systems and time of simulation was also performed to indicate whether the duration of the simulation impact could affect the occurrence of SS. The travel time differed between participants due to individual differences in driving.

The analyses showed a positive correlation between the travel time of the route and the disorientation symptoms only for the PC simulation. Pearson's correlation coefficient (correlation result) is $r = 0.447$; $p < 0.05$ for average simulation time 441,5 s. In other cases, the results were not statistically significant.

5 Results and Conclusions

The analyzes showed that there are significant differences between the symptoms experienced by the PC simulation and those experienced during the simulations using the HTC Vive PRO and Oculus Rift devices. Significant differences appeared both for the overall SSQ score and for particular symptoms groups. Significant differences, however, did not occur in the comparison between goggles. In one case only - for the symptoms of disorientation - the statistical trend of differences was indicated, to the disadvantage of the HTC device.

In the comparison of participants due to their age, significant differences appeared mainly in the field of SS during PC simulation. This may be related to the frequency of using such a simulation - younger people usually use computers more often and more often use virtual reality (e.g. playing computer games). Significant differences between the group of participants between 20–29 y.o. and those after 50 y.o. also appeared in the case of oculomotor symptoms experienced in Oculus goggles. Participants after 50 y.o. felt ailments about three times higher than participants between 20–29 y.o.

The analysis of the correlation between the travel time and the occurrence of symptoms of SS revealed only one statistically significant relationship in this respect. The oculomotor symptoms of the simulator disease using PC simulation largely depended ($p = 0.013$) on the duration of the simulation. For the remaining conditions, no statistically significant results were indicated. The average simulation time was 441,46 s for PC, for Oculus 415,42 and 421,5 s for HTC.

References

1. Anthes, C., Garciá-Hernández, R., Wiedemann, M., Kranzmüller, D.: State of the art of virtual reality technology. In: IEEE Aerospace Conference, pp. 1–19 (2016)
2. Biernacki, M., Dziuda, Ł.: Choroba symulatorowa jako realny problem badań na symulatorach. Med. Pr. **63**(3), 377–388 (2012)
3. Gąsiorek, K., Matysiak, A., Odachowska, E., Pędzierska, M.: Virtual reality technology in professional driver training. In: Comparison of using 2D & 3D simulation (2019, in publication)
4. Grabowski, A.: Wykorzystanie współczesnych technik rzeczywistości wirtualnej i rozszerzonej do szkolenia pracowników (Virtual and augmented reality contemporary techniques in training workers). Bezpieczeństwo Pracy 4/2012, 18–21 (2012)
5. Kennedy, R.S., Fowlkes, J.E.: Simulator sickness is polygenic and polysymptomatic: implications for research. Int. J. Aviat. Psychol. **2**(1), 23–38 (1992)
6. Kruszewski, M., Razin, P., Niezgoda, M., Smoczyńska, E., Kamiński, T.: Analiza efektów oddziaływania symulatora na powstawanie choroby symulatorowej w badaniach kierowców. Systemy Logistyczne Wojsk 44/2016, 188–201 (2016)
7. LaViola Jr., J.: A discussion of cybersickness in virtual environments. SIGCHI Bull. **32**(1), 47–56 (2000)
8. Milanowicz, M.: Koncepcja wykorzystania technik rzeczywistości wirtualnej do prowadzenia badań nad utratą równowagi człowieka. Mechanik **7**, 433–442 (2013)
9. Nooji, S., Pretto, P., Oberfeld, D., Hecht, H., Bülthoff, H.: Vection is the main contributor to motion sickness induced by yaw rotation: implications for conflict and eye theories. PLoS ONE **12**(4), 1–19 (2017)
10. Rebenitsch, L., Owen, C.: Review on cybersickness in applications and visual displays. Virtual Real. **20**(2), 101–125 (2016)
11. Stanney, K., Kennedy, R., Drexler, J.: Cybersickness is not simulator sickness. In: Proceedings of the Human Factors and Ergonomics Society Annual Meeting, vol. 41, no. 2, pp. 1138–1142 (1997)
12. Tiiro, A.: Effect of visual realism on cybersickness in virtual reality. University of Oulu (2018)

13. Weidner, F., Hoesch, A., Poeschl, S., Broll, W.: Comparing VR and non-VR driving simulations: an experimental user study. In: 2017 IEEE Virtual Reality (VR), pp. 281–282. IEEE (2017)
14. Youngblut, C., Johnston, R., Nash, S., Wienclaw, R., Will, C.: Review of Virtual Environment Interface Technology. Institute for Defense Analyses (1996)
15. VRsite. http://vrsite.pl/htc-vive-pro/. Accessed 08 Oct 2018
16. VRsite. http://vrsite.pl/oculus-rift/. Accessed 08 Oct 2018
17. Systel. https://systel.pl/virtual-reality/. Accessed 08 Oct 2018
18. PWC. https://www.pwc.pl/pl/pdf/pwc-raport-rynek-pracy-kierowcow.pdf. Accessed 14 Jan 2019

Optimizing the Data Flow in a Network Communication Between Railway Nodes

Rafał Polak[1], Dariusz Laskowski[2], Robert Matyszkiel[3], Piotr Łubkowski[2(✉)], Łukasz Konieczny[4], and Rafał Burdzik[4]

[1] Transbit, Warsaw, Poland
rafal.polak@transbit.com.pl
[2] Military University of Technology, Warsaw, Poland
piotr.lubkowski@wat.edu.pl
[3] Military Communication Institute, Zegrze Poludniowe, Poland
[4] Silesian University of Technology, Gliwice, Poland

Abstract. An important element in the communications, providing a high level of requirements for transport by rail, is the need to ensure access to and exchange of data in a strictly defined time regime. Another determinant of is to provide a guarantee of reliability. The achievement of these requirements is possible by evaluating current capabilities and adequate to the needs of implementing modern solutions. However, future hardware and software platforms incorporating new technologies and technologies are not always applicable to rail transport communications systems. Therefore, they require detailed analysis from the point of view of meeting the critical indicators i.e. Kg (availability factor), or MTBF (Mean Time Between Failures) and MTTR (Mean time to repair). Another aspect is the use of redundancy. The choice of method depends on the analysis of possible events in the environment in question. Proposed by the authors of the new variant does not require significant investment and is a flexible solution that provides the desired fitness of technical and functional correctness.

Keywords: Optimization · Data · Networks · Railway nodes

1 Introduction

An important element in ensuring the connectivity high level requirements for railway services is the need to ensure the transmission of the data in a strictly defined time regimes and dislocation. This requirement is made possible by the purchase of modern equipment and software from reputable suppliers of equipment with the requisite demonstration of reliability indicators (Mean Time Between Failures - MTBF, Mean time to repair - MTTR, availability factor – Kg [15–17]). The second way, easier to implement, is to use hardware redundancy software preceded by a detailed analysis of possible occurrence of events in the environment. This variant does not require significant investment and is a flexible solution that provides the desired fitness of technical and functional correctness [1–3].

M. Siergiejczyk and K. Krzykowska (Eds.): ISCT21 2019, AISC 1032, pp. 351–362, 2020.
https://doi.org/10.1007/978-3-030-27687-4_35

It is a special requirement that the destruction of one mode of communication does not result in the loss of communication between nodes that are separated from each other. Today's teleinformatic systems (STI) need reliable communication [14, 18] in data transport over the network layer. The constantly developing network of railways contributes to the need to develop the teleinformatic network needed to service the system.

2 Status of the System

Let's assume that a network (e.g. network, *STI*) is the analysis object with the following mathematical model [4–9]:

$$S = \langle G, \{F_z\}, \{f_k\} \rangle \tag{1}$$

where:

- $\{F_z;\ z = \overline{1, Z}\}$ – the set of F_z: $T \rightarrow R^+$ function,
- $\{f_k; k = \overline{1, K}\}$ – the set of f_k: $W \rightarrow R^+$ function,
- G is a complete graph defined by:

$$G = \langle W,\ T,\ P \rangle \tag{2}$$

where:

- $W = \{w_l : l = \overline{1, L}\}$, a scalable set of graph vertices (network nodes),
- $T = \{t_m : m = \overline{1, M}\}$, a scalable set of graph branches (transport resources/ communication lines),
- $P \subset W \otimes T \otimes W$[1], three-term relation of contiguity such that for each line $t_m \subset T$ there is such e pair of nodes $\langle w_i, w_j \rangle$ that $\langle w_i, t_m, w_j \rangle \in P$, where w_i is a starting node and w_j is an ending node. A size of a network component set is a sum of nodes and lines cardinalities:

$$|E| = |W| + |T| = L + M, \qquad E = \{e_i : i = \overline{1, L+M}\} \tag{3}$$

Let's denote the operational state of the component e_i of the network by x_i and assume that it can take one of the two values:

$$x(e_i) = x_i = \begin{cases} 1 & when \quad element \quad e_i \quad is \quad able \\ 0 & when \quad element \quad e_i \quad is \quad unable \end{cases} \tag{4}$$

In this case, the states of network elements can be presented with the vector $\vec{X}$ of components *(L + M)*, which take the values *0* or *1*:

[1] Symbol $\otimes$ denotes a Cartesian product.

$$\vec{X} = [\underbrace{x_1, \ldots, x_L}_{L}, \underbrace{x_{L+1} \ldots, x_{L+M}}_{M}] \tag{5}$$

Let's additionally assume that processes of faults and repairs of elements e_i are mutually independent and the distributions of correct operation time with the parameter λ_i (intensity of faults) and repair time with the parameter μ_i (intensity of repairs) are exponential.

Analogically as for network elements, it can be assumed that a set of network operational states is two-element one. Classification criteria for the network states into the ability or inability ones can be as follows: criterion of structural:

$$\Phi(STI) = \begin{cases} 1 & when \quad network \quad STI \quad is \quad able \\ 0 & when \quad network \quad STI \quad is \quad unable \end{cases} \tag{6}$$

A network is in the ability state if and only if at any time moment t all elements of the network are in the ability state. Technical analysis-level data transport important is the essence of modern network using large-scale all sorts of elements called telecoms-terminals (i.e. end-operator access, hosting and data transport services, etc.) internal architecture like the machines processing data in digital form. In this environment, the realization of the reported demand for the service is associated with:

(1) The execution of threads and processes in terms of a single Terminal.
(2) Functional capabilities of resources: access and transport in terms of data throughput and traffic control (routing, commutation) in data stream control.

Having regard to the size and complexity of the factors, it is proposed to group them in significant collections of determinants in the form of internal properties of the OT (system), the properties of the operator and the harmful effects of the environment, as the components of the formula for the probability of the state of fitness:

$$P_{SZ}(t, e_i) \cong f(P_{wWi}, P_{pOpi}, P_{zOti}) \tag{7}$$

where:

- P_{wWi}: the probability of correctness of functioning of the OT considering internal ownership,
- P_{pOpi}: the probability of correct functioning of the operator OT considering the knowledge and experience in project implementation time regimes (i.e. use or use OT) [10, 11],
- P_{zOti}: the probability of remaining OT able to airworthiness directives considering the environmental effects.

For further analysis will be considered only the probability of correctness of functioning of the OT considering ownership (P_{wWi}), and the remaining probability will be dealt with in subsequent publications. Generalized dependence is:

$$P_{wWi}(t, e_i) \cong f(K_{gi}, P_{sri}, P_{sni}) \tag{8}$$

where:

- K_{gi} – ready indicator structure,
- P_{sri} – the likelihood of blockage in the i-th element of the network (switch, router: access edge or wireframe) considering the routing protocol considering zoning the safe and open,
- P_{sni} – the likelihood of the loss of the data stream in the i-th element of the endpoint (the sender and the recipient, server, etc.) because of mistakenly operating procedures, sorting, processing, data transfer.

Proposes to adopt for the network elements the following assumptions:

- for l-th node (router, central) (P_{wWl}):

$$P_{wW\_l}(w_l) \cong \begin{cases} \omega_l K_{g\_l}(t, e_l)\omega_{Op\_l} R_{Op\_l}(\Delta t/t)\omega_{sr\_l} P_{sr\_l}(t)\omega_{sn\_l} P_{sn\_l}(t) \\ \omega_l R_l(t, e_l)\omega_{Op\_l} R_{Op\_l}(\Delta t/t)\omega_{sr\_l} P_{sr\_l}(t)\omega_{sn\_l} P_{sn\_l}(t) \end{cases} \tag{9}$$

- For m-th transport resource (wireless and cable links) (P_{wWm}):

$$P_{wW\_m}(tr_m) \cong \begin{cases} \omega_m K_{g\_m}(t, e_m)\omega_{Op\_m} R_{Op\_m}(\Delta t/t)\omega_{sr\_m} P_{sr\_m}(t) \\ \omega_m R_m(t, e_m)\omega_{Op\_m} R_{Op\_m}(\Delta t/t)\omega_{sr\_m} P_{sr\_m}(t) \end{cases} \tag{10}$$

where:

- $\omega_{Kg\_i}/\omega_{Rn\_i}$ – the weight of the impact indicator/function reliability:
  - $\omega_{Kg\_l}/\omega_{Rn\_l}$ – l-th node,
  - $\omega_{Kg\_m}/\omega_{Rn\_m}$ – m-th transport resource,
- ω_{Op_i} – the weight of the impact of the software according to the model:
  - ω_{Op_l} – l-th node,
  - ω_{Op_m} – *m-th* transport resource,
- ω_{sr_i} – the weight of the impact of ownership model leads the model:
  - ω_{sr_l} – *l-th* node,
  - ω_{sr_m} – *m-th* transport resource,
- ω_{sn} – the weight of the impact of synchronization.

The value of readiness indicators ((11), (12)) and non-readiness ((13), (14)) e_i, depending on the implementation of monitoring and repair processes for more common situations in operational practice, is determined from the mathematical form [1–3]:

(1) Readiness:

- non-stationary:

$$K_{gi}(t, e_i) = \frac{\mu_i}{\lambda_i + \mu_i} + \frac{\lambda_i}{\lambda_i + \mu_i} e^{-(\lambda_i + \mu_i)t} \tag{11}$$

– stationary:

$$K_{gi}(e_i) = \frac{\mu_i}{\lambda_i + \mu_i} e^{-(\lambda_i + \mu_i)t} \leq K_{gi}(t, e_i) \tag{12}$$

(2) Non-readiness:

– non-stationary:

$$Q_{gi}(t, e_i) = \frac{\lambda_i}{\lambda_i + \mu_i} [1 - e^{-(\lambda_i + \mu_i)t}] \tag{13}$$

– stationary:

$$Q_{gi}(e_i) = \frac{\lambda_i}{\lambda_i + \mu_i} \tag{14}$$

where:

- λ_i – Damage intensity of the i-th element of the repaired OT expressed by the formula:

$$\lambda_i = f(\lambda_b, \pi_Q, \pi_E, \pi_A, \pi_R, \pi_S, \pi_C, \pi_T, \pi_L) \tag{15}$$

where:

- λ_b – Base damage intensity, usually expressed by the model corresponding to the influence of electrical and temperature loads;
- $\pi_{Q,\ E,\ A,\ R,\ S,\ C,\ T,\ L}$ – Other coefficients modifying the baseline vulnerability intensity for the environmental category of applications and other parameters affecting non-enforceability.

For example, for integrated circuits:

$$\lambda_{US} = \pi_L \pi_Q (C_1 \pi_T + C_2 \pi_E) \tag{16}$$

where:

- π_L – considers the degree of mastery of the manufacturer's production,
- π_Q – takes quality into account,
- π_T – considers the influence of temperature,
- π_E – considers the environmental impact,
- C_1, C_2 – considers the complexity of the integrated circuit.

3 Reaction Mechanism

The most relevant information for network administrators and users the information about the State of network services and optimize the use of resources. Check the network status is possible through accurate monitoring of selected network elements (i.e. computers, servers, switches, routers, etc.). To detect the State of emergency as soon as there is a need to implement mechanisms to maintain and restore the system. Shielding subsystem (SSS) solves this problem because it is a set of activities and measures to prevent activation of the processes leading to system failure. An example of this type of solution is the OAM (Operations, administration and management) mechanism. External OAM functionality is therefore essential in any carrier-class technology and is indicated in intelligent Ethernet endpoint devices. Ethernet link OAM specifies the following procedures for monitoring link OAM: discovery, remote indication of failure, including link error, blanking and other critical events, remote feedback tests, recovery of the variables from the database management (MIB) and features specific for the organization.

OAM covers the range of Ethernet services on any path, regardless of whether it is a single link or end, which allows full monitoring of Ethernet services without regard to the layers that support the service, network route Ethernet traffic. OAM is also referred to as "connectivity fault management" (CFM). Sharing network on the domain maintenance in the form of levels of hierarchy, which are then allocated to users, service providers and operators. Technical specification of specifying the requirements arising from the mechanisms of OAM services, particularly elements of the network affects the following service functions OAM:

- error management (including the detection, verification, localization and notification),
- performance monitoring (including performance measurement),

auto-detect (including detection of network elements in the service networks).

4 Network Architecture and Testbed

To ensure the reliability of the network, it is proposed to use hardware and software redundancy. The TCP/IP and UDP/IP stack are a good solution for network services in rail transport. Communication between the units is provided by the following media transmission (Fig. 1.) [12]: digital subscriber lines, fibre optic trunk and trunk radio.

The well-known and available materials and research papers on the issues discussed in the article point to various and uneven solutions in terms of determining the potential and effectiveness of teleinformatic networks. The multitude of solutions is closely linked to the scientific interests represented by individual members of institutes and research teams. From the analysis of existing materials, it is concluded that the research problem in question is still in the modelling and design phase, i.e. that there are no developed and implemented and widely accepted methods to provide a reliable assessment of the network.

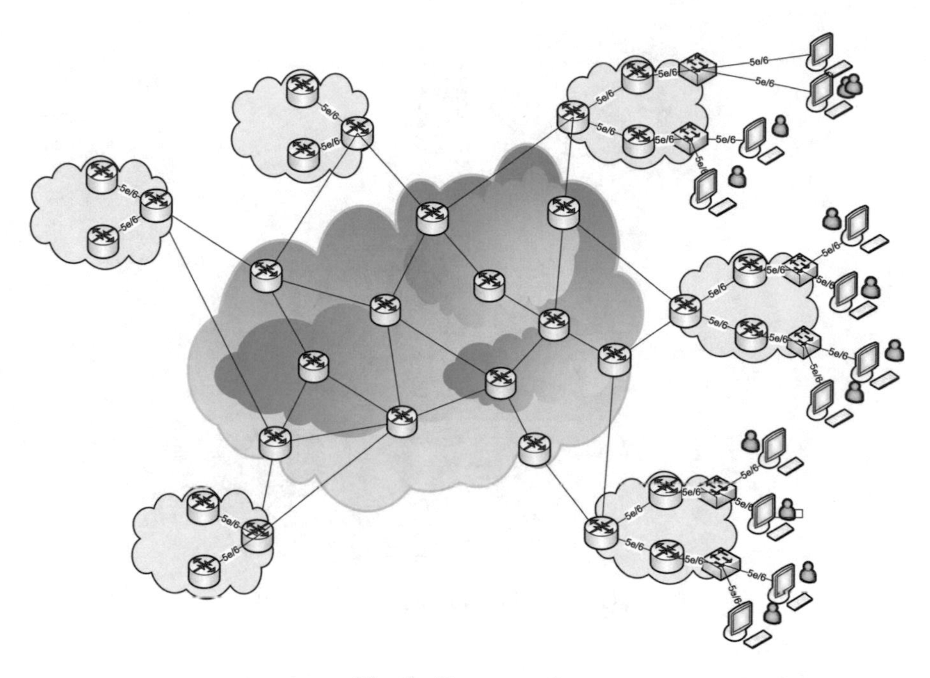

Fig. 1. Test network.

5 Analysis of the Results Simulation Research

The object of research and modelling is a teleinformatic network built for the article about a complex architecture and usual place of operation. The ICT network simulation model reflects the key elements of the actual network, from the point of view of the realization of network services. Network architecture consists of several parts, among which stands out the end users' part of access, aggregation, distribution, and a skeleton network. Below is an overview of the network structure (Fig. 2) and hardware-software platform [13].

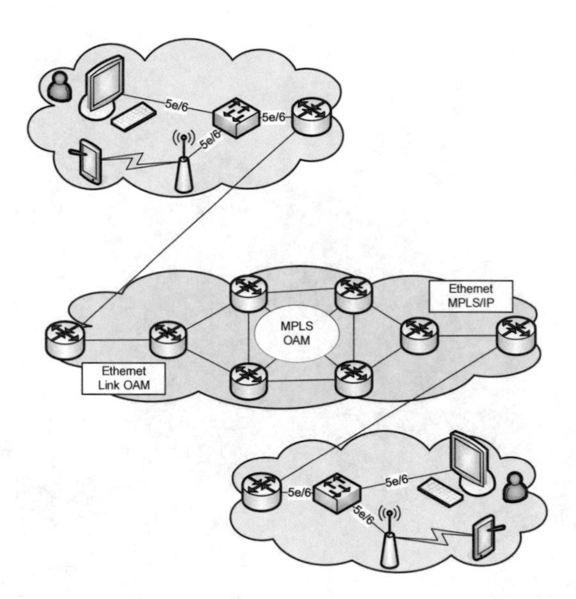

Fig. 2. General network structure.

The following graphs test results of network bandwidth for connections using the OAM mechanism and Ethernet without OAM mechanism. This parameter specifies the maximum number of data that can be sent by the test skeleton network at a given time (Fig. 3).

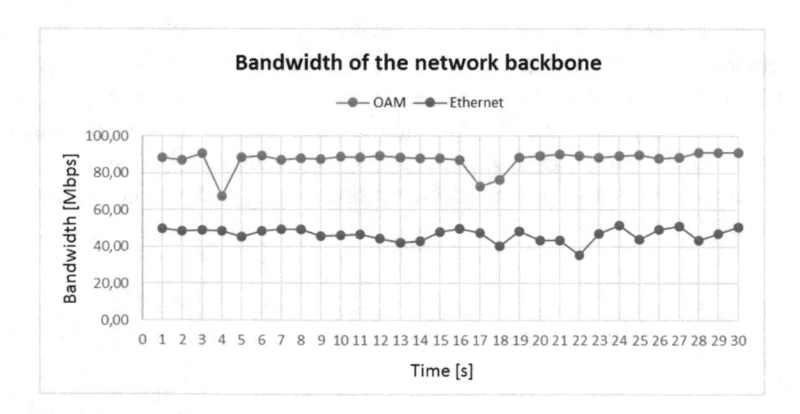

Fig. 3. Bandwidth of the network backbone.

The following graphs test results of network jitter for connections using the OAM mechanism and Ethernet without OAM mechanism. This parameter specifies the maximum delay difference between those in flowing per unit time by a link (Fig. 4).

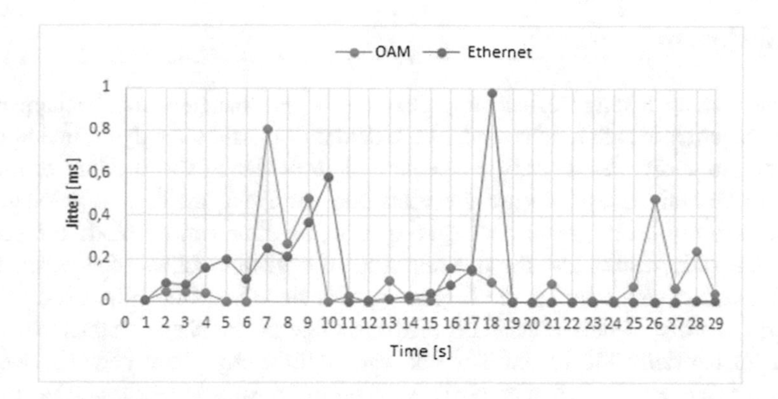

Fig. 4. Jitter.

The following graphs test results of network delay for connections using the OAM mechanism and Ethernet without OAM mechanism. This parameter determines the time required for delivery of packets from the client device to the server. The final study was to determine the rate of a computer network providing data transport. The network used the Fast Ethernet technology with a theoretical value of 100 Mbps. During testing, considering the operating conditions of the classical value obtained from the interval 18÷70 Mbps. Results from network research have made it possible to determine the probability of an ability condition (Fig. 5).

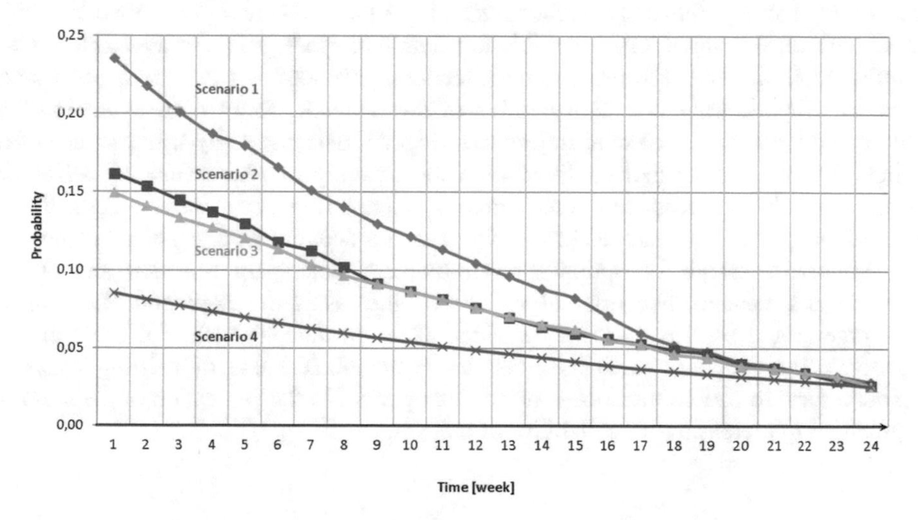

Fig. 5. Probability of an ability condition (scenario 1 … 4).

6 Conclusion

By implementing a range of scenarios examined test that network mechanism shall meet the requirements in reliable and secure data transmission. Analysis, including how to respond to anomalies, i.e. instantaneous failure or damage to the link or node has proved to be entirely and during the transition from one medium to another transmission have not been loss of packages, or stop communication. Made the scenarios provided reliable results and the development of statistics indicate the optimal pallet network mechanisms offering the highest performance providing the expected reliability and security of data transmission. Knowledge of the shape characteristics, designated for the daily life of the network, can significantly affect network operations process management for telematics. To sum up the results of research, the following conclusions can be drawn:

(1) throughout the day takes the variable value dependent on, inter alia, the number of users, their working hours and possible to use the value potentiality obtainable passage network. This leads to periodic network shrinkage in the perception of the number of users because of the sheer volume of or occurring environmental exposures may not meet needs in a timely manner;
(2) the delay in the implementation of the services is subject to similar fluctuations as load on the network and is different for different services (for email and ftp has the level of acceptable throughout the day, and for browsing the WEB are time periods in which to implement this service is not possible).

Measures proposed in the form of probabilities of network components reliability can be used at any network operational step to restore or increase the network ability state with considering stresses possible to occur. Therefore, it is also advisable not to confine to design and establish an object rendering services, but to carry a permanent process of monitoring operational status of the network. Knowledge related to the shape and location of potential trajectories (Fig. 5), determined by using values of a reliability and security indices distribution for a twenty-four-hour (mouth) period of using network resources, can considerably influence improving management of the operational process for each technical object, and a network with separated channels.

Proposed methods of optimization (multiple agents, path selection and OAM technologies) enhance network and, as a result, increasing the likelihood of complete coverage on the service. It has a beneficial effect on the satisfaction of users of the network. The proposed methods allow for more efficient use of existing network resources and to handle more users in a shorter period. In the perception of the users is equivalent to increasing the reliability of the STI.

References

1. Barlow, R.E., Proschan, F.: Mathematical Theory of Reliability. Wiley, New York (1965)
2. Misra, K.B.: Reliability Analysis and Prediction. Elsevier, New York (1992)
3. Ireson, W.G.: Handbook of Reliability Engineering and Management. McGraw-Hill, New York (1996)
4. Laskowski, D., et al.: Anthropo-technical systems reliability. In: Safety and Reliability: Methodology and Applications - Proceedings of the European Safety and Reliability Conference, ESREL 2014, pp. 399–407, 883–888. CRT Press, A Balkema Book (2015). https://doi.org/10.1201/b17399-58. Print ISBN: 978-1-138-02681-0
5. Kowalski, M., et al.: Exact and approximation methods for dependability assessment of tram systems with time window. Eur. J. Oper. Res. **235**(3), 671–686 (2014)
6. Kowalski, M., et al.: Analysis of transportation system with the use of Petri nets. Maint. Reliab. **49**(1), 48–62 (2011)
7. Nowakowski, T., Werbińka, S.: On problems of multicomponent system maintenance modelling. Int. J. Autom. Comput. **6**(4), 364–3784 (2009)
8. Walkowiak, T., Mazurkiewicz, J.: Hybrid approach to reliability and functional analysis of discrete transport system. Comput. Sci. **3037**, 236–243 (2004)
9. Werbinska-Wojciechowska, S., Zajac, P.: Use of delay-time concept in modelling process of technical and logistics systems maintenance performance. case study. Maint. Reliab. **17**(2), 174–185 (2015)
10. Butlewski, M., Sławińska, M.: Ergonomic method for the implementation of occupational safety systems. In: Occupational Safety and Hygiene II - Selected Extended and Revised Contributions from the International Symposium Occupational Safety and Hygiene, SHO 2014, pp. 621–626 (2014)
11. Jasiulewicz-Kaczmarek, M., Drożyner, P.: Social dimension of sustainable development – safety and ergonomics in maintenance activities. In: Stephanidis, C., Antona, M. (eds.): Universal Access in Human-Computer Interaction. Design Methods, Tools, and Interaction Techniques for eInclusion, UAHCI/HCII 2013, Part I, LNCS 8009, pp. 175–184. Springer, Switzerland (2013)
12. Lubkowski, P., et al: Provision of the reliable video surveillance services in heterogeneous networks, safety and reliability: methodology and applications. In: Proceedings of the European Safety and Reliability Conference, ESREL 2014, pp. 883–888. CRT Press, A Balkema Book (2015). https://doi.org/10.1201/b17399-58. ISBN 978-1-138-02681-0
13. Łubkowski, P., Laskowski, D.: Selected issues of reliable identification of object in transport systems using video monitoring services. In: Communication in Computer and Information Science, vol. 471, pp 59–68, Springer, Switzerland (2014). https://doi.org/10.1007/978-3-662-45317-9_7. ISSN 1865-0929
14. Rychlicki, M., Kasprzyk, Z.: Increasing performance of SMS based information systems. In: Proceedings of the Ninth International Conference Dependability and Complex Systems DepCoS-RELCOMEX. Given as the monographic publishing series – Advances in Intelligent Systems and Computing, vol. 286, pp. 373–382 (2014)
15. Siergiejczyk, M., et al.: Reliability assessment of integrated airport surface surveillance system. In: Proceedings of the Tenth International Conference on Dependability and Complex Systems DepCoS-RELCOMEX", given as the monographic publishing series – Advances in intelligent systems and computing", vol. 365, pp. 435–443, Springer (2015)
16. Siergiejczyk, M., Paś, J., Rosiński, A.: Issue of reliability–exploitation evaluation of electronic transport systems used in the railway environment with consideration of electromagnetic interference. IET Intell. Transport Syst. **10**(9), 587–593 (2016)

17. Siergiejczyk, M., Rosiński, A., Krzykowska, K.: Reliability assessment of supporting satellite system EGNOS. In: Zamojski, W., Mazurkiewicz, J., Sugier, J., Walkowiak, T., Kacprzyk, J. (eds.) New Results in Dependability and Computer Systems, Given as the Monographic Publishing series – Advances in Intelligent and Soft Computing, vol. 224, pp. 353–364. Springer (2013)
18. Stawowy, M., Dziula, P.: Comparison of uncertainty multilayer models of impact of teleinformation devices reliability on information quality. In: Podofillini, L., Sudret, B., Stojadinovic, B., Zio, E., Kröger, W. (eds.) Proceedings of the European Safety and Reliability Conference ESREL 2015, pp. 2685–2691. CRC Press, Balkema (2015)

Improving of Selected Routing Strategies for Order Picking

Radosław Puka, Jerzy Duda, and Marek Karkula(✉)

AGH University of Science and Technology, Faculty of Management, Krakow, Poland
{rpuka, jduda, mkarkula}@zarz.agh.edu.pl

Abstract. The problem of effective order picking gains in importance, among other things, due to the growing interest in Internet shopping. This type of shopping allows wholesalers to shorten their distribution channels by omitting intermediaries and selling directly to the end customer. However, it results in servicing a much larger number of smaller orders with a very diverse number of goods. Therefore, it is important that the picking routes are as short as possible, because they greatly affect the overall time of order processing. In order to determine the routes of movement of the order pickers, heuristics are used, which in a short time allow to determine the picking routes. The authors proposed a new heuristic to designate picking routes and an improvement, which can be implemented in such heuristics as midpoint and largest gap and the proposed new routing method. In order to analyze the effectiveness of the proposed solutions, the authors used the computer simulation method. The results showed that the proposed improvement allows for shortening the picking routes by about 1.5–2%, with the value of improvement being strongly dependent on the number of goods to be picked up and the size of the warehouse.

Keywords: Order picking · Heuristics · Midpoint · Largest gap

1 Introduction

Since the beginning of the century, we can observe the continuous development of e-commerce. Currently online shopping accounts for almost 9% of total retail spending worldwide, while in United Kingdom it is nearly 20% (Millenians already make more than 50% of their purchases online). The increase in the number of online purchases results, among others, in the fact that wholesalers are forced to handle a much larger number of smaller orders with a very diverse number of goods. This, in turn, forces them to look for more and more effective order completion strategies. It is widely known that the efficiency of order picking process is the main cost component of the entire logistics processes in the warehouse [6].

Despite the ongoing automation majority of warehouses are still operated mainly manually (according to [12] – 80% of warehouses in Western Europe employs pickers who manually collect items for orders). For manually order picking systems, the most time consuming activity is travelling between different locations in the warehouses,

M. Siergiejczyk and K. Krzykowska (Eds.): ISCT21 2019, AISC 1032, pp. 363–373, 2020.
https://doi.org/10.1007/978-3-030-27687-4_36

while the travel time is an increasing function of the travel distance (see e.g. [7, 9]). In consequence a method of determining the shortest possible route is usually the first objective in optimization of processes in the warehouse. Determining the optimal picking route can be reduced to the traveling salesman problem with additional constraints resulting, for example, from the layout of the warehouse space [3]. Since TSP is an NP-hard problem and it not always can be directly used in practice, many heuristics have been identified for the problem of determining picker routes. Several the most popular heuristics are distinguished in [1, 7, 9, 10] and [11]: S-shape, return, midpoint, largest gap and some combinations of them. Each of the listed heuristics has its advantages and its efficiency depends on the type of the warehouse, the strategy of placing the goods, available paths for movement. In work [5] the authors presented an extensive simulation analysis of routing and storage policies in order-picking systems. In turn, Chen et al. [2] presented the nonparametric heuristic method addressed to the real-time online order problems. Nevertheless, in many situations, further improvements are possible, which may result in additional shortening of the routes traveled by order pickers.

In this paper, the authors propose a modification of the midpoint and largest gap heuristics as well as a new one called d-midpoint, that when combined with the proposed modification is better than the original midpoint heuristic.

The diagram of the warehouse that will be used in the further part of the study is shown in Fig. 1.

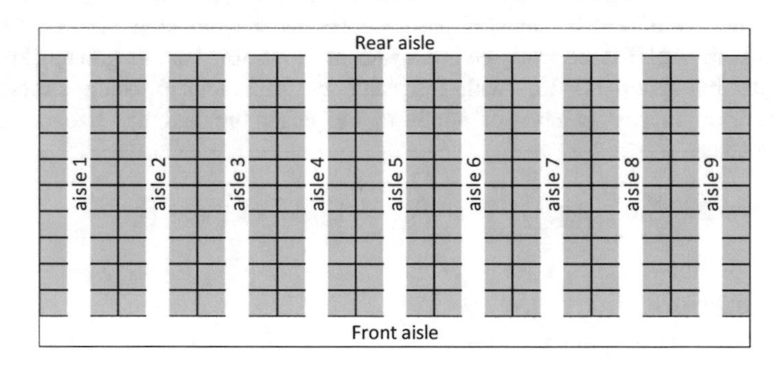

Fig. 1. The structure of an exemplary warehouse.

2 Midpoint Heuristics

Midpoint heuristics consists in dividing a warehouse into two equal parts (with accuracy to one rack location for racks with an odd number of rack locations). Each part is assigned to the operator from one of the alleys: front aisle or rear aisle respectively. Figure 2 shows the division of the warehouse into two parts, with an

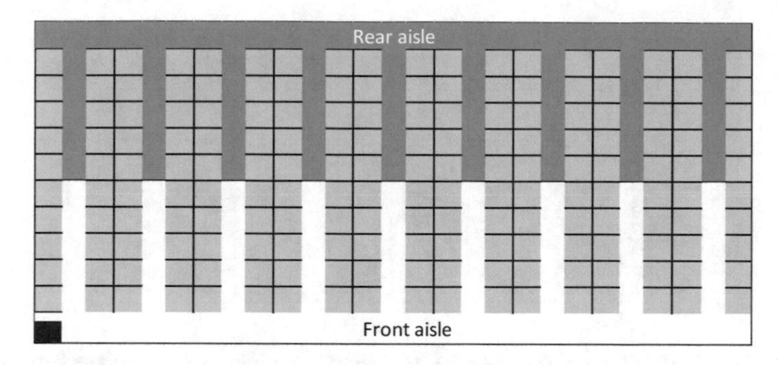

Fig. 2. Division of the warehouse into front and rear areas.

indication of which part of the warehouse is serviced from which alley. In the lower left corner, in black, the starting point for picking, which is the same as the picking point for a completed order, is marked.

The operator starts order picking and in the first alley containing the goods (located in any half of the warehouse) moves to the rear area (if there is at least one goods to be picked in the rear area) [4]. He then collects all the goods in the rear area. In each alley visited, the operator picks the goods in front of the "centre point" – once all the goods have been picked from the alley, the picker leaves it with the same entrance as he entered the alley. The last alley containing the goods (also located in any half of the warehouse) returns to the front area of the warehouse. Going to the order picking point, the operator picks up the goods in the front area. Also in this case the operator compiles the goods in front of the middle point, only this time from the front aisle. Figures 3 and 4 show examples of order picking routes using the midpoint heuristics. The shelves marked in black indicate the goods to be picked.

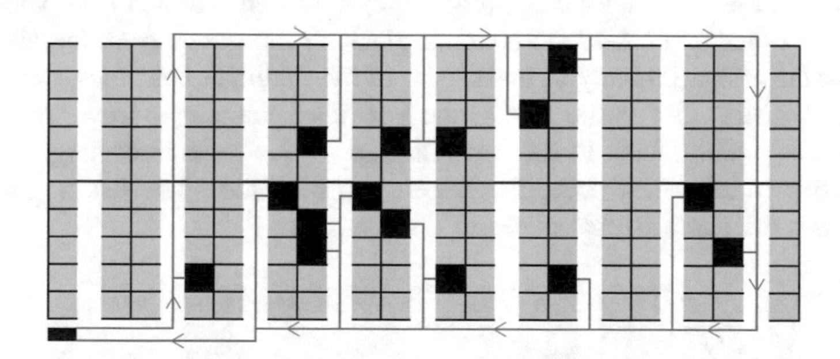

Fig. 3. Route calculated with the help of midpoint heuristics for an exemplary location of goods (1).

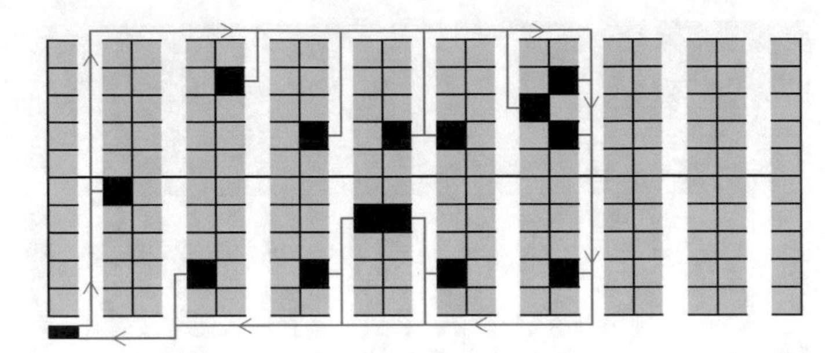

Fig. 4. Route calculated with the help of midpoint heuristics for an exemplary location of goods (2).

3 Largest Gap Heuristics

Largest gap heuristics was proposed by Petersen [9] as a modification of the midpoint heuristics. The main difference between these heuristics is the determination of the place where each of the alleys is divided into front and rear areas. In the case of midpoint heuristics, the division is made at a fixed point for each alley. In heuristics, the largest gap point of division is determined separately for each of the alleys, which in an extreme case may cause the situation that each alley is divided into front area and rear area in a different place. The point of division of a given alley is determined on the basis of the largest distance selected from a list of three values [8]:

1. Maximum distance between two adjacent items to be picked in an alley.
2. The distance between the front aisle and the nearest goods to be picked in the alley.
3. Distance between the rear aisle and the nearest item to be picked in the alley.

Only if the value of the greatest distance is equal to the value of the first one, the operator enters the same alley twice (once with the front aisle and once with the rear aisle). In the other two cases, the operator picks all the goods from the alley in one entrance to the alley. Similarly as in the case of the midpoint heuristics, except for the case when a given alley is the first or the last of all the alleys, the operator always leaves the alley going back to the corridor from which he entered. Figures 5 and 6 show the routes calculated using the largest gap heuristics. The list of goods to be collected was the same as presented in Figs. 3 and 4.

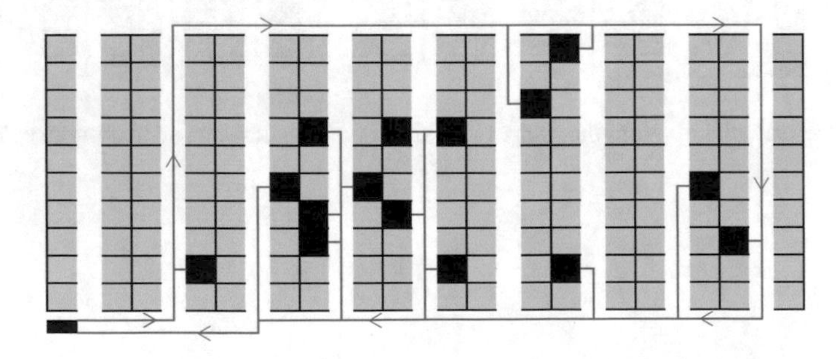

Fig. 5. Route calculated with the help of largest gap heuristics for an exemplary location of goods (1).

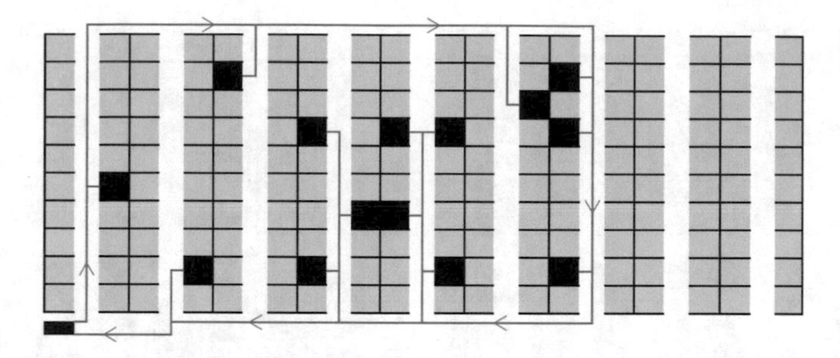

Fig. 6. Route calculated with the help of largest gap heuristics for an exemplary location of goods (2).

4 D-Midpoint Heuristics

The proposed d-midpoint heuristics is a dynamic version of midpoint heuristics. It consists in checking where the alleys should be divided into front area and rear area to ensure the shortest/fastest picking route. For this purpose, for each possible "middle point" position (between front aisle and rear aisle) subsequent picking routes are determined and the most advantageous solution is selected from among the solutions selected.

Figure 7 shows the picking routes analysed within the framework of the d-midpoint heuristics of the picking route (for the distribution of goods to be picked, see Figs. 3 and 5). The "middle point" position in the first analysed case is at the rear aisle. The middle point is then moved towards the front aisle until the front end of the rack is reached. As can be seen from the diagram in Fig. 7, the d-midpoint heuristics can act as midpoint, return and, in some cases, largest gap heuristics.

Figure 8 shows a block diagram used for calculation of picking route with d-midpoint heuristics.

D-midpoint heuristics seems to be easier to implement than largest gap heuristics. However, it is more complex in terms of calculation – in order to obtain a result, calculations for the d-midpoint heuristics have to be made: the number of *rack places* + 1 times more than in the case of the largest gap or midpoint heuristics.

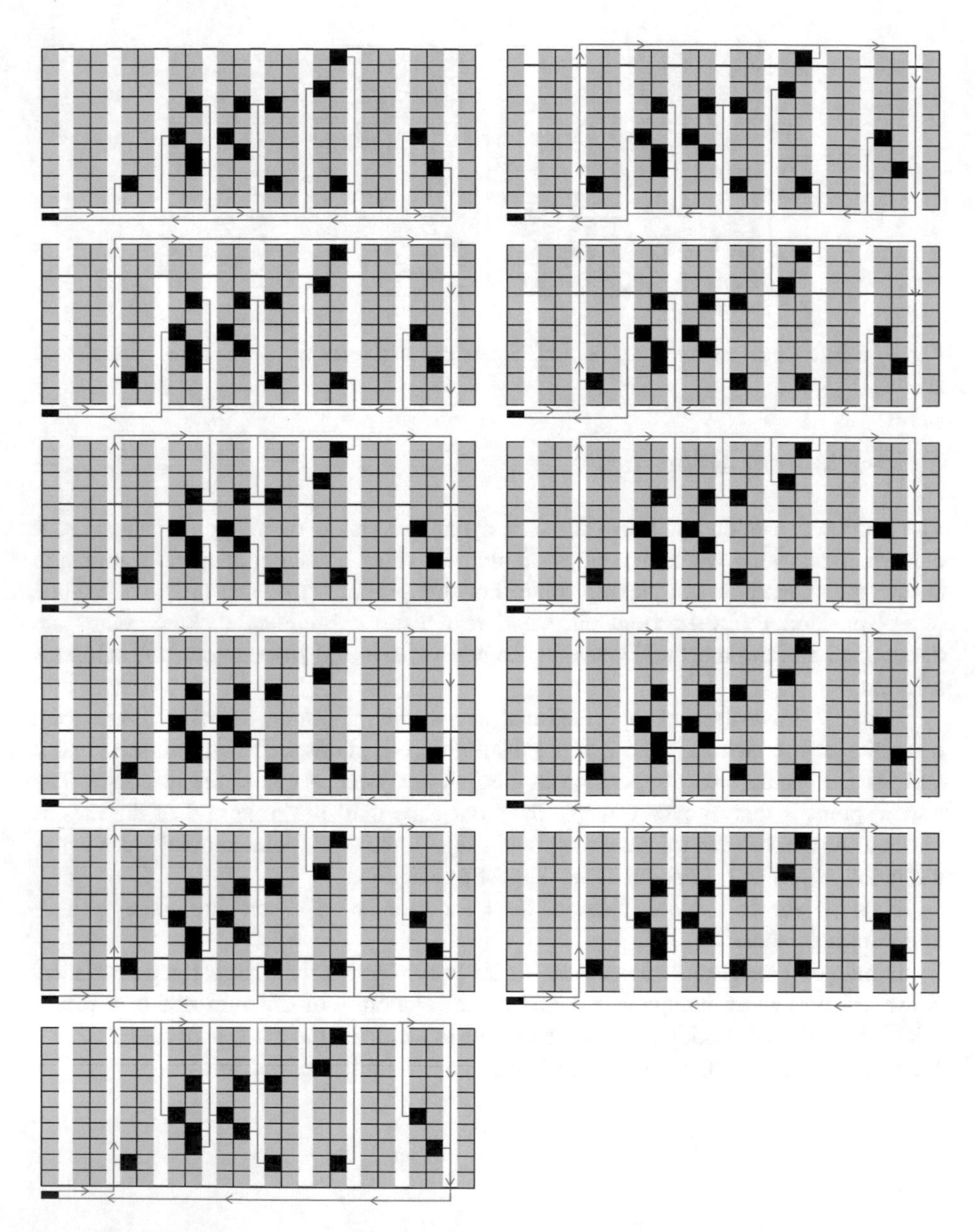

Fig. 7. Diagram of the middle point positions analysed for d-midpoint heuristics.

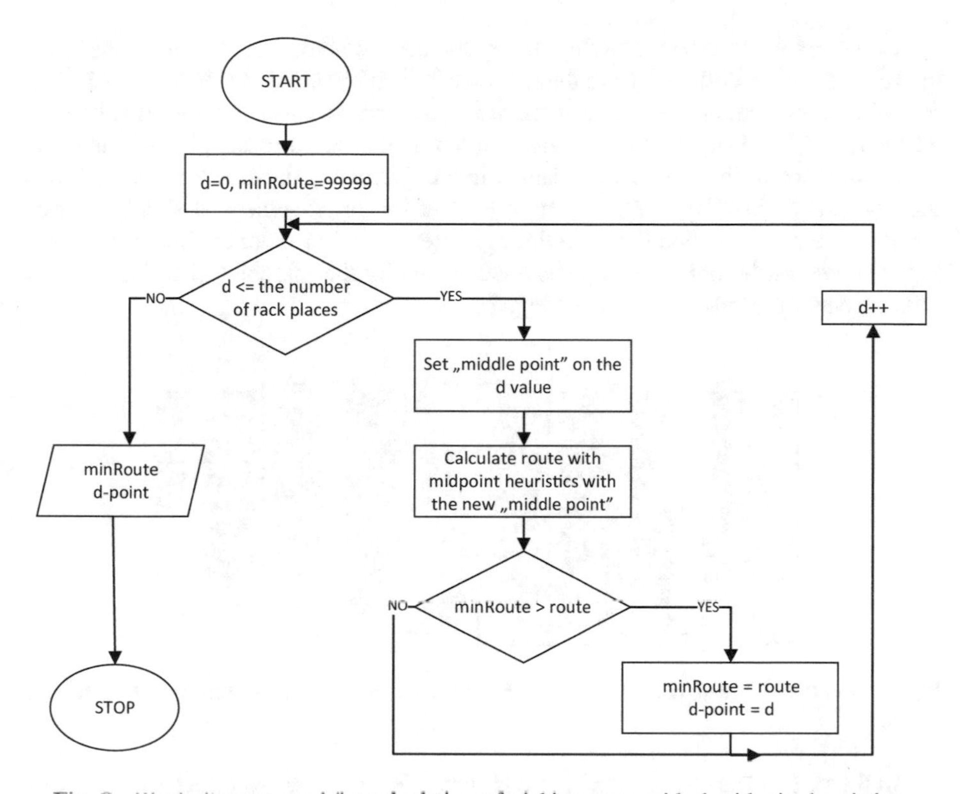

Fig. 8. Block diagram used for calculation of picking route with d-midpoint heuristics.

5 Description of Improvement

The proposed improvement of the described heuristics consists in checking whether the transition to the warehouse rear area through alleys that are not extreme will not allow to shorten the route of order picking. The authors have adopted two constraints concerning the transition to and from the rear area:

1. The nearest (to the picking point) goods alley in the rear area is the last one under consideration, which the picker should go to the rear area. If the first alley contains goods from the rear area, the picker must use it to move to the rear area.
2. The furthest alley containing goods in the rear area is the first considered alley that the picker can return to the front area. If the last alley contains goods from the rear area then the operator must go to the front area with it.

Because all heuristics described in this paper are based on the division of the warehouse into two areas (front and rear), the described improvement for all heuristics will look the same. The authors will use midpoint heuristics and examples of the arrangement of goods in Figs. 3 and 4 to illustrate the improvement. Heuristics with the described improvement will contain a "+" sign attached at the end of the name (e.g. midpoint+).

In the example presented in Fig. 3, the first item to be collected is in alley 2, while the first item to be collected from the rear area is in alley 4. According to point 1 of the described improvement, this means that it is necessary to check whether it is better to go through alley 3 or 4 to the rear area. As far as the return of the picker to the front area is concerned, the last downloadable product is in alley 9, while the last product in the rear area is in alley 7. Based on point 2 of the improvement, it should checked whether the operator should not go through alley 7 or 8 in order to shorten the route. Figure 9 shows the order picking route calculated by the midpoint+ heuristics for the arrangement of goods analogous to Fig. 3.

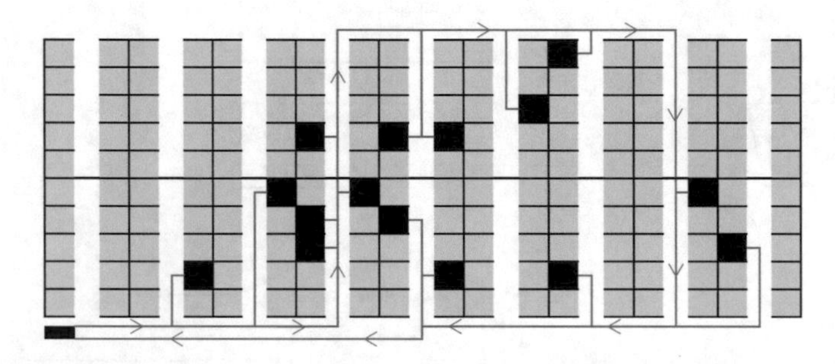

Fig. 9. Route calculated with the help of midpoint+ heuristics for an exemplary location of goods (1).

In the case of the distribution of goods shown in Fig. 4, the route determined by the midpoint+ heuristics will be the same as the midpoint route because:

- the first alleyway to the rear area is more advantageous than the second or third alley,
- in the last alley containing the goods, there are also two goods in the rear area, so according to point 2 of the described improvement, the warehouseman passes this alley to the front area.

6 Effectiveness Study of the Improvements

The effectiveness of the proposed improvement and the d-midpoint heuristic described in the previous sections has been verified on the basis of simulation test. Data for the simulation came from one of the automotive parts warehouses with the following characteristic of the storage space:

- width of a single place on the rack: 1 m,
- width of the rack: 0.75 m,
- widths of the aisle: 0.9 m.

The simulation has been carried out for different rack numbers, different number of storage places in a rack and finally with different numbers of items that have to be collected. In each case 100 000 randomly chosen placements of items have been considered. The results of the simulations have been gathered in Tables 1, 2 and 3. The values in columns 2–4 represent an average percentage change of the picking route compared to the length of the route determined by the heuristic without the improvement, while the values in columns 5–7 represent an average length of the route.

The tables also include information about the average length of the route for heuristics implementing the described improvement. The shortest average route lengths are marked in grey. The number of goods in each iteration was selected randomly from the range from 4 to 27 (see Table 3 for details).

Table 1. Average change in the length of the picking route depending on the number of racks.

Number of racks	%Mid-point+	%D-Midpoint+	%Largest Gap+	Mid-point+	D-Mid-point+	Largest Gap+
10	-1,67%	-3,07%	-2,34%	158,2	148,0	148,1
14	-1,84%	-2,65%	-2,26%	194,9	184,7	183,5
18	-1,85%	-2,38%	-2,14%	222,1	212,6	210,7
22	-1,81%	-2,19%	-2,01%	243,9	235,1	233,0
26	-1,75%	-2,04%	-1,90%	262,5	254,3	252,1
30	-1,69%	-1,92%	-1,81%	278,8	271,2	269,0
34	-1,63%	-1,82%	-1,73%	293,5	286,5	284,3
38	-1,57%	-1,74%	-1,66%	307,2	300,7	298,6
AVG	-1,73%	-2,23%	-1,98%	245,1	236,6	234,9

Table 2. Average change in the length of the picking route depending on the number of places in the rack.

The number of places in the rack	%Mid-point+	%D-Mid-point+	%Largest Gap+	Mid-point+	D-Mid-point+	Largest Gap+
6	-0,58%	-0,64%	-0,64%	112,0	108,7	107,5
12	-1,13%	-1,44%	-1,27%	150,5	146,0	144,6
18	-1,44%	-1,83%	-1,62%	188,5	182,4	180,9
24	-1,64%	-2,07%	-1,85%	226,4	218,7	217,1
30	-1,78%	-2,23%	-2,01%	264,2	255,0	253,2
36	-1,89%	-2,36%	-2,13%	302,0	291,2	289,3
42	-1,97%	-2,46%	-2,23%	339,8	327,4	325,3
48	-2,04%	-2,53%	-2,31%	377,7	363,7	361,4
AVG	-1,56%	-1,94%	-1,76%	245,1	236,6	234,9

Table 3. Average change in the length of the picking route depending on the number of items.

Number of items	%Mid-point+	%D-Mid-point+	%Largest Gap+	Mid-point+	D-Mid-point+	Largest Gap+
<4, 6>	-2,97%	-2,61%	-2,87%	133,7	132,5	133,0
<7, 9>	-2,75%	-2,96%	-2,78%	173,4	170,1	170,6
<10, 12>	-2,27%	-2,72%	-2,42%	208,0	202,4	202,5
<13, 15>	-1,89%	-2,41%	-2,10%	238,8	231,0	230,3
<16, 18>	-1,61%	-2,14%	-1,86%	266,4	256,5	254,9
<19, 21>	-1,38%	-1,91%	-1,66%	291,5	279,8	276,9
<22, 24>	-1,20%	-1,71%	-1,49%	314,2	300,8	296,6
<25, 27>	-1,05%	-1,53%	-1,35%	335,0	320,0	314,4
AVG	-1,89%	-2,25%	-2,07%	245,1	236,6	234,9

On the basis of the presented results it can be concluded that for the vast majority of cases the best results are obtained using the largest gap+ heuristic. Slightly worse results have been achieved by d-midpoint+ heuristic (less than 0.6% in favour of the largest gap+ heuristic). The worst results have been reported for the midpoint+ heuristic that was on average 3.9% worse than largest the gap+ heuristic and 3.3% worse than d-midpoint+ heuristic.

7 Conclusions

The article proposes a new heuristics for calculating order picking routes called by authors as d-midpoint. The proposed heuristics is an extension of the midpoint heuristics by the possibility of dynamic "middle point" calculation. In the implementation it is simpler than the largest gap heuristics, but it is also more computationally complex.

The authors also presented a method of improving the described heuristics (midpoint, d-midpoint and the largest gap). In order to analyze the effectiveness of the proposed heuristics and the presented improvement, simulation studies were carried out, covering in total over 50 million locations of various numbers of goods in warehouses of various sizes.

On the basis of the presented results of computer simulation it can be concluded that the d-midpoint+ heuristics allows to obtain results slightly worse than the largest gap+ heuristics (and even better for very specific examples) and significantly better than the midpoint+ heuristic. The improvement proposed in the article allows to obtain better results (compared to the heuristics not using the improvement) by 1.5% to 2.2% on average. The best percentage improvement of the proposed solutions can be observed in a small number of long alleys, from which few products are taken.

Acknowledgements. This study was conducted under a research project funded by a statutory grant of the AGH University of Science and Technology in Krakow for maintaining research potential.

References

1. Cano, J.A., Correa-Espinal, A.A., Gómez-Montoya, R.A.: An evaluation of picking routing policies to improve warehouse efficiency. Int. J. Ind. Eng. Manag. **8**(4), 229–238 (2017)
2. Chen, F., Wei, Y., Wang, H.: A heuristic based batching and assigning method for online customer orders. Flex. Serv. Manuf. J. **30**(4), 640–685 (2018)
3. De Koster, R., Le-Duc, T., Roodebergen, K.J.: Design and control of warehouse order picking: a literature review. Eur. J. Oper. Res. **182**(2), 481–501 (2007)
4. De Koster, R., Van Der Poort, E.S., Wolters, M.: Efficient order batching methods in warehouses. Int. J. Prod. Res. **37**(7), 1479–1504 (1999)
5. Dukić, G., Oluić, Č.: Order-picking routing policies: Simple heuristics, advanced heuristics or optimal algorithm. J. Mech. Eng. **50**(11), 530–535 (2004)
6. Habazin, J., Glasnović, A., Bajo, I.: Order picking process in warehouse: case study of dairy industry in croatia. Promet Traffic Transp. **29**(1), 57–65 (2017)
7. Hall, R.W.: Distance approximation for routing manual pickers in a warehouse. IIE Transact. **25**, 77–87 (1993)
8. Le-Duc, T.: Design and control of efficient order picking processes, ERIM Ph.D. Series Research in Management. Erasmus University Rotterdam (2005)
9. Petersen, C.G.: An evaluation of order picking routing policies. Int. J. Oper. Prod. Manag. **17** (11), 1098–1111 (1997)
10. Roodbergen, K.J., Vis, I.F.A., Taylor, G.D.: Simultaneous determination of warehouse layout and control policies. Int. J. Prod. Res. **53**(11), 3306–3326 (2015)
11. Roodbergen, K.J., De Koster, R.: Routing order pickers in a warehouse with a middle aisle. Eur. J. Oper. Res. **133**(1), 32–43 (2001)
12. Zulj, I., Glock, C.H., Grosse, E.H., Schneider, M.: Picker routing and storage-assignment strategies for precedence-constrained order picking. Comput. Ind. Eng. **123**(2018), 338–347 (2018)

Subjective Assessment of the Process of Taking Over Vehicle Control with Conditional Autonomy

Paula Razin[1(✉)] and Iwona Grabarek[2]

[1] Motor Transport Institute, Transport Telematics Center, 80 Jagiellonska St., 03301 Warsaw, Poland
paula.razin@its.waw.pl

[2] Faculty of Transport, Warsaw University of Technology, 75 Koszykowa St., 00662 Warsaw, Poland

Abstract. The article presents a subjective assessment of the takeover of steering control by drivers in vehicles with conditional autonomy. The assessment of both the process of taking control and the feeling of safety during autonomous driving was made on the basis of surveys. The questionnaires were supplemented by drivers of a passenger car after the end of the journey. The key role in the case of test scenarios was played by signals of various modalities informing the driver about the need to take control of the vehicle. The research results presented in the article refer to the effectiveness of signals with different modalities in the subjective assessment of the subjects. The obtained results indicate a varied reception of modality among the age groups and people with different levels of driving skills. High safety level assessment during autonomous driving may indicate a high level of confidence in autonomous vehicles, especially for people over 55 and experienced drivers.

Keywords: Autonomous vehicles · Takeover process · Signal modality · Surveys

1 Introduction

The dynamic development of highly automated transport affects not only the emergence of modern technological solutions, but also has a direct impact on the driver using the proposed solutions. Human Machine Interface (HMI) systems dedicated to vehicles with conditional autonomy are designed to quickly and accurately convey information to the driver, who in critical traffic situations may be forced to take control of the vehicle [1]. The research results presented in the article refer to the effectiveness of signals with different modalities in the subjective assessment of the subjects.

While driving a vehicle with conditional autonomy, the driver may at any time regain control of the vehicle or the vehicle system may generate such a request. Control transfer is the phase of driver activity in which the driver must meet the requirements of taking manual control of the vehicle by performing an activity appropriate to driving. The main difference for control transfer is between level 2 and level 3 systems. Level 2 assumes that the driver has to take control of the vehicle immediately after deactivation

M. Siergiejczyk and K. Krzykowska (Eds.): ISCT21 2019, AISC 1032, pp. 374–380, 2020.
https://doi.org/10.1007/978-3-030-27687-4_37

of the system. In levels 3 and 4, on the other hand, it is assumed that the driver should have sufficient time to take control of the vehicle.

The subjective assessment of the takeover of control by the driver in vehicles with conditional autonomy required simulator tests. The test procedure is described in more detail in [7]. In these studies scenarios were used that took into account the need for drivers to take over selected road situations.

2 Transfer of Control in Vehicles with Conditional Autonomy

The driver can be informed about the current traffic situation, possible threats and the need to transfer the control through dedicated HMI interfaces. The warning signals can be emitted through the auditory, visual or tactile channel (vibrations). The necessary messages can be shown on the displays in the control panel or be emitted as information on the basic indicators on the vehicle's dashboard. Interpretation of signals should take place quickly and cause the driver to react properly to the situation [8]. Another situation is taking control at the driver's request, i.e. switching from autonomous to manual mode. There are the following ways for the driver to intervene to regain manual control of the vehicle:

- deactivating the system by depressing the brake pedal,
- deactivate the system by pressing the accelerator pedal,
- deactivation of the system by a significant movement of the steering wheel,
- deactivate the system by pressing the corresponding light.

Only the first three ways to take control of the vehicle were allowed in the test experiment.

3 Research Methods

3.1 Characteristics of the Research Group

The questionnaire survey was conducted on a group of 30 adults who had a category B license. Access to the tests was conditioned by possession of a driving license for at least two years and active car travel every day. The subjects represented three parallel groups classified by age. The first group consisted of drivers aged 20–25, the second group - drivers aged 30–40, and the third drivers aged 55 and over. The average age of drivers was 38 years (SD = 14.04), the youngest driver was 20 years old and the oldest was 64 years old. The research group was also classified on the basis of active driving time (given in years). Persons who had a driving license and actively traveled under 3 years were defined as inexperienced drivers. In turn, a group of experienced drivers have been characterized as possessing licenses and using a car every day or almost every day for over 3 years.

3.2 Research Procedure

In order to subjectively assess of the transfer of control in a vehicle with conditional autonomy, it was necessary to conduct simulation tests. The testing procedure included driving in accordance with the assumed scenario reflecting the road conditions, which included the necessity for the drivers to take control in selected road situations. The subjects were driving a vehicle with conditional autonomy (at the level of L3 according to SAE [9]). The assessment of both the transfer of control and the sense of safety during autonomous driving was made on the basis of surveys. The questionnaires were supplemented by drivers of a passenger car equipped with a multisensory stand directly after the end of the journey. The use of the survey method allowed to standardize the respondents' answers, which enabled the subsequent analysis of the results obtained.

A survey with a 20-point scale was used for the study, which was created on the basis of the modified NASA-TLX sheet [2–4]. The survey included three questions:

- How do you rate the situation of taking control (from manual driving to autonomous mode)?
- How do you rate the situation of taking over control (from autonomous driving to manual mode)?
- How did you feel when driving autonomously?

The respondents gave answers by marking the "X" in the right place. The first grid from the left on the scale was negative, meaning "Very Uncomfortable" (1st and 2nd question) or "Very Dangerous" (3rd question). In turn, the last grid on the right was a positive grade, meaning "Very Comfortably" (1st and 2nd question) or "Very Safe" (3rd question). Figure 1 shows an example of a scale of grades.

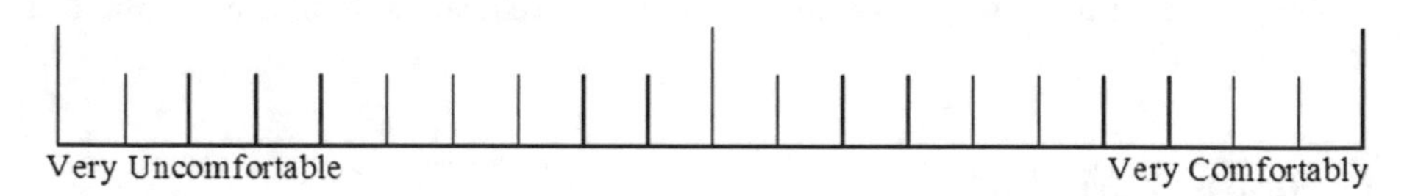

Fig. 1. The scale of grades in questionnaire surveys. Source: own study based on [2].

4 Experiment Results

4.1 Subjective Evaluation of the Process of Taking Control

The driver in manual mode can activate the autonomous mode at any time by selecting the appropriate button on the ADS console. After activating the system, the driver can take his hands off the steering wheel and legs off the pedals. The driver in autonomous mode, on the other hand, can deactivate the system either at his own request or by the on-board system initiating the process. In the experiment, the drivers evaluated the process of handing over control to the vehicle (e.g. after a safe avoidance of a road accident) and the process of taking control of the vehicle caused by the existing road situation. The need to take control of the vehicle was communicated via HMI interfaces located inside a driving simulator with a multisensory station. The subjects received visual, audible and haptic signals.

The research group was divided into 3 parallel subgroups, the first of which was informed only by means of visual stimuli, the second - by means of visual and auditory stimuli, and the third - by visual, audible and haptic stimuli in the form of seat vibration. The subjects were given the opportunity to take control from autonomous to manual mode by depressing the accelerator or brake pedal as well as by significantly moving the steering wheel. After completing the test scenario, the drivers filled out a survey, the results of which are shown in Figs. 2, 3, 4 and 5. For the purposes of the analysis, the research group was additionally divided according to age (< 25 years, 30–40 years, > 55 years) and due to experience in driving (inexperienced - actively leading to 3 years, experienced - drivers actively over 3 years).

Subjective Assessment of the Process of Taking Control from Autonomous to Manual Mode. The graph (Fig. 2) shows the results of the subjective evaluation of the takeover of control from autonomous to manual mode, divided into groups of inexperienced and experienced drivers. Active drivers for more than 3 years considered the most comfortable situation of taking control as the one preceded by a visual and audible signal (average 14.29 points). On the other hand, the combination of three modalities (18 points on average), i.e. visual, auditory and haptic stimuli (in the form of seat vibrations), was considered the best way to inform about the need to take control by inexperienced drivers who drove vehicles under 3 years old.

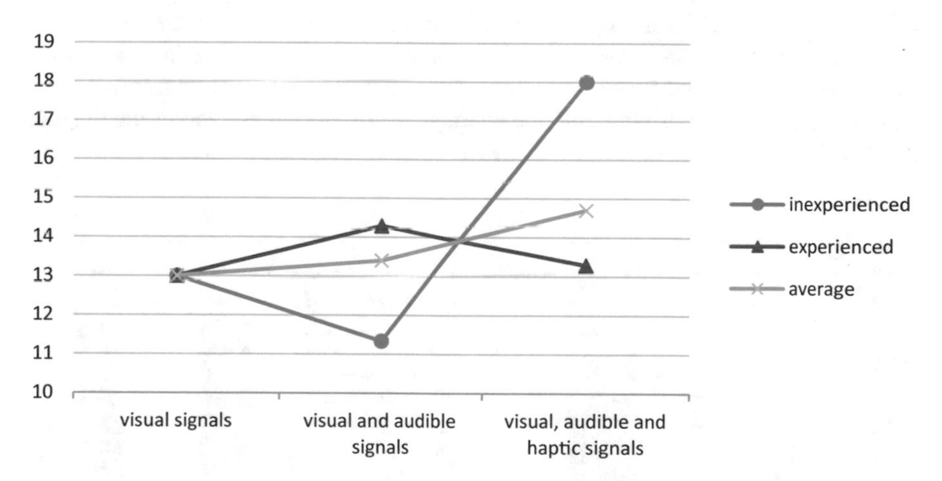

Fig. 2. Evaluation of the comfort of taking control from autonomous to manual mode (drivers' division due to experience).

The division of subjects into age groups showed other differences in the reception of signals with different modalities. Both the group up to 25 years of age and people aged 30–40 years indicated a combination of three types of stimuli as the most comfortable in reception. On the other hand, the group over 55 years of age rated the visual stimulus as the lowest and the combination of visual and auditory signals the highest, as shown in Fig. 3.

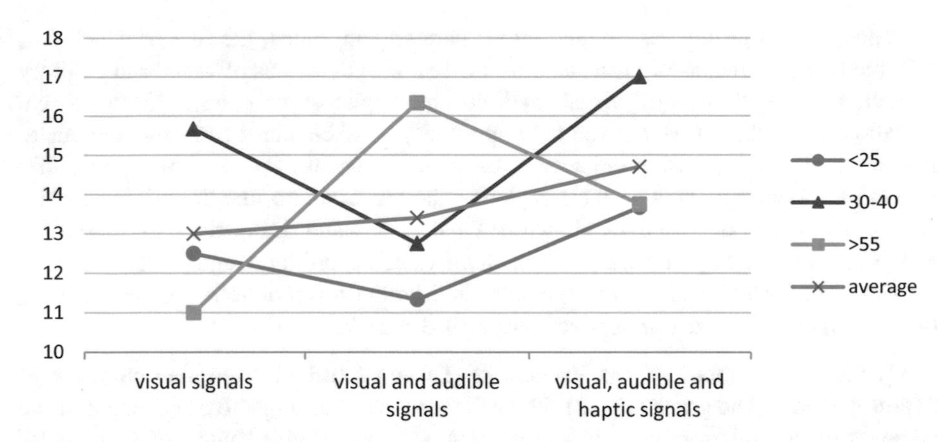

Fig. 3. Assessment of the comfort of taking control from standalone to manual mode (division of drivers by age).

Subjective Assessment of the Process of Taking Control from Manual to Autonomous Mode. By the term comfort of transition from manual mode to autonomous mode, the drivers understood, among other things, the convenience and intuitive use of the ADS console. The transfer of steering control and switching to the autonomous mode was performed by the drivers after bypassing a dangerous road section (e.g. a road accident). The mode was started after pressing the appropriate indicator on the ADS console [7]. In the survey, the drivers evaluated the comfort of transition from manual to autonomous driving, and the average results were presented in the diagrams (Fig. 4).

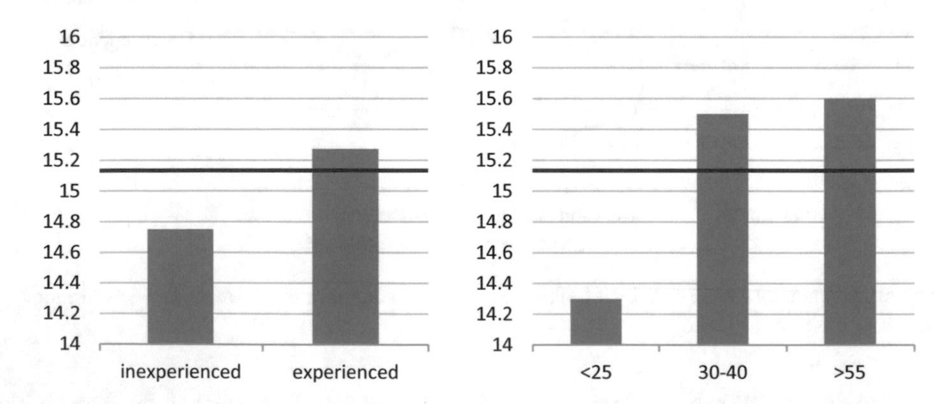

Fig. 4. Assessment of the comfort of taking control from manual to autonomous mode (the division of drivers due to experience - on the left, the division of drivers due to age - on the right).

Among all participants in the study, the comfort of handing over the control to the vehicle was assessed as the lowest by inexperienced drivers and young people (14.75

and 14.3 points respectively). During this process, the most comfortable people felt active drivers for over 3 years and people over 55 years of age (15.27 and 15.6 points respectively). Slightly lower results were obtained by people aged between 30 and 40 years (15.5 points). The average of all respondents was 15.13 points with a standard deviation of 5.37.

4.2 Subjective Evaluation of Safety Level During Autonomous Driving

During the survey, results were also obtained on subjective evaluation of the level of safety during autonomous driving. The experiment took place in simulated traffic conditions and included motorway driving. The subjects evaluated the safety level on a 20-degree scale. The average results are presented in the diagrams below (Fig. 5).

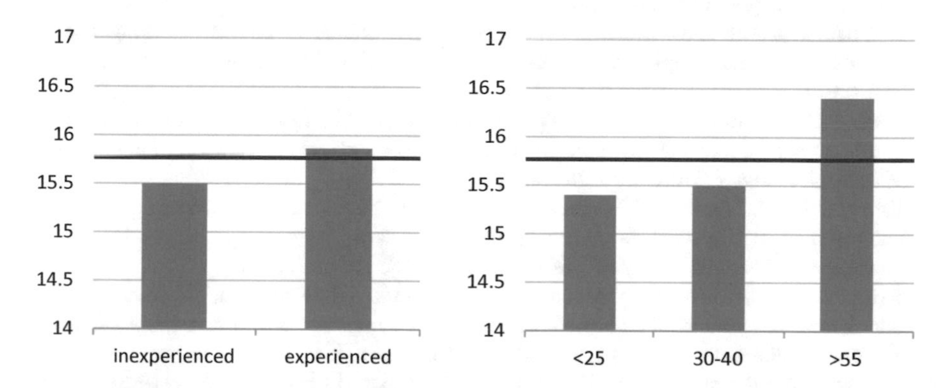

Fig. 5. Subjective assessment of the level of safety during autonomous driving (the division of drivers by experience - on the left, the division of drivers by age - on the right).

The obtained results indicate a higher level of safety during autonomous driving for experienced drivers and persons over 55 years of age (15.86 and 16.4 points respectively). The results were below the average, which amounted to 15.77 points. (SD = 5.53) were scored by inexperienced drivers (15.5 points), drivers under 25 years of age (15.4 points) and people aged 30–40 (15.5 points).

5 Summary and Conclusions

The article presents a subjective assessment of both the process of transfer control in vehicles with conditional autonomy, as well as the feeling of safety while driving in autonomous mode.

The obtained results indicate a varied reception of signal modalities among age groups and people with different levels of driving skills. Drivers experienced above assessed the use of visual and sound signals than inexperienced drivers, who in turn rated higher the addition of a haptic stimulus in the form of seat vibration. In the case of switching the vehicle from manual to autonomous mode, both inexperienced drivers and people below 25 years of age felt less comfortable than other test subjects.

High assessment of the perceived level of safety during autonomous driving may indicate a high level of confidence in autonomous vehicles, especially for people over 55 and experienced drivers.

This article is the result of the PhD thesis and the dean's grant at the Faculty of Transport of the Warsaw University of Technology.

References

1. Choromański, W., Grabarek, I., Spirzewska, A.: Systemy human machine interface (HMI) dedykowane samochodom poziomów L2/L3. Prace Naukowe Politechniki Warszawskiej – Transport **115**, 35–45 (2017)
2. Enquist, J., Rudin-Brown, C.M., Lenne, M.G.: The effects of on-street parking and road environment visual complexity on Ravel Speer and reaction time. Accid. Anal. Prev. **45**, 759–765 (2012)
3. Hart, S.G., Staveland, L.E.: Development of NASA-TLX (Task Load Index): results of empirical and theoretical research. In: Hancock, P.A., Meshkati, N. (eds.) Human Mental Workload. Elsevier Science, Amsterdam (1988)
4. Kruszewski, M., Razin, P., Niezgoda, M., Nader, M.: Model oceny dodatkowego obciążenia poznawczego u kierowcy na podstawie badań z wykorzystaniem symulatora jazdy. Prace Naukowe Politechniki Warszawskiej - Transport **118**, 155–166 (2017)
5. Kruszewski, M., Razin, P., Niezgoda, M., Smoczyńska, E., Kamiński, T.: Analiza efektów oddziaływania symulatora na powstawanie choroby symulatorowej w badaniach kierowców. Systemy Logistyczne Wojsk **44**, 188–201 (2016)
6. Matysiak, A., Razin, P.: Analysis of the advancements in real-life performance of highly automated vehicles' with regard to the road traffic safety. In: MATEC Web of Conferences (2018)
7. Razin, P., Grabarek, I.: Koncepcja oceny przejęcia kontroli sterowania przez kierowcę w pojazdach z warunkową autonomizacją, Prace Naukowe Politechniki Warszawskiej – Transport **121**, 319–327 (2018)
8. Razin, P., Matysiak, A., Kruszewski, M., Niezgoda, M.: The impact of the interfaces of the driving automation system on a driver with regard to road traffic safety. In: MATEC Web of Conferences (2018)
9. SAE International, Sufrace Vehicle Recommended Practice, Taxonomy and Definitions for Terms Related to Driving Automation Systems for On-Road Motor Vehicles, J3016 (2016)

The Role and Importance of Electric Cars in Shaping a Sustainable Road Transportation System

Ewelina Sendek-Matysiak(✉)

Department of Mechatronics and Machine Construction,
Kielce University of Technology,
al. Tysiąclecia Państwa Polskiego 7, 25-314 Kielce, Poland
esendek@tu.kielce.pl

Abstract. Road transportation is an important area of research within the framework of sustainable development, which has had the greatest destructive impact on the environment for many years. Among the negative effects of this mode of transportation on the natural environment presented in the literature on the subject, attention is drawn primarily to air pollution by harmful emissions from transport means, climate change caused by CO_2 emissions and noise pollution. Despite the fact that new vehicles emit much less harmful substances, road transportation (including passenger vehicles) remains one of the few industrial sectors for which total emissions are increasing. This is due to increasing freight transportation, growing car fleets and increasing mileage. As a result, there is a constant need for new solutions which would be more environmentally friendly and linked to the implementation of the concept of sustainable development. One of them, currently strongly promoted, is the large-scale deployment of purely battery-powered electric cars, the so-called BEVs (Battery Electric Vehicle). The purpose of this article is to demonstrate whether such vehicles can actually contribute to reducing the negative impact of road transportation on the environment and consequently to the balancing of transportation systems.

Keywords: Electric car · Sustainable development · Noise pollution

1 Introduction

Since the industrialization period, natural resources, especially raw materials, have been intensively exploited, which has had a negative impact on the environment (with emissions of harmful gases, air and water pollution, etc., being byproducts of the process). Parallel to the process of industrialization and population growth, agricultural production was intensified, the development of which, over time, became an increasing threat to ecosystems, e.g. through acidification of soils and pollution of water with artificial fertilizers. As a result, some species of flora and fauna became extinct. According to the belief and theory of economics, the intensification of production processes was supposed to lead to an increase in prosperity. The assumption that social development is based on the pursuit of material needs was revised over time and

M. Siergiejczyk and K. Krzykowska (Eds.): ISCT21 2019, AISC 1032, pp. 381–390, 2020.
https://doi.org/10.1007/978-3-030-27687-4_38

considered incomplete. The need of inclusion in economic activity, in addition to the economic dimension, of the role of social and environmental factors in order to meet the needs of not only the current generation, but also those of future generations, has begun to be noticed. Such an approach corresponds to the concept of sustainable development. The idea of sustainable and lasting development is nowadays the binding postulate of human activity in all developed and developing countries of the world. Sustainable and lasting development aims at the sustainable improvement of the quality of life of present and future generations through the appropriate shaping of the proportions between the various types of capital: economic, human and natural. One of the basic assumptions of this concept is the principle: "think globally – act locally".

Currently, the environmental friendliness aspect, i.e. the necessity to limit the negative impact on the natural environment, is the main factor leading to the development of technology and organization in all areas of the economy.

Meanwhile, according to the studies, transportation is the most environmentally harmful sector of the economy. Unfortunately, apart from numerous benefits evident in the progress and economic development, transportation also carries a very high risk in the form of external costs, i.e. noise pollution and vibrations, surface and ground water pollution, soil pollution, microclimate changes, occupation of land for the construction of transportation route networks, accidents and, above all, air pollution (including CO_2 and other greenhouse gases) through the exhaust emission (see Table 1).

Table 1. Environmental impact of transportation [1]

Environmental impact	Air pollution	Water pollution	Land and forests	Health and safety
1	2	3	4	5
constant			land use, damage to the landscape, migration of people	
during the exploitation	CO_2 and other greenhouse gas emissions (including HC_2)	spread of pollution	acid rains, adverse effects	accidents, noise pollution, vibrations
risk during transhipment	spread of dangerous substances	spread of dangerous substances	spread of dangerous substances, risk of fire	toxic leaks, risk of fire and explosion
congestion			compared to normal exploitation, congestion exerts greater impact due to time losses and restricted energy efficiency	
Ranking of impact by environmental elements and modes of transportation*				
roads	***	*	***	***
railroad	*b		**	*
waterways		**	*	
maritime transportation	*	**c	*	
air transportation	*		*	*

a - * low influence, ** important influence, *** very important influence; b - plus activity of power plants; c – the influence can be significant, when an accident happens

Next to the energy sector, transportation has been the main source of greenhouse gas emissions in the European Union for many years (see Fig. 1), and in the transportation sector itself, road transportation is the largest generator responsible for greenhouse gas emissions (see Fig. 2).

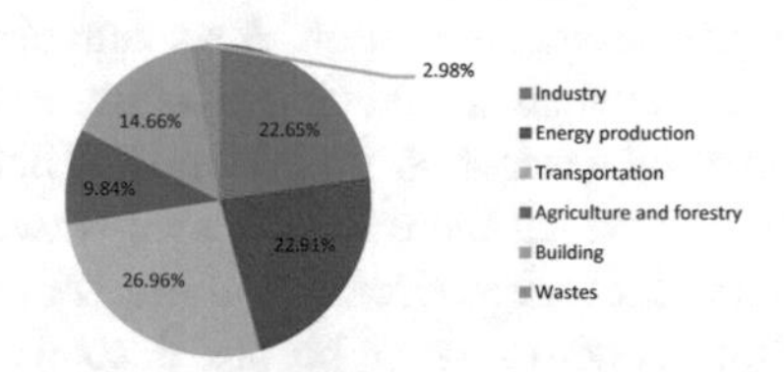

Fig. 1. Greenhouse gas emissions in the EU in 2016 [2].

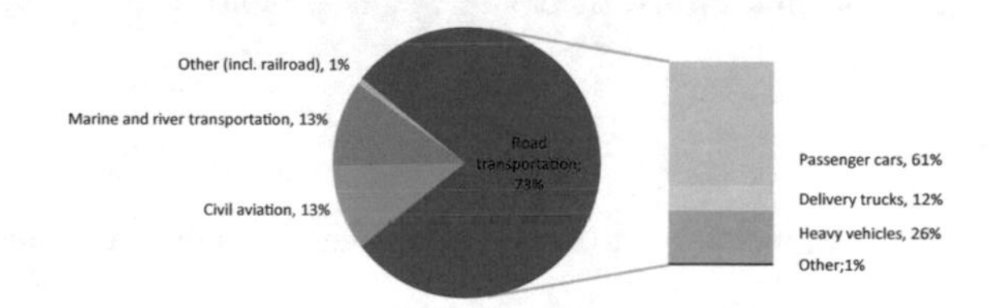

Fig. 2. Greenhouse gas emissions in 2015 in the European Union in the transportation sector [author's own study based on: Eurostat: 3].

Another equally serious problem related to road traffic is excessive traffic noise. According to the EU report [4], approximately 40% of the European population is exposed to traffic noise levels exceeding 55 decibels, while 20–30% to levels above 65 decibels during the day and 55 decibels at night.

Figures 3 and 4 show the approximate number of people in the European Union exposed to harmful noise.

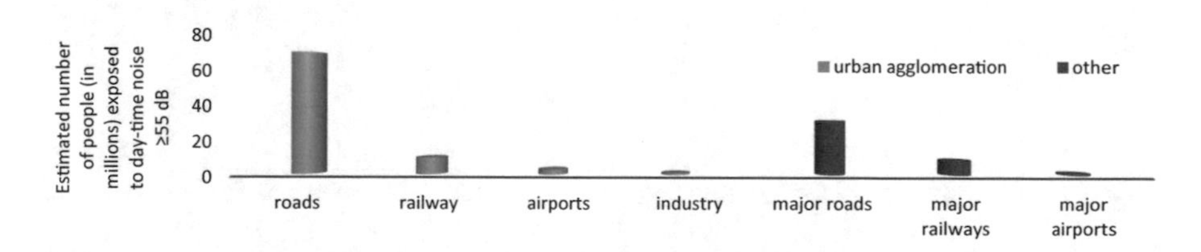

Fig. 3. Number of people exposed to day-time noise (Lden) $\geq$ 55 dB in EU-28 in 2017 [5].

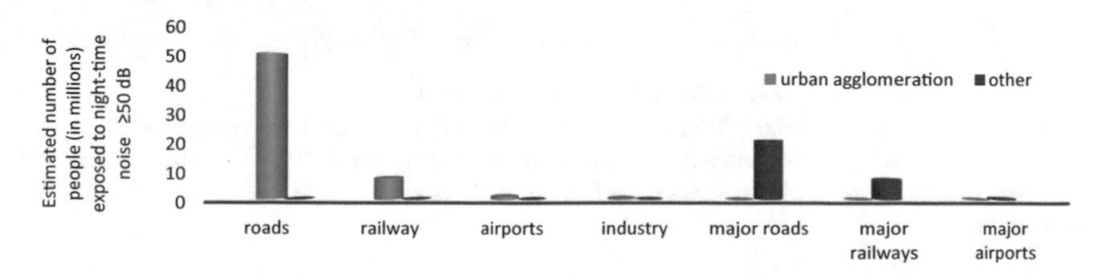

Fig. 4. Number of people exposed to night-time noise (Lnight) $\geq$ 50 dB in EU-28 in 2017 [5].

At present, at least 10,000 Europeans die prematurely as a result of excessive noise [6] and the costs resulting from the impact of road noise on public health are estimated at €40 billion per year [7].

Therefore, the European Union has been taking coordinated actions to promote and implement a single sustainable transportation system within its territory for many years, paying particular attention to road transportation. A sustainable transportation system is one that strikes a balance between social and economic factors, and the spatial development and environmental protection in a country. Therefore, the right form of transportation system is one in which there is a balance between economic and social aspects, and spatial development and environmental protection. This suggests that the design of the transportation system cannot be based solely on economic or social factors, but must take into account environmental concerns [8]. Therefore, the European Union's documents on the common transport policy specify objectives, priorities and corresponding measures aimed at ensuring sustainable development in this area (see Table 2).

Table 2. EU progress in meeting selected transportation goals on energy and climate [9]

	Source	Target	Unit	Where we were		Where we want to be		Where we are (current trends vs. target paths)																	Latest annual trend
				Base year		Target		2000	2010		2011		2012		2013		2014		2015		2016		2017		
				Year	Value	Year	Value	Observed	Target	Observed	Target	Observed	Target	Observed	Target	Observed	Target	Observed	Target	Observed	Target	Observed	Target path	Observed	
Indicative goals	European Commission's 2011 Transport White Paper (EC, 2011; EC, 2016)	Transport GHG (including international aviation, excluding international maritime shipping) (a)	$MtCO_2$	1990	856	2030 2050	920 (+8 %) 335 (-60 %)	1 039	1 113	1 067	1 115	1 058	1 117	1 027	1 119	1 021	1 122	1 031	1 124	1 052	1 106	1 080	1 088	1 096	**1.5 %**
Indicative goals	European Commission's 2011 Transport White Paper (EC, 2011)	EU CO_2 emissions of maritime bunker fuels	$MtCO_2$	2005	163		98 (-40 %)	136	155	162	154	164	153	151	151	143	150	140	148	141	147	147	145	n.a.	**4.3 %**
Indicative goals	Impact assessment accompanying document to the 2011 Transport White Paper (EC, 2011)	Reduction of transport oil consumption	million TJ	2008	17.3	2050	5.2 (-70 %)	15.9	16.8	16.4	16.5	16.3	16.2	15.6	15.9	15.4	15.6	15.6	15.3	15.8	15.0	16.3	14.7	n.a.	**3.1 %**
Indicative goals	Renewable Energy Directive 2009/28/EC	10 % share of renewable energy in the transport sector final energy consumption for each Member State (here EU-28 as a proxy) (b) (c)	%	2010	5.20	2020	10.0		5.22	5.22	5.70	3.95	6.18	5.56	6.66	5.94	7.13	6.52	7.61	6.6	8.09	7.1	8.6	7.2	**0.8 %**
Binding targets	Passenger car CO_2 EC regulation 443/2009	Target average type-approval emissions for new passenger cars (a)	gCO_2/km	2010	140	2015 2021	130 95	172	140	140	138	136	136	132	134	127	132	123	130	120	124	118	118	118.5	**0.3 %**
Binding targets	Van CO_2 EC regulation 510/2011	Target average type-approval emissions for new light commercial vehicles (a)	gCO_2/km	2012	180	2017 2020	175 147						180	180	179	173	178	169	177	168	176	164	175	156	**-4.6 %**

([a]) Preliminary date for 2017

([b]) EU-28 excl. Croatia until 2013. Eu-28 from 2014 onwards. Data for 2017 is provisional

([c]) In the case of Renewable Energy Directive (EU, 2009) target, Eurostat published for the first time (2011 date) the share of biofuels in transport energy use which meet the sustainability criteria of the Directive. The huge increase between 2011 and 2012 (increase by 40.8%) is explained by the fact in previous years the new sustainability criteria were not fully applied. The system for certifying sustainable biofuels is increasingly operational across all MS.

The currently in force 2011 White Paper entitled "Roadmap to a single European transport area – Towards a competitive and resource-efficient transport system" states, among other things, that by 2050 the EU Member States should reduce emissions of harmful substances into the atmosphere by 60% compared to 1990 levels and phase out combustion vehicles from cities. This opens up enormous opportunities for the development of electric cars, which use exclusively electricity stored in the on-board battery that must be recharged regularly by connecting to a charging point connected to the local electricity grid. The aim of the considerations presented below was to conduct multifaceted evaluation of the use of electric cars in modeling a sustainable road transportation system with current technologies and manufacturing techniques.

2 The Role of Electric Cars in Sustainable Development

Electric car enthusiasts believe that the increased market share of such vehicles in the automotive market will contribute to:

- the reduction of emissions of harmful substances generated by the transportation sector,
- the reduction of traffic noise,
- the improvement of the energy security of the European Union due to the reduction of import of fuels.

However, studies shows that although electric cars do not generate emissions at their place of use, in order for them to become an alternative to ICE vehicles in terms of reducing pollution, they must have lower emissions during their use. Nowadays, when comparing the production process of a BEV to an internal combustion engine vehicle, the former produces higher emissions, which is related to the production of batteries installed in such cars. Apart from the production of batteries for electric cars, the level of harmful substance emissions in the production of electric cars and those with conventional engine is comparable (see Fig. 5).

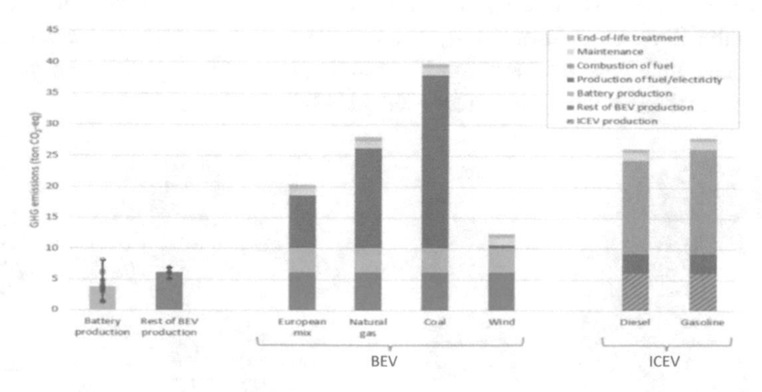

Fig. 5. Lifecycle GHG emissions of mid-sized 24 kWh battery electric (left) and internal combustion engine (right) vehicles [10]

Therefore, in order for electric cars to have lower lifecycle emissions than conventional engine vehicles, they need to have lower emissions in the use phase.

Lifecycle assessment studies concluded that whether or not BEVs can compensate for higher emissions during production is largely dependent on the carbon intensity of the energy sources used to recharge the batteries. Studies [11–13] prove that electricity obtained solely from coal results in electric cars producing more greenhouse gas emissions than ICE vehicles (see Fig. 6).

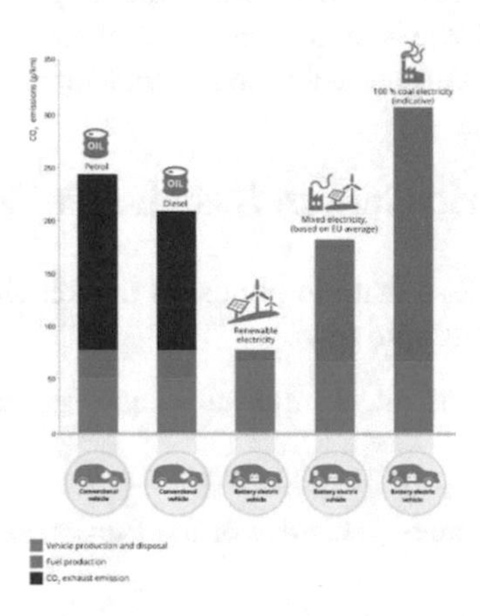

Fig. 6. Range of life-cycle CO_2 emissions for different vehicle and fuel types [14].

In countries where electricity generation is based on a variable mix of different energy sources, the lifecycle emissions gap between the two powertrain technologies is relatively small. The greatest benefits in shaping a green transportation system, due to the use of electric cars, can be obtained in countries where the share of renewable or nuclear energy is high.

2.1 Use of Electric Cars in the European Union

According to a report prepared by Agora Energiewende and Sandbag, in 2017, for the first time in the European Union wind, sun and biomass were responsible for producing more electricity than hard coal and lignite combined (see Fig. 7). An even greater share of energy from these sources was recorded in 2018. (see Fig. 7).

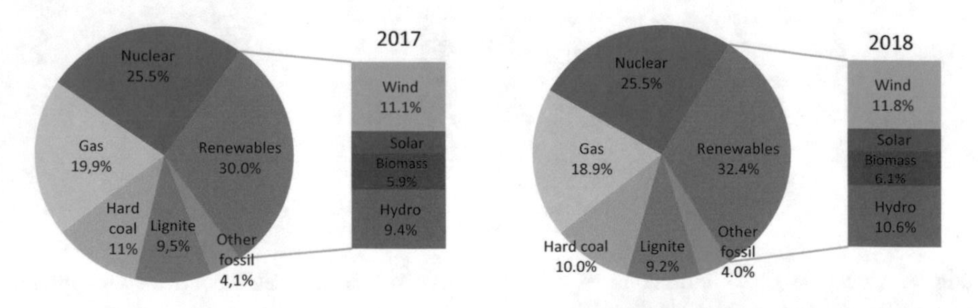

Fig. 7. Generation mix in 2017 and 2018 in the European Union [15].

This means that BEVs powered by electricity from the average European basket, albeit slightly, will reduce greenhouse gas emissions compared to vehicles powered by both diesel and gasoline engines, assuming a vehicle life of 150,000 km [11–13].

However, there are countries in the EU where the share of coal in the production of electricity is still significant and in 2017 it was as high as 77% (see Fig. 8).

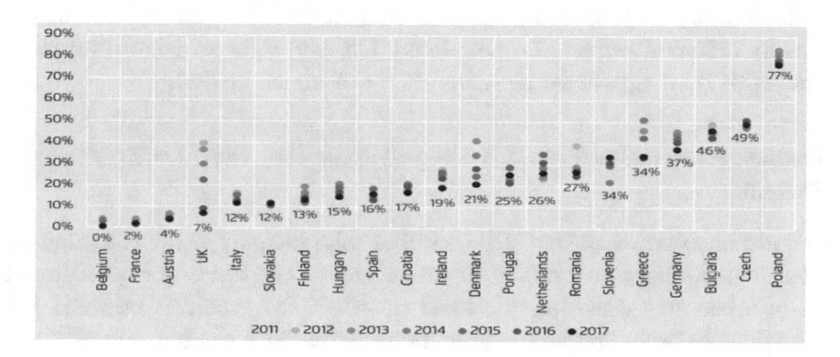

Fig. 8. Hard coal and lignite as percentage of national electricity production [3].

However, in paper [16] it was pointed out that even for an energy mix in which about 85% of electricity is produced by burning coal, the replacement of a combustion engine car with an electric one is more favorable for climate protection. The research conducted by [16] also prove that diesel-powered cars cause higher greenhouse gas emissions than those propelled by electricity produced by power plants with an efficiency of up to 40% (e.g. in Poland) (see Fig. 8).

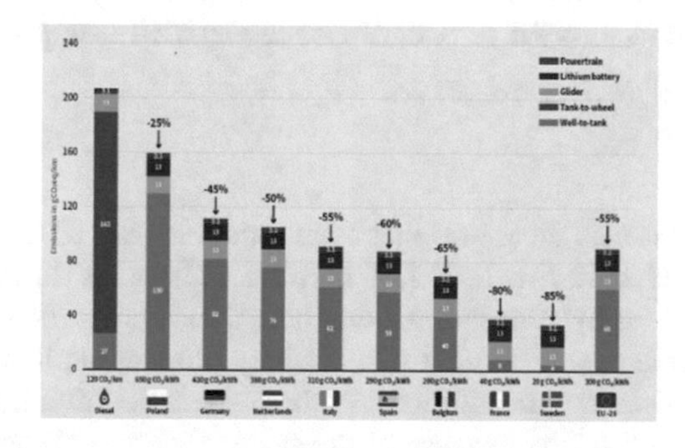

Fig. 9. Electric vehicles' climate impact in different energy mixes [16].

Paper [17] found that the electric Nissan Leaf (80 kW, 280 Nm) emits less, although this difference is slight, CO_2 per kilometer than Toyota Auris with an internal combustion engine (1.4 diesel, 66 kW, 205 Nm). Converting the energy consumption of a vehicle into the air pollution emissions of Polish power plants gives emissions of 118 g/km, whereas for Toyota Auris it is 124 g/km. However, this data shows the advantage of an electric car, which should increase over time (Fig. 9).

According to a Swiss study, a combustion engine-powered car would have to consume three to four liters of fuel in order to be as environmentally friendly as an electric car (with the same characteristics) powered by a power plant that emits an average European amount of pollutants. The data is even more in favor of electric cars if they use renewable or zero-emission energy. Using energy from natural resources can even reduce their emissivity by at least 50% compared to ICE vehicles (see Fig. 5) [11, 18, 19].

In addition, by implementing electric cars, it will be possible to reduce CO_2 emissions by reducing demand for petroleum. Currently, its extraction and processing is accompanied by a significant emission factor of about 110 g/l.

2.2 Reducing Excessive Traffic Noise and Balancing the Electric Power System

Increased use of electric cars can also contribute significantly to the reduction of noise from road transportation, especially in urban areas. The noise generated by electric vehicles in urban traffic reaches the level of 40–60 dB (vehicles powered by combustion engines in urban traffic: a passenger car – 60–70 dB, a bus – 90 dB, a truck – 100 dB [20]). The paper [21] indicated that electric cars generate much lower noise level than their combustion equivalents, especially in the speed range up to 50 km/h. Measurement of noise for two identical vehicles differing only in the power unit showed differences of 3 to 7 dB(A) depending on the speed of the vehicle, which means 2- to 5-fold reduction in noise level. For the quietest electric car used in the test (Nissan Leaf) and the noisiest combustion vehicle (BMW E30), the maximum difference was approx. 17 dB for speeds up to 50 km/h and 14 dB for higher speeds.

Undoubtedly, electric cars that require charging from the grid can become a very important link in the power grid. In addition to their transport function, electric cars could, in the future, act as mobile batteries, which not only consume, but also feed electricity into the grid in order to better balance electricity supply and demand.

3 Summary

The research carried out so far shows that even today, in case of the average European Union energy mix, electric cars emit less harmful substances during their entire lifecycle than cars powered by conventional engines. Considering the EU policy aimed at intensifying renewable energy sources in electricity production, it can be assumed that their emissions will decrease. In addition, another benefit contributing to decreasing the emissions of BEVs will be the increase of the share of renewable sources in electricity production in the countries producing battery cells.

In addition, due to the fact that batteries used in BEVs are a relatively new technology, they have considerable potential for improving production efficiency and technological improvements such as increased gravimetric energy density and longer lifespan. Such improvements can also contribute to reducing the environmental impact of electric cars during their lifecycle.

In conclusion, despite the difficulties associated with electric cars (purchase cost, short range, long battery life), these vehicles are promoted as a key factor in building a sustainable mobility system. Everything seems to indicate that they will break the European Community's long-standing dependence on internal combustion engines and petroleum products as a means of meeting its transportation needs. Increased use of electric vehicles, in particular when powered by renewable energy sources, can play an important role in achieving the EU's objective of reducing greenhouse gas emissions by 80–95% by 2050 and in the transition to a low-carbon economy in the future.

Under the current conditions, electric cars, through zero emissions of toxic compounds and CO_2 during their use, can contribute to improving air quality near heavy traffic roads and urban agglomerations. This is important, as it is in cities that the external costs of road transportation development are most perceivable in terms of local air pollution, smog and noise. It should be noted, however, that the implementation of electric cars is currently associated with risks associated with the use of rare earth elements or the disposal of batteries.

References

1. European Commission: The State of the Environment in the European Community, COM (23) final, vol. 3, p. 71, Brussels (1992)
2. European Environment Agency: EEA greenhouse gas – data viewer. ww.eea.europa.eu/data-and-maps/data/data-viewers/greenhouse-gases-viewer#tab-based-on-data. Accessed 02 Apr 2019
3. Eurostat. http://ec.europa.eu/eurostat/data. Accessed 29 Mar 2019
4. World Health Organization: Data and statistics. http://www.euro.who.int/en/health-topics/environment-and-ealth/noise/data-and-statistics. Accessed 03 Apr 2019
5. European Environment Agency. Data and Maps. https://www.eea.europa.eu/data-and-maps. Accessed 25 Mar 2019
6. European Environment Agency: Signals 2016 – Towards clean and smart mobility. https://www.eea.europa.eu/publications/signals-2016. Accessed 27 Mar 2019
7. European Commission: Report from the Commission to the European Parliament and the Council on the implementation of the Environmental Noise Directive in accordance with Article 11 of Directive 2002/49/EC. COM (2011) 321 final, Brussels (2011)
8. Żak, J., Jacyna-Gołda, I., Merkisz-Guranowska, A., Sivets, O.: Rola i znaczenie transportu drogowego w kształtowaniu efektywnego zrównoważonego systemu transportowego, Logistyka 6 (2014)
9. European Environment Agency. https://www.eea.europa.eu/themes/transport/term/term-briefing-2018/table1. Accessed 20 Mar 2019
10. Directorate-General for Internal Policies of the Union, Directorate for Structural and Cohesion Policies: Research for TRAN Committee: Battery-powered electric vehicles: market development and lifecycle emissions Transport and Tourism (2018)
11. Ellingsen, L.A.-W., Singh, B., Strømman, H.: The size and range effect: lifecycle greenhouse gas emissions of electric vehicles. Environ. Res. Lett. **11**, 054010 (2016)
12. Miotti, M., Hofer, J., Bauer, C.: Integrated environmental and economic assessment of current and future fuel cell vehicles. Int. J. Life Cycle Assess. **22**, 94–110 (2015)

13. Notter, D.A., Kouravelou, K., Karachalios, T., Daletou, M.K., Haberland, N.T.: Life cycle assessment of PEM FC applications: electric mobility and μ-CHP. Energy Environ. Sci. **8**, 1969–1985 (2015)
14. TNO: Energie- en milieu-aspecten van elektrische personenvoertuigen. http://www.nederlandelektrisch.nl/file/download/33742992. Accessed 19 Mar 2019
15. Sandbag, The European Power Sector in 2017 – Agora Energiewende. https://sandbag.org.uk/wp-content/uploads/2018/01/EU-power-sector-report-2017.pdf. Accessed 02 Apr 2019
16. Le Petit, Y.: Life Electric cars have significantly lower climate impact than diesels over their lifetime – study. https://www.transportenvironment.org/news/electric-cars-have-significantly-lower-climate-impact-diesels-over-their-lifetime-%E2%80%93-study. Accessed 26 Mar 2019
17. Solaris. http://solaris18.blogspot.com/2009/12/czy-samochody-ektryczne-sa.html. Accessed 02 Mar 2019
18. Bauer, C., Hofer, J., Althaus, H.-J., Del Duce, A., Simons, A.: The environmental performance of current and future passenger vehicles: Life Cycle Assessment based on a novel scenario analysis framework. Appl. Energy **157**, 1–13 (2015)
19. Hawkins, T.R., Singh, B., Majeau-Bettez, G., Strømman, A.H.: Comparative environmental life cycle assessment of conventional and electric vehicles. J. Ind. Ecol. **17**, 53–64 (2012)
20. Záskalický, P.: Study of current space phasor trajectory of the three-phase asynchronous motor with one phase open circuit fault. http://www.komel.katowice.pl/ZRODLA/FULL/109/ref_24.pdf. Accessed 24 Mar 2019
21. Łebkowski, A.: Electric vehicles - the sound of silence. Maszyny Elektryczne - Zeszyty Problemowe (Electrical Machines - Transaction Journal), No. 1/2016

Analysis and Assessment of Risk in Digital Railway Radio Communication System

Mirosław Siergiejczyk(✉)

Faculty of Transport, Warsaw University of Technology, Warsaw, Poland
msi@wt.pw.edu.pl

Abstract. This paper presents an analysis of potential threats to information transmission in a GSM-R system, including technical factors, human factors and organizational factors. Modern telecommunications systems are exposed to a variety of hazards variable in time. In this paper, they are divided into three underlying groups: hardware and software related, intentional and unintentional as well as internal and external.

The role of risk assessment in information security management is presented. Distinct phases of risk assessment process are characterized. Methods used in the process of risk assessment and risk planning are discussed, taking into account possible responses to threats and opportunities. Lastly, an original method for identifying threats and detecting vulnerabilities in railway radio communication systems is presented.

Keywords: Threats · Vulnerability · Risk · Assessment

1 Introduction

GSM-R (Global System for Mobile Communications - Railway) is the European standard for digital radio communication, developed and used for the needs of rail transport. It provides digital voice communication and digital data transmission. It offers an extended functionality of the GSM system. It is characterized by infrastructure located only near the railway lines. GSM-R is designed to assist systems introduced in Europe: ERTMS (European Rail Traffic Management System) and ETCS (European Train Control System) which is designed to continuously collect and transmit data concerning rail vehicle, such as its speed and geographical location [4, 7]. Implementing the above-mentioned systems improves rail traffic safety significantly, enables vehicle diagnostics in real time and monitoring of consignments and carriages. Moreover, track utilisation can be increased considerably thanks to precise train distance data for GSM-R enabled railway routes [1, 8, 20].

One of the key issues in implementation of GSM-R is planning and configuring the system to make it highly resistant to corruption of sensitive data, thus minimizing the likelihood of system failure.

GSM-R is a telecommunications system, which must exhibit upmost availability and provide a high level of data transmission security in the railway environment. The key issues affecting the security of the system is reliability of the radio interface and the elements directly related to it (e.g. transponders). Dependability of data feed via GSM-R is achieved by providing adequate radio coverage along the route,

M. Siergiejczyk and K. Krzykowska (Eds.): ISCT21 2019, AISC 1032, pp. 391–400, 2020.
https://doi.org/10.1007/978-3-030-27687-4_39

corresponding to train speeds and offered services [18]. The level of radio coverage for the service of data transmission is determined by the maximum permitted line speed, while voice services, regardless of the maximum permissible speed should have radio signal strength of at least −98 dBm. Reliable access to telecommunications services is of paramount importance for the railway infrastructure administration, since it has a direct impact on safety and flow of rail traffic. Due to GSM-R's substantial role in railway traffic control and railway traffic safety, it is imperative to ensure the uppermost reliability of the system. High availability of the system is important due to the nature of the ERTMS system solutions, which operates autonomously to a large extent, based on continuous data feeds, acquired among others, through GSM-R. System outage can lead to errors in communication between devices or staff that may contribute to a railway accident. The aim of this paper is to identify threats and system vulnerabilities in railway radio communication and attempt to assess the risk in information security management system. Main steps of the risk assessment process are also demonstrated. The methods used in the process of risk assessment and risk planning are discussed, taking into account possible responses to threats and opportunities.

2 Identifying Threats and Detecting Vulnerabilities in Railway Radio Communication System

Information Security Management System (ISMS) needs to ensure compliance of the system with the requirements of the latest standard, which at the moment is the PN ISO/IEC 27001 [11]. One of the key requirements it stipulates, is an obligation to conduct a risk assessment process. One could argue that in the process of safety management, one of its components plays a prominent role, namely - risk management. This is hardly a new perspective, but due to the fact that contemporary organizations of all types and sizes must address a wide spectrum of potential threats capable of hampering their day-to-day, it became apparent that all of those risks must be managed one way or another.

A risk can be characterized as an uncertain event or set of events, which if materialized, would affect how an objective is realized. It is customarily measured as the product of likelihood that a perceived risk or opportunity arises and the magnitude of its impact on the goals.

Risk assessment involves three steps:

- risk identification,
- risk description,
- risk measurement.

The first step in the process of risk assessment, according to the standard PN ISO/IEC 27001 [11], is risk identification. The principal purpose of risk identification is to recognize the threats and opportunities that may exert influence over the outcomes of a project or a venture. It is recommended that a few tasks are performed during this stage, i.e.:

- identification of assets,
- identification of threats,

- determination of each threats' likelihood,
- determination of assets' vulnerability to threats,
- determination of risks' impact on the project.

The key aspect of the risk assessment process are the assets. When using a descriptive way to evaluate their significance, it is important to determine the importance of individual items. For instance, the significance of an asset can be represented by the following classification:

- critical - loss or breach of asset's security interrupts processes related to task execution,
- significant - loss or breach of asset's security may affect execution of tasks,
- medium - loss or breach of asset's security results in suboptimal performance of the system,
- negligible - loss or breach of asset's security does not affect system's performance.

The next step is threat identification. In order to effectively protect assets or resources of a system, each of them needs to be individually analysed in terms of exposure to potential threats and risks. A good practice is to list primarily realistic threats, that is those that can and do occur in the system (specific failures, power outages, accidental information disclosure, theft, etc.), as opposed to those that are easily named (terrorist attacks, deliberate human action, environmental factors and the like). At the same time, from the viewpoint of risk assessment's comprehensiveness, as many threats as possible should be considered (including those unlikely). At this point, let's be mindful of the fact that an important part of the process of risk analysis is availability of timely and reliable results.

The next step in the process of risk assessment is determining threats' likelihood. Risks tend to materialize irregularly, hence the concept of threat likelihood was introduced. Equipment failures, or power outages are definitely more commonplace than fires or natural disasters. Different levels of threat likelihood will depend on the structure of the system, its functionalities and size. A good practice, which had proven itself useful in the process of risk assessment of transport safety, is the following classification:

- very high - likely to occur/occurred frequently (e.g. once a month) or regularly at fixed intervals,
- high - likely to occur/occurred sporadically (occurred in the last year or happens irregularly),
- average - probably would not occur, but it is possible or infrequent,
- low - did not occur even once last year,
- very low - very unlikely to occur, there is no record that it occurred at all.

Determining assets' vulnerability to threats is the next stage of risk identification. Once assets' role in maintaining service continuity is identified, possible threats recognized and subsequently an approach for assets' protection developed, gaps in its security need to be uncovered i.e. characteristics and/or properties that could be exploited.

A stage of risk identification that is last in order of mention but not of importance, is determining risk's impact on objectives of a project and identifying an adequate level of protection. It covers issues related to impact assessment of threats to information security in accordance with latest standards and determining applicable level of existing security measures. With regard to the GSM-R system, the following classification of the impact/effect of threats on system availability, privacy and integrity of transmitted information can be presented:

– critical - risk's materialization poses a security threat to train traffic, damage to equipment and infrastructure, incapacity to perform tasks,
– severe - serious safety hazard - large losses in the infrastructure, restricted ability of the system to perform its tasks,
– substantial - a significant safety hazard, a serious incident - loss of equipment, a major impediment to operation of the system,
– small - negligible impact on safety - incident, impediment, constraints,
– negligible - threat's materialization does not cause any disruptions to the business or they are marginal.

Risk management entails a systematic implementation of policies, procedures and management practices in tasks related to determining the context of risk, its identification, analysis, impact assessment, treatment and communication (defined by PN-IEC 62198 [10]). One of the first steps is to identify threats and vulnerabilities. It is recommended for identified threats [5, 6] to be classified into the following groups:

– strategic (affecting company's long-term goals),
– operational (affecting company's day-to-day operations),
– financial (related to direct financial activities and company's capital),
– informational (affecting security of corporate information resources),
– regulatory (affecting compliance with applicable regulations).

In terms of identifying threats to ICT systems used in digital railway radio communication, the following threat classification could be assumed as the starting point for their detailed identification and description [3, 4, 6]:

– force majeure (natural disasters, financial disasters, changes in legislation, etc.) - possible consequences: destruction of information and physical resources, outage of network, lowered level of protection.
– unauthorized and criminal activities, including:
– risks associated with physical theft and loss of hardware, software and documents - possible consequences include network outage and breach of information confidentiality,
– risks associated with different types of wiretaps on (including use of "classical" techniques of espionage) hardware and software - possible consequences: breach of information confidentiality,
– unauthorized staff actions - possible consequences: network outage, loss of integrity and confidentiality of information, reduced level of protection,
– unauthorized actions of third parties - possible consequences: network outage, loss of integrity and confidentiality of information, reduced level of protection.
– mistakes by staff operating computer systems at control centers and OMC (Operations and Management Center) - possible consequences: network outage, loss of integrity and confidentiality of information, reduced level of protection,
– results of bad work organization, including threats associated with errors in physical and technical protection - possible network outage, loss of integrity and confidentiality of information,

– damage to equipment and software bugs - possible consequences: mainly loss of information availability and reduced level of protection.

When determining scope (environmental) of risk analysis for an ICT system's security, it is also important to identify vulnerabilities as a starting point for initiating future risk mitigation activities. Vulnerabilities can be related to:

ICT System

– hardware components that constitute a digital railway radio communication system, along with equipment communicating subsystems together,
– software components: - operating system - application software, especially in the OMC,
– staff operating digital radio communication system: operators, administrators,
– operating procedures (particularly procedures describing access to the system and information processed and stored in that system).

Security System

– elements of physical security i.e. mechanical and construction elements, security staff, physical security procedures,
– electronic components of the security system, including hardware, software components, staff operating the computerized system of technical security,
– human supervision system i.e. people performing supervisory functions and enforcing supervisory procedures,
– built-in hardware and software components or added to the system of digital radio communication in order to ensure adequate protection of information. Those measures include, among others, firewalls, IPS, authentication and authorization systems part of logical access control, antivirus software, cryptographic devices and software etc.

Operating environment of the digital railway radio communication and associated security system, especially power supply systems, air conditioning systems, buildings and enclosed spaces.

Presented classifications particularize and demonstrate the scope of responsibilities put in place to ensure ICT system's security. It shows what elements should be included in case of risk analysis for ICT security. Analysis of internal damage log and data from the market are the usual sources of information on potential risks and resultant damage. It can be performed through review and synthesis of own damage data and reports, by studying available company archives and press materials covering incidents suffered by competitors. A good source of information are also analyses of damage identified immediately after commissioning the network, then after few years of network operation (3–5 years) and further down the line (e.g. about 10 years).

To assess acceptable risks, the likelihood-impact matrix was used. The likelihood should be considered as the likelihood of specific risk to occur, and the impact represents estimated impact associated with a risk. Matrix "Likelihood/impact" (Table 1) illustrates areas of different risk levels.

Table 1. Likelihood-impact matrix in the digital railway radio communication system.

LIKELIHOOD	0.9	71–90%	0.0045	0.09	0.18	0.36	0.72
	0.7		0.035	0.07	0.14	0.28	0.56
	0.5		0.025	0.05	0.10	0.20	0.40
	0.3	11–30%	0.015	0.03	0.06	0.12	0.24
	0.1	<10%	0.005	0.01	0.02	0.04	0.08
			Negli-gible	Small	Sub-stan-tial	Severe	Critical
			0.05	0.10	0.20	0.40	0.80
			IMPACT				

In terms of assessing information security in the digital railway radio communication system, the items with highest combined value (product of likelihood and impact) are most important.

3 Risk Assessment in the Digital Railway Radio Communication System

Risk assessment of the railway radio communication system (Table 2) shows identified threats and likelihoods of their occurrence. At the time of writing this paper, very scant information about risk assessment of railway radio communication is available. Therefore the author had subjectively assigned those values, based on analysis of security levels recommended and used in information management systems, as well as on the basis of information obtained from departmental and public network operators. The environment in which GSM-R system operates was taken into account when estimating those risks, namely the environment of railway transport. Needless to say, once railway radio communication systems in Poland will have been used for several years, both the threats identified in this paper and their likelihoods might have different values. Nonetheless the methodology of risk analysis and assessment in digital railway radio communication system is quite flexible and thus suitable for use under changing circumstances.

Table 2. Risk assessment of digital railway radio communication system.

ID	Threat	L	I	R
1.1	Fire, flooding	0.1	0.8	0.08
1.2	Construction disasters	0.1	0.8	0.08
1.3	Extreme environmental factors (temperature, humidity, dust, smog)	0.1	0.4	0.04
1.4	Natural disasters (earthquakes, hurricanes, epidemics, floods)	0.1	0.8	0.08
1.5	Legal changes	0.7	0.4	0.28
2.1	Theft of workstations and other hardware that stores data	0.3	0.1	0.03
2.2	Theft of paper or electronic documentation	0.3	0.1	0.03
2.3	Wiretapping	0.1	0.1	0.01
2.4	Corruption and blackmail in order to extract specific information from company employees	0.3	0.1	0.03
2.5	Hacking into the system - impersonation of an authorized user	0.5	0.8	0.40
2.6	Fraud, counterfeit documents, access cards, passwords, etc.	0.1	0.4	0.04
2.7	Unauthorized attempt to modify the event log	0	0	0.00
2.8	Unauthorized installation of devices phishing for confidential information	0.1	0.1	0.01
2.9	Attacks that affect availability of resources (DoS, DDoS, computer sabotage)	0.1	0.4	0.04
2.10	Unauthorized, intentional modification of software installed on a computer by other users	0.1	0.1	0.01
2.11	Loss or copying of information by unauthorized personnel during repairs or maintenance	0.1	0.2	0.02
2.12	Unauthorized copying of data from a hard drive or other storage medium	0.7	0.1	0.07
2.13	Theft of a computer hard drive	0.1	0.1	0.01
3.1	Permanent loss of processed information	0.1	0.1	0.01
3.2	Unauthorized misuse of software	0.3	0.1	0.03
3.3	Preview of documents processed by previous user	0.5	0.1	0.05
ID	**Threat**	**L**	**I**	**R**
3.4	Negligence on the part of data processing personnel	0.3	0.8	0.24
3.5	Accidental alteration of configuration settings	0.1	0.4	0.04
3.6	Errors in using software	0.3	0.2	0.06

3.7	Incorrect use of procedures	0.3	0.4	0.12
4.1	Accidental loss of documentation or permanent loss of processed information due to a breakdown, fire, flooding, etc.	0.1	0.4	0.04
4.2	Use of unlicensed software	0.5	0.1	0.05
4.3	Unreliable audit of recorded system events	0.1	0.1	0.01
4.4	Preview of documents processed by previous user	0.5	0.1	0.05
4.5	Unauthorized access to data processing	0.5	0.1	0.05
4.6	Reuse of waste materials - printouts and CDs (instead of destroying them)	0.1	0.1	0.01
4.7	Exploitation of irregularities during duplication of classified documents, including duplication of classified documents outside the security zone	0.1	0.1	0.01
4.8	Lack of clarity regarding roles and responsibilities in the process of planning risk and day-to-day operations	0.1	0.1	0.01
5.1	Failure of heating, lighting, power or air conditioning system	0.1	0.8	0.08
5.2	Damage to equipment at base stations, control centers, OMC and ICT network elements	0.3	0.8	0.24
5.3	Failure of operating system or application software bug	0.1	0.1	0.01
5.4	Aging data storage mediums	0.1	0.1	0.01

where:

L - Likelihood,

I - Impact,

R - Risk.

In terms of business continuity and availability of digital railway radio communication, the most important threats (product of likelihood and impact) will have the highest values. Those include the following risks:

- Damage to equipment at base stations, control centers, OMC and ICT network elements,
- Hacking into the system - impersonation of an authorized user,
- Negligence on the part of data processing personnel in control centers and OMC.

GSM-R is a vast and territorially distributed system. Damage to elements of its subsystems (for example damage to or outage of a base transceiver station, or even a group of base stations or base station's controller on one of the routes) must not affect the availability of services provided by GSM-R on other routes. Therefore it is advisable that for purposes of GSM-R's availability analysis, single routes are considered as opposed to the whole system. Only the NSS subsystem could be considered in terms of availability of the entire system.

Legislative changes are a substantial risk too, however, in case of a system which is responsible for safety of railway traffic, it does not directly affect the security of information transmission in railway radio communication.

4 Summary

Any telecommunications network is intended to send information at a specified time and at specified error rate. GSM-R is a communication system, which needs to be highly reliable and secure for data transmitted in the railway environment. Reliable access to telecommunications services is paramount for the railway infrastructure administration, since it directly impacts safety and flow of rail traffic.

Because ETCS (*European Train Control System*) level 2 relies on GSM-R for continuous data transmission, ERTMS (*European Railway Traffic Management System*) imposes maximum outage of the GSM-R communication network as follows:

- for ETCS level two and level three - 4 h per 10 years (99.995% availability).
- for other voice and data services - 8 h per year (99.91% availability).

Key issues affecting security of the system is reliability of the radio interface and the elements directly related to it (e.g. transponders). Any information transmitted by radio is exposed to eavesdropping and interception. Therefore, all connections should be encrypted so that their content is kept confidential and unavailable to a accidental user. Encryption, however, does not apply to Railway Emergency Call (REC), where shortest possible time to make the call is required. Encryption requires the use of an cryptographic algorithm by both the network and the mobile station. Nevertheless, before information is encrypted, the network needs to identify the user by performing an authorization procedure, also referred to as authentication.

References

1. Xun, D., Xin, C., Wenyi, J.: The analysis of GSM-R redundant network and reliability models on high-speed railway. In: 2010 International Conference on Electronics and Information Engineering (ICEIE 2010), Kyoto, Japan (2010)
2. Kowalewski, M., Kowalewski, J.: Information security policy in practice. Library IT Professional (2014)
3. Kowalewski, M., Kowalewski, J.: Threats of Information in Cyberspace Cyberterrorism. Publishing House of Warsaw University of Technology, Warsaw (2017)
4. Krzykowska, K., Siergiejczyk, M.: Selected aspects of risk analysis and assessment of the use of air navigation satellite systems. In: Smith, I. (ed.) Logistics Magazine 6/2014. Institute of Logistics and Warehousing, Conference TransComp (2014). ISSN 1231-5478
5. Lehrbaum, M.: GSM-R Disaster Recovery, GSM-R Business Operations, Warsaw, October 2009

6. Liderman, K.: Risk analysis and protection of information in computer systems. PWN Warsaw (2009)
7. Pawlik, M.: The European Rail Traffic Management System, an overview of the features and technical solutions - from concept to implementation and operation. KOW, Warsaw (2015)
8. Pawlik, M., Siergiejczyk, M., Gago, S.: European rail transport management system mobile transmission safety analysis. In: Walls, L., Revie, M., Bedford, T. (eds.) Risk, Reliability and Safety: Innovating Theory and Practice © 2017. Taylor & Francis Group, London, (2017)
9. Perzynski, T., Lewinski, A., Lukasik, Z.: The Concept of Emergency Notification System for Inland Navigation. Monograph Information, Communication and Environment, Radom (2015)
10. ISO/IEC 17799: 2003 Information technology. Practical information security management
11. PN-ISO/IEC 27001:2007. Information security management systems. Requirements
12. Project Management Institute: A guide to the Project Management Body of Knowledge (PMBOK Guide). Ingram International Inc. (2013)
13. Siergiejczyk, M.: Analysis of information secure transmission methods in the intelligent transport systems. Arch. Transp. Syst. Telematics **10**(3) (2017)
14. Siergiejczyk, M.: Critical assessment of ERTMS systems reliability based on the example of the GSM-R system. J. KONES **23**(4) (2016)
15. Siergiejczyk, M., Gago, S.: Ensure information security problems in the GSM-R network. In: Siergiejczyk, M. (ed.) Scientific Papers of Warsaw University of Technology. Transport, Z. 92. Publishing House of Warsaw University of Technology, Warsaw (2013)
16. Siergiejczyk, M., Gago, S.: Safety and security, availability and certification of the GSM-R network for ETCS purposes. Arch. Transp. Syst. Telematics **7**(1) (2014)
17. Siergiejczyk, M., Gago, S.: Selected problems of reliability and security of data transmission in the GSM-R system. Railway Probl. **58**(162) (2014)
18. UIC project EIRENE. System Requirements Specification, GSM-R Operators Group, System Requirements Specification (SRS) Version 16.0.0, December 2015
19. UIC project EIRENE. Functional Requirements Specification (FRS) Version 8.0.0, December 2015
20. Winter, P.: International Union of Railways, Compendium on ERTMS. Eurail Press, Hamburg (2009)

Analysis of Transport Process' Costs with Use Various Technologies in Terms of Last Mile Delivery Problem

Jakub Starczewski(✉)

Cracow University of Technology, Str. Warszawska 24, 31155 Kraków, Poland
starczewski.jakub@onet.pl

Abstract. The research includes the calculation and comparison of the cost of cargo delivery process using a car and a cargo bicycle without electric drive. The methodology of calculation was indicated and the case study was considered, assuming the assumptions consistent with the bicycle transport characteristics. The costs of invested money, energy consumption, using of vehicle, using of infrastructure and external costs of transport were taken into account. A calculation of delivery costs was made when the capacity of vehicles is on maximum level, in relation to the weight of the load and the distance of transport. An equilibrium point was determined as the distance of carriage, for which the delivery costs of both vehicles have comparable values and then a simulation of delivery costs in relation to different load mass when the distance equals the point of equilibrium, was made. The research was crowned with conclusions where the directions of future researches were indicated.

Keywords: Cargo bike · Cost of transport · Last mile

1 Introduction

As a market analysis show, with each year an increase in the importance of e-commerce with reference to traditional trade is observed. This contributes to the generation of new transport needs and the need to modernize existing supply chains, where direct distribution, neighborhood distribution, and "to the collection point" distribution are becoming more and more important. The difficulty, which is at the same time a consequence of this trend, is the increase in the importance of the last mile delivery problem as a highly cost-intensive transport stage.

As of today, there are many examples of practical implementations that attempt to reduce the negative consequences of last mile deliveries. They can be divided into groups referring to: the organization of the system and improving the efficiency of its operation and the usage of alternative means of transport. Both groups are closely related, because usually the use of different transport technologies makes it necessary to modernize the organization of the current delivery system. However, the example of a second group of solutions can be mentioned as: use of cargo bikes, including conventional and electrically assisted drive.

M. Siergiejczyk and K. Krzykowska (Eds.): ISCT21 2019, AISC 1032, pp. 401–410, 2020.
https://doi.org/10.1007/978-3-030-27687-4_40

Because in Poland, electricity is produced in the majority from fossil fuels and non-renewable sources (89,06% of installed energy as at 31/12/2018 [6]), the author omits those solutions that rely on energy from the battery. However, the highly ecological (in terms of the product life cycle) mean of the transport (as a cargo bicycle powered by human muscle power) and conventional technology (as an internal combustion vehicle) was compared. The study includes an analysis of operational costs and also the environmental costs of the transport process. The results were compared basing of the adopted efficiency criterion relating to the policy of sustainable transport development.

2 Methodology for Calculating the Cost of Cargo Delivery

This analysis shows the cost's differences created during the transport process using different transport technologies. Costs of the entire supply chain and related to the need to modernize the current system to introduce a different transport technology are not taken into account. Also, the aspect of driver's salary is omitted due to the lack of the possibility of generalizing wages, based on real data, but it may be the direction of further research. However the study takes into account the following costs related to transport costs of: freezing capital related to the purchase of a vehicle, energy and consumables, vehicle wear, using the logistics infrastructure and environmental costs.

The calculation methodology is based on the time in which the vehicle is involved in the transport process (1) and the rate for one hour operating of the device in relation to all listed costs. The time of involvement is calculated as follows:

$$t_Z = \frac{M \cdot t_N}{60} \cdot \frac{n \cdot L}{V} \cdot \frac{M \cdot t_W}{60} \cdot \frac{n \cdot t_M}{60} \quad (1)$$

t_Z - time of vehicle involvement in the transport process expressed in [h],
M - mass of load [kg],
L - distance of cargo transportation [km],
V - technical speed [km/h],
n - number of courses [-],
t_N - loading time [min],
t_W - unloading time [min],
t_M - time of stops and maneuvering [min].

2.1 Costs of Freezing Capital Related to the Purchase of the Vehicle

The costs of freezing capital related to the purchase of the vehicle were calculated on the basis of the following (2). In both cases, the purchase of means of transport from own resources and monthly depreciation write-offs (amortization using the straight-line method), were assumed.

$$K_K = \frac{W_P \cdot \alpha_K}{12} \cdot \min\left\{\left[\frac{12}{\alpha_A}\right]; N \cdot 12\right\} \cdot \left(1 - \min\left\{\left[\frac{12}{\alpha_A}\right] - 1; N \cdot 12 - 1\right\} \cdot \frac{\alpha_A}{2 \cdot 12}\right) \cdot \frac{t_Z}{N \cdot t_R} \quad (2)$$

K_K - cost of freezing capital related to the purchase of a vehicle [PLN],
W_p - initial value of the vehicle used for transport [PLN],
α_K - annual cost of freezing capital [%/year],
α_A - annual depreciation rate for a vehicle [%/year],
t_R - average annual vehicle engagement time [days/year],
N - number of years of vehicle operation expressed in years.

2.2 Costs of Energy Consumption and Consumables

The costs of energy and consumables consist of three components: cost of energy used by the human, cost of energy consumed by the machine and cost of consumables (3):

$$K_E = K_{EL} + K_{EM} + K_M \tag{3}$$

K_E - cost of energy and consumables [PLN],
K_{EL} - cost of energy consumed by the human [PLN],
K_{EM} - costs of energy consumed by the machine [PLN],
K_M - cost of consumables [PLN].

The Costs of Energy Consumed by Human
In order to calculate the energy consumed by the human for carrying out the transport task, the methodology of Harris and Benedict was used [2]. First, the Basal Metabolic Rate was calculated - i.e. the minimum number of calories required to maintain vital functions by an individual during one day. The formula (4) and (5) were used:

$$z_{LP}(m) = 66,4730 + 13,7516 \cdot W(m) + 5,033 \cdot H(m) - 6,7550 \cdot A(m) \tag{4}$$

$$z_{LP}(k) = 655,0955 + 9,5634 \cdot W(k) + 1,8496 \cdot H(k) - 4,6756 \cdot A(k) \tag{5}$$

$z_{LP}(m)$ and $z_{LP}(k)$ - Basic Metabolism Rate for men and women [kcal],
$W(m)$ and $W(k)$ - weight of the person for men and women [kg],
$H(m)$ and $H(k)$ - height of the person for men and women [cm],
$A(m)$ and $A(k)$ - age of the person for men and women [years].

At the next step the Total Energy Expenditure were calculated. It is defined as a number of calories covering the total daily energy demand of the organism, including its physical activity (6).

$$z_{LC}(m,k) = a \cdot z_{LP}(m,k) \tag{6}$$

$z_{LC}(m,k)$ - Total Energy Expediture for men and women [kcal],
a - activity coefficient determined according to the assumptions (based on: [7]): $a \in \langle 1,2;1,3 \rangle$ - for a sick person lying in bed; $a = 1,4$ - for low physical activity; $a = 1,6$ - for moderate physical activity; $a = 1,75$ - for high physical activity; $a = 2$ - for very high physical activity; $a \in \langle 2,2;2,4 \rangle$ - for professional sports activities.

Finally, TEE was calculated as follow (7):

$$K_{EL} = \frac{c_L \cdot z_{LC} \cdot t_Z}{24} \tag{7}$$

c_L - unit price of energy kcal in [PLN/kcal].

The Costs of Energy Consumed by the Machine

The energy costs consumed by the machine were calculated as follows:

$$K_{EM} = \frac{c_M \cdot z_M \cdot n \cdot L}{100} \tag{8}$$

c_M - unit price of fuel for the vehicle [PLN/l],
z_M - average fuel consumption by the vehicle [l/100 km].

The Costs of Consumables

In many literature sources, the costs of consumables of combustion vehicles are calculated as a percentage of fuel consumption (about 20%). In an addition, there is often a separate factor such as tire wear, which generally makes up about 5% of fuel consumption. However that methodology would cause substantive errors in the case of a delivery process using a cargo bicycle without electrical support, because the fuel consumption for the machine is zero. On the other hand, it is problematic to find an appropriate dependence of the costs of consumables in relation to (for example) the mileage of the vehicle, due to the lack of researches of exploitation of the cargo bikes and the less availability of results in the scientific literature. For this reason, a simplification has been made by taking the percentage factor of a value of the vehicle as the cost of consumables. The adopted coefficient (14.85%) gives similar values of results as coefficients relating to the fuel consumed. These results are similar for the size of the mileage and the time of vehicle engagement taken into account in the research. It should be noted that the exact value of the coefficient should be found experimentally for the real object and the real operating conditions. The costs of consumables were calculated in the following way:

$$K_M = \frac{14,85\% \cdot [W_p - W_k]}{N \cdot t_R} \cdot t_Z \tag{9}$$

W_k - the final value of the vehicle used for transport [PLN].

2.3 Costs of Wear

According to the proposed approach in the source [11], the cost of wear of the vehicle, was calculated as the cost of amortization the vehicle using the line method (10):

$$K_A = [W_p - W_k] \cdot \frac{t_Z}{t_R \cdot N} \tag{10}$$

K_A - the cost of vehicle wear [PLN].

2.4 Costs of Using Logistic Infrastructure and Other Costs

In road transport, there are cost items resulting from the use of logistic infrastructure. These may include: road fees, fees for entry into specific zones, parking fees for vehicles, taxes on means of transport and other fees resulting from separate local and state law. The analysis takes into account the costs of using logistic infrastructure, such as:

- permission to enter the city center - due to the fact that the distribution of loads by cargo bikes should be a response to the last mile delivery problem, which clearly increases in such areas as the historic and/or the strict city center.
- the cost of parking - resulting from the need to pay a fee by the driver who enters the paid parking zone for unloading the goods,
- the cost of obligatory vehicle insurance and technical inspections as a necessary condition for allowing the vehicle to traffic.

The analysis does not include tax on means of transport due to the fact that both tested vehicles are exempt from this obligation. The idea of calculating costs of using the infrastructure and the rest, is to calculate their value for one hour. Together with the time of engagement it will be a component of total costs. The formula (11) was used for calculations:

$$K_I = \left(\sum c_I + \sum c_P\right) \cdot t_Z \tag{11}$$

K_I - the cost of using the logistics infrastructure and other costs [PLN],
c_I - the cost of fees and taxes per one hour [PLN/h],
c_P - other fees resulting from the transport process per hour [PLN/h].

2.5 Environmental Costs

External transport costs, otherwise environmental transport costs, are defined as negative (expressed in monetary) external effects of the movement of people and goods that are not reflected in the price of the process itself [5]. This study is based on the results of European and Polish researchers [5, 8, 10] and takes into account the total external costs in which the following components are included: costs of accidents, air, soil and water pollution, biodiversity losses, noise, urban effects, climate change, land defragmentation, changes in the landscape, up- and downstream. The calculation of environmental costs was made on the basis of formula (12).

$$K_S = M \cdot n \cdot L \cdot c_S \tag{12}$$

K_S - the environmental cost for the vehicle used for transport [PLN].
c_S - unit environmental cost of 1 tkm in a country [PLN/tkm].

2.6 The Criterion for Efficiency

To compare transport processes, the following criterion for efficiency was defined as a function of transport length (13):

$$K(L) = \frac{K_K(L) + K_E(L) + K_A(L) + K_I(L) + K_S(L)}{M} \tag{13}$$

$K(L)$ - the criterion of efficiency for the transport process of length L [PLN/t].

3 Case Study

The following input data regarding the motor vehicle and the cargo bicycle was adopted, to analyze the issue of cargo transport costs - Table 1. The choice of a specific model was random, but was taken into account the set of vehicles available on the market and used in existing cargo distribution systems.

Table 1. Vehicles input parameters.

Feature	Symbol	Unit	Value for car	Value for bike
Name	-	-	Fiat Ducato	Long John
Drive	-	-	Diesel engine 2999 [cm^3]	Human muscles without electrical support
Fuel consumption	z_M	l/100 km	11,0	0
Fuel price	c_M	PLN/l	5,14	Not applicable
Technical speed	V	km/h	20	15
Capacity	Q	t	1,2	0,1
Time to loading 1 ton	t_N	min	15	10
Time of unloading 1 ton	t_W	min	15	10
Time of stops and maneuvering	t_M	min	15	5
Initial value	W_p	PLN	100 000	12 000

(*continued*)

Table 1. (*continued*)

Feature	Symbol	Unit	Value for car	Value for bike
Final value	W_k	PLN	30 000	1 500
The annual cost of freezing capital	α_K	-	7%	7%
Annual depreciation rate	α_A	-	20%	18%
Number of years of operation	N	-	8	8
Time of involvement	t_R	day/year	250	250
The cost of inspections	-	PLN/year	99	Not applicable
Insurance	-	PLN/year	1485	Not applicable
Fee for entering the city center	-	-	Not applicable	Not applicable
Parking fee	-	PLN/h	3	Not applicable

In addition, in order to calculate the energy consumption needed by the car's driver and the cargo bicycle's driver, the following values were used - Table 2. Due to the greater effort of the bicycle's driver, a higher rate of activity than the car's driver was assumed.

Table 2. Input parameters of the drivers.

	Symbol	Unit	Value for car	Value for bike
Sex	-	-	M	M
Weight	W	[kg]	75	75
Growth	H	[cm]	175	175
Age	A	[years]	25	25
Activity coefficient	a	[-]	1,75	2,00

For the analysis of the price of energy consumed by a man, the parameters included in Table 3 were adopted. 15 food products considered as consumer products were taken into account. According to [9], their average gross price in PLN for the fourth quarter of 2018 in the Małopolskie voivodeship was adopted. Moreover, according to [3] energy value was adopted too. Then, the average price per unit of energy, in [PLN/kcal], was calculated.

Table 3. Parameters used to calculate the average price per 1 kcal.

No.	Product name	Quantity	Unit	Price for quantity [PLN]	Energy value [kcal/100 g(ml)]	Energy price [PLN/kcal]
1	Wheat roll	50	g	0,52	277	0,0038
2	Pork ham cooked	1	kg	24,84	233	0,0106
3	Fresh butter ok. 82.5%	200	g	6,18	735	0,0042
4	Milk chocolate	100	g	4,02	551	0,0073
5	Half-curd cheese	1	kg	13,26	133	0,0100
6	Fresh chicken eggs	60	g	0,60	139	0,0072
7	Rice	1	kg	3,71	349	0,0011
8	Beef meat with bone	1	kg	27,68	152	0,0182
9	Barley	0,5	kg	2,65	339	0,0016
10	Cow milk 3–3,5% sterilized	1	l	2,73	62	0,0044
11	Potatoes	1	kg	1,64	80	0,0020
12	Apple juice	1000	ml	3,87	42	0,0092
13	Natural ground coffee	250	g	7,73	21	0,1472
14	White crystal sugar	1	kg	2,09	405	0,0005
15	Apples	1	kg	1,77	50	0,0036
The average price for 1 kcal:						0,0154

To simulate the calculation of environmental costs a source [5] was used. Their total value for transporting cargo was: in the min. scenario: 5 264,70 mln EUR; in the max. scenario: 7 374,10 mln EUR. However the author did not provide the share of light vehicles in the cost structure. Therefore, using the [1] (for the analogous period: 2014), in which it was included that: 0,8% of the transport work was carried out with vehicles with a capacity to 4,5t and at the same time, a transport work of 96 627 mln tkm was carried out, it was calculated that light trucks made a transport work of value of: 773,016 mln tkm. In this study, the author assumed the value of external costs proportional to the share of light commercial vehicles in total transport, which amount to EUR 263,235 mln (min. scenario) and EUR 368,705 mln (max. scenario). It is therefore assumed that delivery trucks with a capacity to 4,5 t generated external costs at the level of 0,0021794 EUR/tkm and 0,0030526 EUR/tkm respectively. With the average EUR/PLN exchange rate in 2014: 4,184494 PLN [4], and the arithmetic mean of received values, the c_{S1} = 0,01095 PLN/tkm was calculated. Due to the lack of scientific research on external costs of transport using cargo bikes and at the same time, taking into account that the bike itself does not emit harmful substances and greenhouse gases during the exploitation phase, a value of price for external costs for cargo bike was assumed for calculations as a 30% of the rate for the delivery vehicle: c_{S2} = 0,00328 PLN/tkm.

It was assumed that in the case of both vehicles, deliveries are made during the hours allowed for loading operations, so charges for entering the city center are not required. The simulation was made assuming a homogeneous load, which can be divided arbitrarily. The mass in the first case was respectively: 0.1t for the cargo bicycle and 1.2t for the delivery vehicle, which was the maximum use of the vehicles' capacity. The length of the transport - that is the route from the consignor to the consignee, was selected from the range of 0 km to 70 km, due to the fact that the larger distance becomes unreachable for a cargo bicycle without electrical support. Based on calculations an equilibrium point was determined, in which the levels of the defined efficiency criterion for vehicles have the smallest absolute difference. This point for the assumed initial conditions was: 16.0 km and the time of engagement of vehicles was: 1.18 h for a bicycle and 1.65 h for a car. The results of the analysis are presented in Fig. 1.

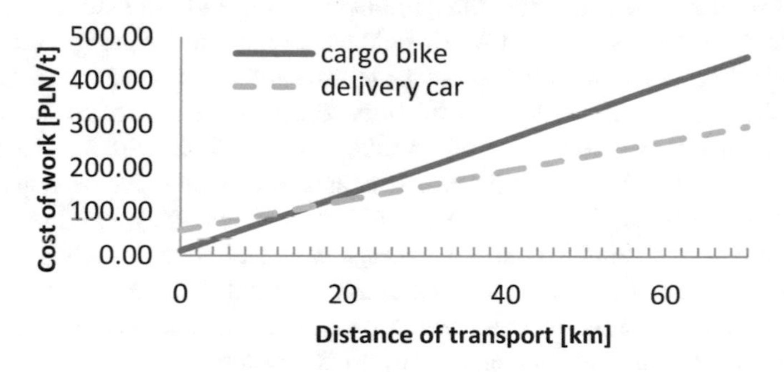

Fig. 1. The cost of vehicle's work as a function of transport distance. Source: own study.

In the second case, it was tested how the transport costs would change, for the same weight of the load (for a maximum capacity of larger mean of transport) taking into account the constant distance of transport, i.e. the previously calculated point of balance. The necessity of returning the cargo bicycle for the rest of load, due to its lower capacity was taken into account. The results are presented in Fig. 2.

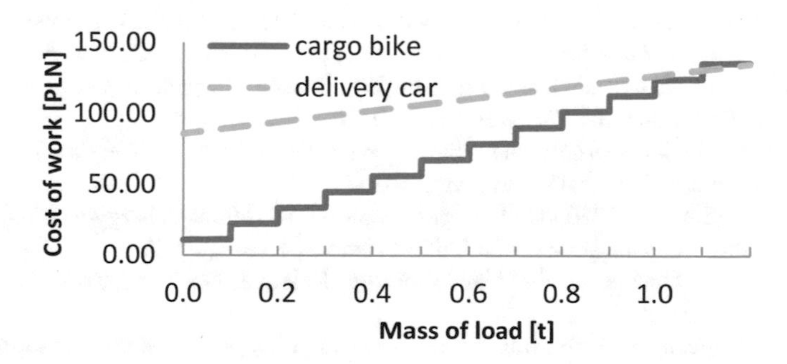

Fig. 2. The cost of vehicle's work as a function of cargo weight. Source: own study.

The above figure shows that, the transport costs of cargo bike are lower than in the case of a delivery truck. It should be mentioned that the time of engagement of vehicles for transporting a cargo with a mass of 1.2t and a distance of 16.0 km are: 14.2 h for a bicycle and 1.65 h for a car. Therefore such carriage, despite the theoretical basis, can not be applied in practice due to the employee's daily working time according to the Labor Code, but this may be a starting point for further research.

4 Conclusions

The conducted research shows that the carriage of load by cargo bikes is an alternative for the transport by delivery cars. This is especially for short distances, but only selected loads with a specific transport susceptibility will be suitable for transporting by these vehicles. It should be assumed that the demonstrated dependencies would be change in favor of the cargo bicycle, when it would achieve a higher technical speed than the car. The analysis does not exhaust the subject but only indicates some cost dependencies. Therefore the analysis of the demand for transport in the aspect of transport susceptibility of load, justifying the chose a cargo bike, should become the next step to further research. Moreover a case of study which characterized by a lot of stops and small distances between them, should be analyzed. Times of: reloading activities, maneuvers and stops, should be examined and the average technical speed which is characteristic for a given area should appointed and taken into account. Further, distant directions of research are: taking into account all company's costs, collecting data of the operational costs and the cost analysis of the entire cargo bike system.

References

1. Główny Urząd Statystyczny: Transport – wyniki działalności 2014r. Zakłady Wydawnictw Statystycznych, Warszawa (2015)
2. Harris, J.A., Benedict, F.G.: A biometric study of human basal metabolism. In: Proceedings of the National Academy of Sciences of the United States of America, Washington (1918)
3. Kunachowicz, H., Nadolna, I., et al.: Wartość odżywcza wybranych produktów spożywczych i typowych potraw. Wydawnictwo Lekarskie PZWL, Warszawa (2012)
4. Narodowy Bank Polski: Archiwum kursów średnich – Tabela A/2014. www.nbp.pl
5. Pawłowska, B.: Koszty zewnętrzne transportu w Polsce. Przegląd Naukowy – Inżynieria i Kształtowanie Środowiska, Warszawa (2018)
6. Polskie Sieci Elektroenergetyczne: Raport: zestawienie danych ilościowych dotyczących funkcjonowania KSE w 2018 roku. www.pse.pl
7. Report of a Joint FAO/WHO/UNU Expert Consultation: Human energy requirements. Food and Agriculture Organization of the United Nations, Rome (2004)
8. Ricardo AEA: Update of the Handbook on External Costs of Transport. European Commission (2014)
9. Urząd Statystyczny w Krakowie: Biuletyn statystyczny województwa małopolskiego IV kwartał 2018, Kraków (2019)
10. Van Essen, H., Schroten, M., Otten, M.: External Costs of Transport in Europe Update Study for 2008. CE Delft, Delft (2011)
11. Wasiak, M., Jacyna-Gołda, I.: Transport drogowy w łańcuchach dostaw. Wyznaczenie kosztów. PWN, Warszawa (2016)

Application of Mathematical Evidence to Estimate the Demand for Parking Spaces for Motorways

Marek Stawowy[(✉)], Krzysztof Perlicki, Tomasz Czarnecki, and Grzegorz Wilczewski

Warsaw University of Technology, Warsaw, Poland
marek.stawowy@pw.edu.pl

Abstract. This article presents the application of uncertainty modelling based on DS (Dempster and Shafer) mathematical evidence in order to estimate the factors which enable to determine the parameter defining the attractiveness of a car park. The mathematical evidence theory renders the synthesis of information from many independent sources possible. Thus it appears to be the best choice, because the attractiveness of a car park for the user depends on several independent qualities. A simulation for the chosen modelling was conducted. The presented results demonstrated a dependency between the final values and the factor values of the qualities denoting facilities applied at the parking place.

Keywords: Rest area · Demand for parking spaces · Mathematical evidence

1 Introduction

A huge demand for parking spaces for trucks alongside main transportation routes in all Europe and the USA appears to be in recent years a major concern [8, 9]. It is related to the drivers' work regulations and their need to rest especially during the night. Currently the car park capacity is insufficient. Due to this car parks are congested at night, often from evening hours. Car park occupancy research has been conducted in many countries yet infrequently and the results convey an incomplete study of the problem [3, 4]. The results derive from random measurements, often flawed by measurement uncertainty. Only constant measurements allow valid conclusions. In Poland in the years 2015–2017 a continuous research was undertaken regarding rest area parking occupancy during peak periods. The measurements were conducted for the "Parking spaces in rest areas" project and financed by NCBiR/GDDKiA (The National Centre of Research and Development/The General Directorate for National Roads and Motorways) within the joint project "RID" (Road Innovations Development) contract DZP/RID-I-44/8/NCBR/2016 partially with the participation of the authors of this article [2].

The analysis of the results led to the development of a method for determining parking spaces demand, yet one of the equation parameters, namely car park attractiveness, is difficult to observe. That is why in this article the uncertainty modelling based on DS (Dempster and Shafer) mathematical evidence is applied in order to

M. Siergiejczyk and K. Krzykowska (Eds.): ISCT21 2019, AISC 1032, pp. 411–419, 2020.
https://doi.org/10.1007/978-3-030-27687-4_41

estimate the factors enabling the determination of the parameter defining car park attractiveness [1, 5]. The mathematical evidence method allows a synthesis of information coming from many independent sources, which is why it seems the most appropriate choice, because the car park attractiveness for the user often relies on several independent features, which affect the decision to use the car park parallelly, i.e., independently of other features.

2 State of the Art

In the final report of the RID 4E project [2], two equations are presented, which allow determining parking space demand. The first equation grants calculating the number of parking spaces, which meets the demand. It should be possible on the basis of a couple of parameters, namely:

1. traffic volume,
2. defining the factor of the number of cars parking in respect to the traffic volume,
3. car park attractiveness consisting of features such as: proximity to an agglomeration or a frontier and car park facilities.

Parking space demand[1]:

$$N_{RS} = OC_{max} * SF_{max} * AADT_{d} * RAA \tag{1}$$

where the following stand for:

N_{RS} - the number of required parking spaces,
OC_{max} - car park occupancy factor (or car parks, if more than one car park is considered),
SF_{max} - seasonal factor, different for various periods (for calculating the maximum demand the maximum value product of the year and of a week has to be adopted),
$AADT_{d}$ - annual average daily traffic volume (for an existing motorway) or presumed (e.g. a planned motorway) in one direction,
RAA - rest area attractiveness factor.

The second formula defines the car park occupancy factor resulting from the traffic volume (the number of cars on the motorway passing the car park).

Occupancy factor:

$$OC_{max} = \frac{max(N_{os_t})}{TV} \tag{2}$$

where the following stand for:

OC_{max} - car park occupancy factor (or car parks, if more than one car park is considered),

[1] Source: final report of "RID" 4E project [2].

N_{ost} - the number of occupied parking spaces as a function of time,
TV - traffic volume on the motorway in one direction.

The traffic volume TV in formula (2) can be easily calculated and next linked to the parking spaces demand. The second parameter of the formula (1) SF_{max} should be assessed on the basis of traffic seasonality. From the results published on the websites of The General Directorate for National Roads and Motorways an annual change of traffic volume can be verified. Figure 1 presents the average daily traffic relationship in respective months.

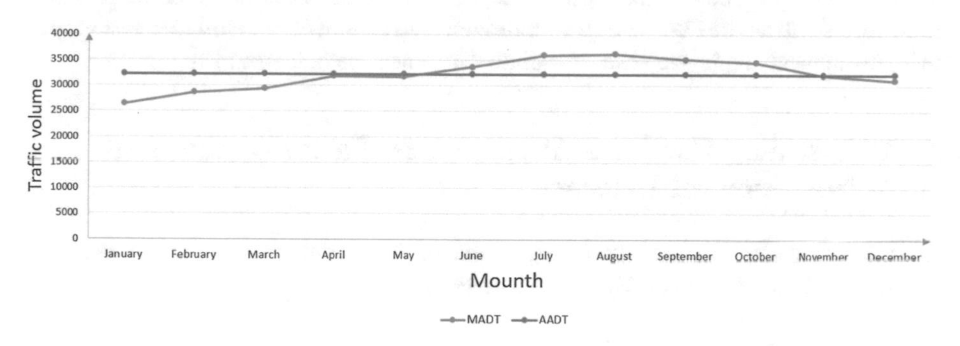

Fig. 1. Traffic volume seasonal variability. Devised on the basis of [7].

Table 1 shows the relationship $SF = MADT/AADT$ (monthly average daily traffic/annual average daily traffic) for the whole year.

Table 1. Annual traffic volume variability [7].

Month	MADT	MADT/AADT
January	26402	0.819
February	28601	0.888
March	29345	0.911
April	31835	0.988
May	31741	0.985
June	33678	1.045
July	35988	1.117
August	36194	1.123
September	35153	1.091
October	34590	1.073
November	32037	0.994
December	31102	0.965

It is possible to calculate the third parameter $AADT_d$ from formula (1) assuming that there is a sufficient number of variously located car parks. The final parameter *RAA* from formula (1) is the most problematic to estimate, because it derives from the car park facilities, but also the distance from cities or a frontier. This article has applied uncertainty modelling based on DS (Dempster and Shafer) mathematical evidence in order to estimate the factors which enable to determine the parameter defining the attractiveness of a car park. The mathematical evidence theory renders the synthesis of information from many independent sources possible. As it has been stated in the introduction this appears to be the best choice for this solution. The mathematical evidence theory is more complex than the adopted variate difference method presented in [4, 6]. Yet it can be expected to be more universal and will not require a vast number of measurements as in the case of variate difference method from [2].

3 Determining Car Park Attractiveness Factor on the Basis of Mathematical Evidence

In order to determine the attractiveness factor *RAA* from formula (1) a model of elements must be designed, which constitute it.

3.1 The Model of Attractiveness Factor

The impact on the attractiveness factor value and, consequently deducing from formula (1), on the parking space demand, can derive from many factors. Such as:

1. Facilities:
   a. filling station,
      i. fuel price,
      ii. filling station brand,
   b. restaurant,
      i. restaurant brand,
      ii. opinion about the restaurant,
   c. hotel,
   d. play ground,
   e. guarded car park,
   f. Wi-Fi.
2. Location:
   a. the proximity of a big city,
   b. en route to holiday destinations,
   c. the proximity of a major transportation hub,
   d. location related to the periodicity of drivers working hours,
   e. proximity to a frontier,
3. Time:
   a. during the day,
   b. during the year.

As presented above many parameters influence the attractiveness factor value and consequently the required number of parking spaces next to a motorway. Further discussion will embrace the following factors (the parentheses enclose symbols of observation related to the parameter influencing the attractiveness factor).

1. Facilities (e1):
   a. filling station (e1a):
      i. fuel price (e1ai),
      ii. filling station brand (e1aii),
   b. restaurant (e1b):
      i. restaurant brand (e1bi),
      ii. opinion about the restaurant(e1bii),
   c. guarded car park (e1c).
2. Location (e2):
   a. proximity of a big city (e2a),
   b. en route to holiday destinations (e2b),
   c. proximity to a frontier (e2c).

Figure 2 presents a model of elements of car park attractiveness factor taking into account the above listed parameters. All dependents in the model are parallel (independent). This enables the application of mathematical evidence based on Dempster-Shafer theory [1, 5]. Symbol h means hypothesis evaluation from observations.

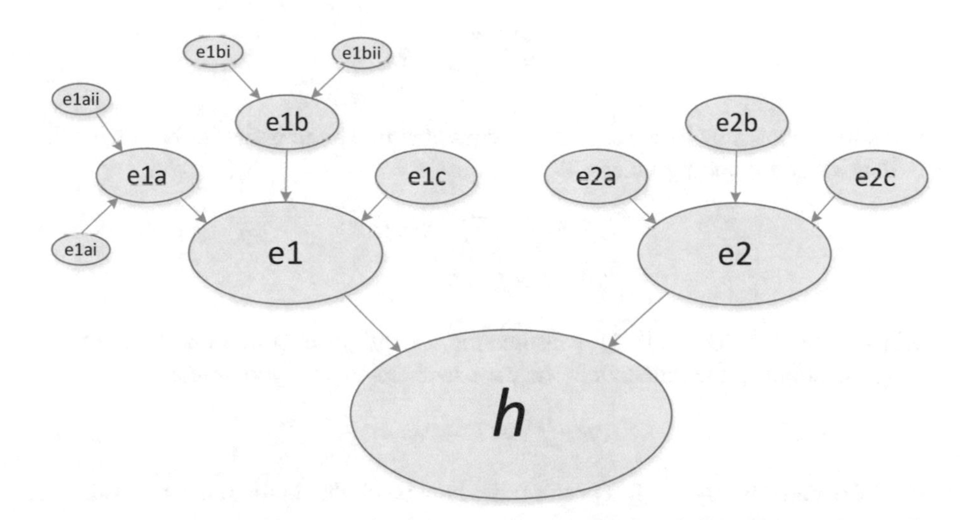

Fig. 2. Attractiveness factor model of elements.

3.2 Uncertainty Modelling on the Basis of the Theory of Mathematical Evidence

Mathematical theory of evidence [1, 5] implies that information synthesis for individual elementary probability measures m is possible. Information can be synthesized even if its elements are contradictory or derive from different sources. Such a synthesis can be described with the following formula:

$$m_3(C) = \frac{\sum\limits_{A_i \cap B_j = C} m_1(A_i)m_2(B_j)}{1 - \sum\limits_{A_i \cap B_j = O} m_1(A_i)m_2(B_j)} \quad (3)$$

Where the following stand for:

A, B i C – are the sources of observation which represent the subset of Θ set;
m_1 i m_2 – sets of masses;
m_3 – new set of mass.

The synthesis is known as the Dempster's rule of combination [1]. The basic belief assignment function (BBA) can be formed in the following way.

$$\begin{aligned} & m : 2^{\Theta} \rightarrow [0, 1] \\ & m[\varnothing] = 0 \\ & \sum_{A \subseteq \Theta} m(A) = 1 \end{aligned} \quad (4)$$

Belief denoted by Bel ∈ [0, 1] measures the strength of acquired observations supporting the belief in the authenticity of the examined set of hypotheses.

$$Bel(A) = \sum_{B \subseteq A} m(B) \quad (5)$$

Plausibility denoted by Pl ∈ [0, 1] determines how much the belief in the authenticity A is limited by supporting evidence ¬A.

$$\begin{aligned} & Pl(A) = \sum_{B \cap A \neq 0} m(B) \\ & Pl(A) = 1 - Bel(\neg A) \end{aligned} \quad (6)$$

Doubt denoted by Dou ∈ [0, 1] measures the strength of acquired observations supporting the doubt in the authenticity of the examined set of hypotheses.

$$Dou(A) = 1 - Bel(A) \quad (7)$$

Disbelief denoted by Dis ∈ [0, 1] determines how much the doubt in the authenticity A is limited by supporting evidence ¬A.

$$Dis(A) = 1 - Pl(A) \quad (8)$$

The combination rule affects the belief function and can be presented in the following way:

$$Bel1 \oplus Bel2(A) = \sum_{B \subseteq A} m_1 \oplus m_2(A) \tag{9}$$

3.3 An Example of Determining the Car Park Attractiveness Factor

A model of car park attractiveness factor *RAA* was presented in Fig. 2. The model included two observations of the third level e1ai – fuel price and e1aii – filling station brand, connected with the second level e1a – filling station. Apart from that two observations from the third level were included e1bi – restaurant brand and e1bii – opinion about the restaurant, connected with the observation of the second level e1b – restaurant.

Table 2 shows values, which are ascribed to the third level observations connected with the observation of the second level e1a – filling station.

Table 2. Values ascribed to the third level observations connected with the observation of the second level e1a. Own elaboration.

Observation	Value
Fuel price (e1ai)	0.2
Filling station brand (e1aii)	0.1

Having assigned the value with the use of the suggested method e1a = 0.069.

Table 3 shows values, which are ascribed to the third level observations connected with the observation of the second level e1b – restaurant.

Table 3. Values ascribed to the third level observations connected with the observation of the second level e1b. Own elaboration.

Observation	Value
Restaurant brand (e1bi)	0.8
Opinion about the restaurant (e1bii)	0.05

Having assigned the value with the use of the suggested method e1b = 0.648.

Tables 4 and 5 show values, which are ascribed to the second level observations connected with the observation of the first level respectively e1 – facilities, e2 – location.

Table 4. Values ascribed to the second level observations connected with the observation of the first level e1. Own elaboration.

Observation	Value
Filling station (e1a)	0.069[a]
Restaurant (e1b)	0.648[a]
Guarded car park (e1c)	0.02

[a]Value calculated in the previous stage of modelling.

Having assigned the value with the use of the suggested method e1 = 0.434.

Table 5. Values ascribed to the second level observations connected with the observation of the first level e2. Own elaboration.

Observation	Value
Proximity of a big city (e2a)	0.3
En route to holiday destinations (e2b)	0.02
Proximity to a frontier (e2c)	0.2

Having assigned the value with the use of the suggested method e2 = 0.144. Table 6 shows values, which are ascribed to the first level observations

Table 6. Values ascribed to the first level observations. Own elaboration.

Observation	Value
Facilities (e1)	0.434[a]
Location (e2)	0.144[a]

[a]Value calculated in the previous stage of modelling.

Having assigned the value with the use the suggested method $RAA = 1 + h = 1.779$.

This value will be the attractiveness factor, which can be substituted directly into the formula (1).

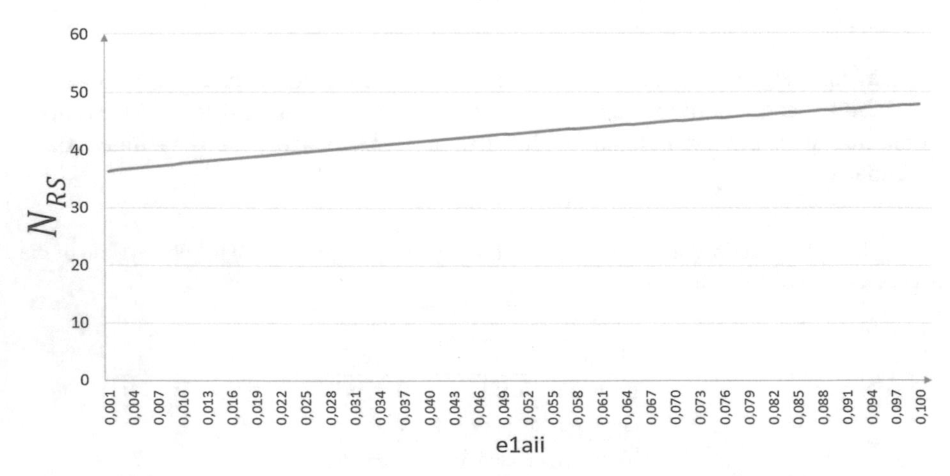

Fig. 3. IQ dependency graph as a function of one of the IQ features. Own elaboration.

In order to obtain a visualisation of dependency change N_{RS} as a value function of one of the factors of the third level e1aii was adopted with the range between 0 and 0.1. For the purpose of the simulation the following value of $OC_{max} = 0.001$ was adopted [2]. From Table 1 the value $SF_{max} = 1.123$ from the month August was adopted. Value $AADT_d = 32222$ was adopted from the graph in Fig. 1. The results of this simulation are presented in the form of a graph in Fig. 3.

4 Conclusion and Summary

The suggested method enables determining car park attractiveness factor with the use of uncertainty modelling based on Dempster-Shafer evidence theory. The model can comprise observations consisting of many levels. The article contains an example and a simulation of the influence of the lowest level observations on the car park attractiveness factor value. The next stage in the development of this method might be the application of other ways of modelling uncertainty, because the theory of mathematical evidence is of little practical use in multi-element models. This results from the fact that complications in calculations grow rapidly with regards to the number of independent objects in the model.

References

1. Dempster, A.P.: Upper and lower probabilities inducted by a multi-valued mapping. Ann. Math. Stat. **38**, 325–339 (1967)
2. Final report of project "Parking spaces in rest areas" financed by NCBiR/GDDKiA (The National Centre of Research and Development/The General Directorate for National Roads and Motorways) within the joint project "RID" (Road Innovations Development) contract DZP/RID-I-44/8/NCBR/2016. Consortium MOPS Warsaw (2018)
3. Lüttmerding, A.: Calculation model for estimating the demand of truck parking areas along motorways. In: Proceedings of the 12th International Conference on Transport Science, Portorož, Slovenija (2009)
4. Rodrigues, V.S., Stantchev, D., Potter, A., Naim, M., Whiteing, A.: Establishing a transport operation focused uncertainty model for the supply chain. Int. J. Phys. Distrib. Logist. Manag. **38**(5), 388–411 (2008)
5. Shafer, G.: A Mathematical Theory of Evidence. Princeton University Press, Princeton (1976)
6. Stawowy, M., Perlicki, K., Mrozek, T.: Application and simulations of uncertainty multilevel models for estimating the number of motorway parking spaces. In: Cepin, M., Bris, R. (eds.) Proceedings of the European Safety and Reliability Conference, ESREL 2017, pp. 2653–2657. CRC Press/Balkema, London (2017)
7. The General Directorate for National Roads and Motorways. Archiwizacja i analiza danych ze stacji ciągłych pomiarów ruchu z 2017 roku. Archiwizacja i analiza roczna danych z 2017SCPR nr 02627. Warszawa (2018)
8. United Kingdom Government, Department for Transport. The Strategic Road Network and the Delivery of Sustainable Development 2014. www.gov.uk. Accessed 05 Jan 2019
9. USA New York State Department of Transport. Guide for the Development of Rest Areas on Major Arterials and Freeways 2001. www.dot.ny.gov. Accessed 05 Apr 2018

Methodical Approach to Assessing Interference in GSM-R Network

Marek Sumiła(✉)

Railway Research Institute, 04275 Warsaw, Poland
msumila@ikolej.pl

Abstract. The article presents a cross-section of methods used to identify and assess the probability of interference radio communication in GSM-R network. The article proposes different approaches in identifying areas susceptible the interferences by conducting research in three areas: theoretical (using geographical method and propagation models), simulation (using advanced measuring equipment and software) and measuring (based on field tests). The article was enriched with additional knowledge of the practical application of methods.

Keywords: GSM-R · Radio interference · Interference identification methods

1 Introduction

The GSM-R system is a radio-based communication system dedicated in the EU to interoperable railway communication. This was done under Decision 1692/96/EC [4] and later Council Directives 96/48/EC of 23 July 1996 [5] on the interoperability of the trans-European high-speed rail system and Directive 2001/16/EC of 19 March 2001 on the interoperability of the trans-European conventional rail system [8][1]. A one of the key role in the idea of interoperable railway system plays a uniform radio communication system that aroused from public mobile radio communication standard GSM 2G. New GSM-R system enables voice communication between groups of railway employees and is also used to data transmission for the European Train Control System (ETCS).

In recent years, there have been report a number of occurrence radio disturbances in this network. As a result of this, the flow of railway traffic was disturbed. The UIC (International Union of Railways) has taken action to assess the causes and scale of the problem. ERA (EU Agency for Railways), ETSI (European Telecommunications Standards Institute), CEPT (European Conference of Postal and Telecommunications Administrations), GSM-R network operators and research centers were involved in the case. As a result of the carried out works, a number of studies was issued confirming the occurrence and sources of interferences [1, 2, 13, 17, 19–22]. On this basis, can be indicated the sources of interference which are BTS of public mobile radiocommunications systems.

[1] Both directives were later amended.

M. Siergiejczyk and K. Krzykowska (Eds.): ISCT21 2019, AISC 1032, pp. 420–429, 2020.
https://doi.org/10.1007/978-3-030-27687-4_42

In the documents [1, 2, 13] three main categories of radio interferences have been identified:

1. Intermodulation products are self-generated by the GSM-R receiver itself in its reception band in presence of two or more mobile networks signals in the adjacent band.
2. Blocking is a coexistence phenomenon that can be caused by either insufficient selectivity of the GSM-R receiver (filter discrimination), saturation of the front-end (LNA and/or mixer) or reciprocal mixing (with local oscillator phase noise).
3. Out of Band emissions from radio services in the adjacent frequency band falling into the native receive band of the GSM-R receiver can also cause a degradation of the communication service.

There is no evidence that weather events affect the frequency of interference. One of the important conclusions of the performed analyzes was the statement that the GSM-R network has low immunity and is not prepared to coexist with newer wideband technology in the same 900 MHz band.

Although, CEPT developed guidance to support a more efficient usage of spectrum, and ECC published in Report 229 [2], providing views from all interested parties and guidance to enable a better coexistence between GSM-R and mobile networks, the problem exist and need to be solved in particular GSM-R networks.

The next parts of the article will present a methodical approach to assess the possibility of interference in the GSM-R network.

2 Different Approaches to GSM-R Interference Testing

2.1 Geographical Identification

Indication of disruption of railway radio communication network requires an analysis of hundreds or thousands of kilometers of lines of railway within the covcrage of GSM-R network. It is not a simple task and it is not always possible to do it through field tests. Author of the article proposed the method of identifying places of interference using digital maps. He presented it in articles [17, 18].

This method consists in searching of public networks transmitters (BTS) located a short distance from railway lines. For this purpose, the author use digital maps with the coordinates of public BTS transmitters. The conducted simulation tests prove that the biggest threat is transmitters located within 700 m from the railway line [19]. The base stations of public operators located at distances smaller than specified may pose a threat to GSM-R receivers.

In addition to this, the method focused on BTS transmitting signals next to the vicinity of the GSM-R band and use technology GSM 3G or GSM 4G. Such systems produce broadband IM3 intermodulation in GSM-R radio receivers. In Poland, these are P4 (925.1–930.1 MHz) and Aero2 (930.1–935.1 MHz) public operators. Therefore, these are two premises that suggest us areas threatened by disturbances of GSM-R networks.

The following figure shows a map containing the location of BTS public transmitters and measured distance from the railway line in Ostrów Wielkopolski town (Fig. 1).

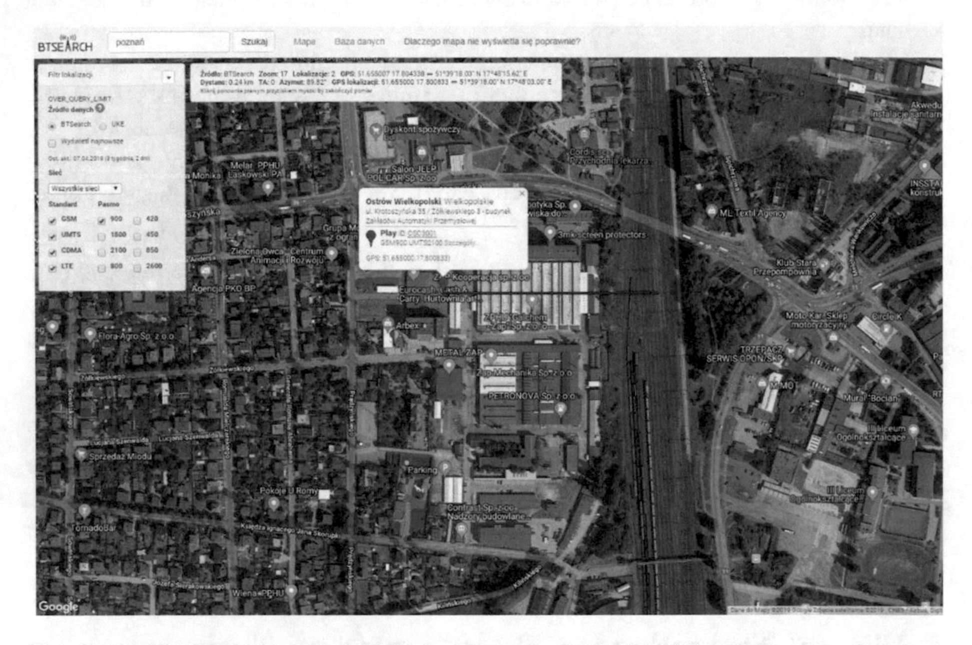

Fig. 1. Satellite map with a marked distance between the public BTS and the railway line (own study based online tool http://beta.btsearch.pl).

The satellite map shown the coordinates of the public BTS transmitter working in the band 925–930 MHz. It have been marked in purple. The distance to the nearby railway line was marked with a red line. In this case, the measured distance was about 260 m and can be considered as a dangerous.

2.2 Use of Propagation Models

There are currently available a large number of propagation models, which aim is to estimate the level of radio signal or attenuation from the transmitter. The use of models requires a number of standard parameters like bandwidth, distance transmitter – receiver, antenna suspension height, terrain type (urban, suburban, rural), etc. Use of some models allow to introduce a specific radio propagation conditions for example terrain obstacles, weather conditions, multi-path signaling and many others.

Some specific parameters for the GSM-R system are available in its specifications [6, 7, 9, 12]. For examples when we talk about the place of installation elements network (BTS, the terminal handhelds and train radios: Cab Radio, EDOR). Typically, handheld terminals work at a height of 1.5–2.0 m with an output power of 2 W. Terminals installed on vehicles have a suspended antenna at a height of about 4 m and work with output power up to 8 W. On the other hand the BTS transmitters are suspended at heights of approx. 30–40 m above the ground and depending on the class their output power not exceeding EIRP 2 kW.

Many of useful propagation models are available on ETSI websites, i.e.:

- ITU.R P.525 (Free Space Loss) – band: >30 MHz, range: LOS-limited,
- Extended Hata – band: 30 MHz–3 GHz, range: up to 40 km,
- COST 231-Walfish-Ikegami – band: 800 MHz–2 GHz, range: 0.02–5 km,
- ITU-R P.370 – band: 30 MHz–1 GHz, range: up to 1000 km,
- ITU-R P.452 – band: 700 MHz–50 GHz, range: up to 10000 km,
- ITU-R Rec. P.528 – band: 125 MHz–15.5 GHz, range: 0–1800 km,
- ITU-R Rec. P.1411 – band: 300 MHz–100 GHz, range: <3 km,
- ITU-R P.1546 – band: 30 MHz–3 GHz, range: 1–1000 km,
- JTG5-6 – band: 30 MHz–3 GHz, range: 0–1000 km,
- Longley Rice (ITM) – band: 20 MHz–40 GHz, range: 1–2000 km.

The models can be compared because each model give different results. Proper estimating the signal levels at the front of the GSM-R terminal give more reliable answer about interference issue. Figure 2 compares the signal attenuation results for the two models. The results of the calculations indicate differences in results for different distances from the transmitter.

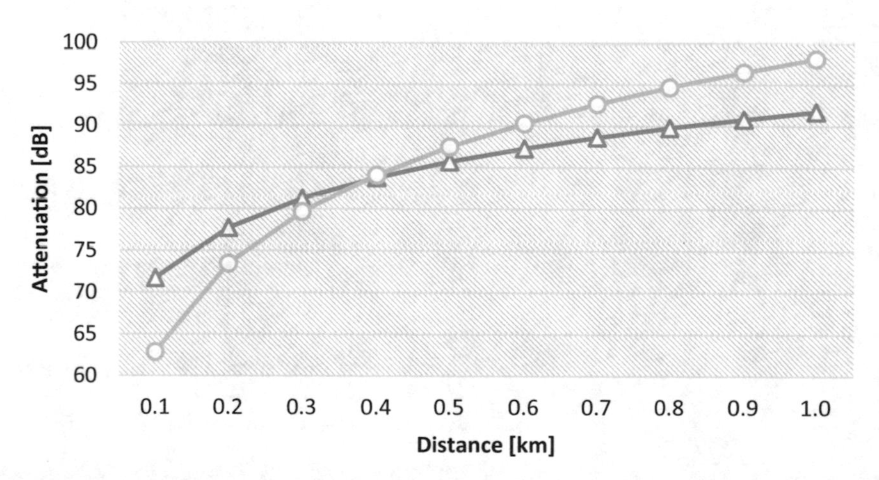

Fig. 2. Comparison of two radio propagation models for the carrier 925,1 MHz. (triangles – Okumura-Hata Model (open space), circles – Free Space Loss) (own study).

2.3 Simulation in Software Tools

Each of the radio signal propagation models presented in the previous section is described by more or less complex mathematical equations. Nowadays, computational tools can improve the calculation process. In the scope of the considerations, it is worth mentioning especially the tool developed by the SEAMCAT Technical Group (STG) in the European Communications Office (ECO) who is the Secretariat of the CEPT.

Project SEAMCAT (Spectrum Engineering Advanced Monte Carlo Analysis Tool) is a free tool to use a statistical simulation model that uses a method of analysis called Monte Carlo to assess the potential interference between different radiocommunication systems such as broadcasting, point to point, point to multipoint, radar, mobile networks, aeronautical and satellites.

A Monte Carlo simulation is a statistical technique based upon the consideration of many independent events in time, space and frequency. Using this technique each event, or simulation trial is built up using a number of different random variables that define the system to be simulated (e.g. the location of the interferers with respect to the victim, the victim link's wanted signal strength, the frequencies of the victim and interferer use). As a result of simulation the probability of a certain event occurring can be evaluated with a high level of accuracy by taking the average result over all trials.

In the simulation process is necessary to define the distributions for all relevant parameters of the radiocommunications systems to be modelled (e.g. antenna heights, powers, operating frequencies, positions of the transceivers, etc.). An exemplary result of the simulation of coexistence two GSM 2G and GSM-R networks is shown in the Fig. 3.

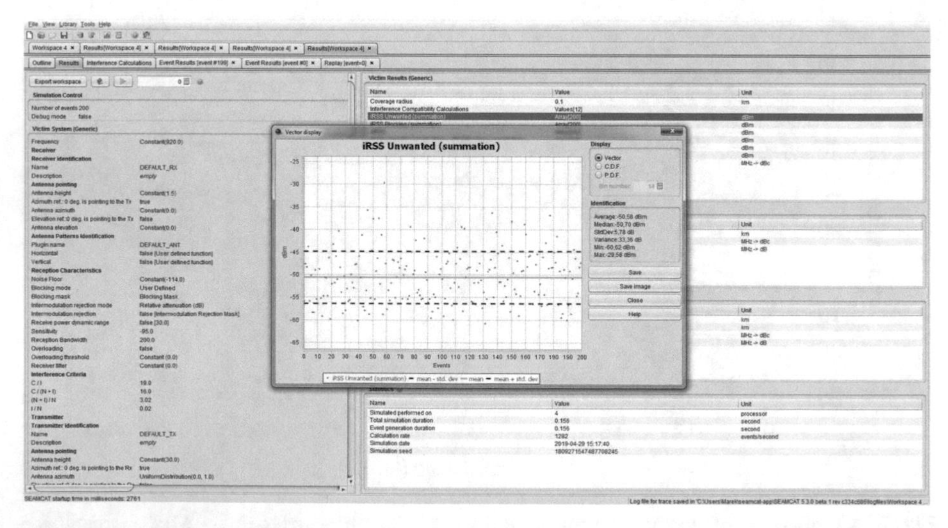

Fig. 3. Analysis of the simulation results of interference in the unwanted 900 MHz band in SEAMCAT tool (own study).

In this example, it was calculated that the average level of wanted signal will be around −81 dBm. The estimated value of the unwanted signal (iRSS unwanted) was −50 dBm. The presented results of 200 trials were marked with dots in red, the average value of interference with the green line, standard deviation with a black dotted line. The obtained test results can be displayed in many different forms (tables, graphs, figures). The tool is so universal that it allows you to simulate not only GSM-R networks and can be used as a tool to conformance the coexistence hypotheses of radio networks.

The second tool is an intermodulation simulator, It is available online on the website of the Triorail radio equipment manufacturer [15]. The tool allows to simulate a wideband and narrowband transmitters in the 900 MHz band by determining their channel parameters. Despite the great simplicity of this tool, it is possible to map many different sources and check the probability of occurrence the intermodulation products in the receiver. An example of such a simulation is shown in the Fig. 4.

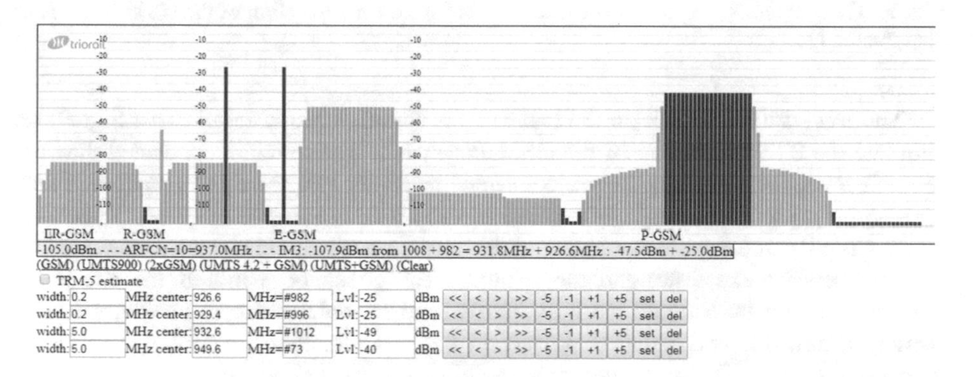

Fig. 4. The result of a complex simulation of 4 networks operating in the 900 MHz band (own study base on online tool https://www.triorail.com/trio-im3-sim/).

The simulator calculates the values for each channel in the 900 MHz band and allows to determine the intermodulation products and pointing its source signals.

2.4 Laboratory Measurements

In addition to methods based on computer simulations, it is possible to use real measurement systems for the reconstruction of specific network conditions in a research laboratory. The such apparatus like CMU200 and SMU200 allows to simulate interferences on the available on the market GSM-R terminals.

Such tests as before are performed for both signals (the desired and the interfering) occur [14, 16, 24]. The effects of simulated interferences are observed directly during the operation on the GSM-R terminal and via the spectrum analyzer (see Fig. 5). The quality of voice calls are assessed using the RxQual parameter. The higher the value of the RxQual means better quality of connection. The authors of the documents [2, 3, 13, 24] indicated various guidelines for the proper value of this parameter, however, most of them indicate that it should be greater than 4.

The report [24] provides an example of a laboratory stand for interference phenomena simulation presented as the block diagram.

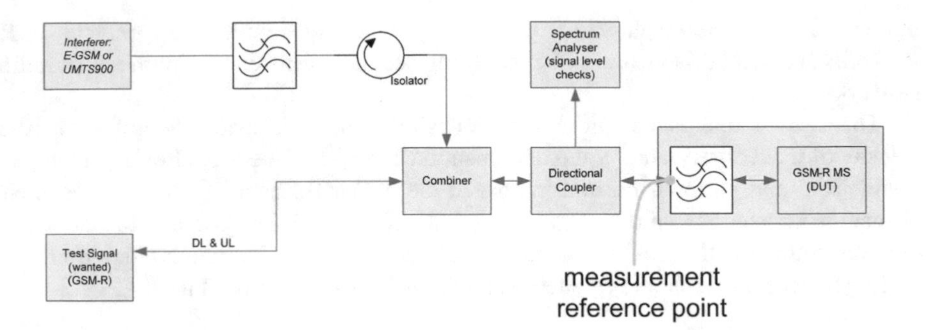

Fig. 5. General block diagram of the measurement stand for simulation of the GSM-R receiver disturbance [24].

The test stand was built on the basis of the communication tester (Test Signal) to simulate the BTS transmitter of the GSM-R network. The terminal connected directly to such tester through the antenna connector reacts as if it were in a separate GSM network. Such terminal can be subjected to various tests confirming correct operation in the basic RF and application layers. A RF signal generator (Interferer) is included in the link between the tester and the terminal. The generator works as the source of interference signals. It could work as narrowband or wideband system with selected carrier, bandwidth and radiated power. In case of simulation a larger number of interfering sources, it may be necessary to use more such generators.

As a result, we can map the conditions presented in a given area, but also check the resistance of selected GSM-R terminals in present to various types of interference. The power of the signals (desired and interfering) can be adjusted by using additional attenuators. All signals are summed in the Combiner. A Spectrum Analyzer is added to the stand to allows observation of parameters and measurement signals at the antenna input connector. This is possible by the additional Directional Coupler element. Finally, the signal comes to the radio input of the GSM-R terminal whose behavior is evaluated in the laboratory test process.

Tests carried out according to the above method can confirm the correct operation of GSM-R terminals for compliance with technical specifications TS 102 933-1 [10] and TS 102 933-2 [11]. Simultaneously, they can be used to test the improved parameters of the old types terminals, after applying additional filters in accordance with recommendations O-9760 [23].

2.5 Radio Field Tests

Field tests are the most popular methods among the network operators and are used to assess the occurrence of strong out-of-band signals and their impact on the generation of interference in GSM-R network receivers. The GSM-R network operators are obliged to carry out the tests twice a year on the basis of the Railway Safety Directive 2016/798 [6].

The tests are performed as measuring systems mounted on trains, as service systems or "on attendant". The tests are usually performed on the main tracks for both travel directions. This means that for each railway line the signal level in the 900 MHz band is measured at least twice. The measurement is usually made at night due to logistical reasons, but it should also be performed during the daytime.

Due to high costs of testing, GSM-R network operators limit the number of measurement campaigns to minimum. However, in the case of reported and confirmed the occurrence of problems with communication in the GSM-R network, these tests are performed in the indicated area immediately to confirm the causes and take action as indicated in the ECC report 229 [2].

3 Applying Methods

In Sect. 2 was presented a number of methods for assessing the probability of interference in the GSM-R network. The choice of method is determined by the purpose of research and the state of knowledge.

The method presented in Sect. 2.5 requires a minimum level of knowledge. Field tests enable the measurement and assessment of the occurrence of interference based on the analysis of signals in the 900 MHz band.

The reverse method to the mentioned is the geographical method presented in Sect. 2.1. It does not require field tests. To use it, you need to know about BTS transmitters of nearby networks and to determine the distance from the railway area. This method makes it possible to gain knowledge about the areas susceptible to interference. Complementing mentioned method is the use of one or more radio propagation models. The method using such models was the content of Sect. 2.2.

The computer tools described in Sect. 2.3 allows to estimate the probability of interference based on a limited number of data. This type of estimation allows to obtain broad radio analysis by creating many different scenarios of network configuration in the reviewed area.

The last of the presented in Sect. 2.4 method is mainly used for testing GSM-R receivers and their reactions to various out-of-band signals occurring in the 900 MHz band. Laboratory tests carried out with the method allows to gain knowledge about the various GSM-R terminals or the quality of used methods to mitigate the effects of interference in these devices.

4 Conclusions

The GSM-R system is the first coherent radio communication system dedicated to railways in the EU. Idea of GSM-R system solve many of the previous existing problems associated with radio train communication on cross border international trains. In addition to this, it enables the data transmission that supports the management of train traffic and it is a milestone in the development of railway control systems. Unfortunately, as it was mentioned in the introduction, the development of public mobile technologies, increase in the number of subscribers, direct adhesion of GSM-R and public bands has become the cause of mutual interferences of the networks and reported problems with communication in GSM-R network.

Disruptions in communication do not have a direct impact on safety, but it cause perturbations in the movement of trains caused by increasing delays or stopping traffic. The awareness of this causes that currently in many EU countries great importance is attached to the problem of GSM-R network interference and methods of their detection. The ways of dealing with the phenomenon of GSM-R interference have been presented in the ECC report 229 [2]. The report does not contain methods for detecting affecting by interference areas, but only shows means to mitigate them.

The article presents various methods for identifying GSM-R interferences. They were divided into categories and the scope of their use. Some of them can be use during a new GSM-R radio network is planned or during the reconfiguration an existing network. The GSM-R network operator can use this knowledge and act in an active way preventing interference, instead of in a reactive manner, as is often the case today, that is acting after a problem in a given area.

The experience gained in the field of GSM-R network interference has also been helpful in the work on the development of the Future Railway Mobile Communication System (FRMCS) [16].

References

1. CEPT ECC Report 162, Practical mechanism to improve the compatibility between GSM-R and public mobile networks and guidance on practical coordination (2011)
2. CEPT ECC Report 229, Guidance for improving coexistence between GSM-R and MFCN in the 900 MHz band (2015)
3. CEPT ECC Report 231, Mobile coverage obligations (2015)
4. Decision No 1692/96/EC of the European Parliament and of the Council of 23 July 1996 on Community guidelines for the development of the trans-European transport network
5. Council Directive 96/48/EC of 23 July 1996 on the interoperability of the trans-European high-speed rail system
6. Directive 2004/49/EC of the European Parliament and of the Council of 29 April 2004 on safety on the Community's railways and amending Council Directive 95/18/EC on the licensing of railway undertakings and Directive 2001/14/EC on the allocation of railway infrastructure capacity and the levying of charges for the use of railway infrastructure and safety certification (Railway Safety Directive)
7. EIRENE System Requirements Specification. European Integrated Railway Radio Enhanced Network. GSM-R Operators Group. UIC CODE 951. Version 16.0.0, Paris (2015)
8. ETSI TS 100 910, Technical Specification. Digital cellular telecommunications system (Phase 2 +). Radio Transmission and Reception. (3GPP TS 05.05 version 8.20.0 Release 1999) European Telecommunications Standards Institute (2005)
9. Directive 2001/16/EC of the European Parliament and of the Council of 19 March 2001 on the interoperability of the trans-European conventional rail system
10. ETSI TS 102 281 (V3.0.0), Railways Telecommunications (RT). Global System for Mobile communications (GSM). Detailed requirements for GSM operation on Railways
11. ETSI TS 102 933-1 (V2.1.1), Railway Telecommunications. GSM-R improved receiver parameters. Part 1: Requirements for radio reception (2015)
12. ETSI TS 102 933-2 (V2.1.1), Railway Telecommunications (RT). GSM-R improved receiver parameters. Part 2: Radio conformance testing (2015)

13. ETSI EN 301 515 Global System for Mobile communication (GSM); Requirements for GSM operation on railways. V2.3.0 (2005)
14. FM(13)134 GSM-R Measurement Report - BNetzA Germany, Radio Spectrum Committee. Updated version of the working document on GSM-R interferences and coexistence with public mobile networks, taking into consideration observations made by RSC delegations. Working Document. RSCOM15-60 rev3, Brussels (2016)
15. Hasenpusch, T.: Compatibility measurements UMTS/LTE/GSM -> GSM-R. Federal Network Agency, Germany/ECC (2013)
16. Intermodulation simulator. www.triorail.com/trio-im3-sim. Accessed 30 Apr 2019
17. LS telcom AG. Coexistence of GSM-R with other Communication Systems ERA 2015 04 2 S.C. Made for European Union Agency for Railways. Version 3, Lichtenau (2016)
18. Sumiła, M.: Risk analysis of interference railway GSM-R system in Polish conditions. In: Advanced in Intelligent Systems and Computing, Proceedings of the Eleventh International Conference on Dependability and Complex Systems DepCoS-RELCOMEX, Brunów, Poland, 27 June–1 July 2016, vol. 470. Springer, Heidelberg (2016)
19. Sumiła, M.: Risk analysis of railway workers due to interference into GSM-R system by MFCN. In: Communications in Computer and Information Science, Challenge of Transport Telematics Proceedings of 16th International Conference on Transport Systems Telematics, Katowice-Ustroń, Poland, 16–19 March, vol. 640. Springer, Switzerland (2016)
20. Sumiła, M., Miszkiewicz, A.: Analysis of the problem of interference of the public network operators to GSM-R. In: Mikulski, J (ed.) Tools of Transport Telematics, pp. 76–82. Springer, Heidelberg (2015)
21. UIC O-8700-2.0, Report on interferences to GSM-R. UIC (2012)
22. UIC O-8736-2.0, UIC Assessment report on GSM-R current and future radio environment. UIC (2014)
23. UIC O-8740, Report on the UIC interference field test activities in UK (2013)
24. UIC O-8760-1.0, GSM-R RF filter requirements. UIC (2015)
25. Red-M UMTS 900 - GSM-R Interference Measurements. A Report prepared for OFCOM (2011)

Assessment of Availability of Railway Video Surveillance System

Piotr Szmigiel[1](✉) and Zbigniew Kasprzyk[2]

[1] PKP Polskie Linie Kolejowe S.A., Targowa 74, Warsaw, Poland
piotr.szmigiel@plk-sa.pl
[2] Politechnika Warszawska, Wydział Transportu, Koszykowa 75, Warsaw, Poland

Abstract. The article describes both existing and planned video surveillance systems on railway infrastructure. Problems regarding the creation of a centralized video surveillance system covering the entire country have been identified. An assessment of the emerging system in terms of its availability has been made. The emerging system is one of the most modern systems of this type in Europe, however, due to certain compromises made during the development of the system, it may not provide the availability on the level of other auxiliary systems used in railway transport, in particular due to insufficient availability of server infrastructure.

Keywords: Video surveillance · Railway safety systems

1 Introduction

In recent years, surveillance systems of passenger infrastructure are playing an increasingly important role in ensuring passenger safety and property protection. The growing challenges related to terrorism force infrastructure managers to take steps to ensure the ability to identify potential threats and crises. Technical progress in the field of image sensors creates a constant decline of prices of solutions that make visual identification of a person possible, which in turn encourages the development of such systems.

Although the first thing that comes to mind when dealing with a terrorist threat are airfields, it is actually railway infrastructure that is particularly vulnerable to acts of terrorism and vandalism. The significant density of people during rush hours, especially in larger cities, effectively hinders the possibility of providing safety for passengers. A major challenge lies in geographical dispersion of railway infrastructure and the sheer number of stations and other passenger-related railway objects. The Polish National Railway Manager – PKP Polskie Linie Kolejowe S.A. – manages a total of over three thousand passenger stations. Their geographical dispersion, location (especially when it comes to smaller stations – often located far away from urban areas) and number basically makes it impossible for Polish Railway Protection Service (SOK – Służba Ochrony Kolei) to actively supervise the railway areas on a daily basis.

Video surveillance systems have become a solution to the problems described above. Such systems, in addition to making it possible for operators to notice potential

M. Siergiejczyk and K. Krzykowska (Eds.): ISCT21 2019, AISC 1032, pp. 430–439, 2020.
https://doi.org/10.1007/978-3-030-27687-4_43

terroristic threats, are also noticeably contributing to crime rate reduction, on the example of the Stockholm subway, where Closed Circuit Television (CCTV) systems contributed to a 25% decrease of criminal events [7].

2 CCTV Systems for Railway Passenger Infrastructure - Existing and in Development

In 2015, only 108 passenger stations managed by PKP Polskie Linie Kolejowe S.A. had a CCTV system installed. From 924 cameras, 48.4% (447 cameras) were obsolete analog devices[1]. Police units and local authorities in many parts of Poland contributed to the change of safety policy of the National Railway Infrastructure Manager due to the increase of possible threats and frequent inquiries about safety systems made by the passengers. The debut of the Dynamic Passenger Information System (pol. SDIP – Systemy Dynamicznej Informacji Pasażerskiej) also played an important role, as the system's executive elements are composed of massive LCD displays, particularly vulnerable to vandalism. The National Railway Infrastructure Manager did not possess any regulations regarding CCTV dedicated for passenger infrastructure at that point. Development of guidelines for systems that could provide the ability to detect suspicious events required a new approach. Previous solutions applied in the railway have been implemented in the form of dispersed groups of devices without any form of remote access, usually handled by traffic administrators. Such solutions cannot be used for anything else than post-factum analysis, hence the decision to create a nation-wide CCTV system. The development of that system's core – the central server room, an integrating platform and a security center – are one of the tasks of the project titled "Projekt, dostawa i instalacja elementów prezentacji dynamicznej informacji pasażerskiej oraz systemu monitoringu wizyjnego wraz z infrastrukturą techniczną na dworcach, stacjach i przystankach kolejowych", which is a part of the Polish National Railway Program [3], designated as 5.2-11.

In 2017, guidelines titled "Ipi-4 Wytyczne dotyczące projektowania i budowy Systemów Monitoringu Wizyjnego (SMW) na obiektach obsługi pasażerskiej" have been published by PKP Polskie Linie Kolejowe S.A. The Ipi-4 guidelines defined the method of device selection, the basics of their integration, the use of image analysis algorithms (known as VCA – Video Content Analysis) as well as general guidelines for the design, installation and operation of the CCTV system. In 2018 a suplmentary guidelines regarding network and data transmission requirements were released as "Ie-122 Wymagania na transmisję danych systemów SMW, SPA i SDIP oraz integrację z siecią teletransmisyjną PKP Polskie Linie Kolejowe S.A.". Since 2018, development of new CCTV systems takes place only according to these guidelines. In 2018, the number of stations under video surveillance increased to 268, the number of cameras increased over four-fold (from 924 to 3832), and the percentage of obsolete analog devices dropped to 20,46% (to a total of 784)[2].

[1] Source: own study. As of 16.07.2015.

[2] Source: own study. As of 29.05.2018.

The guidelines [12] assume that each passenger station will be equipped with a Video Surveillance System, excluding those categorized as D, D- and E class stations (respectively: objects of local significance, objects of low local significance and objects with low number of train stops per day) for which the average daily number of train stops is less than or equal to 20. This essentially means that the target state will be a Video Surveillance System covering more than one thousand passenger stations, which would make it one of the largest security systems in Europe.

In addition to the development of the video surveillance system, PKP Polskie Linie Kolejowe also plan to install a system enabling two-way communication with the passenger in an emergency situations (pol. System Przywoławczo-Alarmowy – SPA), operated from the security center (pol. Centrum Bezpieczeństwa Infrastruktury Pasażerskiej – CBIP), based on videophones using Voice over Internet Protocol (VoIP).

3 Architecture of the Central Video Surveillance System of PKP Polskie Linie Kolejowe S.A.

The Central Video Surveillance System of PKP Polskie Linie Kolejowe S.A. can be divided into three major parts:

– station section (picture 1) – devices within individual passenger stations,
– network section (picture 2) – telecommunication infrastructure covering the area of whole Poland,
– server section (picture 3) – hardware platforms hosting applications that manage and integrate all elements of the system.

The station section is composed of cameras (in accordance with [12], divided into classes of different functionalities – with image sensors from at least 3 Mpix to at least 8 Mpix), communication modules and auxiliary devices (i.e. infrared radiators). These devices are connected to network switches, which themselves are connected to form a ring topology. Video transmission to the Network Video Recorder (NVR) might be local or utilizing the network section (in case of stations which are aggregated i.e. within a single railway line – in which case cameras from several stations are supported by a single NVR). The image is processed on NVRs by the VCAs, which are meant to detect events such as left luggage, zone intrusion (i.e. on railway tracks) or loitering (i.e. by the homeless). The operators stationed in the safety center (CBIP) are notified about those events through alarm calls.

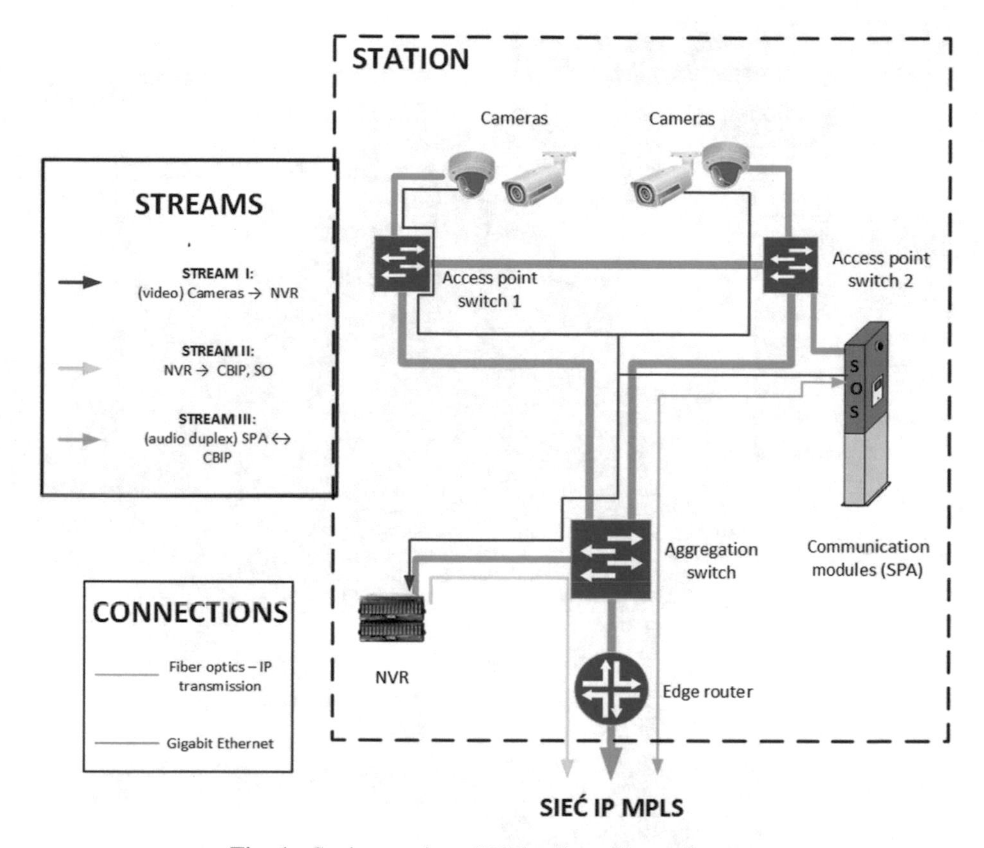

Fig. 1. Station section of Video Surveillance System

The network section is based on the IP-MPLS network owned by PKP Polskie Linie Kolejowe S.A. The same network is the backbone of all railway ICT systems, GSM-R and SDIP in particular. The MPLS backbone network is based on multiple ring topology. The whole network is supervised through SNMP packets from the Network Management Centre (NMS) of PKP Polskie Linie Kolejowe S.A. in Sosnowiec, Poland (Fig. 2).

The server section is composed of hardware platforms in the central server room (pol. Główna Serwerownia – GS), hosting the following applications:

- software integration platform – PSIM (Physical Security Information Management),
- VoiP software platform VoIP for the SPA system,
- data repository (disk matrices).

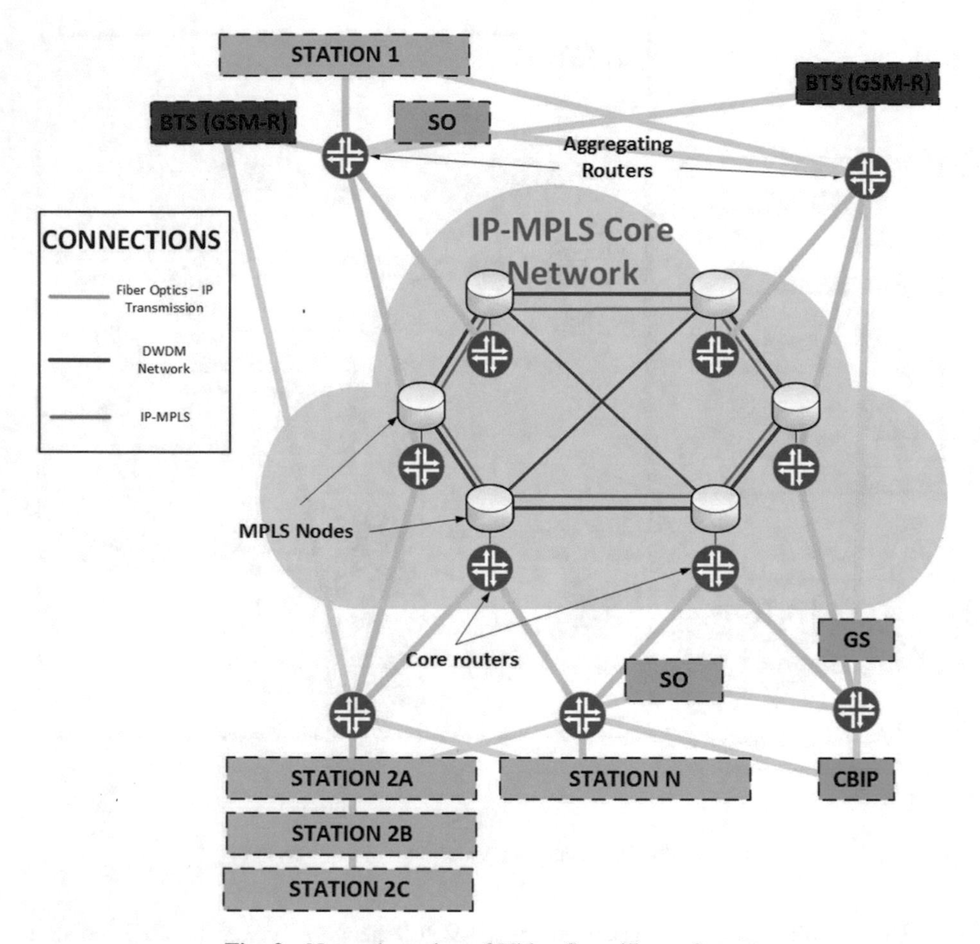

Fig. 2. Network section of Video Surveillance System

The server section includes a security center (CBIP) used to view the recordings, for management of the network and to carry out two-way communication with passengers utilizing the SPA system. The server section also includes sets of workstations (pol. Stanowiska Oglądowe – SO), deployed within railway buildings, allowing authorized entities (such as SOK or the Police) to monitor chosen stations, without administrative rights (Fig. 3).

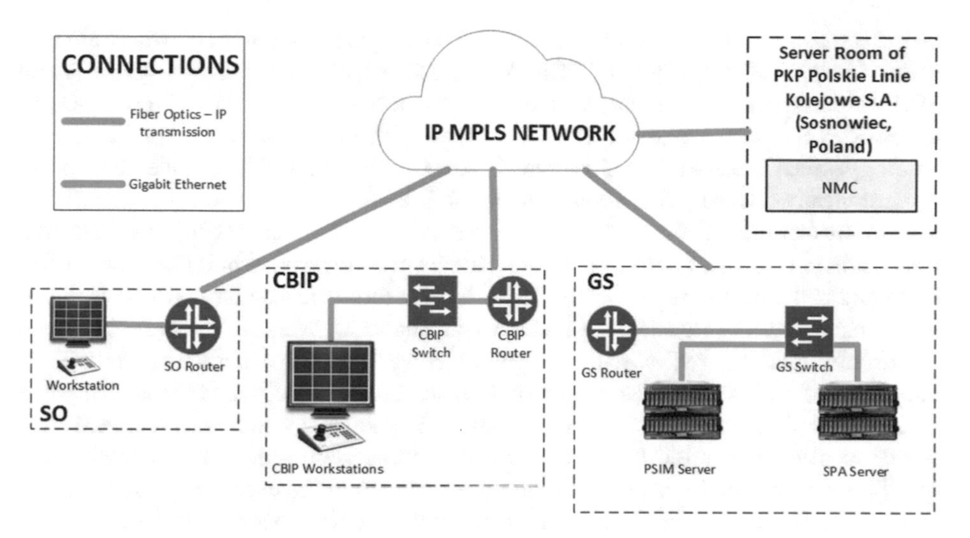

Fig. 3. Server section of Video Surveillance System

4 Assessment of Availability of the Video Surveillance System

The Video Surveillance System is not intended to be used for railway traffic control directly, hence it's not subject to national [9] and international requirements regarding railway traffic control. Elements of the Video Surveillance System can however help in traffic control, i.e. providing the dispatchers with the ability to check the situation on the station and to alarm them about passengers intruding the tracks, providing indirect utility. Thus, the Video Surveillance System can be considered an auxiliary operation system. For such a system, the availability requirements for GSM-R auxiliary voice transmission can be used as reference levels. The required availability for such transmission, when it is not used directly for traffic control, is 99.91% [8]. Fulfilling the requirements for such a vast system in practice comes down to assuring redundancy for each of the system's elements, groups and services.

In case of the station section, the redundancy for executive elements can be granted through following the design recommendations – in accordance with [12] the pairs of cameras should be placed in a way that enables them to have each other in their field of view, thus ensuring that if one camera breaks down, the visibility in its field of view will be maintained by the other device. The cameras have a guaranteed power supply, thus in case of the switch breaking down - the video does not stop recording. The cameras are provided with SD cards, which makes it possible for them to record the image even if the connection to the NVR gets severed. Recordings are downloaded to the NVR from the SD card after the connection gets re-established. The access point switches themselves are connected in a ring topology, which provides protection in case of physical damage to the fiber optic cable, i.e. during renovation work on railway tracks.

The network section is equipped with numerous solutions that ensure high availability. The Edge routers are connected in a ring topology i.e. within a single railway line. Each access point is connected to at least two aggregation routers, which are also nodes for other ICT systems of PKP Polskie Linie Kolejowych S.A. – GSM-R and SDIP in particular. Each aggregation node is connected to at least two core routers of the IP-MPLS network and the IP-MPLS network itself is based on a multiple ring topology. Such a network architecture ensures high availability despite the system geographically covering the whole country of Poland.

The availability of the server section seems to be the weakest link in the chain. According to [13, 14], PKP Polskie Linie Kolejowe S.A. plans to build only one server room and only one security center (CBIP) – both of those are also meant to be deployed in the same building (located at Chodakowska street 50, Warsaw, Poland). It's worth mentioning that the PSIM platform, update server and SPA central server will be implemented on a virtual platform hosted on server hardware, which should improve their availability, however, there is no physical redundancy implemented. Also, the choice to place both critical central elements of the system – the server infrastructure and the security center – in one location raises concern in the event of a terrorist attack or a natural disaster. An occurrence of those could effectively paralyze the system in the whole country.

If additional server rooms/security centers are not provided eventually, changing the requirements for "maximum time of reaction to damage" and "maximum time of damage repair" could provide a solution, lowering the Mean Time To Repair (MTTR) of the server section. At the moment, PKP Polskie Linie Kolejowe S.A. doesn't have any regulations regarding central components of the system, as [12] does not include guidelines for those. Those maximum times would have to exceed those in the highest category given in [12], which is 2 h maximum for reaction to damage and 8 h maximum to repair the damage.

The access network of the system (Fig. 1) can be presented as a mixed structure of series-parallel type for the availability of the access route displayed on Fig. 4.

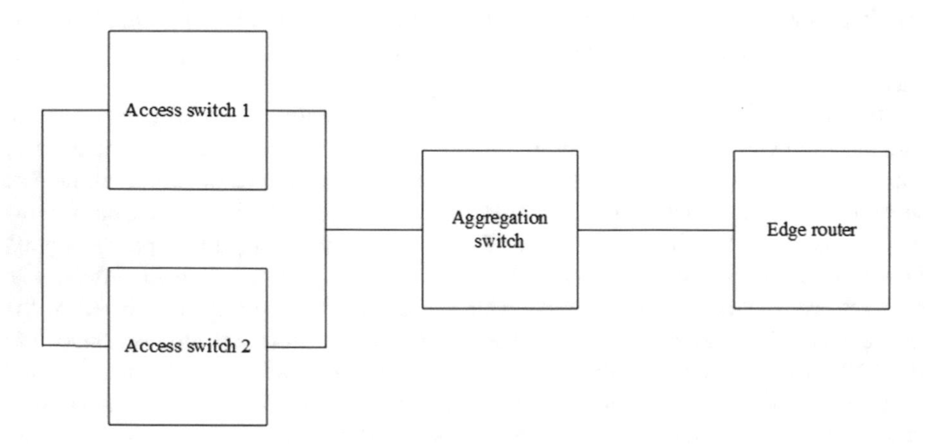

Fig. 4. General reliability scheme for the station section of the network

Damage to any of the access point switches in the parallel structure causes the transition from full efficiency state $R_O(t)$ to safety risk state $Q_{ZB}(t)$. Damage to the aggregating switch or the edge router in the series structure causes the transition from full efficiency state $R_O(t)$ to safety failure state $Q_B(t)$. Those relations are displayed on Fig. 5 in the context of system safety.

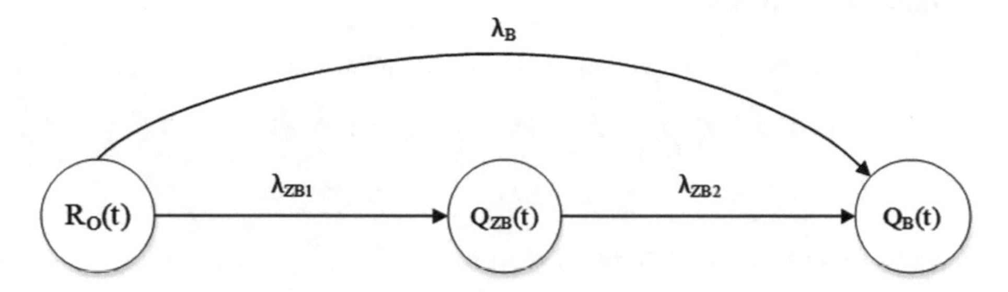

Fig. 5. Access network relations

Figure 5 designations:

$R_O(t)$ – probability function of access network being in full efficiency state,
$Q_{ZB}(t)$ – probability function of access network being in safety risk state,
$Q_B(t)$ – probability function of access network being in safety failure state,
λ_B – failure intensity of the elements of series structure of the access network (aggregating switch and edge router),
λ_{ZB1}, λ_{ZB2} – failure intensity of the elements of parallel structure of the access network (access point switches).

System shown on Fig. 5 can be described with the following Chapman-Kolmogorov equations:

$$\begin{aligned} R_0'(t) &= -\lambda_B \cdot R_0(t) - \lambda_{ZB1} \cdot R_0(t) \\ Q_{ZB}'(t) &= \lambda_{ZB1} \cdot R_0(t) - \lambda_{ZB2} \cdot Q_B(t) \\ Q_B'(t) &= \lambda_B \cdot R_0(t) + \lambda_{ZB2} \cdot Q_{ZB}(t) \end{aligned} \tag{1}$$

Assuming the initial conditions:

$$\begin{aligned} R_0(0) &= 1 \\ Q_{ZB}(0) &= Q_B(0) = 0 \end{aligned} \tag{2}$$

Utilizing Laplace's Transformation, we get the following system of linear equations:

$$
\begin{aligned}
s \cdot R_0^*(s) - 1 &= -\lambda_B \cdot R_0^*(s) - \lambda_{ZB1} \cdot R_0^*(s) \\
s \cdot Q_{ZB}^*(s) &= \lambda_{ZB1} \cdot R_0^*(s) - \lambda_{ZB2} \cdot Q_B^*(s) \\
s \cdot Q_B^*(s) &= \lambda_B \cdot R_0^*(s) + \lambda_{ZB2} \cdot Q_{ZB}^*(s)
\end{aligned} \tag{3}
$$

Through transformations we get:

$$
\begin{aligned}
s \cdot R_0^*(s) + \lambda_B \cdot R_0^*(s) + \lambda_{ZB1} \cdot R_0^*(s) &= 1 \\
s \cdot Q_{ZB}^*(s) - \lambda_{ZB1} \cdot R_0^*(s) + \lambda_{ZB2} \cdot Q_B^*(s) &= 0 \\
s \cdot Q_B^*(s) - \lambda_B \cdot R_0^*(s) - \lambda_{ZB2} \cdot Q_{ZB}^*(s) &= 0
\end{aligned} \tag{4}
$$

Using the reverse transformations, we get:

$$R_0(t) = e^{-(\lambda_B + \lambda_{ZB1}) \cdot t} \tag{5}$$

$$Q_{ZB}(t) = \lambda_{ZB1} \cdot \left[\frac{e^{-(\lambda_B + \lambda_{ZB1}) \cdot t} - e^{-\lambda_{ZB2} \cdot t}}{\lambda_{ZB2} - \lambda_B - \lambda_{ZB1}} \right] \tag{6}$$

$$
\begin{aligned}
Q_B(t) = {} & \frac{\lambda_B}{\lambda_B + \lambda_{ZB1}} \cdot \left[1 - e^{-(\lambda_B + \lambda_{ZB1}) \cdot t} \right] + \\
& \lambda_{ZB1} \cdot \lambda_{ZB2} \cdot \left[\frac{e^{-(\lambda_B + \lambda_{ZB1}) \cdot t}}{(\lambda_B + \lambda_{ZB1}) \cdot (\lambda_B + \lambda_{ZB1} - \lambda_{ZB2})} - - \frac{e^{-\lambda_{ZB2} \cdot t}}{(\lambda_B + \lambda_{ZB1} - \lambda_{ZB2}) \cdot \lambda_{ZB2}} + + \frac{1}{(\lambda_B + \lambda_{ZB1}) \cdot \lambda_{ZB2}} \right]
\end{aligned} \tag{7}
$$

The designated relationships allow to determine the probability of the network based on the elements displayed on Fig. 4 to be in full efficiency state R_O, safety risk state Q_{ZB} and safety failure state Q_B.

5 Summary

The centralized Video Surveillance System in polish railway is one of the most modern systems of its kind in Europe. It will noticeably improve the safety of passengers and property of the Polish National Railway Manager. The station section of the system has been developed based on the ring topology. In accordance with [12] at least two or three cameras should be deployed for each scene to ensure the expected pixel density (either 62.5 pix/m or 125 pix/m in the area of basic utility, depending on station class) is reached. The network design can be rated positively regarding its availability. However, for the system as a whole to meet the expectations regarding the availability the server section should be improved by providing additional redundancy points, as well as a partial correction of Mean Time To Repair (MTTR) – the maximum time of reaction and maximum time of damage repair – should be made. A single security center and a single server room (located in close vicinity) raises concern when it comes to the risks of possible terroristic threats and natural disasters.

References

1. Zhang, H., Shao, X., Zhang, J., Liu, B., Si, J.: Design of mechatronic-hydraulic product dynamic and multi-state reliability simulation platform. In: 2013 International Conference on Quality, Reliability, Risk, Maintenance, and Safety Engineering (QR2MSE) (2013)
2. Kołowrocki, K.: Reliability of Large Systems. Elsevier Ltd., Amsterdam (2004). https://doi.org/10.1016/B978-0-08-044429-1.X5000-4
3. Krajowy Program Kolejowy do 2023 r, Infrastruktura kolejowa zarządzania przez PKP Polskie Linie Klejowe S.A. Załącznik do uchwały nr 162/2015 Rady Ministrów z dnia 15 września 2015 r
4. PN-EN 50126: 2002 Zastosowania kolejowe – Specyfikacja niezawodności, dostępności, podatności utrzymaniowej i bezpieczeństwa
5. PN-EN 62676-1-1:2014-06 – wersja angielska – Systemy dozorowe CCTV stosowane w zabezpieczeniach – Część 1-1: Wymagania systemowe – Postanowienia ogólne
6. PN-EN 62676-4: 2015-06 – Systemy dozoru wizyjnego stosowane w zabezpieczeniach – Część 4: Wytyczne stosowania
7. Priks, M.: The effects of surveillance cameras on crime: evidence from the stockholm subway. Econ. J. **125**(588), F289–F305 (2015)
8. Siergiejczyk, M., Gago, S.: Wybrane problemy niezawodności i bezpieczeństwa transmisji w systemie GSM-R. Problemy Kolejnictwa – Zeszyt 162, pp. 111–124, Warszawa 2014 r
9. Ustawa z dnia 28 marca 2003 r. o transporcie kolejowym (Dz. U. z 2015 r., poz. 1297 z późn. zm.)
10. Werbińska-Wojciechowska, S.: Modele utrzymania systemów technicznych w aspekcie koncepcji opóźnień czasowych. Oficyna Wydawnicza Politechniki Wrocławskiej – Wrocław 2018 r
11. Wymagania na transmisję danych systemów SMW, SPA i SDIP oraz integrację z siecią teletransmisyjną PKP Polskie Linie Kolejowa S.A. Ie-122, PKP Polskie Linie Kolejowe S. A., Warszawa 2018 r
12. Wytyczne dotyczące projektowania i budowy Systemów Monitoringu Wizyjnego (SMW) na obiektach obsługi pasażerskiej Ipi-4, PKP Polskie Linie Kolejowe S.A., Warszawa 2018 r
13. Załącznik nr 5 "Wymagania dot. Zabudowy Stanowisk Oglądowo-Administracyjnych" do Opisu Przedmiotu Zamówienia dla projektu 5.2-11, PKP Polskie Linie Kolejowe S.A., Warszawa 2019 r
14. Załącznik nr 7 "Wytyczne dla kontenera telekomunikacyjnego dla potrzeb serwerowni GS" do Opisu Przedmiotu Zamówienia dla projektu 5.2-11, PKP Polskie Linie Kolejowe S.A., Warszawa 2019 r
15. Chołda, P., Jajszczyk, A.: Ocena gotowości w sieciach telekomunikacyjnych. Przegląd Telekomunikacyjny i Wiadomości Telekomunikacyjne, Wyd. SIGMA-NOT, Warszawa 2003, nr 2–3, pp. 66–71 (2003)

Impact of Airfield Pavement's Operability on Its Anti-skid Properties

Mariusz Wesołowski, Krzysztof Blacha, and Paweł Iwanowski(✉)

Air Force Institute of Technology, Warsaw, Poland
pawel.iwanowski@itwl.pl

Abstract. The article focuses on presenting the impact of the airfield pavements operability on the obtained values of the friction coefficient. The problem was captured in the context of the surface contamination's increase during the half-year measurement period. In addition, a reduction of the pavement surface anti-skid parameters in the touchdown zone on the runway, in relation to its central part was presented.

Anti-skid properties are characterized by two main parameters: coefficient of friction at wheel/surface contact zone and depth of surface texture. There are proven and used for years methods for measuring the macro- and micro-texture of the surface and the roughness of the surface.

In order to check the impact of operating processes on the anti-skid properties of airfield pavements, appropriate tests have been carried out. Changes resulting from arise of rubber amounts on the selected runway's (DS1) and taxiway's (DK) surfaces starting from late winter till late summer were assessed. In addition, a different runway was assessed (DS2) for which a map of friction coefficient values was made in order to show areas with worse anti-skid properties.

The article confirmed the necessity of constant control of anti-skid parameters of airport pavements in order to ensure safety conditions at an appropriate level. Maintenance services and proper operation of airfield pavements also is necessary to minimize the effects of surface deterioration and the accumulation of contaminants in its area.

Keywords: Anti-skid properties · Friction coefficient · Operability · Airfield pavement

1 Introduction

Anti-skid properties of airfield pavements play a key role in the safety of conducting air operations in the ground maneuvering area. The correct trajectory of the aircraft's movement during acceleration, braking and turning maneuvers is possible only if the pavement surface is rough enough. This is important in the case of unfavourable weather conditions, in particular during rainfall or freezing air temperature. Anti-skid properties are characterized by two main parameters: coefficient of friction at wheel/surface contact zone and depth of surface texture. In fact, the texture depth parameter determines certain value of friction coefficient.

M. Siergiejczyk and K. Krzykowska (Eds.): ISCT21 2019, AISC 1032, pp. 440–449, 2020.
https://doi.org/10.1007/978-3-030-27687-4_44

The operational pavement surface is exposed to constant deterioration, due to atmospheric factors and factors related to its daily maintenance services. The deterioration process changes the texture of the pavement surface, thereby changing the friction coefficient values obtained, and thus changing the anti-skid properties of the surface. That is the reason why operational airport pavements shall be regularly checked in terms of its roughness.

Changes in the pavement surface texture are also possible due to the maintenance treatments. The winter maintenance of the surface and the process of removing rubber deposits, coming from the aircraft's wheels, from the pavement surface is of great importance here. Both groups of treatments may change, and usually change, not necessarily in negative way, the anti-skid properties of the pavement surface.

2 Airfield Pavement's Operability

Airfield pavements during operations constantly deteriorate. The construction type, atmospheric conditions and the way they are operated, and in particular care for their technical condition have an impact on such a process. Each surface is characterized by specific surface parameters, due to which it achieves its own anti-skid properties. Abovementioned surface parameters change over time due to external influences, of which aircraft loads, atmospheric conditions and maintenance services are the most significant. The atmospheric conditions also affect the degradation process of airfield pavement surfaces. Cyclical freezing and thawing of water in the voids causes microcracks, which later on transform into other types of deteriorations. High sun exposure of the surface and heating of the upper surface layer cause stress in the construction, which can consequently cause damage both in the upper and lower layers of the construction structure.

Part of the abovementioned deteriorations is extremely important for the anti-skid properties of the airfield pavement surface. These are in particular the shallow and deep scaling, as well as hairline and frost cracks. The view of map cracking is presented in the photograph (Fig. 1). Surface changes its texture, and thus the parameter of the friction coefficient due to these deteriorations.

The surface condition of the airport pavement is also affected by maintenance procedures, in particular concerning winter maintenance [2]. Solutions of acetates and potassium formates, as well as acetates and sodium formates are commonly used for airfield pavement's de-icing purposes. The use of de-icing agents in large amounts has a devastating effect on airfield pavements. This is especially important in case of construction that has been built in the technology of cement concrete not resistant to de-icing agents. Cyclical freezing and thawing of the airfield pavement surface in the presence of de-icing agents leads to scaling formations on the construction's, resulting in changes of texture and anti-skid properties.

An important factor that may have an impact on changing the anti-skid properties of airfield pavements is rubber remaining on the pavement surface, arised from aircraft tires. The view of rubber arisen on the runway pavement surface is shown in the photograph (Fig. 2). The rubber seals the macrotexture of the pavement surface and affects its microtexture. Taking into account the fact that micro- and macrotexture has

Fig. 1. Surface scaling of airfield pavement. *Source: Author's study on the basis of Air Force Institute of Technology archives*

an impact on the friction coefficient values obtained, the presence of rubber changes the anti-skid properties of the pavement surface.

Fig. 2. Rubber arisen on the runway pavement surface. *Source:* https://www.airport-suppliers.com/supplier-press-release/airport-runway-marking-removal-runway-rubber-removal-15/, *19 April 2019*

3 Airfield Pavement's Anti-skid Properties

3.1 Pavement Surface Texture

The anti-skid properties of the pavement's surface depends in particular on the surface texture. The obtained values of the friction coefficient depend in a significant part on the characteristics of the micro- and macros texture of the pavement's surface.

Microtexture refers to a depth of texture in the range of up to 0.5 mm and determines surface irregularities in the wavelength range of less than 0.5 mm. This parameter is closely related to the aggregate surface texture of which pavement consists of. It is essential for maintaining the proper anti-skid properties of the pavement surface at low speeds [5]. The microstructure of the surface undergoes changes during the surface operability due to the aggregate polishing phenomenon or collection of aircraft rubber sediments in the touchdown zone. That is a reason why it is so important to remove rubber from the surface of the runway pavement regularly. The friction coefficient measurements shall be performed on the previously cleaned surface after each procedure of rubber removal.

A texture with a depth above 0.5 mm is referred to as macrotexture and characterizes irregularities in the wavelengths exceeding 0.5 mm [5]. Macrotexture allows water disposal at the wheel and pavement contact zone at high speeds, thus improving the anti-skid properties of the pavement surface during precipitation conditions. Macrotexture is closely related to the size of the aggregate used for the road surface layer or the surface construction technology (so-called roughing). The rate of coefficient of friction decreases along with the test speed measurement increases depends mostly on the macrotexture [5].

Airfield pavement's texture requirements are not straightforward and various values are specified in different documents. The average depth of texture is provided as 1.0 mm in most cases [10, 15, 16, 17].

There are various methods of measuring the surface texture, whereas the basic volumetric method is included in PN-EN 13036-1 [13]. Often referred to as the "calibrated sand method" or the "sand patch method", it has always been used all over the world. It involves spreading the previously measured amount of calibrated sand with an appropriate graining in a circular manner on the evaluated surface. Glass beads may also be used for testing. The material is spread over the pavement surface with the use of a metal disc with a diameter of 65 mm and a thickness of about 25 mm. The test result is mean surface texture depth MTD calculated from formula:

$$MTD = \frac{4V}{\pi D^2} \tag{1}$$

where:

V – volume of granular material spread over the surface [mm^3]

D – mean diameter of circle covered with spread material [mm].

Due to the spot character of the measurement, with a small number of test points, test does not accurately reflect the entire surface in terms of the macrotexture parameter. For continuous macrotexture depth measurements, the profilometric method described in the PN-EN ISO 13473-1 [14]. In contrary to the method discussed earlier, the profilometric measurement allows to capture the texture in the range from 0 to 5 mm (the sand patch method has a range of 0.25 to 50 mm). The method enables to evaluate both macro and micro texture [5]. It uses point or linear laser sensors. As a result the surface profile, that is further the basis for analysis in order to determine the average depth of the MPD profile, is obtained. The value of MPD derived after transformation into the estimated depth of the texture ETD is a reference value to the results obtained by the traditional sand patch method. Transformation is performed using the formula:

$$ETD = 0,2 + 0,8MPD \tag{2}$$

The error between the MTD and ETD results is assessed as significantly smaller in comparison with the dispersion of sand patch method results, and related to the measurements performance by different operators. One of the devices used to evaluate the surface profile is an RSP laser profilograph mounted on a car.

3.2 Friction Coefficient

The basic parameter characterizing the airport pavement surface in terms of roughness is the friction coefficient. Friction is a phenomenon occurring at the contact zone of the surface and material bodies. The friction force at rest is always equal to the resultant external forces acting in the parallel direction to the contact surface of the objects. This is due to the first principle of Newton's dynamics [4].

The friction force depends on two parameters: the value of the one body's pressure on the other and the coefficient of friction. In the case of static friction, when the objects remain stationary in relation to each other, the coefficient of friction is constant [5].

The general formula for the maximum friction force is:

$$T = \mu N \tag{3}$$

where:

T - friction force [N],

μ - friction coefficient value,

N - force one body is pressing on another [N].

The value of the aircraft wheel's coefficient of friction on the airfield pavement's surface is determined interchangeably as the roughness of the surface. The surface roughness depends on many factors, such as the type of material from which the surface pavement was built, the appropriate macrotexture, moisture, the presence of contaminants or de-icing agents during its winter maintenance. In order to ensure the safety and reliability of aircraft during the flight operations at the highest level, periodic tests of roughness of airfield pavements shall be carried out. This is especially important during the winter maintenance services of airfield pavements, during atmospheric precipitation and freezing air temperatures. Because of that the documents regulating study were created.

International aviation documents present values of the coefficient of friction, which shall correspond to the airfield's functional elements surface, so that it is possible to perform flight operations maintaining the safety and reliability of aircraft. The International Civil Aviation Organization (ICAO) has developed Doc document. 9137 AN/898 Airport Service Manual Part 2 - Pavement Surface Conditions [6] that specifies the requirements for the airfield pavement in terms of friction coefficient. These requirements are also given in ICAO Annex 14 [7], in AC 150/5320-12C FAA [8] and in Polish military standard NO-17-A501: 2015 [9]. The requirements specify minimum friction coefficient values for newly designed airfield pavements and friction coefficient values for operational airfield pavements, above which repair actions shall be taken, and minimum friction coefficient values below which the surface is unusable for flight operations in the movement area.

The frequency of friction coefficient measurements shall be specified at the level enabling the identification of airfield functional elements in terms of minimum friction coefficients, in order to take repair actions. The frequency shall depend on the aircraft type to be taken, the number of air operations to be carried out, the type of airfield pavement surface, weather conditions and requirements in the scope of pavement operability. The frequency of measurements shall not be less than one test per every three months. In addition, tests shall be carried out whenever there is a suspicion that the runway surface may be slippery, especially in the event of unfavourable weather conditions, such as heavy rainfall. What is more, the roughness test of the airfield pavement shall be carried out each time after rubber from aircraft's tire had been removed.

The most popular devices for measuring the coefficient of friction at Polish airports are the ASFT unit mounted on an autonomous vehicle and the ASFT T-10 trailer towed behind the vehicle. The view of both devices is shown in the photograph (Fig. 3).

Fig. 3. ASFT device for coefficient of friction measurements. T-10 trailer is visible at the forefront, in the background - mounted on Skoda Octavia vehicle. *Source: archives of Air Force Institute of Technology*

4 Friction Coefficient Test Results at Selected Airfield Pavements

In order to check the impact of operating processes on the anti-skid properties of airfield pavements, appropriate tests have been carried out. Changes resulting from arise of rubber amounts on the selected runway's (DS1) and taxiway's (DK) surfaces starting from late winter till late summer were assessed. In addition, a different runway was assessed (DS2) for which a map of friction coefficient values was made in order to show areas with worse anti-skid properties. The DS1 runway was made of asphalt concrete, while the runway DS2 and taxiway DK were constructed in cement concrete technology. Selected airport facilities perform high number of air operations (over 3 million passengers served per year) and each of the airfield's functional elements assessed was in constant use. The Swedish company's ASFT T-10 device built in on the trailer, was used in the friction coefficient measurements tests.

The tests on DS1 and DK pavements were performed at the measuring speed of both 65 and 95 km/h. The measurement was carried out with the disposal of water to the measuring wheel due to create a water film with a thickness of 1 mm. Two selected sections (one at DS1, the other at DK) were assessed, with a length of 200 m each. 10 measurements were taken on each section for each of the measurement speeds. The test was carried out three times, at similar time intervals, every 3 months. That resulted in 120 test results.

4.1 Measurements on DS1 and DK Pavements

In order to illustrate the subject problem, mean values of the coefficient of friction, results standard deviation, minimum and maximum values of the coefficient of friction obtained in the diagrams (Figs. 4 and 5) were presented.

As it is shown in the graphs beyond, the mean value of the friction coefficient decreased regardless of the measurement conditions and the assessed airfield's functional element. The inspection of the evaluated pavement surfaces proved a significant increase in rubber contamination from the aircraft tires on the DS1 element. On the other hand, no rubber was found on the DK element, although in winter - spring period the pollination of the pavement surface increased.

In the summer, maintenance services in form of concrete hydrophobation was carried out. The treatments were performed directly before the test. It shall be assumed that it had a direct impact on the anti-skid properties of the assessed section.

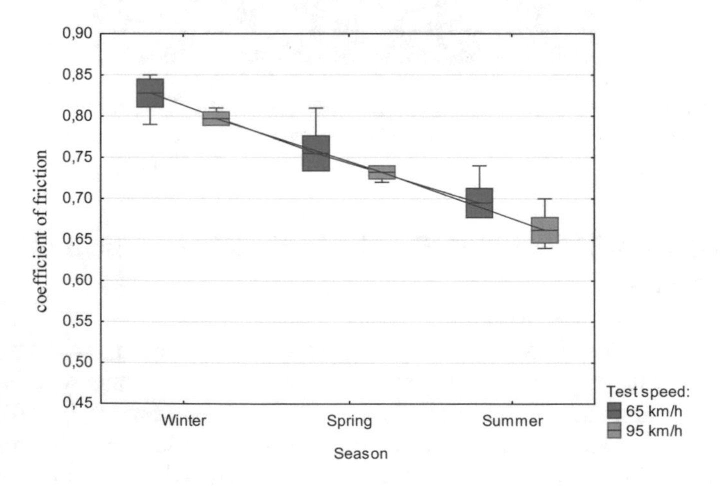

Fig. 4. Friction coefficient results from the DS1 pavement

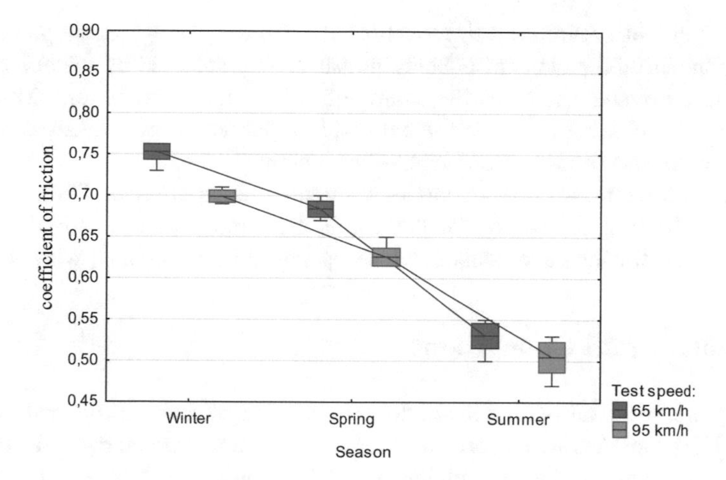

Fig. 5. Friction coefficient results from the DK pavement

4.2 Measurements on DS2 Pavement

The pavement surface of the whole runway, designated as DS2 was assessed in order to show the deterioration of anti-skid properties in the touchdown zone. 10 runs were made along the entire length of the runway, excluding sections for the run up and safe braking of the measurement equipment. The measurement routes were located parallel to the DS2 axis, west-east (EC) at a distance of 5 m. The average friction coefficient value for every 10 m measuring route was taken into account, which resulted in approximately 280 friction coefficients for each measuring route and over 2800 results for the entire DS2. On thc basis of the population of results collected, a map of friction coefficient values presented in figure (Fig. 6) was developed.

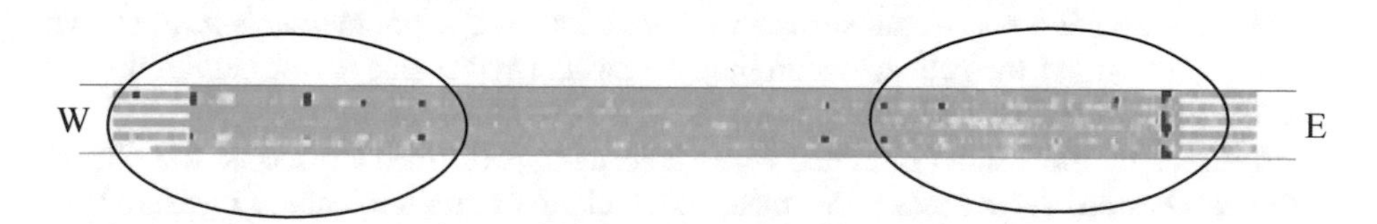

Fig. 6. Map of DS2 pavement surface friction coefficient values with touchdown zones marked

In the illustration above, the areas in which the mean coefficient of friction fell below the required limit value of 0.4 were highlighted in red, whereas yellow areas were of friction coefficient between 0.4 and 0.55.

The analysis shows that in the touchdown zones the mean friction coefficient decreased significantly, comparing to the central part of the runway. It may be assumed that this phenomenon is directly related to the presence of contaminants in the form of rubber coming from aircraft tires. Tires at the moment of touchdown are pressed towards the surface, thus contaminating the surface.

At the moment of aircraft's tire touching the surface, the aggregate present on the surface of the airfield pavement is being polished. As explained by the authors in [5], the aggregate polishing reduces the average depth of the microtexture. This process may have a significant impact on the anti-skid properties of the evaluated pavements and, as a result, the friction coefficient values obtained.

Despite the decreased coefficients of friction in some areas of DS2, the average value of the friction coefficient for the whole measuring route was higher than the minimum value for the pavements being in operation in accordance with [12].

5 Summary and Conclusions

The safety of aircraft air operations in the terrain maneuvering area depends strictly on the anti-skid properties of airport pavements. These are characterized by two basic, dependent parameters. We are talking about the surface texture and coefficient of friction at the wheel/surface contact zone. Changes in the surface texture lead to changes of anti-skid properties, and thus obtained values of the coefficient of friction.

There are proven and used for years methods for measuring the macro- and micro-texture of the surface and the roughness of the surface. The article presents the most important of them, commonly used on Polish airport facilities. The requirements for these parameters were also specified in accordance with the cited normative documents, which, as it turns out, are not always consistent.

Texture of the surface during the life and operability of the airfield pavement undergoes constant changes. This is due to the surface deterioration, resulting from many factors, including the atmospheric conditions, the pavement's operability and maintenance services.

During the exploitation of the surface pavement, the surface contaminants arise and the texture changes. Rubber coming from aircraft tires is the main pollution, mainly referring to the runway pavement.

The tests carried out on the surface of selected runways and taxiways have shown that the problem of the rubber occurrence on airfield pavement is not negligible and significantly affects the anti-skid properties of the pavement surface.

Analysis of the results obtained from field tests performed at one of the airport facilities showed decreases in the mean coefficient of friction values measured on selected sections during subsequent tests performed for half a year at time intervals of 3 months. Examination of the surface of the assessed sections showed a noticeable increase in rubber contamination on the surface of the functional elements of the airport. The newly hydrophobized concrete treatment and the pollination of the paved road also affected the degradation of anti-skid properties.

The presence of rubber on the surface of the runway in the touchdown zones and its anti-skid properties degradation was confirmed by tests carried out on the whole surface of this element. The map of the average values of the friction coefficient presented in Sect. 4 shows the noticeable differences between the touchdown zones and the central part of the runway. At critical points, the individual friction coefficient values obtained were lower than the required limit value in normative documents of 0.4.

The article confirmed the necessity of constant control of anti-skid parameters of airport pavements in order to ensure safety conditions at an appropriate level. Maintenance services and proper operation of airfield pavements also is necessary to minimize the effects of surface deterioration and the accumulation of contaminants in its area. This is of key importance in the aspect of the safety of air operations performance in the terrain maneuvering area.

The authors, in cooperation with a wider research team, are working on an innovative method of assessing anti-skid properties on airfield pavements, enabling continuous assessment of the surface texture.

References

1. Barszcz, P., Wesołowski, M.: Assessment of the deterioration stage in the process of operation of airfield pavement functional elements made of cement concrete. J. KONBiN **32**, 45–56 (2014)
2. Kowalska, D., Misztal, A.: Winter maintenance of airport pavements and the safety of aircraft operations. Autobusy **17**, 278–285 (2016)
3. Iwanowski, P., Wesołowski, M.: Characteristics of friction of airport surfaces in the aspect of the safety of flight operations. Autobusy (2016)
4. Iwanowski, P., Blacha, K., Wesołowski, M.: Review of modern methods for continuous friction measurement on airfield pavements. In: IOP Conference Series: Materials Science and Engineering, vol. 356, p. 012002 (2018)
5. Wesołowski, M., Blacha, K.: Evaluation of airfield pavement micro and macrotexture in the light of skid resistance (friction coefficient) measurements. In: MATEC Web Conference, vol. 262, Art. No. 05017 (2019)
6. Hall, J.W., Smith, K.L., Titus-Glover, L., Wambold, J.C., Yager, T.J., Rado, Z.: Guide for pavement friction. Contractor's final report for NCHRP Project (2009)
7. Habich, E.: Vehicle mechanics. Technical car guide. Cz.1, Warszawa (1962)
8. Mataei, B., Zakeri, H., Zahedi, M., Nejad, F.M.: Pavement friction and skid resistance measurement methods: a literature review. Open J. Civ. Eng. **6**, 537–565 (2016)
9. Nita, P.: Construction and maintenance of airport pavements. WKŁ, Warszawa (2008)
10. Doc. 9137 ICAO AN/898 Airport Service Manual Part 2 - Pavement Surface Conditions, ICAO, 4th Edn. (2002)
11. ICAO Annex 14 - Volume 1 - Aerodromes - Aerodrome Design and Operations, 7th Edn. (2016)
12. Advisory Circular no: 150/5320-12C, U.S. Department of Transportation, Federal Aviation Administration (FAA). Accessed 18 Mar 1997
13. NO-17-A501:2015 Airport pavements. Roughness test
14. PN-EN 13036-1 Surface features of road and airport pavements. Test methods. Part 1: Measurement of depth of macrotexture by volumetric method
15. PN-EN ISO 13473-1 Characteristics of the surface structure using surface profiles. Part 1: Determination of the average depth profile
16. ED Decision 2017/021/R Issuing Certification Specifications and Guidance Material for Aerodrome Design (CS ADR-DSN)
17. Doc. 9157 ICAO AN/901 Aerodrome Design Manual Part I – Runways, ICAO, 3rd edn. (2006)
18. Advisory Circular no: 150/5320-12C, U.S. Department of Transportation, Federal Aviation Administration (FAA) (1997)

Comparative Analysis of the Electric Vehicle Charging Costs Using DC and AC Charging Stations Throughout the Country, Based on Own Research

Piotr Wiśniowski(✉) and Maciej Gis

Motor Transport Institute, Jagiellońska 80, Warsaw, Poland
piotr.wisniowski@its.waw.pl

Abstract. The Act on electromobility and alternative fuels presupposes the introduction of a million electric vehicles on the Polish market by 2025. Market analyzes assume that with an optimistic scenario, it will be possible to supply such a number of vehicles with electricity. Own analyzes resulting from the conducted research presented in the article confirm this fact.

According to social surveys, the majority of potential electric vehicle users assume charging these types of vehicles using a public charging station network. On the Polish market, the number of such points is gradually increasing. Both DC and AC chargers are available. The article focuses, inter alia, on presenting costs related to the use of electric vehicles when charging this type of car. The presented analysis is aimed at demonstrating the real operating costs associated with electric vehicles and confirming the energy possibilities for the introduction of one million vehicles of this type. In order to determine the actual current consumption of the car in road traffic conditions, the authors of the article carried out RDE type tests on the designated measurement route. A six-channel YOKOGAWA WT1806E measuring device was used in the research. Vehicle loading costs have been calculated on the basis of data received from the electricity supplier.

Keywords: Electromobility · Electric vehicle · Charging costs

1 Introduction

Historically, the first attempts to introduce an electric vehicle began in 1800, so the electric vehicles are not a novelty in the history of motoring. In later years, they were replaced by vehicles powered by conventional fuels [11].

The ongoing return to the electric vehicle is the result of the increase in greenhouse gas emissions caused by the combustion of conventional fuels.

According to the IEA report, the number of electric vehicles (EV), hydrogen vehicles (FCEV) and plug-in hybrid vehicles (PHEV) on global roads in 2017 reached 3.1 million (an increase of 54% y/y). In 2017, more than 1 million electric cars were sold in total (54% more than in 2016), of which half (580,000) in China. The US came second with 280,000 units.

M. Siergiejczyk and K. Krzykowska (Eds.): ISCT21 2019, AISC 1032, pp. 450–456, 2020.
https://doi.org/10.1007/978-3-030-27687-4_45

There are currently 1.23 million electric vehicles on the Middle Kingdom roads (40% of the global EV fleet), 0.82 million in Europe, 0.76 million in the US, and 0.3 million in the rest of the world. In 2018, the most EVs were observed in Norway (39.2%), Iceland (11.7%) and Sweden (6.3%). Next places were taken by China (2.2%), Germany (1.6%), USA (1.2%) and Japan (1.0%) [1].

Electric vehicles are currently referred to as the main pillars of the future automotive industry. By 2030, they are to account for around 30–35% of total vehicle sales in the world [3, 4].

2 Road Tests of the Electric Vehicle

2.1 Development of the RDE Research Route

To better determine the actual impact of electric vehicles on the environment, it was necessary to determine the actual energy consumption in real road conditions, in accordance with the real emission test (RDE).

For this purpose, it was necessary to designate a test route that would be representative of the tests and meet the requirements imposed by the WLTP procedure.

The requirements imposed by the legislator are presented in Table 1.

Determination of the test route (Fig. 1) was made by the Motor Transport Institute. Its development was a priority due to further research possibilities [12, 17].

Fig. 1. RDE research route [7]

The designated transit route meets the requirements imposed by the legislator. It complements the WLTP testing procedure. The transit route is representative for Warsaw and allows performing tests in the field of emission tests in accordance with the requirements of the currently applicable WLTP procedure [13, 15, 16] (Fig. 2; Table 2).

Table 1. Requirements for the RDE test route

Parameter	Requirement
Outdoor temperature (T_z)	Normal range: 0 °C $\leq$ T_z < 30 °C
	Extended bottom range: −7 °C $\leq$ T_z < 0 °C
	Extended upper range: 30 °C < T_z $\leq$ 35 °C
	Normal range: 0 °C $\leq$ T_z < 30 °C
The topographical value of conducting the research (h)	Normal range: h $\leq$ 700 m n.p.m.
	Extended range: 700 < h $\leq$ 1300 m n.p.m.
Assessment of the impact of external weather and road parameters, as well as driving style	Total height increase: less than 1200 m/100 km
	Relative positive acceleration (RPA): bigger than RPA_{min} (for all driving conditions)
	Coefficient of velocity and acceleration ($v \cdot a_{pos}$): less than $v \cdot a_{pos\ min}$ (for all test conditions)
Thermal condition of the vehicle before the test	Cold start of the vehicle: cooling liquid below 70 °C Time at least 300 s
	No counting of emissions from a cold start to the RDE test
One time vehicle stop	Not longer than 180 s
Operation of driving comfort systems	Normally used as intended (e.g. operation of the air conditioning system)
Test requirements	Duration 90–120 min
Requirements for the urban part of the test	Share 29–44% the length of the entire test
	Distance: greater than 16 km
	Vehicle speed (v): v $\leq$ 60 km/h
	Average speed: 15–40 km/h
	Staging: 6–30% of the time of the urban part
Requirements for the extra-urban part of the test	Share of 23–43% of the total length of the test
	Distance: greater than 16 km
	Vehicle speed (v): 60 km/h < v $\leq$ 90 km/h
Requirements for the test section of the highway	Share of 23–43% of the total length of the test
	Distance: greater than 16 km
	Vehicle speed (v): v > 90 km/h
	Driving speed over 100 km/h for at least 5 min
	Driving speed over 145 km/h for up to 3% of the time

Table 2. Selected technical data of the tested EV vehicle [2]

Parameter	Unit	Value
Length	mm	4140
Width	mm	1800
Height	mm	1593
Wheelbase	mm	2570
Engine		Synchronous with permanent magnets
Power	kW	80,2
Maximum engine rotational speed	rpm	10500
Range	km	Around 200
Battery type		Lithium-polymer
Number of links	pcs	192
Number of modules	pcs	8
Weight of links	kg	135

Fig. 2. An electric research vehicle

2.2 Measuring Equipment

Specialist test equipment was used to test the electric vehicle. The YOKOGAWA WT1806E device (Fig. 3) is a six-channel analyzer capable of measuring, among other things, power or energy consumption collected and recovered from batteries and subassemblies of electric and hybrid cars, among others

- voltage measurement range 0–1000 V,
- current measurement range: 3 channels 0–5 A, 3 channels 0–50 A,
- accuracy not lower than 0.05% of reading, +0.05% of the measuring range,
- frequency of updating the measurement (voltage, current, power) at least 1 ms,
- sampling rate 2 Msa/s, 16-bit converter.

Fig. 3. Yokogawa research apparatus

3 Test Results and Analysis

The tests were repeated several times in order to determine the average of the obtained results.

The results from the electric vehicle tests on the RDE route are shown in Table 3.

Table 3. The average energy consumption of the EV vehicle and the average energy taken from the electricity grid during its charging

Parameter	Unit	Value
Energy consumed during research in RDE	kWh	17.3
Energy collected from the mains while charging	kWh	14.8

On the basis of own tests of an electric vehicle in real traffic conditions, it was possible to examine its average energy consumption and determine how much electricity is needed to charge the batteries after passing the RDE test. Then it was determined with the assumption of daily runs, what would be the energy demand for one vehicle and energy costs related to it (Table 4).

Table 4. Annual energy consumption needed to recharge the batteries of one vehicle and charging costs depending on the tariff

Parameter	Unit	Value
Daily mileage on business days	km	60
Daily mileage on Sunday and holidays	km	60
Annual mileage on business days	km	15 060
Annual mileage on Sundays and holidays	km	6840
Total mileage	km	21 900

(continued)

Table 4. *(continued)*

Parameter	Unit	Value
Price of 1 kWh for a household for different tariffs:		
G11	zł/kWh	0.58
G12 - peak	zł/kWh	0.57
G12 - off the peak	zł/kWh	0,25
G12w - peak	zł/kWh	0.62
G12w - off the peak	zł/kWh	0.25
Annual energy consumption needed to recharge the batteries of one vehicle	kWh	3602
Annual cost of charging one vehicle for different tariffs:		
G11	zł	2089
G12 - peak	zł	2053
G12 - off the peak	zł	900
G12w - peak	zł	2233
G12w - off the peak	zł	900

The peak value described in Table 4 applies to the current price from 6:00 to 13:00 and from 15:00 to 22:00. The remaining period of 24 h for the so-called off-peak period, which takes place between 13:00 and 15:00 and at night.

4 Conclusions

The result of the research, carried out by the authors of the article, in real traffic in RDE tests was to determine the consumption of electricity during test cycles. On this basis, it was possible to estimate the average daily and annual electricity demand in an electric vehicle that is in the possession of the average car user. The next stage of consideration was to calculate the total amount of money needed, with the estimated demand for electricity, to cover the costs of charging the vehicle. These amounts depend on the selected tariff for electricity settlement and its delivery by the supplier. The most advantageous solution from among those presented is charging the vehicle in the off-peak period, in the G12 and G12w electricity tariffs. Then the cost of driving 100 km is 4.1 PLN and therefore it is lower than 1 L of gasoline. Charging the vehicle in the other tariff variants is more expensive, the most expensive would be if the vehicle is charged at tariff G12w during the peak period. Then the cost of traveling 100 km would be 10.2 PLN which is equivalent to about 2 L of gasoline [8–10].

References

1. Wu, B., Offer, G.: Environmental impact of hybrid and electric vehicles. In: Environmental Impacts of Road Vehicles: Past, Present and Future, vol. 44. ISBN-13 978-1782628927
2. http://kia.com
3. https://dieselnet.com/news/2018/05iea2.php
4. Guo, C., Yang, J., Yang, L.: Planning of electric vehicle charging infrastructure for urban areas with tight land supply. Energies **11**(9), 2314 (2018)
5. JEC (JRC-Eucar-Concawe): Tank-To-Wheels Report Version 4a – Well-to Wheels Analysis of Future Automotive Fuels and Powertrains in the European Context. https://ec.europa.eu/jrc/en/jec
6. Climate Change 2014: Synthesis Report. Contribution of Working Groups I, II and III to the Fifth Assessment Report of the Intergovernmental Panel on Climate Change. In: Pachauri, R. K., Meyer, L.A. (eds.) Core Writing Team, 151 pp. IPCC, Geneva, Switzerland (2014)
7. Wei, P., Hsieh, Y., Chiu, H., Yen, D., Cheng, Y., Ting, T.: Absorption coefficient of carbon dioxide across atmospheric troposphere layer. Heliyon **4**(10), e00785 (2018)
8. mib.gov.pl
9. Moro, A., Lonza, L.: Electricity carbon intensity in European Member States: impacts on GHG emissions of electric vehicles. Transp. Res. Part D: Transp. Environ. **64**, 5–14 (2018). https://doi.org/10.1016/j.trd.2017.07.012
10. Li, B., Li, Q., Lee, F.C.Y., Liu, Z.: Bi-directional on-board charger architecture and control for achieving ultra-high efficiency with wide battery voltage range. In: 2017 IEEE Applied Power Electronics Conference and Exposition (APEC) (2017)
11. Rajashekara, K.: Present status and future trends in electric vehicle propulsion technologies. IEEE Power Energy Soc. **1**, 3–10 (2013)
12. Commission Implementing Regulation (EU) 2018/1002 of 16 July 2018 amending Implementing Regulation (EU) 2017/1153 to clarify and simplify the correlation procedure and to adapt it to changes to Regulation (EU) 2017/1151
13. Wiśniowski, P., Ślęzak, M., Niewczas, A., Szczepański, T.: Selection of the vehicle speed waveforms filtering method. **9**, 24–29 (2017)
14. UN ECE. Regulation 101, Rev.3. United Nation Economic Commission for Europe (2013). http://www.unece.org/trans/main/wp29/wp29regs101-120.html
15. Implementation of the test procedure for conducting real driving emissions tests (RDE). Statutory work 06/18/ITS/008. Motor Transport Institute
16. Wu, X., Freese, D., Cabrera, A., Kitch, W.: Electric vehicles' energy consumption measurement and estimation. Transp. Res. Part D Transp. Environ. **34**, 52–67 (2015)
17. Malling, N., Heilig, M., Weiss, C., Chlond, B., Vortisch, P.: Modelling the weekly electricity demand caused by electric cars. Proc. Comput. Sci. **52**, 444–451 (2015)

Assessment of Modernization of Electric Multiple Units - Case Study

Jerzy Wojciechowski(✉), Zbigniew Łukasik, and Czesław Jakubowski

Faculty of Transport and Electrical, Kazimierz Pulaski University of Technology and Humanities in Radom, Malczewskiego 29, 26-600 Radom, Poland
{j.wojciechowski,z.lukasik,c.jakubowski}@uthrad.pl

Abstract. The development of passenger rail transport is an element of the development of our civilization. The requirements set by passengers as to the quality and comfort of driving are very high. This mainly applies to travel conditions, i.e. the means of transport itself. The second important factor is the cost of travel, resulting, inter alia, from the energy consumption of the means of transport used. The purchase of new railway units is expensive. A competitive solution is the modernization of units already in use. The authors presented in the article an analysis of the modernization of the group of electric multiple unit vehicles (EMU). This modernization involved, among other things, the reconstruction of almost the entire drive unit, the reconstructed driver's cab, the change of the external appearance and the reconstruction of the interior. The article presents technical solutions used in the modernization process. The authors focused on the energy issues of vehicles before and after modernization, presenting the results of their own research. Measurements are carried out by the authors at EMU functioning on railway lines connecting Łódź with: Warsaw, Tomaszów Mazowiecki and Częstochowa (Poland). Data regarding electricity consumption by EMU before and after modernization were presented. The data on energy recuperation by the analyzed EMU group were presented. Comparative results of electric energy consumption by a selected group of drivers are presented. In the final part, the authors referred to the economic issues of the modernization.

Keywords: Electric multiple unit · Modernization · Energy consumption

1 Introduction

Electric multiple unit (EMU) is a rail, self-propelled, multi-segment traction vehicle designed for passenger transport [1–10]. They are used both in urban agglomerations, in suburban traffic, as well as in long-distance traffic. They are supplied with electricity from the traction network of a given transport system. In many European countries, it is a 3 kV DC network.

Progress in transport systems forces, among others, replacement or modernization of rolling stock [3, 4, 7, 10]. This also applies to EMU. The decision to buy a new rolling stock or to modernize the old one should be made on the basis of the economic

M. Siergiejczyk and K. Krzykowska (Eds.): ISCT21 2019, AISC 1032, pp. 457–465, 2020.
https://doi.org/10.1007/978-3-030-27687-4_46

and technical balance. The articles below will present the results of modernization for the EMU - EN57 unit group. This modernization involved, inter alia, the reconstruction of almost the entire drive unit, the reconstructed driver's cab, the change of the external appearance and the interior remodelling.

2 EMU Modernization

Modernized EMUs must comply with applicable legal and passengers requirements. The latter are primarily: the shortest travel time, a sense of security, driving comfort (silence and lack of vibrations), comfort in the passenger area, access to information (media, Internet, free hotlines), vehicle sound system (messages about stops and connections), air conditioning system with air heating supply, system of platforms, toilet for disabled people and wheelchair users, external and internal monitoring, 230 V sockets for each pair of seats.

As a result of the analyses carried out, the EMU group - type EN 57 was modernized. The modernization concerned mainly the following elements:

- drive system with three-phase AC asynchronous motors - EMU - EN 57 drive system after modernization consists of two sets which include: inverter and two asynchronous LKm 450X6 motors; inverters were made in HV IGBT 6,5 kV technology; the inverter meets all UIC and EN standards in the field of safety and electromagnetic compatibility; the drive system is characterized by a very low level of low frequency noise passed to the traction network;
- drive system control - control and monitoring system TCMS (Train Control and Monitoring System), is in compliance with the EN 50155 standard,
- motor protection - GEL speed sensor, shielded PT resistance thermometers (installed in each motor phase and on the bearing discs).
- external and internal monitoring - cameras in: driver's cab, at the head of the vehicle for observation of the pre-field, pantographs and passenger compartments;
- voltage converter - static converter (U_{in} = 3 kVDC, U_{out} = 3 × 0,4 kVAC 50 Hz, U_{out1} = 230 VAC 50 Hz, U_{out2} = 110 VDC, U_{out3} = 24 VDC, S_{out} = 14 kVA, THD < 5%);
- brakes - electrodynamic brakes with the possibility of energy recuperation to the traction network;
- information systems - information boards made in LED technology, have the ability to display from one to several lines of text,
- door opening system - a system of individual door opening with automatic closing (central opening from the driver's desk or individually by the passenger).

The view of the stator of the asynchronous motor and of the used asynchronous motor LKm-450X6 with the wheel set after EMU modernization is shown in Fig. 1.

Fig. 1. View of the stator of the asynchronous motor after drying and of the used asynchronous motor LKm-450X6 with the wheel set (Source: Own research).

EMU parameters obtained thanks to the modernization are presented in Table 1.

Table 1. List of parameters before and after modernization EN57.

	Before modernization	After modernization
Wheel arrangement	2′2′ + Bo′Bo′ + 2′2′	2′2′ + Bo′Bo′ + 2′2′
Supply voltage	3000 V DC	3000 V DC
Maximum speed	110 km/h	110 km/h
Mass	125 t	125 t
Rating	608 kW	1000 kW
One hour rating	740 kW	1200 kW
Maximum tractive force	98 kN	127 kN
Acceleration 0–40 km/h	0,5 m/s^2	1,0 m/s^2
Acceleration at Vmax	0 m/s^2	0,3 m/s^2
Acceleration time up to 100 km/h	120 s	42 s
Boot type	Resistor	IGBT VVVF
Boot control	Mechanical	Microprocessor
Traction motor type	LKf-450	LKm-450X6
Number of traction motors	4	4

(Source: [11])

3 Assessment of Electricity Consumption in the Modernized EMU

Prior to modernization EMUs hey were driven by the use of a series of direct current resisters. The use of such a drive is associated with high costs of electricity consumption. They result from the loss of electric energy converting into thermal energy in the starter resistors. In these drives there is a loss of kinetic energy generated during braking, which in some part could be recovered by recuperation. To reduce the electricity consumption caused by the start-up, the starting frequency is minimized and the engine switching is used. In the first step, a serial connection is used later in parallel. All these complications do not occur with the drive with AC motors.

Data regarding the electricity consumption of the selected EMU group before modernization is shown in Table 2. The authors have carried out these tests at EMU on railway lines connecting Łódź with: Warsaw, Tomaszów Mazowiecki and Częstochowa (also results from Tables 3 and 4). The measurements were made with EM 3000 meters installed in EN 57 vehicles with DC motor drive.

Table 2. Energy consumption in the EMU group before modernization.

	Numer EMU	Distance driven [km]	Energy consumed [MWh]	Consumption per 1 km [MWh/km]	Number of kilometers driven per 1 MWh
1.	EMU-1	065717	00428,813	0,00653	153,3
2.	EMU-2	138539	00816,638	0,00589	169,6
3.	EMU-3	133480	00846,505	0,00634	157,7
4.	EMU-4	035968	00345,736	0,00961	104,0

(Source: Own research)

As a result of the modernization, a drive was used with asynchronous motors and power electronics converters. It enables starting without losses and braking with energy recovery. The use of an asynchronous motor allows for greater starting and braking efficiency, and thus a lower energy consumption. Data regarding the electricity consumption of the selected EMU group after modernization is shown in Table 3 (listed in Table 3 EMU will be the same as in Table 4).

Table 3. Energy consumption in the EMU group after retrofitting

	Numer EMU	Distance driven [km]	Energy consumed [MWh]	Energy given [MWh]	Consumption per 1 km [MWh/km]	Number of kilometers driven per 1 MWh
1.	EMU-11	14932	108,636	16,304	0,00728	137,4
2.	EMU-12	7898	52,660	1,408	0,00667	150,0
3.	EMU-13	66567	473,479	73,524	0,00711	140,6
4.	EMU-14	13562	86,820	18,911	0,00640	156,2
5.	EMU-15	17404	109,612	17,861	0,00630	158,8
6.	EMU-16	8816	61,787	9,455	0,00701	142,7
7.	EMU-17	6386	46,256	5,714	0,00724	138,1

(Source: Own elaboration)

On the basis of the data from Tables 2 and 3, the traction energy consumption of the electrically conductive traction units it can be found that the upgraded EN57 consume less energy than the EMU with DC motor. However, comparing the energy consumption per 1 km and the number of kilometres driven per 1 MWh, the results are practically the same. However, if you consider engine power, acceleration, maximum traction (Table 1), then the savings are significant. There is a better driving comfort in upgraded EN57, resulting from supply heating, air conditioning, the possibility of using 230 VAC voltage.

Minimizing energy intensity in rail passenger transport forces the necessity to apply the latest technological solutions, concerning, among the others, recovery and recuperation of electricity [12, 13]. Recuperation is a process that involves recovering energy in a rolling stock with an electric drive and using it to power the drive motor of another traction unit. Most often, energy is recovered in vehicles powered by asynchronous traction engines, using the kinetic energy of the speeding vehicle. It is converted into electrical energy during electrodynamic braking. In this way, the recovered energy can be delivered directly to the power supply traction network, stored in a storage tank or cut off on resistors.

One of the factors influencing the energy recovery is the number of braking, resulting from the characteristics of the route and the number of stops. According to the analysis of the energy consumption of the EMU used, vehicles operating in different directions have different effectiveness of recuperation. The second factor limiting the possibility of energy recovery from braking is the unit power of the rolling stock. In the case of trains pulled by a locomotive, the power of traction motors and the strength of the locomotive's adhesion in relation to the mass of wagons is too small to obtain adequate effects in braking.

In EN57, there are four driving and braking axes as well as eight rolling axles. The weight of the wagon is additionally increased by mounted engines. The high mass allowed to achieve high efficiency of electrodynamic braking. Table 4 shows the percentage of energy consumed to be recovered. It illustrates the desirability of using recuperation.

Table 4. The ratio of the energy taken to the donated.

	Numer EMU	Energy collected [MWh]	Energy given [MWh]	Balance [%]
1.	EMU-11	652,648	94,963	14,55
2.	EMU-112	479,179	66,579	13,89
3.	EMU-113	1061,386	145,281	13,69
4.	EMU-114	426,141	83,887	19,68
5.	EMU-115	2003,851	312,368	15,59
6.	EMU-116	1351,200	152,842	11,31
7.	EMU-117	1567,558	240,137	15,32

(Source: Own elaboration)

The efficiency of recuperation depends to a large extent on the way the driver conducts the EMU. This is best evidenced by the measurement data obtained during test and technical rides. They were conducted as part of the unit repair cycle. Technical test runs took place after the completion of a specific repair cycle and were intended to check the operation of electrical, pneumatic circuits and other mechanisms, in general confirmation of the vehicle's efficiency. The journeys took place on the same route with no passenger load, at similar speed. Figure 2 presents (a graph of the number of miles driven by the EMU for 1 MWh of collected electricity, for test and operational rides (data come from the authors' of the article own research). There were units of the same series, i.e. EN 57 taken for comparison.

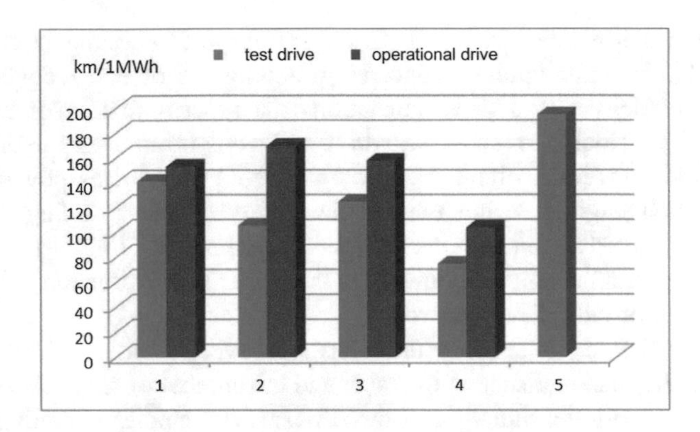

Fig. 2. The number of kilometres driven by EMU for 1 MWh of electricity consumed for test and operational rides (Source: Own elaboration).

Figure 3 shows the unit consumption of electric energy for test and operational drives of EN 57 units.

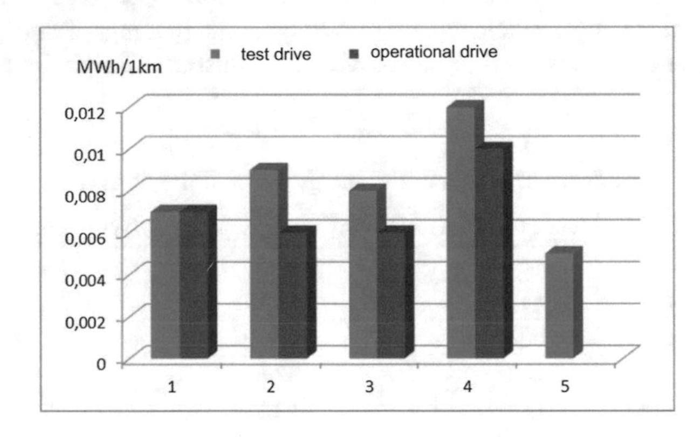

Fig. 3. Electrical power consumption for test and operational drives of EN 57 units (Source: Own elaboration).

The data presented above determine the value of electricity consumption during test and operational rides. It can be concluded from them that the energy consumption for a 1 km drive is greater for test drives than for operating conditions. This happened despite the fact that the test drives were not laden with passengers. The increased energy consumption was due to: not the same conditions as the rehearsal tests, the high rigiveness of the running parts of the chassis bogies, the lapping of the brake pads, the friction of the wheel set flanges. In addition to the increased energy consumption, the heating effect was attached (regardless of the temperature outside). This was due to the need to confirm the efficiency of all vehicle heaters. The heating capacity of EMU is within 76,0 kW. The test drive is always carried out at a maximum speed of 110 km/h. The variation in electricity consumption for 1 km and 1 MWh during the exploitation traffic was due to different EMU speeds, the technical condition of the tracks, the track line profile (there is a higher friction when driving the track arch). Also the form of terrain tracks (hills, mountains) has a very high impact on energy consumption.

It is important to mark a very large role of the personnel operating the vehicle (that is the driver) in the energy consumption. The driver conducting a vehicle more dynamically consumes more traction energy. The measurements show large differences in the handling of the same routes by different drivers. The differences in long-distance driving are up to 18%, while in the regional trains - up to 30%. This shows how large the opportunities of savings can be achieved thanks to the proper operation of the driver. However, there must be an incentive system to reward such behaviour, e.g. bonuses for saving the electric energy of the vehicle.

The basic question when deciding whether to modernise the EMU relates to its costs. The cost of upgrading is compared with the cost of purchasing of new units. From an economic point of view, the size of the contract that the co-operants are implementing based on the list of modernised elements should be decisive. An example of the number of units based upgrade cost is shown in the following statistics - Enterprise X (according to press reports):

- 21 EMU - 1,72 million Euro per vehicle,
- 27 EMU - 1,72 million Euro per vehicle,
- 4 EMU - 1,93 million Euro per vehicle,
- 1 EMU - 1,94 million Euro per vehicle.

The alternative could be new vehicles bought from manufacturers. The average cost of such a unit is about 4 million euros per vehicle. Compared to modernized units, the cost is more than doubled. The purchase of new vehicles brings also additional costs which are: maintenance and servicing, upgrading the qualifications of technical staff. Modernized vehicles inherit old problems, although modernization contractors ensure that they are practically new units. However, failures of upgraded units do not confirm this claim.

4 Conclusions

The scope of EN57 modernization and the applied constructional solutions bring measurable financial benefits both in operation and maintenance. Replacing the resistor start with a frequency controlled start leads to a reduction in energy consumption both at start-up and during driving. Less wear while driving, especially at low speeds, translates into fewer inspections and repairs caused by the lack of mechanical components (cog wheel, adjusters, contactors). The lack of resistor start-up caused the operation time of the gear transmission to be extended.

The use of asynchronous motors has resulted in less energy consumption with increased efficiency. It reduced the labor consumption of repairs, inspections of the traction engine. It enabled the use of energy recuperation during electrodynamic braking. Extensive diagnostics and protection systems have allowed to minimize the effects of a failure.

The limitation of energy consumption in rail passenger transport forces the constructors to use the best technologies in the field of using devices that enable the storage of electricity. The number of tests and tests of rail vehicles equipped with energy storage devices during train braking is increasing.

The industry working for railway rolling stock has a high chance of development and the prospects of the reconstruction of the rolling stock park in Poland and Eastern European countries. Abandonment of purchases (lasting 15 years) caused a large gap in the market, attractive to producers, but difficult for railway undertakings to manage. Taking into account the economic situation of entities operating in the passenger transport sector, it is necessary to confirm the sense of modernization of railway traction units.

Using the experience of Western countries, it can be assumed that in order to be competitive in the sphere of passenger transport, electric multiple units need to be modernized every 12–15 years, i.e. twice in the period of use of the vehicle.

References

1. Wang, J., Rakha, H.: Electric train energy consumption modeling. Appl. Energy **193**, 346–355 (2017)
2. Jong, J.-C.: Models for estimating energy consumption of electric trains. J. East. Asia Soc. Transp. Stud. **6**, 278–291 (2005)
3. Liudvinavičiusa, L., Jastremskas, V.: Modernization of diesel-electric locomotive 2M62 and TEP-70 locomotives with respect to electrical subsystem. Procedia Engineering **187**, 272–280 (2017). In: 10th International Scientific Conference Transbaltica; Transportation Science and Technology. Elsevier
4. Wanga, R., Kudrot-E-Khudab, M., Nakamuraa, F., Tanakaa, S.: A Cost-benefit analysis of commuter train improvement in the Dhaka Metropolitan Area, Bangladesh. Proc. - Soc. Behav. Sci. **138**, 819–829 (2014)
5. Wang, J., Lin, B., Jin, J.: Optimizing the shunting schedule of electric multiple units depot using an enhanced particle swarm optimization algorithm. Comput. Intell. Neurosci. **2016**, 11 (2016)
6. Chen, R., Zhou, L., Yue, Y.-X., Tang, J., Lu, C.: The integrated optimization of robust train timetabling and electric multiple unit circulation and maintenance scheduling problem. Adv. Mech. Eng. **10**, 1–16 (2018)
7. Armstrong, A.: Electrical multiple unit testing: commissioning to overhaul. In: IET 13th Professional Development Course on Electric Traction Systems, London, pp. 1–11 (2014)
8. Li, W., Nie, L., Zhang, T.: Electric multiple unit circulation plan optimization based on the branch-and-price algorithm under different maintenance management schemes. PLoS ONE **13**(7), e0199910 (2018)
9. Wu, J., Lin, B., Wang, H., Zhang, X., Wang, Z., Wang, J.: Optimizing the high-level maintenance planning problem of the electric multiple unit train using a modified particle swarm optimization algorithm. Symmetry **10**, 349 (2018)
10. Cadarso, L., Marín, Á.: Robust rolling stock in rapid transit networks. Comput. Oper. Res. **38**(8), 1131–1142 (2011)
11. Biliński, J., Buta, S., Gmurczyk, E., Kaska, J.: Modern asynchronous drive with regenerative braking produced by MEDCOM for modernized electric multiple units of the EN57 AKL series. Rail Transp. Tech. **4**, 20–25 (2012). [in Polish]

12. Kawałkowski, K., Młyńczak, J., Olczykowski, Z., Wojciechowski, J.: A case analysis of electrical energy recovery in public transport. In: Sierpiński, G. (ed.) Advanced Solutions of Transport Systems for Growing Mobility, TSTP 2017. Advances in Intelligent Systems and Computing, vol. 631, pp. 133–143. Springer, Cham (2017)
13. Wojciechowski, J., Lorek, K., Nowakowski, W.: An influence of a complex modernization of the DC traction power supply on the parameters of an electric power system. In: MATEC Web of Conferences, 13th International Conference "Modern Electrified Transport", MET 2017, vol. 180, pp. 1–6 (2018)

Application of Graph Theory for Description of the Infrastructure of the Signalling System

Paweł Wontorski(✉) and Andrzej Kochan

Faculty of Transport, Warsaw University of Technology, Warsaw, Poland
pawel.wontorski.dokt@pw.edu.pl, ako@wt.pw.edu.pl

Abstract. The article presents the concept of using graph theory to describe the infrastructure of the signalling system, in accordance with the assumptions of the *MAP-WP* design automation method. The principles of formalization of the system structure based on *TU* and *TS* graphs describing the structure of the track system division into areas controlled by internal signalling device sets and the structure of external connections between internal signalling device sets are presented. The *MAP-WP* method generates many variants of the system structure depending on the number of controlled areas, network topology, location of the control room relative to the station track system, and then selects a variant that meets the requirements for the availability and reliability of the structure. Confirmation that the structure meets specific requirements is possible thanks to the analysis of edge and tip coherence of the *TU* and *TS* graphs representing it.

Keywords: Graphs · Signalling system · Accessibility · Reliability · Infrastructure

Abbreviations

A(TU)	availability of the signalling system structure
BIM	Building Information Modelling
ICT	Information and Communication Technologies
MAP-WP	the method of automating the design of the infrastructure of the computer signalling system [8]
MS	model of the signalling system structure
ob	controlled area, defined part of the track system
R(TS)	reliability of the signalling system structure
TS	graph of the structure of the external connections and signalling devices
TU	graph of the structure of the track system and external signalling devices
$\mathbf{Z}_{UW}$	a set of internal devices controlling one set $\mathbf{Z}_{UZ}$ of external devices
$\mathbf{Z}_{UZ}$	set of external devices in *ob* area

1 Introduction

The automation of design requires a clear description of the design subject. In the case of signalling systems, it is necessary to formalize the description of their infrastructure. The structure of the signalling system MS is a way of organizing elements of the system, their attributes and relations between them.

M. Siergiejczyk and K. Krzykowska (Eds.): ISCT21 2019, AISC 1032, pp. 466–475, 2020.
https://doi.org/10.1007/978-3-030-27687-4_47

To describe the MS structure we can successfully apply the graph theory. It should be emphasized, however, that it does not have to be only a schematic diagram of elements, presented, for example, in [11] and [12].

The graphs allow you to map the structure of the system on a different scale and level of detail, depending on the needs and purpose of modeling. They are used, for example, in the *RailTopoModel* railway infrastructure description standard [1] and in other methods of description of structure of the signalling system [6, 10, 13].

This article proposes the use of graphs for the description of the signalling system in accordance with the assumptions of the method *MAP-WP* of automating the design of the infrastructure of the computer signalling system [8], which refers to BIM.

The method enables automatic generation of schematic plans for the location of srk devices on the station and the route along with the choice of the system structure variant. The chosen variant meets the designer's requirements by the regarding availability, reliability and costs. The method includes the design of interoperable infrastructure, including deployment of ETCS eurobalises and lineside electronic units (LEU).

The reliability *R(TS)* of the structure of the system is a measure of the system's resistance to damage that could cause part of the station to be out of service.

The availability *A(TU)* of the system structure can be defined as the ability of the traffic system to safely implement a traffic process at the station, also in conditions of limited efficiency and functionality (e.g. shutting down some parts of the station). This is a measure of the resistance of the physical process to the exclusion of some parts of the station from exploitation, caused by damage to the structure of the system.

2 Formal Description of the Signalling System Structure

The MS model of the signalling system structure can be presented as:

$$\mathrm{MS} = \{TU, TS\} \tag{1}$$

where:

TU – graph of the structure of the track system and external signalling devices,
TS – graph of the structure of the external connections and signalling devices.

2.1 Structure of the Track System and External Signalling Devices

The set $\boldsymbol{Z_{UZ}}$ of external devices is defined as a set of external devices, which is controlled by internal devices that form exactly one set $\boldsymbol{Z_{UW}}$ of internal devices.

Controlled area *ob* is a contractually defined part of the track system with external signalling devices forming exactly one set $\boldsymbol{Z_{UZ}}$ of external devices.

The structure of the track system and external signalling devices can be saved as a *TU* graph (see Fig. 1), whose vertexes are control areas *ob*, and the edges of the track system elements *ut* connecting these areas.

$$TU = \langle OB, UT \rangle \quad (2)$$

$$\boldsymbol{OB} = \{ob_1, ob_2, \ldots, ob_{NOB}\} \quad (3)$$

$$\boldsymbol{UT} = \{ut_1, ut_2, \ldots, ut_{NUT}\} \quad (4)$$

$$\boldsymbol{UT} \subseteq \{\{ob_i, ob_j\} : ob_i, ob_j \in \boldsymbol{OB}, ob_i \neq ob_j\} \quad (5)$$

where:

OB – set of vertexes of the graph having an interpretation of controlled areas,
UT – set of graph edges having the interpretation of elements of the track system
NOB – cardinality of the ***OB*** set,
NUT – cardinality of the ***UT*** set.

The ***OB*** set is divided into subsets of elements by element type.

$$\boldsymbol{OB} = \{\boldsymbol{OB_P}, \boldsymbol{OB_W}, \boldsymbol{OB_S}\} \quad (6)$$

where:

$\boldsymbol{OB_P}$ – set of areas controlled directly by the relay room (one area at the station)
$\boldsymbol{OB_W}$ – set of internal controlled areas (within a station)
$\boldsymbol{OB_S}$ – set of controlled areas adjacent to the station

The *TU* graph is consistent, unchecked and is not a multigraph. Each vertex of the *obi* graph denoting the controlled area from the ***OB*** set can be described by:

$A(ob_i)$ – availability of the *i*-th controlled area,

The subject of the publication is not the availability of a single area, but the availability of the structure of the signalling system as a whole, in particular the track system together with external signalling devices capable of implementing the physical process. The structure of the track system in the form of a *TU* graph can be described by:

A(TU) – availability of the signalling system structure.

It has been assumed that the internal controlled areas *obw* will be taken into account for analyzes of the availability of the system structure. They will be the subject of analyzes of the availability of the system structure. No availability of *obp* means no access to the entire structure. The areas controlled by adjacent *obs* are shown as illustrative.

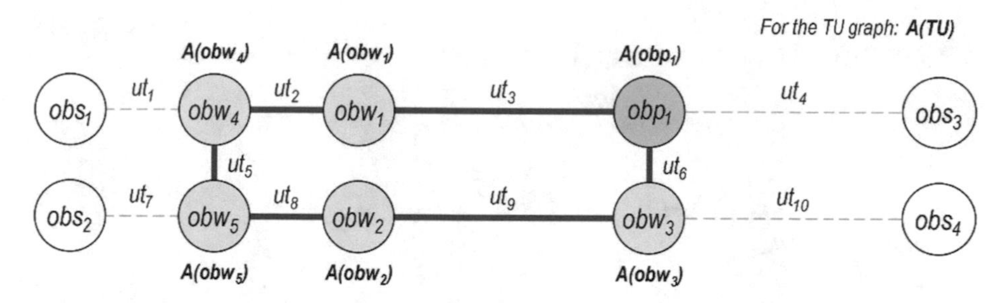

Fig. 1. Example of a track system structure and external signalling devices in the form of a *TU* graph (own elaboration based on: [8])

2.2 Structure of the External Connections Network and Signalling Devices

The structure of connections and signalling devices can be saved as a *TS* graph (see Fig. 2) whose vertices are distribution points (node) *pr*, and the edges are external connections *pz*.

$$TS = \langle PR, PZ \rangle \tag{7}$$

$$\boldsymbol{PR} = \{pr_1, pr_2, \ldots, pr_{NPR}\} \tag{8}$$

$$\boldsymbol{PZ} = \{pz_1, pz_2, \ldots pz_{NPZ}\} \tag{9}$$

$$\boldsymbol{PZ} \subseteq \{\{pr_i, pr_j\} : pr_i, pr_j \in \boldsymbol{PR}, pr_i \neq pr_j\} \tag{10}$$

where:

PR – set of vertexes of the graph having the interpretation of distribution points,
PZ – set of graph edges having an interpretation of external connections,
NPR – cardinality of the ***PR*** set,
NPZ – cardinality of the ***PZ*** set.

The *TS* graph is consistent, unchecked and is not a multigraph. Each edge of the graph $pz_k = \{pr_i, pr_j\}$ denoting the external connection from the ***PZ*** set between two distribution points from the ***PR*** set can be described by:

$R(pz_k)$ – reliability of the *k*-th connection between the *i*-th and *j*-th point of the distribution.

The subject of the publication is not the reliability of a single connection, but the reliability of the structure of the signalling system as a whole, in particular the structure of the external connection network between the distribution points. The structure of the system in the form of a *TS* graph can be described by:

$R(TS)$ - reliability of the signalling system structure.

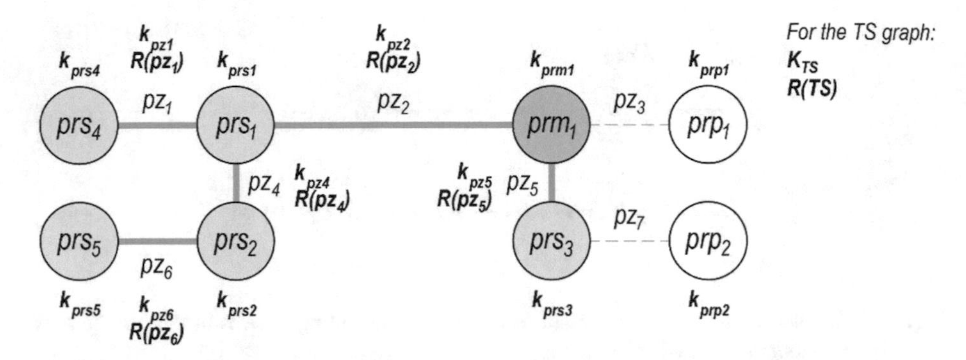

Fig. 2. Example of an external connection structure model and signalling devices in the form of a *TS* graph (own elaboration based on: [8])

It was assumed that for further analysis, external connections *pzf* and *pzz* between the distribution points constituting sets of internal devices, i.e. *prm* and *prs*, will be taken into account. Connections to *prp* distribution points are shown as illustrative.

All *pz* connections will be the subject of analysis of the reliability of the signalling system structure.

3 Variants of the MS Structure

The MS structure can be designed in many variants defined by the designer or generated by the design automation system.

The *vms* variant of structure of the signalling system is every way of organizing this system distinguished on the basis of a combination of three variables, also occurring in several variants:

- the type of cable network structure,
- number of controlled areas,
- the location of the control room in relation to the track's circuit layout

The $\boldsymbol{V_{MS}}$ set of variants is the Cartesian product of sets: $\boldsymbol{TY_S}$, $\boldsymbol{L_{ob}}$ and $\boldsymbol{X_n}$

$$\boldsymbol{T_{YS}} \times \boldsymbol{L_{ob}} \times \boldsymbol{X_n} = \{(tys, lob, xn) \colon tys \in \boldsymbol{T_{YS}}, lob \in \boldsymbol{L_{ob}}, xn \in \boldsymbol{X_n}\} \tag{11}$$

$$\boldsymbol{TY_S} = \{tys_1, tys_2, \ldots, tys_{NTYS}\} \tag{12}$$

$$\boldsymbol{L_{ob}} = \{lob_1, lob_2, \ldots, lob_{NLOB}\} \tag{13}$$

$$\boldsymbol{X_n} = \{xn_1, xn_2, \ldots, xn_{NXN}\} \tag{14}$$

On the Cartesian product $\boldsymbol{TY_S} \times \boldsymbol{L_{ob}} \times \boldsymbol{X_n}$, the mapping σ_{vms} are given, which to the elements of this set of this product assigns elements from the set {0, 1}, i.e.:

$$\sigma_{vms}\colon \boldsymbol{TY_s} \times \boldsymbol{L_{ob}} \times \boldsymbol{X_n} \rightarrow \{0,1\} \tag{15}$$

$$\sigma_{vms}\colon (tys_j, lob_k, xn_m) = 1, \quad \text{if variant exists} \tag{16}$$

$$\sigma_{vms}\colon (tys_j, lob_k, xn_m) = 0, \quad \text{if variant does not exist} \tag{17}$$

Therefore, the set $\boldsymbol{V_{MS}}$ of variants is defined as follows:

$$\boldsymbol{V_{MS}} = \{(tys, lob, xn)\colon \sigma_{vms}(tys, lob, xn) = 1, (tys, lob, xn) \in \boldsymbol{TY_s} \times \boldsymbol{L_{ob}} \times \boldsymbol{X_n}\} \tag{18}$$

$$\boldsymbol{V_{MS}} = \{vms_1, vms_2, \ldots, vms_{NVMS}\} \tag{19}$$

where:

$\boldsymbol{V_{MS}}$ – set of variants of the structure of the signalling system
$\boldsymbol{TY_S}$ – set of variants (types) of the *TS* network structure (network topology)
$\boldsymbol{L_{ob}}$ – set of variants of division into *ob* controlled areas
$\boldsymbol{X_n}$ – set of variants of control room locations
NVMS – cardinality of the $\boldsymbol{V_{MS}}$ set
NTYS – cardinality of the $\boldsymbol{TY_S}$ set
NLOB – cardinality of the $\boldsymbol{L_{ob}}$ set
NXN – cardinality of the $\boldsymbol{X_n}$ set
vms_i – *i*-th variant distinguished in the $\boldsymbol{V_{MS}}$ set
tys_m – *m*-th variant distinguished in the $\boldsymbol{TY_s}$ set
lob_k – *k*-th variant distinguished in the $\boldsymbol{L_{ob}}$ set
xn_j – *j*-th variant distinguished in the $\boldsymbol{X_n}$ set

In the $\boldsymbol{TY_S}$ set, basic variants (types) of *TS* network structures (network topology) have been distinguished:

$$\boldsymbol{TY_S} = \{\text{MAG}, \text{2GWI}, \text{RING}\} \tag{20}$$

where:

MAG – variant with bus,
2GWI – variant with two stars connected with a bus,
RING – variant with ring

The $\boldsymbol{L_{ob}}$ set has several variants of division into *ob* controlled areas:

$$\boldsymbol{L_{ob}} = \{1, 2, 3, 4, 5, 6, 7, 8\} \tag{21}$$

The cardinality of $\boldsymbol{X_n}$ set has been limited in accordance with the number of variants of the control room locations relative to the track system:

$$\boldsymbol{X_n} = \{\text{NW}, \text{NC}, \text{NS}, \text{SW}, \text{SC}, \text{SE}\} \tag{22}$$

where:

NW – location of the control room to the north-west of the track system,
NC – location of the control room to the north of the track system
NE – location of the control room to the north-east of the track system
SW – location of the control room to the south-west of the track system
SC – location of the control room to the south of the track system
SE – location of the control room to the south-east of the track system

The $\boldsymbol{V_{MS}}$ set of variants can be presented in the form of ***MVMS*** matrix. Figure 3 shows a part of an exemplary ***MVMS*** matrix.

$\boldsymbol{MVMS} = [mvms_{tys,lob,xn}]$ is three-dimensional matrix with dimensions.

$$tys\,x\,lob\,x\,xn \tag{23}$$

$$tys \in \{1, 2, \ldots, NTYS\}, \quad lob \in \{1, 2, \ldots, NLOB\}, \quad xn \in \{xn1, 2, \ldots, NXN\} \tag{24}$$

The matrix contains information about the existence of *vms* variants of the MS system structure. Each element of this matrix:

$$mvms_{tys,lob,xn} = 1, \quad \text{if variant exists} \tag{25}$$

$$mvms_{tys,lob,xn} = 0, \quad \text{if variant does not exist} \tag{26}$$

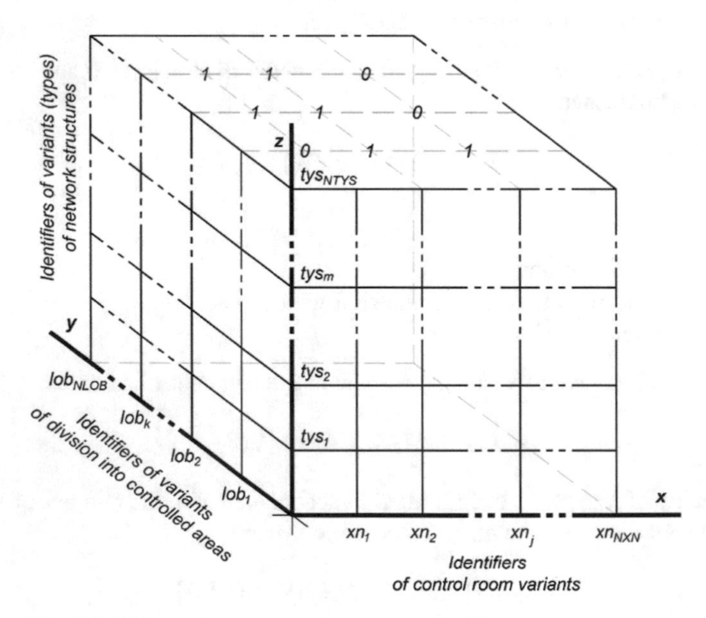

Fig. 3. The ***MVMS*** matrix of MS structure variants (own elaboration based on: [8])

4 Availability of the Signalling System Structure

In relation to the structure of signaling system, the availability of the *A(TU)* system structure is defined as the ability of the signalling system structure to safe implementation the traffic process of the signalling system at the station, also in conditions of reduced efficiency and functionality (limited availability) [9]. These limitations result from the temporary decommissioning of certain elements of the system structure and, as a result, specific controlled areas *ob*, without which it is possible to continue traffic.

$A_{obwi,obwj}(TU)$ means the minimum number of internal control areas *obw* that must be removed from the *TU* graph in order to prevent a train from passing at the station (or route) between the other two different vertices *obwi* and *obwj* in *TU* graph.

The availability of the *A(TU)* system structure is the minimum number of vertices *obw* (internal control areas) that must be removed from the *TU* graph to make it impossible for a train to pass through a station (or track) for any pair *obwi* and *obwj* in the *TU* graph. Therefore:

$$A(TU) = min_{obw}(A_{obwi,obwj}(TU)) \tag{27}$$

Assuming that the graph *TU* is *k*-consistent it should be assumed that the availability of the graph *A(TU)* is equal to the consistency of the vertex *k* of the graph *TU*.

The concept of the *articulation point* may also be used to explain the concept of availability of the graph. The point of articulation is the *v* vertex, for which there exists a pair of *x*, *y* vertexes such that each path joining *x* and *y*, such that $x \neq v$ and $y \neq v$ goes through the *v* vertex [7]. Removing an articulation point from the graph will increase the number of coherent components (it will make the graph inconsistent).

Figure 4 shows an example of a structure from which one of the indoor units has been removed, resulting in the complete exclusion of one control area from motion *obw*. The graph presented in the example is 2-consistent (vertex), so the measure of its availability is 2. This means that only after any two sets of internal devices have been removed will the structure no longer be available.

It can be argued that the division of internal equipment into assemblies and the modular construction of signalling systems increases their availability [5].

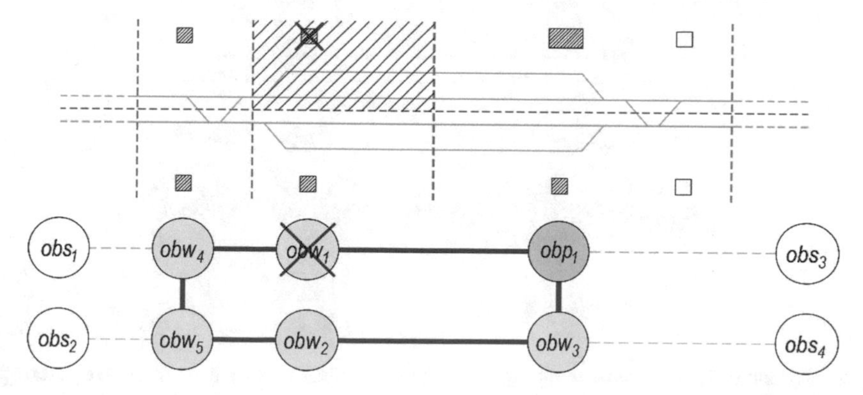

Fig. 4. Example of mapping of *TU* graph and *A(TU)* availability: structure with one interrupted connection (own elaboration based on: [8])

5 Reliability of the Signalling System Structure

The analysis of the reliability of the *TS* graph structure was carried out in relation to the shaping of the structure of devices and signalling connections. Vertex coherence of the graph can be expressed as a minimum number of failures at vertexes that cause the failure of the entire network of connections [3]. Edge consistency of the graph can be expressed as the minimum number of failures of connections between vertexes, which will cause a failure of the whole network of connections [4].

$R_{pri,prj}(TS)$ indicates the minimum number of edges to be removed from the graph to break the connection on all paths between *pri* vertex and *prj* in the *TS* graph. The deterministic reliability of the *R(TS)* graph is the minimum number of edges that must be removed from the *TS* graph to break all possible connections between at least one pair of different vertices of the graph. Therefore:

$$R(TS) = min_{pri,prj}(R_{pri,prj}(TS)) \tag{28}$$

Assuming that *TS* graph is *k*-consistent, it should be assumed that the deterministic reliability of *R(TS)* graph is equal to the edge coherence of *k* of *TS* graph.

The concept of a bridge can also be used to explain the reliability of a graph. A bridge is an edge *e* for which there is a pair of vertices *x* and *y* such that each path connecting *pri* and *prj* passes through the edge *e* [7]. Removing the bridge in the graph causes the graph to be inconsistent.

Figure 5 shows an example of a structure from which one of the connections has been removed, which resulted in the total exclusion of two distribution points, including the point constituting the assembly of internal devices *prs*. The graph presented in the example is 1-consistent (edge-to-edge), so the calculated measure of its reliability is 1.

Further information on the mathematical foundations of graph research algorithms can be found in [7] and sample algorithms in [2] and [5].

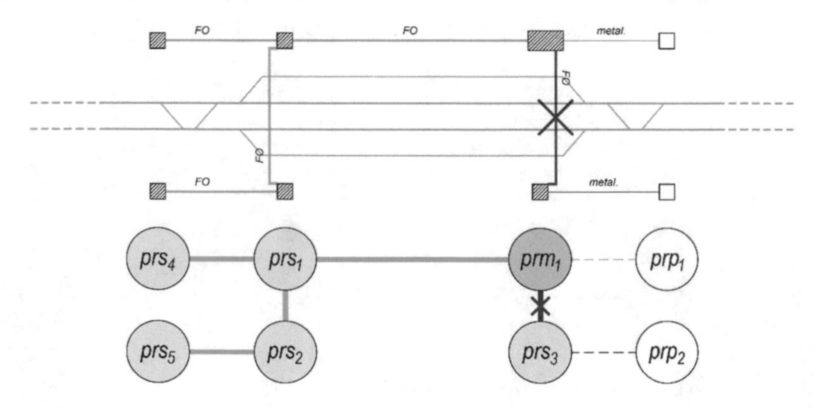

Fig. 5. Example of *TS* graph mapping and *R(TS)* reliability: structure with one interrupted connection (own elaboration based on: [8])

6 Conclusion

The paper presents the issue of description of infrastructure of signalling system with the use of graph theory. Formalization of the system structure realized in this way allows to determine the availability and reliability of the signalling system structure by conducting graphs cohesion analyses. Such an assessment is necessary in the case of automated design with implemented algorithms for generating optimal or suboptimal design solutions and supporting the designer's decisions.

The possibility of generating further variants of the system structure with different parameters and requirements enables the designer to carry out analyses and conceptual works in the pre-design phase in a short time. The formal description presented in the article can also be used for analyses of other structures with distinguished nodes and edges, e.g. ICT or power networks.

References

1. Augele, V.: Comparative Analysis of Building Information Modelling (BIM) and RailTopoModel/railML in View of their Application to Operationally Relevant Railway Infrastructure, Technische Universität Dresden, Dresden (2017)
2. Byrne, H.R., Feddema, T.J., Abdallah, T.C.: Algebraic Connectivity and Graph Robustness. Sandia Report, Sandia National Laboratories, Albuquerque and Livermore (2009)
3. Jungnickel, D.: Graphs, Networks and Algorithms. Springer, Heidelberg (2013)
4. Problem niezawodności sieci. http://www.mini.pw.edu.pl/MiNIwyklady/grafy/prob-sieci.html. Accessed 02 June 2018
5. Wałaszek, J.: Algorytmy, struktury danych. http://eduinf.waw.pl/inf/alg/001_search/0130a.php. Accessed 02 June 2018
6. Wang, D., Chen, X., Huang, H.: A graph theory-based approach to route location in railway interlocking. Comput. Ind. Eng. **66**(4), 791–799 (2013)
7. Wojciechowski, J., Pieńkosz, K.: Grafy i sieci, PWN, Warsaw (2013)
8. Wontorski, P.: Metoda automatyzacji projektowania infrastruktury komputerowego systemu sterowania ruchem kolejowym, Phd Thesis, Faculty of Transport, Warsaw University of Technology, Warsaw (2019)
9. Wontorski, P., Kochan, A.: Projektowanie zdecentralizowanych struktur komputerowych systemów sterowania ruchem kolejowym. In: Zeszyty Naukowo-Techniczne Stowarzyszenia Inżynierów i Techników Komunikacji w Krakowie, Cracow, vol. II, no. 2(113), pp. 173–185 (2017)
10. Xiangxian, C., Yulin, H., Hai, H.: A component-based topology model for railway interlocking systems. Math. Comput. Simul. **81**(9), 1892–1900 (2011)
11. Zabłocki, W.: Modelowanie stacyjnych systemów sterowania ruchem kolejowym. In: Prace Naukowe Transport, No. 65, Oficyna Wydawnicza Politechniki Warszawskiej, Warsaw (2008)
12. Zabłocki, W.: Modelowanie systemów sterowania ruchem kolejowym - struktury informacji i elementy opisu formalnego. In: Prace naukowe Transport, No. 57, Oficyna Wydawnicza Politechniki Warszawskiej, Warsaw (2006)
13. Zafar, N.A.: Formal model for moving block railway interlocking system based on undirected topology. In: 2006 International Conference on Emerging Technologies, pp. 217–223 (2006)

The Optimization of Railway Transition Curves with an Emphasis on Initial and End Zones

Krzysztof Zboinski and Piotr Woźnica(✉)

Faculty of Transport, Warsaw University of Technology, Warsaw, Poland
{kzb, pwoznica}@wt.pw.edu.pl

Abstract. The aim of this article is to clarify, why the curvatures' bends at the terminal points of optimum railway transition curves (TCs) obtained by the authors have a relatively small negative impact on vehicle dynamics while running along such curves. In this context the authors put forward the following research hypothesis. Maybe bigger length of the middle zone causes that shape of the terminal zones has become less important. The idea of the research was to modify quality function calculation so that initial and end zones have bigger weights (importance) than the middle zone. At present stage the central part of TC was completely ignored in the quality function calculation. Two dynamical quantities being the results of simulation of railway vehicle dynamical model were exploited in the determination of the quality functions.

Keywords: Railway transition curves · Optimization · Simulation

1 Introduction

In last years an increase in number of the publications that deal with railway transition curves (TCs) can indeed be observed [1, 2, 4, 6, 7, 9–20]. The same touches the publications, that deal with the railway vehicle dynamics, e.g. [5, 8].

In many works, e.g. [19], the authors of this article showed, that for the railway polynomial transition curves of odd degrees (5th, 7th, 9th and 11th) the best properties (the smallest values of quality function applied) had curves with the biggest possible number of their terms. For such curves of 5th degree the number of terms was 3, for curves of 7th degree – 5, for curves of 9th degree – 7, and for curves of 11th degree – 9. The curvatures of the obtained optimum TCs did not have tangence (G^1 continuity) in the terminal points of the curve, however.

The aim of current article is to clarify, why the curvatures' bends at the terminal points (G^0 continuity) of optimum railway transition curves obtained by the authors have a relatively small negative impact on vehicle dynamics while running along such TCs.

The idea used in present work was the modification in calculation of the quality function (*QF*), so that the initial and final curve zones were of greater weight (importance), than the central zone. The authors asked, if the greater length of the middle zone of the transition curve can cause that the shape of shorter terminal zones becomes less important. In the method adopted, the calculation of *QF* was performed for the first and last *p%* of the transition curve length and the whole circular arc length. This meant,

M. Siergiejczyk and K. Krzykowska (Eds.): ISCT21 2019, AISC 1032, pp. 476–483, 2020.
https://doi.org/10.1007/978-3-030-27687-4_48

that the QF was not calculated for (100–2p) % of the length in the middle part of transition curve. In this work the percentage of p was equal to 1%, 5%, 10%, 20% and 50%. In the further part of the work p will be presented as a fraction, e.g. $p = 0.05$. The value $p = 0.5$ means, that QF was calculated for the whole transition curve.

In all optimizations done for needs of the current work, the authors adopted polynomial of 5th and 7th degree as the TC. The radii of the circular arc were equal to 600, 1200 and 2000 m. The values of cant in the circular arc were ranged from 45 mm to 150 mm.

2 The World Literature Survey

The increase in the number of publications that relate to railway transition curves can be observed. Certain qualitative change in their content can also be noticed. The authors of the current work mean here the attempts to resign from the traditional approach and to find new methods of assessing of the properties of railway transition curves. Despite these, some limitations visible in the earlier works still exist, in the present authors' opinion.

The authors selected 4 group of works and these groups are represented by: Tari and Baykal, Ahmad and Ali, Long et al. and Kufver. The works which belong to first three groups show how much the shape of railway transition curves can influence on its properties. The last group reveals that there are some works in the world literature, where view on the problem of TCs evaluation and formation is close to that represented by the present authors. In this article the authors made the short description of 4 mentioned groups of works.

Tari and Baykal [14] specified continuity of function describing lateral change of acceleration (LCA), in the domain of the curve current length l, as the most important criterion, which should be satisfied while negotiating TC. They found only the transition curves which satisfied this criterion.

The example of the work that treats transition curves mainly as the mathematical object is the work of Ahmad and Ali [1]. Mathematical properties of the curves located between two straight lines or two circles are described there. The curves are defined with the Bezier curves that possess the continuity of type G^2 in each point. The example of the work belonging to the second group let be e.g. [2].

The next reference worthy of discussion is work of Long et al. [11]. These authors use an advanced model of a railway vehicle and the corresponding simulation software to evaluate dynamic properties of the TCs. They compare TCs of six different shapes. The another example of the work belonging to the third group let be e.g. [13].

Kufver optimised length of TC taking passenger comfort and track–vehicle dynamical interactions into account for one type of the curve. In [10] he exploited the European standard [3] dealt with the passenger ride comfort.

Generally in many works, the approach to track–vehicle interactions is traditional [6]. It is limited to treating all vehicles collectively and studying only the selected quantities in the vehicle body. The authors of this paper do not know the method applied in practice, which uses the complete dynamical model of vehicle to form railway TCs. Many of the methods in use refer to a very simple vehicle model.

3 Method of the Analysis Used for Needs of the Current Research

3.1 The Railway Vehicle and the Corresponding Model

Generalised approach to the modelling was used, as explained in [17]. Basically, dynamics of relative motion is used in that approach. This means, that description of motion (dynamics) is relative to track-based moving reference frames. The quality function is calculated as a result of the numerical simulation of motion of the dynamical mechanical system as described in [17].

3.2 The Optimization Method and the Objective Functions

The optimization problem, which is solved in the current studies is to find the A_i polynomial coefficients, that define TC's shape. Type of a TC chosen for optimisation is the polynomial TC of degree $n \geq 4$:

$$y = \frac{1}{R}\left(\frac{A_n l^n}{l_0^{n-2}} + \frac{A_{n-1} l^{n-1}}{l_0^{n-3}} + \frac{A_{n-2} l^{n-2}}{l_0^{n-4}} + \frac{A_{n-3} l^{n-3}}{l_0^{n-5}} + \ldots + \frac{A_4 l^4}{l_0^2} + \frac{A_3 l^3}{l_0^1}\right), \quad (1)$$

where y defines curve lateral co-ordinate. The R, l_O, and l define curve minimum radius (at its end), total curve length, and curve current length, respectively. In the polynomial coefficients A_i one has $i = n, n - 1, \ldots, 4, 3$, while n is the polynomial degree. In this work, $n = 5$ and 7.

For needs of the current paper the authors utilised two quality functions QF. These functions concerned a minimisation of integral of vehicle body lateral acceleration and its change with respect to the time:

$$QF_1 = L_C^{-1} \int_0^{L_C} |\ddot{y}_b| dl, \quad (2)$$

$$QF_2 = L_C^{-1} \int_0^{L_C} |\dddot{y}_b| dl, \quad (3)$$

where:

$\ddot{y}_b$ – lateral acceleration of vehicle body,
$\dddot{y}_b$ – change of lateral acceleration of the vehicle body,
L_C – whole TC and the adjacent CC of 100 m length.

The authors determined lengths of transition curves according to the algorithm given in [9].

3.3 General Look at the Software

Scheme of the software used in optimisation TCs shape is shown e.g. in [19]. The major objects within this scheme are two iteration loops. The first one is the integration loop. This loop is stopped when distance l_{lim}, being the length of route (usually compound route ST, TC and CC or CC, TC and ST), is reached by the model. The second one is the optimisation process loop. It is stopped when number of iterations reaches limit value i_{lim} or optimum solution is found.

4 The Results of the Research for 5th and 7th Degrees

The main aim of this article is the optimization of the shape of railway TCs of 5th and 7th degree using 3 radii of the circular arc – R = 600 m, 1200 m and 2000 m. As the initial TCs the authors took standard TCs of 5th and 7th degrees [11].

In Table 1 the authors presented the conditions of optimizations adopted in the current chapter: degree of polynomial n, velocity of vehicle v [m/s], curve radius R [m]

Table 1.

No.	n	v [m/s]	R [m]	H [mm]	l_0 [m]	QF	p
1	5	24.26	600	150	97.47	2	0.01
2	5	30.79	600	150	123.74	2	0.01
3	5	24.26	600	150	97.47	1	0.05
							0.1
4	5	30.79	600	150	123.74	1	0.05
							0.1
5	5	24.26	1200	75	48.73	2	0.05
							0.1
							0.2
							0.5
6	5	24.26	2000	45	29.24	2	0.05
							0.1
							0.2
							0.5
7	5	36.17	1200	75	72.67	1	0.05
							0.1
							0.2
							0.5
8	7	36.17	1200	75	90.84	1	0.05
							0.1
							0.2
							0.5
9	7	24.26	1200	75	60.92	1	0.05
							0.1
							0.2
							0.5

and cant H [mm], length l_0 [m] and quality function QF. For each of 9 sets of the conditions from 1 to 4 optimization processes – each one for the different value of p – were made.

For each of 9 sets of optimization, the authors obtained the shapes of optimum transition curves, which improved the values of quality functions in comparison with the initial transition curves. In all 25 out of 26 analyzed cases the program found transition curves having the curvatures very close to linear curvatures of 3rd degree parabola. Qualitatively similar results for degrees 5th and 7th the authors obtained in the past, e.g. in [15], but here the novelty is the fact that such transition curves were obtained for different values of p.

In Fig. 1 the authors showed the curvatures of the obtained optimum polynomial transition curves of 7th degree (point no. 7 in Table 1 and the superelevation ramp slopes corresponding to them compared with the courses for initial transition curve. The courses for initial transition curve are marked INI TC. Figure 1a reveals, that the program found in practice the optimum curvatures very similar for each p value.

All the curvatures do not have a tangence (G^1 continuity) in the terminal points, although theoretically, it was possible to find the transition curves with such a tangence.

For each of 9 sets of optimum transition curves the obtained results were very similar one to another. For the case no. 7 from Table 1 the shapes of the optimum transition curves were simply identical.

Such situation means here, that:

- there is one shape of transition curve and one course of acceleration of the vehicle body mass centre (arising from the passage of the vehicle through the route consisted of straight track, optimum transition curve and circular arc), which minimizes quality functions assumed,
- different values of p do not influence on the shape of the optimum transition curves significantly,

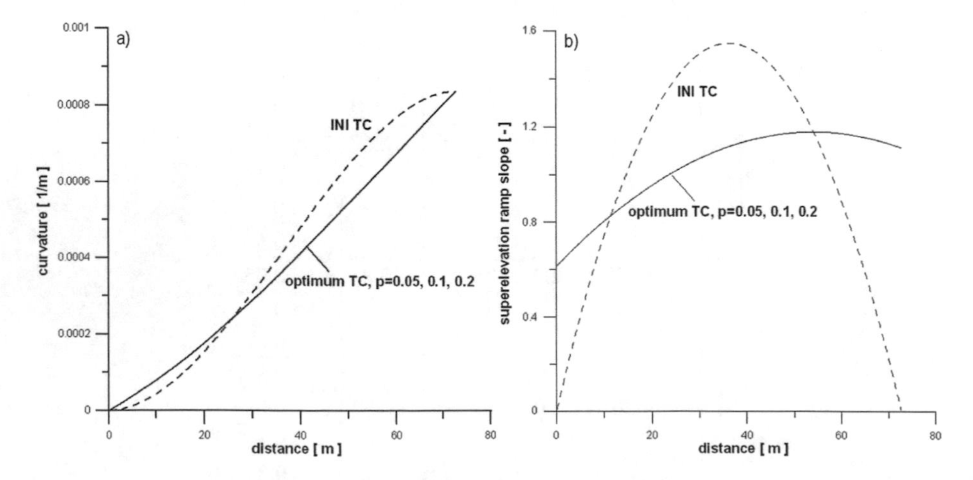

Fig. 1. (a) curvatures, (b) superelevation ramp slopes of the optimum TCs

– for the cases, where the lengths l_0 are shorter than 90 m and transition curves have curvature very close to the linear one, the vehicle dynamics is milder in the whole route (see Fig. 2). The lack of G^1 continuity in the starting point of the curve has positive influence on the vehicle dynamics at the beginning of the TC.

Figure 2 shows both the lateral displacements and lateral acceleration of the mass centre of the vehicle body. In majority of cases performed the general betterment of the dynamical properties of the system for the optimised transition curves shapes in comparison to the initial curve was confirmed by simulation results: the displacements y_b and accelerations $\ddot{y}_b$ of vehicle body.

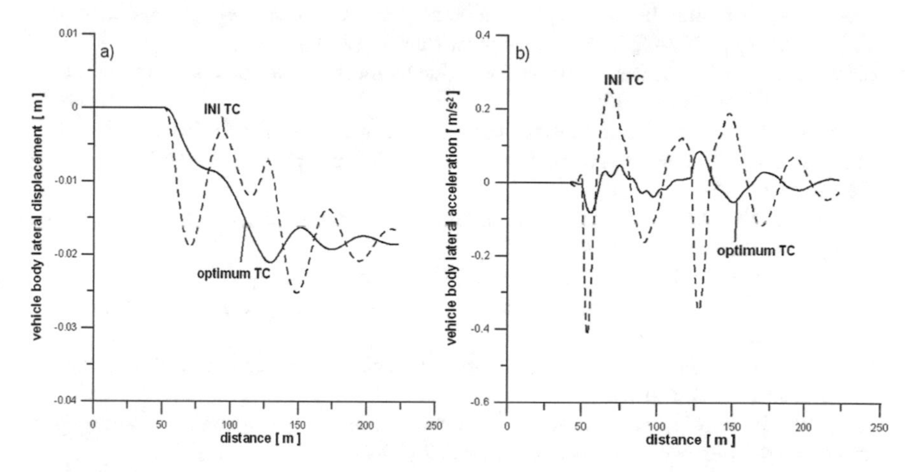

Fig. 2. Dynamical characteristics: (a) lateral displacements, (b) lateral accelerations of vehicle body mass centre

5 Conclusions

As a result of the optimizations performed several original and important conclusions can be drawn. First of all, the idea used in the work – the modification of calculation of the quality function, so that the initial and final curve zone were of greater weight (importance), than the central zone – has not changed both the assessment of TCs, and the results of optimizations. Quite similar results i.e. the optimum shapes of railway polynomial TCs of 5th, 7th degrees the authors obtained in [15] and [18].

Secondly, the methodology used in the work, i.e. calculation of quality function just for *p%* of TC length for the initial and the end part of the curve, did not let to obtain the optimum TCs with curvatures (superelevation ramps) having G^1 continuity in terminal points of a curve in all the cases.

Thirdly, for the optimum TCs of lower degrees, i.e. 5th and 7th ones, with curvatures with G^0 continuity, shorter lengths of TCs, and bigger curve radii, the vehicle dynamics is generally milder, than for the TCs with G^1 continuity in each part of the route. So, if we omit the central zone in *QF* calculation for these degrees, we may expect to obtain optimum TCs of curvatures with G^0 continuity in terminal points of a curve.

References

1. Ahmad, A., Ali, J.M.: G^3 transition curve between two straight lines. In: Proceedings of 5th International Conference on Computer Graphics, Imaging and Visualisation, pp. 154–159. IEEE Computer Society (2008)
2. Ahmad, A., Gobithasan, R., Ali, J.M.: G^2 transition curve using quadratic Bezier curve. In: Proceedings of 4th International Conference on Computer Graphics, Imaging and Visualisation, pp. 223–228. IEEE Computer Society (2007)
3. ENV 12299: 2009. Railway applications - Ride comfort for passengers – Measurement and evaluation. CEN Brussels (2009)
4. Drozdziel, J., Sowinski, B.: Railway car dynamic response to track transition curve and single standard turnout. In: Allan, J., et al. (eds.) Computers in Railways X, pp. 849–858. WIT Press (2006)
5. Dusza, M.: The study of track gauge influence on lateral stability of 4-axle rail vehicle model. Arch. Transp. **30**(2), 7–20 (2014)
6. Esveld, C.: Modern Railway Track. MRT-Productions, Duisburg (1989)
7. Fischer, S.: Comparison of railway track transition curves types, pollack periodica. Int. J. Eng. Inf. Sci. **4**(3), 99–110 (2009)
8. Kardas-Cinal, E.: Selected problems in railway vehicle dynamics related to running safety. Arch. Transp. **31**(3) (2014)
9. Koc, W., Radomski, R.: Analysis of transition curves with nonlinear superelevation ramp. Drogi Kolejowe (J. Polish) **11**, 261–267 (1985)
10. Kuvfer, B.: Optimisation of horizontal alignments for railway – procedure involving evaluation of dynamic vehicle response. Ph.D. thesis. Royal Institute of Technology, Stockholm (2000)
11. Long, X.Y., Wei, Q.C., Zheng, F.Y.: Dynamic analysis of railway transition curves. Proc. IMechE Part F: J Rail Rapid Transit **224**(1), 1–14 (2010)
12. Pombo, J., Ambrosio, J.: General spatial curve joint for rail guided vehicles: kinematics and dynamics. Multibody Syst. Dyn. **9**(3), 237–264 (2003)
13. Tanaka, Y.: On the transition curve considering effect of variation of the train speed. ZAMM – J. Appl. Math. Mech. **15**(5), 266–267 (2006)
14. Tari, E., Baykal, O.: A new transition curve with enhanced properties. Can. J. Civ. Eng. **32** (5), 913–923 (2005)
15. Woznica, P.: Formulation and evaluation of dynamic properties of railway transition curves using the methods of optimization and simulation. Ph.D. in Polish, WUT, Warsaw (2012)
16. Woznica, P., Zboinski, K.: Optymalizacja kształtu kolejowych krzywych przejściowych ze szczególnym uwzględnieniem strefy początkowej i końcowej (in polish). Scientific Works of WUT. Transport, no. 114 (2016)
17. Zboinski, K.: Dynamical investigation of railway vehicles on a curved track. Eur. J. Mech. A-Solids **17**(6), 1001–1020 (1998)
18. Zboinski, K.: Nonlinear dynamics of railway vehicles in the curved track (book in Polish). WNITE, Warsaw-Radom (2012)

19. Zboinski, K., Woznica, P.: Optimisation of railway polynomial transition curves: a method and results. In: Proceedings of the First International Conference on Railway Technology: Research, Development and Maintenance. Civil-Comp Press (2012)
20. Zboinski, K., Woznica, P.: Combined use of dynamical simulation and optimisation to form railway transition curves. Veh. Syst. Dyn. (2018). https://doi.org/10.1080/00423114.2017.1421315

The Issue of Individual Adjustment of Active Manual Wheelchairs to People with Disabilities

Zuzanna Zysk, Sylwia Bęczkowska(✉), and Iwona Grabarek

Faculty of Transport, Warsaw University of Technology, Warsaw, Poland
zuzanna.zysk@gmail.com, sylwiaglow@op.pl

Abstract. The growing number of people with disabilities in Poland forces researchers to take a completely different approach to the issue of individual adjustment of active manual wheelchairs. A properly adjusted wheelchair, and especially an active wheelchair, is the basic equipment of a disabled person. The issues of scientific research focuses mainly on the study of upper extremities demands, then the construction and simulation of kinematic models. The results are the basis for determining guidelines to minimize injuries caused by excessive demands of upper extremities, and to develop optimal ranges of limb movements. The review and analysis of the literature have shown a number of important issues that have not found their place in the field of research on the broadly understood problem of the disability-wheelchair system. It is necessary to increase the amount of research in a given field. This conclusion became the premise for defining the purpose of future research, i.e. developing a methodology for individual adjustment of active manual wheelchairs to the possibilities, limitations and expectations of potential users based on the principles of user-centered design. The implementation of the objective provides for the use of modern measurement techniques, including electromyography (EMG) or wireless analysis of 3D motion kinematics using inertial sensors.

Keywords: Active wheelchair · Disability · Adjustment

1 Introduction

A wheelchair is one of the basic elements of orthopedic (rehabilitation) care for disabled people, as well as, permanently or temporarily, seriously ill and elderly people. According to the Eurostat methodology, there were nearly 7.7 million disabled people living in Poland at the end of 2014, i.e. people who claimed that due to health problems they had limited capacity to perform the activities that people normally perform (serious and less serious restrictions were taken into account). The aspect of having a legal certificate of disability was not taken into account. The population of people with disabilities in Poland at the end of 2014 with legal disability status and/or limited capacity to perform their duties amounted to 4.9 million people, which is about 15% of the total population [14]. Currently, the number of citizens of our country permanently using wheelchairs is estimated at least 100,000 people, and the number of all wheelchair users at least 400,000, with the vast majority (>90%) of users using manual wheelchairs [14]. Active wheelchairs significantly give the term "disabled person" a

M. Siergiejczyk and K. Krzykowska (Eds.): ISCT21 2019, AISC 1032, pp. 484–492, 2020.
https://doi.org/10.1007/978-3-030-27687-4_49

new meaning, because people who use active wheelchairs are most visible in our environment, outside their own home and hospital. These people - despite the lack or limitation of motor skills in the lower limbs - can undertake educational, professional, sports or social activity.

Normalization and standardization in the field of wheelchairs for disabled people, also active wheelchairs, include previous systematics tests included, among others in:

- Polish standards, including PN-ISO 6440,
- standards of the International Organization for Standardization (ISO), including ISO 7176, ISO 16840,
- ANSI/RESNA (American National Standards Institute/Rehabilitation Engineering and Assistive Technology Society of North America) standards, including RESNA WC-1, WC-2, WC-3 and WC-4.

These regulations, however, do not always keep pace with the medical and technical progress in this area. The manufacturers of wheelchairs, basing on the current regulations, offer wheelchairs with different sizes of individual elements in the so-called series of types with a wide range of regulations and a wide range of individual equipment, creating the basis for the proper selection of wheelchairs for a given patient. Despite the modernization of the wheelchair's construction, its proper adjustment (personalization) is not always ensured, which causes pain and injuries to the upper extremities, which limit functionality and lead to a decrease in independence and quality of life [9]. This article discusses the problems caused by incorrect wheelchair adjustment. It includes a review of available studies on wheelchair adaptation/modernization. On this basis, the aim of future research was defined, i.e. the development of a methodology for individual adjustment of active manual wheelchairs to the capabilities, limitations and expectations of potential users based on the principles of user-centered design. The aim of the project is to use modern measurement techniques, such as electromyography (EMG) or wireless analysis of 3D motion kinematics with the use of inertial sensors that allow to define physiological parameters of people with disabilities, necessary in the process of individual adjustment of active manual wheelchairs.

2 Analysis of the Research Problem

The number of people with disabilities is growing from year to year and the problems they face on a daily basis have been the subject of many studies. The following chapter provides an overview of the literature on the health problems faced by wheelchair users and research into constructional changes of wheelchairs.

2.1 Studies on Health Problems

Wheelchair users often experience upper extremities pain that disrupts basic everyday activities such as difficulties in a wheelchair propulsion, decreased physical activity, weakness, reduced independence and quality of life. This high incidence of pain and injury is correlated with the considerable physical demand placed on the upper extremity during wheelchair propulsion as significant intermuscular coordination is needed to generate the mechanical power necessary to propel the wheelchair while maintaining body posture stability [10]. To date, details on the external load during wheelchair propulsion are becoming available [12].

The problem of alleviating the harmful effects of an upper extremity propulsion is the subject of many studies aimed at determining the influence of various propulsion techniques on the exposure and functioning of upper extremities [4, 7, 10–12].

First and foremost, it is essential to appreciate that variability is inherent within all physiological systems. In the past, motor variability was seen as a nuisance to scientific inquiry; something to be experimentally minimized or completely eliminated. However, this approach to variability usually ignores the fact that variability, especially within an individual, can provide important health and function information. The introduction of non-linear dynamics and chaos theory into the study of motion control and rehabilitation has led to the observation that variability (functioning as fluctuations in physiological power within an individual) can provide unique information concerning the control and health of the neuromuscular system [11]. Optimal health of the musculoskeletal system of a wheelchair user is possible due to repeated loads with a certain variable frequency and speed [11]. The lack of differentiation results in insufficient time for muscle regeneration between successive loads.

Muscle fatigue may result from many mechanisms, but is generally referred to as a temporary decrease in muscle strength as a result of long-term physical activity [3]. It may occur at different rates in individual muscles, and the resulting changes in the load on the musculoskeletal system often lead to injuries. However, the overall impact of fatigue on the biomechanics of wheelchair propulsion is not clearly defined. Some study concluded that fatigue may lead to potentially harmful changes in propulsion mechanics, while others have suggested that during an extended period of propulsion, individuals may actually make beneficial adjustments to their propulsion mechanics to mitigate the increased risk of injury [10]. A number of analyses have found that during manual wheelchair propulsion, the highest net joint moments and powers are generated at the shoulder, suggesting that the glenohumeral joint may be the most at risk for overuse injury [10]. The intensity of muscle tension grows with increasing fatigue. This contributes to imbalances between the contact and regeneration phases of the muscles. The influence of fatigue in individual muscles on propulsion mechanics is an issue rarely analyzed by researchers. A study by Slowik, McNitt-Gray, Requejo, Mulroy and Neptune investigates the effects of rotator cuff tears highlighted the injury mechanism using an inverse dynamics-based model (Fig. 1). The investigators found that rotator cuff tears, simulated by eliminating the force-generating capacity of the individual muscles, can lead to increased deltoid activity and a more superiorly directed glenohumeral contact force during wheelchair propulsion. This force vector alteration can initiate an injury mechanism in which the humeral head migrates upward into the subacromial space causing rotator cuff impingement [10]. Muscle weakness may also result from neurological damage caused by injury or disease [2]. However, the extent and scale of these injuries may be different depending on the degree of disability or type of injury.

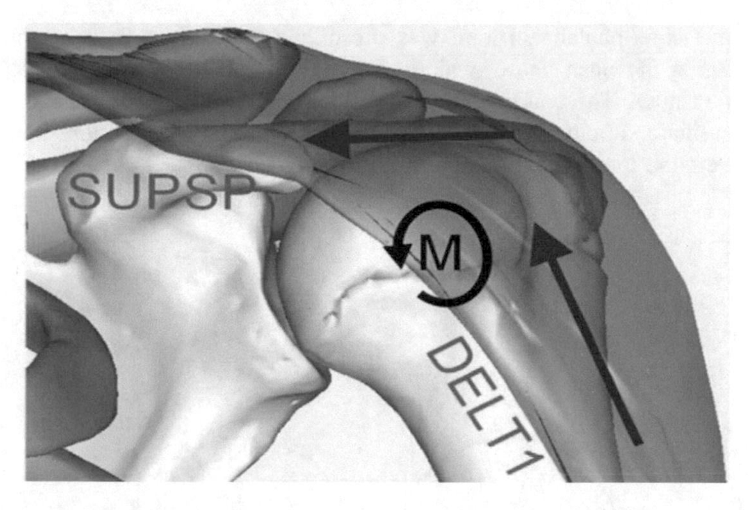

Fig. 1. Moment (M) created by the forces generated by supraspinatus (SUPSP) and anterior deltoid (DELT1). While both muscles can produce an abduction moment, the supraspinatus force draws the humeral head towards the glenoid fossa while the anterior deltoid provides a more superiorly directed force [10].

Research on ways of propulsion of a wheelchair has led, among other things, to the development of guidelines aimed at minimizing injuries caused by excessive muscle stress [4]. The guidelines recommend the use of propulsion techniques that maximize contact angle, such as hemispherical and double loops. These have formed the basis for determining optimal ranges of movement, which has a significant impact on the design of effective training techniques to reduce the muscular demand of upper extremities when moving a wheelchair (Fig. 2) [9].

2.2 Studies on Construction Solutions

The high physical demands placed on the upper extremity during manual wheelchair propulsion can lead to pain and overuse injuries that further reduce user independence and quality of life. Analysis of studies on the construction and exploitation of wheelchairs has shown little interest in individually adjusting the parameters of wheelchairs to users. In this context, it is worth noting that Slowik and Neptune's research, in which it was assumed that the seat position is an adjustable parameter that can influence the mechanical loads placed on the upper extremity. The aim of this study was to investigate the influence of seat position on the demand for muscle strength of upper extremities, including muscle stress and metabolic costs, using a simulation of wheelchair propulsion dynamics [9]. The researchers have generated a model of a standard manual wheelchair with circular handrims and combined it with the musculoskeletal model by defining the position of the hip relative to the rear axle of the wheelchair. This position was kept fixed throughout each simulation of wheelchair propulsion and was systematically varied in both the superior/inferior and anterior/posterior directions to represent a wide range of seat positions. The motion of the third metacarpophalangeal joint was prescribed to follow a path on the circular handrim during the push phase, which was further defined by the contact (θC) and release (θR) angles. These angles were dependent on the seat position, and thus were calculated using a simple 2D model (Fig. 2) [9]. Defining these relationships can help doctors determine the optimal configuration of a wheelchair.

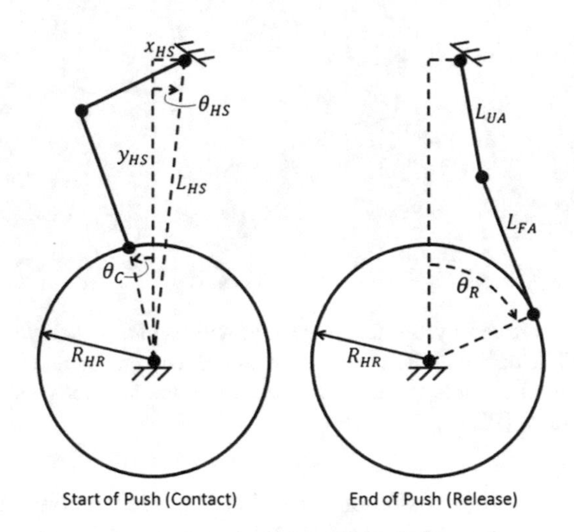

Fig. 2. Kinematic model of wheelchair propulsion used to determine contact and release angles (θ_C and θ_R, respectively). The model consisted of an upper arm (length = L_{UA}), a forearm/hand segment (length = L_{FA}) and a wheel/handrim segment (radius = R_{HR}) with revolute joints at the shoulder, elbow, hand and hub/axle. The shoulder was set at a fixed position relative to the hub/axle (total distance = L_{HS}, anterior/posterior component = x_{HS}, superior/inferior component = y_{HS}).

Determining an optimal seat position reduce upper extremity overuse injuries and pain for wheelchair users. Clinical guidelines suggest that seat position should be adjusted as far posterior as possible without compromising wheelchair stability. The guidelines also recommend superior/inferior positions that correspond to an elbow angle between 100° and 120°. These recommendations have been challenged in recent studies which have shown that with such low, posterior seat positions, the joint ranges of motion and muscle activity levels may be increased, which may adversely affect upper extremity demand [9]. The lack of a clear assessment is due to the fact that it is difficult and time-consuming to study the impact of seat posture on upper extremity muscular strength requirements using experimental techniques (e.g. evaluation of the impact of seat position on metabolic costs).

The aim of the study was to analyze the influence of wheelchair seat position on upper extremity demand, muscle stress and metabolic cost and this goal was achieved. However, many important aspects were not considered in the study, including the diversity of wheelchair propulsion techniques, and future work should therefore include other techniques in the analysis to determine the most effective. Additionally, a model of musculoskeletal system represented by a 50-centile man capable of work was assumed, while wheelchair users are different in age, gender and character of disability. Future work should therefore include the development of thematic models to assess the overall application of the identified trends.

A completely different approach to the problem of upper extremities load when during wheelchair propulsion can be observed in the studies by Wargul, Wieczorek and Kukla [13]. The authors noted that the coefficient of rolling resistance is one of the basic resistances when moving objects. In the case of objects that are not equipped with engine-driven wheels and suspension systems, such as wheelchairs, all resistance is overcome by the strength of the operator's muscles. The study used an innovative method for measuring the rolling resistance coefficient of objects equipped only with wheels and suspension systems, and developed a device for this study. Also presented was an innovative rolling resistance coefficient testing methodology based on patent applications and results of preliminary research on the value of this coefficient for non-pneumatic wheels used in wheelchairs. The research is a part of the work carried out by a research team whose main task is to study the biomechanics of active manual wheelchairs, aimed at the analysis of parameters of the anthropotechnical system of a human wheelchair [13].

Literature reviews on the issue of adjustment of active manual wheelchairs to people with disabilities have shown that research in this area is rather rare. The authors' analysis of literature, both Polish and foreign, showed few examples of research on individual adjustment of active wheelchairs to users. In particular, the number of Polish publications in this area is small. The research focuses mainly on the study of upper extremity load, then the construction and simulation of kinematic models and the results are the basis for the definition of guidelines aimed at minimizing injuries caused by excessive use of upper extremities and developing optimal ranges of movement of extremities.

3 Methodology and Scope for Future Research

The review above revealed a number of important issues which did not find their place in the area of research on the broadly understood issues of the system of a disabled person - wheelchair. This conclusion became a premise for the formulation of the scope of research, which would fill the existing methodological gap. The aim of the research is to develop a original methodology for the adaptation of active manual wheelchairs to people with disabilities, taking into account factors including:

- the user's state of health (including functional state), overall performance and wheelchair driving skills,
- technical parameters of the wheelchair,
- adaptation the wheelchair to the needs, limitations and expectations of the patient/user,
- individual user assessments regarding comfort obtained through self-report surveys.

Considering the above factors in the methodology requires the use of, among others, the technique of measurement of electric muscle signals (EMG) and wireless analysis of the kinematics of body movement of a person using a wheelchair. They are necessary to determine the characteristics and physiological limitations of wheelchair users, which will enable an individual approach to each case. The proposed methodology will take into account the principles of user-centered design. The inclusion of adjustable elements in the process can increase the sense of comfort and independence of users, improve the efficiency of wheelchair propulsion, increase physical activity and increase independence and quality of life. The presented solutions are in line with existing research trends, but put more attention to the personalization of the wheelchair.

The table below shows the parameters that have been adjusted to produce positive effects on the comfort of the wheelchair (Table 1) [6].

Table 1. Parameters and effects of wheelchair adjustment

Adjustable parameter	The effect
Wheelbase (distance between wheels)	Can be adjusted to the width of the shoulder of a person
Wheel size	A larger wheel means less propulsion to travel a distance, but more weight
Backrest angle	Regulates the inclination of the user's body, which affects, among other things, directly its stability
Seat tilt angle	Increases the degree of pelvic anchorage in the seat and indirectly stabilizes the torso
Backrest height	The stability of the user's torso depends on it, which is often correlated with the height of the spinal cord injury
Type of side	Relevant during the transition from and to the wheelchair
Location of footrests	Affects the posture of the body
Type of brakes	Central or lateral - easy to use (control)
The type of strings	The compromise between diameter and material (friction) and low weight, affects the character of the movements performed
Type and diameter of the front wheels	Ease of movement under various conditions, stability

In the area of the selection and use of active wheelchairs, the biggest problems to be solved are:

- lack of uniform procedures for wheelchair selection (prescription criteria, selection and regulation procedures and basic training for the user and his/her caregivers),
- the absence of generally accepted standards, including a series of types of wheelchairs, makes it difficult to achieve a complete fit, which is only possible with sufficiently wide adjustment ranges and a wide range of accessories,
- lack of standards with regard to prevention of secondary changes (bedsores, deformations),
- lack of training programs, both for medical personnel and for wheelchair users and their caregivers [6].

It is also important to remember that an incorrect position in the wheelchair, sometimes perpetuated by hours of use, can lead to contractures, for example, by causing incorrect positioning of particular parts of the body. Such a situation results, among others, in worse results of improvement and/or limits the effectiveness of verticalization and re-education of walking [6].

Nowadays, it is important to use modern scientific research such as automation and robotization [1, 8] and the emergence of a real alternative to wheelchairs, e.g. exoskeletons [5]. However, in the pursuit of modern solutions one cannot forget the possibility of modernization of the already existing ones.

4 Conclusions

Appropriately adjusted wheelchairs, especially active wheelchairs, are basic equipment for disabled, seriously ill and elderly people with lower extremities dysfunctions. The growing number of people with disabilities in Poland forces researchers to adopt a completely different approach to the issue of individual adjustment of active manual wheelchairs. The increasing number of studies and publications from a given area gives hope to improve the situation of people with disabilities and enable them to function at the highest possible level. Improper selection of a wheelchair deprives people of the chance of independence and may also become a source of harmful pathological changes. Personalization of a wheelchair should not be difficult and time-consuming, but will require the knowledge of appropriate adjustment methods. Individual adjustment of the wheelchair to the capabilities and preferences of the user shall be made by a specialist (e.g. physiotherapist) with knowledge of the field:

- the medical condition of the user,
- the possibilities and needs of the patient,
- construction and adjustment of the wheelchairs,
- a methodology for individual adjustment of the wheelchair to the needs of the user.

The use of these procedures in clinical practice is not a rule, but only a goal that science is approaching but has not yet achieved, as witnessed by poorly adjusted wheelchairs visible on Polish streets and hospital corridors.

References

1. Champaty, B., Dubey, P., Sahoo, S., Ray, S.S., Pal, K.: Development of wireless EMG control system for rehabilitation devices. In: International Conference on Magnetics, Machines and Drives (AICERA-2014 iCMMD) (2014)
2. Dimitrijevic, M.R., Kakulas, B.A., McKay, W.B., Vrbova, G.: Restorative Neurology of Spinal Cord Injury. Oxford University Press, USA (2012)
3. Enoka, R.M., Duchateau, J.: Muscle fatigue: what, why and how it influences muscle function. J. Physiol. **586**(1), 11–23 (2008). Epub 16 Aug 2007
4. Jayaramana, C., Moonb, Y., Sosnoff, J.J.: Shoulder pain and time dependent structure in wheelchair propulsion variability. Med. Eng. Phys. **38**, 648–655 (2016)
5. Huang, G., Ceccarelli, M., Huang, Q., Zhang, W., Yu, Z., Chen, X., Mai, J.: Design and feasibility study of a leg-exoskeleton assistive wheelchair robot with tests on gluteus medius muscles. Sensors **19**, 54 (2019)

6. Mikołajewska, E.: Dobór wózków dla niepełnosprawnych w polskich i zagranicznych badaniach naukowych. Annales Academiae Medicae Silesiensis **67** (2013)
7. Rankin, J.W., Richter, W.M., Neptune, R.R.: Individual muscle contributions to push and recovery subtasks during wheelchair propulsion. J. Biomech. **44**, 1246–1252 (2011)
8. Sathish, S., Nithyakalyani, K., Vinurajkumar, S., Vijayalakshmi, C., Sivaraman, J.: Control of robotic wheel chair using EMG signals for paralysed persons. Indian J. Sci. Technol. **9**(1) (2016). https://doi.org/10.17485/ijst/2016/v9i1/85726
9. Slowik, J.S., Neptune, R.R.: A theoretical analysis of the influence of wheelchair seat position on upper extremity demand. Clin. Biomech. **28**, 378–385 (2013)
10. Slowik, J.S., McNitt-Gray, J.L., Requejo, P.S., Mulroy, S.J., Neptune, R.R.: Compensatory strategies during manual wheelchair propulsion in response to weakness in individual muscle groups: a simulation study. Clin. Biomech. **33**, 34–41 (2016)
11. Sosnoff, J.J., Rice, I.M., Hsiao-Wecksler, E.T., Hsu, I.M.K., Jayaraman, C., Moon, Y.: Variability in wheelchair propulsion: a new window into an old problem. Front. Bioeng. Biotechnol. **3**, 105 (2015)
12. Veeger, H.E.J., Rozendaal, L.A., Van der Helm, F.C.T.: Load on the shoulder in low intensity wheelchair propulsion. Clin. Biomech. **17**, 211–218 (2002)
13. Wargula, Ł., Wieczorek, B., Kukla, M.: The determination of the rolling resistance coefficient of objects equipped with the wheels and suspension system – results of preliminary tests. In: MATEC Web of Conferences, vol. 254, p. 01005 (2019)
14. Stan zdrowia ludności Polski w 2014 roku. Główny Urząd Statystyczny, Warszawa (2014)

Author Index

M. Siergiejczyk and K. Krzykowska (Eds.): ISCT21 2019, AISC 1032, pp. 493–494, 2020.
https://doi.org/10.1007/978-3-030-27687-4

Printed in the United States
By Bookmasters